Quantum Optics and Nanophotonics

École de Physique des Houches
Session CI, 5–30 August 2013

Quantum Optics and Nanophotonics

Edited by

Claude Fabre, Vahid Sandoghdar, Nicolas Treps,
Leticia F. Cugliandolo

OXFORD
UNIVERSITY PRESS

OXFORD
UNIVERSITY PRESS

Great Clarendon Street, Oxford, OX2 6DP,
United Kingdom

Oxford University Press is a department of the University of Oxford.
It furthers the University's objective of excellence in research, scholarship,
and education by publishing worldwide. Oxford is a registered trade mark of
Oxford University Press in the UK and in certain other countries

First Edition published in 2017

Impression: 1

Published in the United States of America by Oxford University Press
198 Madison Avenue, New York, NY 10016, United States of America

British Library Cataloguing in Publication Data
Data available

Library of Congress Control Number: 2016949071

ISBN 978–0–19–876860–9

DOI 10.1093/oso/9780198768609.003

Printed and bound by
CPI Group (UK) Ltd, Croydon, CR0 4YY

École de Physique des Houches

Service inter-universitaire commun
à l'Université Joseph Fourier de Grenoble
et à l'Institut National Polytechnique de Grenoble

Subventionné par l'Université Joseph Fourier de Grenoble,
le Centre National de la Recherche Scientifique,
le Commissariat à l'Énergie Atomique

Directeur:

Leticia F. Cugliandolo, Sorbonne Universités, Université Pierre et Marie Curie,
Laboratoire de Physique Théorique et Hautes Energies, Paris, France

Directeurs scientifiques de la session:

Claude Fabre, Sorbonne Universités, Université Pierre et Marie Curie,
Laboratoire Kastler Brossel, Paris, France

Vahid Sandoghdar, Max Planck Institute for the Science of Light,
Erlangen, Germany

Nicolas Treps, Sorbonne Universités, Université Pierre et Marie Curie,
Laboratoire Kastler Brossel, Paris, France

Leticia F. Cugliandolo, Sorbonne Universités, Université Pierre et Marie Curie
Laboratoire de Physique Théorique et Hautes Energies, Paris, France

Previous sessions

Publishers

- Session VIII: Dunod, Wiley, Methuen
- Sessions IX and X: Herman, Wiley
- Session XI: Gordon and Breach, Presses Universitaires
- Sessions XII–XXV: Gordon and Breach
- Sessions XXVI–LXVIII: North-Holland
- Session LXIX–LXXVIII: EDP Sciences, Springer
- Session LXXIX–LXXXVIII: Elsevier
- Session LXXXIX–: Oxford University Press

Preface

Over the last few decades, optics has undergone impressive development, in both its applied and fundamental aspects. The quantum aspects of light have been explored both as test benches for understanding the strangeness of the quantum world (entanglement, quantum aspects of measurement, etc.) and with an eye to their potential applications in the domains of information and metrology. There has been major progress in the understanding of the specific quantum aspects of the interaction between light and matter, with applications to the design of quantum memories, and in the generation of highly non-classical states of light, such as 'Schrödinger cat' states. Single photons are now routinely produced by nano-emitters or micro-emitters such as single molecules on surfaces, vacancies in crystals, and quantum dots. The micrometre and nanometre scale is also the privileged range where fluctuations of electromagnetic fields manifest themselves through the Casimir force, a domain that has been very actively investigated in the last few years. Even the domain of classical optics, ruled by the venerable Maxwell equations, has recently seen many exciting new developments, especially in the area of nano-optics.

Many of these advances have been fuelled by the quest to achieve further developments in nanoscience and nanotechnology. Approaches based on single-molecule detection, scanning probe microscopy, and plasmonics, as well as high-performance numerical simulations, have provided new avenues for exploring light–matter interaction at the nanometre scale. For example, optical nano-antennas have been shown to modify the emission pattern and rate of nearby emitters, and a wealth of tailored 'meta-materials' have brought the science fiction dream of optical cloaking closer to realization. Another flourishing domain is that of 'optomechanics', where light is coupled to the motion of physical objects, which is also rich in terms of its manifestation of basic quantum issues and of its potential for applications. All these topics show a common trend of considering and exploiting smaller and smaller objects, down to the micrometre, nanometre, and even atomic range—a region in which there is a gradually transition from classical physics to quantum physics.

The Summer School held in Les Houches in August 2013 gathered participants from a variety of scientific communities—atomic physics, quantum optics, solid state physics, physical chemistry, optical microscopy, plasmonics, and integrated optics—which have mostly evolved independently over the past three decades. This mixing triggered many discussions that we hope have been fruitful and will be the starting points for future scientific cooperations.

In addition to the lectures represented by the chapters in the present volume, the participants in the summer school enjoyed four lectures given by J. Kimble from Caltech on 'Cavity quantum electrodynamics with trapped atoms', three lectures by P. Grangier from the Institut d'Optique on 'Quantum optics with continuous variables', and five lectures by Oskar Painter from Caltech on 'Optomechanics'. Unfortunately, the lecturers did not find the time to write up the corresponding lecture notes.

There were also a number of short presentations in which participants described their own research, as well as working groups to allow deeper discussion of various subjects covered by the school. In addition, a number of researchers came to Les Houches to give seminars to illustrate and elaborate on the topics covered by the courses:

- J. M. Gerard, Service de Physique des Materiaux et Microstructures, CEA Grenoble, France: 'Quantum optics with photonic nanowires and photonic trumpets: basics and applications';
- S. Götzinger, Max Planck Institute for the Science of Light, Erlangen, Germany: 'Steering and controlling single photons with single molecules';
- P. Hommelhoff, Friedrich Alexander University, Erlangen, Germany: 'Nano-optics meets strong field physics—attosecond phenomena at metal tips';
- F. Jelezko, University of Ulm, Germany: 'Quantum optics in diamond';
- P. Lodahl, Niels Bohr Institute, University of Copenhagen, Denmark: 'Quantum optics in photonic-crystal waveguides and cavities';
- B. Lounis, Institut d'Optique Graduate School, Bordeaux, France: 'Optical detection and spectroscopy of individual nano-objects';
- J. O'Brien, University of Bristol, UK: 'Integrated quantum photonics';
- P. Torma, University of Aalto, Helsinki, Finland: 'Strong coupling phenomena in nanoplasmonics'.

We are very grateful to the University of Joseph Fourier at Grenoble, the French–German University, and the Max Planck Institute for the Science of Light for their financial support, and to all the staff in Les Houches for their kind and efficient assistance during the preparation of this summer school.

C. Fabre
V. Sandoghdar
N. Treps
L. F. Cugliandolo

Contents

List of Participants

ORGANIZERS

FABRE CLAUDE
Laboratoire Kastler Brossel, Paris, France
SANDOGHDAR VAHID
Max Planck Institute, Erlangen, Germany
TREPS NICOLAS
Laboratoire Kastler Brossel, Paris, France

LECTURERS

BARNETT STEPHEN
University of Glasgow, UK
GRANGIER PHILIPPE
Institut d'Optique, Palaiseau, France
GREFFET JEAN-JACQUES
Institut d'Optique, Palaiseau, France
KIMBLE JEFFREY
Caltech, Pasadena, USA
LEONHARDT ULF
University of Saint Andrews, UK
LEUCHS GERD
Max Planck Institute, Erlangen, Germany
LUKIN MIKHAIL
Harvard University, Cambridge, USA
ORRIT MICHEL
Leiden University, The Netherlands
PAINTER OSKAR
Caltech, Pasadena, USA
REYNAUD SERGE
Laboratoire Kastler Brossel, Paris, France
STEINBERG AEPHRAIM
University of Toronto, Canada

VUKOVIC Jelena
Stanford University, USA

Seminar speakers

GERARD Jean-Michel
CEA Grenoble, France

GÖTZINGER Stephan
Max Planck Institute, Erlangen, Germany

HOMMELHOFF Peter
Max Planck Institute, Erlangen, Germany

JELEZKO Fedor
Ulm University, Germany

ÊLODAHL Peter
Niels Bohr Institute, Copenhagen, Denmark

ÊLOUNIS Brahim
Université de Bordeaux, France

O'BRIEN Jeremy
University of Bristol, UK

TÖRMÄ Päivi
Aalto University, Finland

Students

ARCANI Marta
Niels Bohr Institute, Copenhagen, Denmark

BODDEDA Rajiv
Institut d'Optique, Palaiseau, France

BUESE Alexander
Macquarie University, Sydney, Australia

CAI Yin
Laboratoire Kastler Brossel, Paris, France

CAMERON Robert
Strathclyde University, Glasgow, UK

CHILLE Vanessa
Max Planck Institute, Erlangen, Germany

CHRISTENSEN Thomas
Technical University of Denmark, Kongens-Lyngby, Denmark

CHRISTODOULOU SOTIRIOS
Istituto Italiano di Tecnologia, Genoa, Italy

COLOMBIER LÉO
Laboratoire Charles Coulomb, Montpellier, France

CORTI TOMMASO
Università degli Studi dell'Insubria, Como, Italy

CROAL CALLUM
University of St Andrews, UK

DHAWAN AMIT RAJ
Institut des NanoSciences de Paris, France

DHEUR MARIE-CHRISTINE
Institut dÓptique, Palaiseau, France

DUFOUR GABRIEL
Laboratoire Kastler Brossel, Paris, France

GLOPPE ARNAUD
Institut Néel, Grenoble, France

GUILLE ANTOINE
Ghent University, Belgium

HAMSEN CHRISTOPH
Max Planck Institute, Garching, Germany

HEASE WILLIAM
Université Denis Diderot, Paris, France

HUANG KUN
Laboratoire Kastler Brossel, Paris, France

JACQUARD CLÉMENT
Laboratoire Kastler Brossel, Paris, France

KIRSANSKE GABIJA
Niels Bohr Institute, Copenhagen, Denmark

KOMAR PETER
Harvard University, Cambridge, USA

LAFONT OMBLINE
Laboratoire Pierre Aigrain, ENS, Paris, France

LEIJSSEN RICK
FOM Institute AMOLF, Amsterdam, The Netherlands

LOVCHINSKY IGOR
Harvard University, Cambridge, USA

MANCEAU MATHIEU
Laboratoire Kastler Brossel, Paris, France

MARSAULT FÉLIX
Université Denis Diderot, Paris, France

MIATTO FILIPPO
University of Ottawa, Canada

MUNRO EWAN
National University of Singapore, Singapore

NEITZKE OLIVER
Humboldt University, Berlin, Germany

PENJWEINI ROJIAR
Max Planck Institute, Erlangen, Germany

PONTIN ANTONIO
Università di Trento, Povo, Italy

RAHBANY NANCY
Université de Technologie de Troyes, France

RASHID MUDDASSAR
University of Southampton, UK

ROSENKRANTZ DE LASSON JAKOB
Technical University of Denmark, Kongens-Lyngby, Denmark

ROY ANANDA
Yale University, New Haven, USA

RUIZ-RIVAS JOAQUIN
University of Valencia, Spain

SARAVI SINA
Friedrich-Schiller University, Jena, Germany

SCHNOERING GABRIEL
Université de Strasbourg, France

SIPAHIGIL ALP
Harvard University, Cambridge, USA

SUN YONGNAN
Hefei University of Technology, China

TIKHONOV KIRILL
St. Petersburg State University, Russia

TISCHLER NORA
Macquarie University, Sydney, Australia

TÜRSCHMANN PIERRE
Max Planck Institute, Erlangen, Germany

UNDERWOOD DEVIN
Princeton University, USA

VERBIN MOR
Weizmann Institute of Science, Rehovot, Israel

VERHART NICO
Leiden University, The Netherlands

VOGELL BERIT
University of Innsbruck, Austria

VYUNISHEV ANDREY
Kirensky Institute of Physics, Krasnoyarsk, Russia

WANG DAQING
Max Planck Institute, Erlangen, Germany

WANG XIAOLONG
Ecole Polytechnique Fédérale de Lausanne, Switzerland

YU LEO
Stanford University, USA

1

Introduction to quantum information

Stephen M. BARNETT

School of Physics and Astronomy
University of Glasgow, UK

Stephen M. Barnett: *Introduction to Quantum Information*. In: *Quantum Optics and Nanophotonics*. Edited by: Claude Fabre, Vahid Sandoghdar, Nicolas Treps, and Leticia F. Cugliandolo. Oxford University Press (2017), pp. 1–60. © Oxford University Press 2017. DOI: 10.1093/oso/9780198768609.003.0001

Chapter Contents

In these five lectures, I shall give a short introduction to the field of quantum information. The lectures are drawn from my book *Quantum Information* [4]. My aim is to make these notes self-contained, but I cannot hope to cover the totality of what is now a large and active field of research. I aim, rather, to give a taster of the subject to whet the reader's appetite. A more complete introduction may be found in [4] or in any of a now large collection of books and review papers devoted to the topic (far too many for me to attempt to make a list and to risk offence by innocent omission). One text that needs to be mentioned, however, is that by Nielsen and Chuang [75], which did so much both to popularize the field and to lay out its founding principles.

I am grateful to Oxford University Press for their permission and, indeed, encouragement to reproduce some of the material from [4]. I wish to express my gratitude also to Allison Yao for her help in preparing these notes.

1.1 A very short history

Our field starts with the work of the Reverend Thomas Bayes (1702–61) and the celebrated theorem that bears his name (of which more below) [7]. His key idea was that probabilities depend on what you know; if we acquire additional *information*, then this modifies the probabilities. Today, such reasoning is uncontentious and forms part of the prevailing paradigm in much of probability theory [18, 19, 57, 58, 64]. This was not the case, however, for most of the 350 years of its history. An entertaining and informative presentation of this troubled history may be found in [71].

The picture is completed by identifying, or formulating the quantity of information. It was Claude Shannon (1916–2001) who solved this problem and, by using it to devise his two coding theorems, founded information theory in a paper entitled *A Mathematical Theory of Communication* [87]. Perhaps I can give an indication of the magnitude of Shannon's achievement by relating that just a year later the paper was republished as a book [89], in which, apart from the correction of a few typographical errors, there are only two changes, namely the inclusion of a short introductory article by Weaver and a change of title to *The Mathematical Theory of Communication*. The theory was born complete—the numerous textbooks on the topic have greatly broadened the scope and application of Shannon's ideas, but have not departed from the fundamentals as explained by Shannon in his first paper [20, 29, 42, 46, 60, 61].

Shannon's information has a simple and familiar form. For a given set of probabilities, $\{p_i\}$, the information is

$$H = -\sum_i p_i \log p_i . \tag{1.1}$$

Remarkably, this is simply Boltzmann's formula for the entropy (of which more later).

We can sum up the fundamental lessons learned from Bayes and from Shannon as follows: Bayes taught us that probabilities are not absolute but rather depend on available information, while Shannon showed that information itself is a precisely defined function of the probabilities.

Shannon's work was aimed, initially, at the problem of providing a quantitative theory of communications, but any set of probabilities can be associated with a quantity of information and it follows that any probabilistic phenomenon has an associated information theory. This idea has been applied, for example, to statistical mechanics [55, 56]. Quantum theory is a probabilistic theory, of course, and so it was inevitable that a quantum information theory would be developed.

1.1.1 Probabilities and conditional probabilities

Consider an event A with possible outcomes $\{a_i\}$. Everything we know is specified by the probabilities for the possible outcomes: $\{P(a_i)\}$. For tossing a fair coin, for example, the outcomes are 'heads' and 'tails', with $P(\text{heads}) = \frac{1}{2} = P(\text{tails})$. In general, the probabilities satisfy the two conditions

$$0 \leq P(a_i) \leq 1,$$
$$\sum_i P(a_i) = 1. \tag{1.2}$$

If we have two events A and B with outcomes $\{a_i\}$ and $\{b_j\}$, then the complete description is given by the *joint* probabilities $P(a_i, b_j)$. Here the comma is read as 'and', so $P(a_i, b_j)$ is the probability that both $A = a_i$ *and* $B = b_j$. If the two events are independent, then $P(a_i, b_j) = P(a_i)P(b_j)$, but this is *not* true in general. More generally, we have

$$P(a_i) = \sum_j P(a_i, b_j),$$
$$P(b_j) = \sum_j P(a_i, b_j). \tag{1.3}$$

If the events are correlated, then what does learning the value of A tell us about the value of B? If we learn, for example, that $A = a_0$, then $P(b_j)$ is *replaced* by

$$P(b_j|a_0) \propto P(a_0, b_j). \tag{1.4}$$

Here the vertical line is read as 'given that', so $P(b_j|a_0)$ is the probability that $B = b_j$ *given that* $A = a_0$. We can find the constant of proportionality by noting that the sum over j of $P(b_j|a_0)$ must be unity, and this leads to Bayes' rule:

$$P(a_i, b_j) = P(b_j|a_i)P(a_i)$$
$$= P(a_i|b_j)P(b_j). \tag{1.5}$$

Bayes' theorem utilizes this rule to relate the two conditional probabilities:

$$P(a_i|b_j) = \frac{P(b_j|a_i)P(a_i)}{P(b_j)}. \tag{1.6}$$

1.1.2 Entropy and information

The link between information and entropy is a subtle and beautiful one. Shannon himself gave a simple derivation of this as a mathematical theorem, using only a few simple axioms that he argued information must satisfy [89]. We shall not reproduce this here, but rather present a simple argument to make the association plausible.

Suppose that we have a single event A with possible outcomes a_i. If one, say a_0, is certain to occur, then we acquire no information by observing A. By simple extension, if $A = a_0$ is very *likely* to occur, then we might confidently expect it to do so—and therefore when it happens we learn very little. If, however, $A = a_0$ is very *unlikely*, then when it occurs we might need to drastically modify our actions. A very simple example is a two-state communication system that you may not have considered: a fire alarm. A fire alarm is either ringing or not ringing. Its most common state (hopefully not ringing) is so innocuous that we give it no thought. When it does ring, however, we stop what we are doing and leave the building (if we are sensible).

It seems reasonable to conclude that learning the value of A provides a quantity of information, h, that *increases* as the corresponding probability *decreases*:

$$h[P(a_i)] \Uparrow \quad \text{as} \quad P(a_i) \Downarrow .$$

We think of learning something new as *adding* to the available information. Therefore, for independent probabilities for a pair of events, $P(a_i, b_j) = P(a_i)P(b_j)$, it is natural to require that

$$
\begin{aligned}
h[P(a_i, b_j)] &= h[P(a_i)P(b_j)] \\
&= h[P(a_i)] + h[P(b_j)],
\end{aligned}
\tag{1.7}
$$

which immediately suggests logarithms. Hence we set

$$h[P(a_i)] = -K \log P(a_i).\tag{1.8}$$

Here K is a positive constant, yet to be determined, and the minus sign ensures both that h is positive and that it increases as the probability decreases (recall that the logarithm of a number less than unity is negative).

It is convenient to define the information as the *average* of h, which means weighting h for each outcome by its probability of occurring and then summing. This procedure leads us, of course, to the entropy:

$$H = -K \sum_i P(a_i) \log P(a_i).\tag{1.9}$$

We can absorb the prefactor K into the choice of base for the logarithm. A convenient choice is to use base 2, so that

$$H = -\sum_i P(a_i) \log_2 P(a_i) \quad \text{bits},\tag{1.10}$$

which is Shannon's formula for the information. Henceforth in these lectures, we shall drop the subscript 2.

For two events A and B, we can write informations for the joint events or for the single events:

$$H(A, B) = -\sum_{i,j} P(a_i, b_j) \log P(a_i, b_j),$$

$$H(A) = -\sum_{i,j} P(a_i, b_j) \log \sum_{k} P(a_i, b_k), \qquad (1.11)$$

$$H(B) = -\sum_{i,j} P(a_i, b_j) \log \sum_{l} P(a_l, b_j).$$

We can also define information based on the conditional probabilities. An especially useful measure of correlation between the two events is the *mutual information*

$$H(A : B) = H(A) + H(B) - H(A, B). \qquad (1.12)$$

This has the important properties that it is positive or zero and that it takes the value zero if and only if the events are independent:

$$H(A : B) \geq 0,$$
$$H(A : B) = 0 \quad \text{iff} \quad P(a_i, b_j) = P(a_i)P(b_j) \quad \forall i, j. \qquad (1.13)$$

It is the mutual information that provides an upper bound on the rate at which we can communicate. A more detailed (but gentle) discussion of these entropies and exploration of their properties may be found in [4].

1.1.3 Information and thermodynamics

The fact that the mathematical form of the information is also the entropy begs the question as to whether information entropy is the *same* quantity that appears in statistical mechanics. It is!

An important and simple example is the way in which we can obtain the Boltzmann distribution by maximizing the information (what we have yet to discover) subject only to a constraint on the average energy. This is such a nice calculation that I cannot resist the temptation to include it here, even though there was not time to present it in my lectures. Let a bound quantum system have a set of discrete energy levels E_i and be in thermodynamic equilibrium with its environment. Our task is to determine the probability that the system is to be found in any of its given energy levels. Of course, we know how to do this by maximizing the entropy, but information theory gives us a rather different take on the problem; we maximize the information of the system subject to the constraint only that its mean energy is fixed. In this way, we maximize, in an unbiased manner, our uncertainty about the state. The derivation is readily performed as a variational calculation using Lagrange's method of undetermined multipliers [14]. It is convenient to work with the natural base of logarithms in this case, and so we define the information to be

$$H_e = -\sum_i P(a_i) \ln P(a_i) \quad \text{nats}.$$ (1.14)

Our task is to maximize this quantity subject to a fixed mean energy, $\sum_i p_i E_i = \bar{E}$, and the probabilities summing to unity, $\sum_i p_i = 1$. We can achieve this by varying, independently, the probabilities in the function

$$\tilde{H} = H_e + \lambda \left(1 - \sum_i p_i \right) + \beta \left(\bar{E} - \sum_i p_i E_i \right).$$ (1.15)

We find

$$d\tilde{H} = \sum_i \left(-\ln p_i - 1 - \lambda - \beta E_i \right) dp_i.$$ (1.16)

We require this to be stationary (zero) for all dp_i, which leads to the solution

$$p_i = e^{-1-\lambda} e^{-\beta E_i}.$$ (1.17)

We can fix the values of the undetermined multipliers by enforcing the normalization of the probabilities and the value of the average energy, thereby arriving at the familiar Boltzmann distribution

$$p_i = \frac{\exp\left(-E_i/k_B T\right)}{\sum_j \exp\left(-E_j/k_B T\right)}.$$ (1.18)

A more dramatic example was provided by Szilard in his paper *On the Decrease of Entropy in a Thermodynamic System by the Intervention of Intelligent Beings* [96]. This paper is all the more remarkable in that it precedes Shannon's work by nearly 20 years. Here is Szilard's argument. Consider a box of volume V_0 with a movable partition dividing the box into two equal halves. There is a single molecule in one of the two sides, as depicted in Fig. 1.1. An intelligent being can look in the box and determine which side the molecule is in. By attaching a small weight to a pulley, we can extract work from heat in an apparent violation of the second law of thermodynamics, an interesting paradox in the style of Maxwell's famous demon [70].

Applying the ideal gas law,

$$PV = k_B T,$$ (1.19)

to our single-molecule gas allows us to calculate the work extracted from the expanding gas in an isothermal expansion:

$$W = \int_{V_0/2}^{V_0} P \, dV = k_B T \ln 2.$$ (1.20)

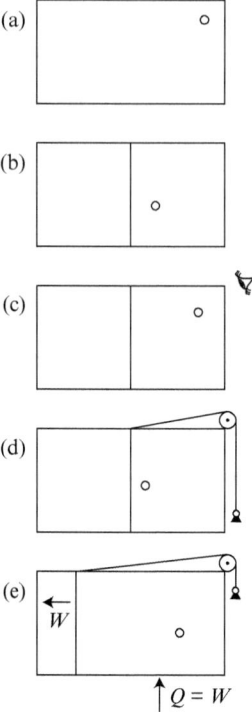

Fig. 1.1 Szilard's illustration of the connection between information and (thermodynamic) entropy. (Reproduced, with permission, from Barnett, S. M. (2009) Quantum information. Oxford University Press, Oxford.)

We have extracted useful work from the reservoir in apparent conflict with the second law of thermodynamics. The second law can be saved, however, if the process of measuring and recording the position of the molecule produces an entropy change of *not less than*

$$\Delta S = k_B \ln 2 \,. \tag{1.21}$$

Szilard concludes, in this way, a direct link between information and thermodynamic entropy.

A more precise statement is that to complete the thermodynamic cycle, we need to include the process of the observer forgetting in which side the molecule was found. The process of forgetting was studied by Landauer [63, 66, 80]. He considered a single-molecule gas, like Szilard's, and proposed using the position of the molecule to represent a logical bit: the molecule being to the left of the partition corresponding to a logical 0, and the molecule being to the right corresponding to a logical 1. Landauer showed that erasing and resetting the bit to 0 requires the dissipation of at least $k_B T \ln 2$ worth of energy as heat. To see this, we can simply remove the membrane (a reversible process) and then push in, slowly, a fresh partition from the right, as

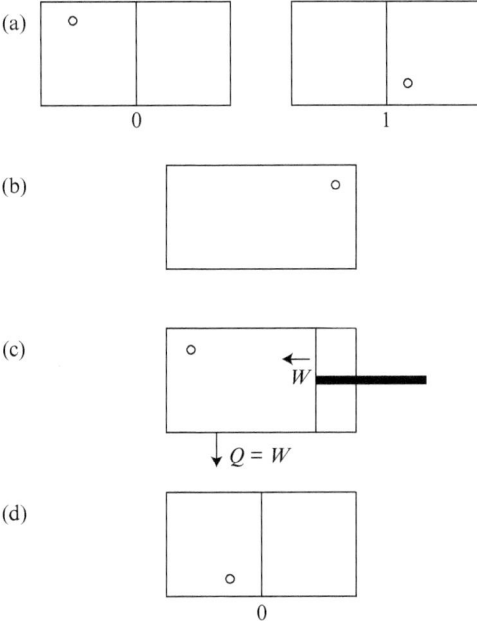

Fig. 1.2 Illustration of Landauer's derivation of the thermodynamic cost of information erasure. (Reproduced, with permission, from Barnett, S. M. (2009) Quantum information. Oxford University Press, Oxford.)

depicted in Fig. 1.2. When the partition has reached halfway, the 'memory' has been reset to the bit value 0 and, of course, all trace of the original bit value has been lost. The work that needs to be done to achieve this is

$$W = -\int_{V_0}^{V_0/2} P\, dV = k_B T \ln 2, \tag{1.22}$$

and this is dissipated, of course, as heat. We can view this as a resolution of the paradox implicit in Szilard's model. This is not the end of the story, however. It has recently been shown that Landauer's limit can be beaten if we pay a cost, not in energy, but in some other quantity such as angular momentum [97]. The *key idea* to take from these models, however, is unaffected: information is physical.

1.1.4 Communications theory

Shannon first introduced his information theory as a/the mathematical theory of communication [87, 89]. In doing so, he introduced the very general model of a communications system depicted in Fig. 1.3. Let us analyse this by introducing two characters, Alice and Bob, who have become very popular in quantum communications. Alice wishes to send a message to Bob. Alice's event, A, is the selection of the message from a set of possible messages $\{a_i\}$. Bob's event, B, is the detection of the message from

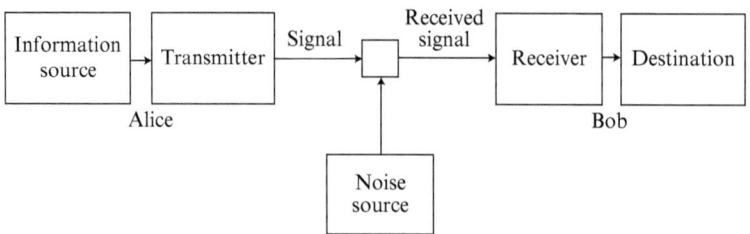

Fig. 1.3 Shannon's model of a communications channel. (Reproduced, with permission, from Barnett, S. M. (2009) Quantum information. Oxford University Press, Oxford.)

the possible set $\{b_j\}$. On Alice's side, there is an information source, which produces the choice of message to be sent (this may be Alice herself, of course) and a transmitter that produces the signal (electrical, optical, acoustical, or even a piece of paper) ready for transmission. The signal propagates from Alice to Bob and, while en route, is subject to noise. Bob receives the noisy signal and his receiving device turns the signal into a message.

The limiting performance of such a communications channel is governed by two theorems derived by Shannon [87, 89]: his noiseless coding theorem, which tells us how much a message can be compressed and still be read, and his noisy channel coding theorem, which tells us how much redundancy is need to correct errors.

1.1.4.1 Noiseless coding theorem

Most messages have an element of redundancy and can be compressed but still be readable. As an example, you might like to try this one:[1]

THS LS HCHS SCHL HS NTRSTNG LCTRS

I compressed the message by removing the vowels. Shannon's noiseless coding theorem quantifies the redundancy in a message. Two simple examples are that in English we mostly don't need to put a 'u' after 'q' and a three-letter word beginning with 'th' will almost certainly be 'the'.

Shannon's theorem shows us the typical number of bits that we need to encode a message efficiently. It is the *entropy* associated with the probabilities for the messages that is the key quantity. If we have one of a set of messages that is N bits long, for example

$$\underbrace{010\cdots 111}_{N \text{ bits}},$$

---

[1] If you are having difficulty, the message is THiS LeS HouCHeS SCHooL HaS iNTeReSTiNG LeCTuReS.

and each of the possible possible messages $\{a_i\}$ is selected with the associated probability $P(a_i)$, then Shannon's noiseless coding theorem tells us that we need only an *average* of $H(A)$ bits, if we can find an optimum coding scheme.

We shall not prove Shannon's noiseless coding theorem, but rather present a simple example from his original paper [87, 89]. Simple derivations of both the noiseless and noisy coding theorems may be found in [4, 95]. Consider a message formed from an alphabet of four letters, A, B, C, and D and let each entry in the message be one of these four with the probabilities

$$P(\mathrm{A}) = \tfrac{1}{2}\,,$$
$$P(\mathrm{B}) = \tfrac{1}{4}\,, \tag{1.23}$$
$$P(\mathrm{C}) = \tfrac{1}{8} = P(\mathrm{D})\,.$$

The information associated with this set of probabilities is

$$H = -\left(\tfrac{1}{2}\log\tfrac{1}{2} + \tfrac{1}{4}\log\tfrac{1}{4} + 2 \times \tfrac{1}{8}\log\tfrac{1}{8}\right)$$
$$= \tfrac{7}{4}\ \mathrm{bits}\,. \tag{1.24}$$

Hence Shannon's noiseless coding theorem says that we should be able (on average) to reduce a message of N characters, or $2N$ bits, to $1.75N$ bits. The key to this reduction is to use short sequences for common elements of the message (here the letter A) and longer ones for the less likely ones (like C and D in our example). Here is a coding scheme that achieves this:

$$A = 0\,,$$
$$B = 10\,,$$
$$C = 110\,, \tag{1.25}$$
$$D = 111\,.$$

A bit of thought will confirm that any message so encoded can be decoded uniquely to recover the original sequence of letters. The average number of bits used to encode a sequence of N letters is then

$$N\left(\tfrac{1}{2} \times 1 + \tfrac{1}{4} \times 2 + 2 \times \tfrac{1}{8} \times 3\right) = \tfrac{7}{4}N = HN\,, \tag{1.26}$$

which is the bound provided by Shannon's theorem. The fact that Shannon's value is reached in this case tells us, moreover, that no better coding is possible: this is the shortest possible length of the message.

1.1.4.2 Noisy coding theorem

The presence of noise on the communications channel will introduce errors in the received signal. We can combat these errors by introducing some redundancy; indeed, this is undoubtedly the reason why language evolved already containing redundancy. As a simple example, let is suppose that any given bit in the message is 'flipped' with probability q and so produces an error. How much redundancy do we need to be able to detect and correct these errors?

Shannon's noisy coding theorem tells us that, on average, we require at least

$$\frac{N_0}{1 - H(q)} \text{ bits} \tag{1.27}$$

to encode, faithfully, one of 2^{N_0} equiprobable messages. Here

$$H(q) = -[q \log q + (1 - q) \log(1 - q)] \tag{1.28}$$

is the entropy associated with the single-bit error probability. In other words, if we first remove all the redundancy to get 2^{N_0} possible optimally compressed messages, we need to *put back* this much redundancy to combat errors.

The general statement is based on the mutual information. It says that the greatest number of messages that can be sent from Alice to Bob on a noisy channel, using N bits, *and be reconstructed by Bob* is

$$2^{NH(A:B)}. \tag{1.29}$$

Any more is impossible in that an attempt to do so will inevitably produce ambiguities in Bob's decoding process.

We conclude with a simple illustration of the principle of using redundancy to combat errors. Try to read the following message:

WNTM NARMQN THRS S FN

You probably didn't manage to do so. The reason for this is that I first compressed the message by removing the vowels and then added in errors. Because much of the redundancy was removed, the message has become unreadable. If I had left the full message (complete with vowels) and then added the errors, we might have

WUANTFM INAORMAQION THEORS US FUN

Hopefully, after a bit of thought, you should be able to read the message.[2] Note that decoding the message is possible even though the errors affect both the characters in the compressed message and the redundant characters added to combat the errors.

1.2   Quantum communications and quantum key distribution

Quantum communications differ from their classical counterpart in that the transmitted signal is carried by a quantum system. At its simplest, classical information is expressed in *bits*; and is carried by a physical system with two distinct states (for example a high or low voltage or the presence or absence of an optical pulse). One state is used to represent the logical 0 and the other the logical 1.

---

[2] If you are struggling, the message is qUANTuM INfORMAtION THEORy iS FUN.

We can take the same approach in quantum communications and use two orthogonal quantum states to represent 0 and 1. We label these, naturally enough, as $|0\rangle$ and $|1\rangle$. The additional element brought in by using a two-state *quantum* system is that we can also prepare any superposition state of the form

$$|\psi\rangle = \alpha|0\rangle + \beta|1\rangle, \tag{1.30}$$

where α and β are complex probability amplitudes. A quantum bit, or *qubit*, is such a two-state quantum system.

1.2.1 Qubits

Before getting into the theory of quantum communications, we pause to elaborate on the idea of a qubit and the mathematical tools used to describe it. A qubit can be any quantum system with two orthogonal states. We choose a basis, often called the computational basis in quantum information, and label the two states in this basis as $|0\rangle$ and $|1\rangle$. It is convenient to think of this as a spin-half particle; this is not literally true in most cases, but it has the benefit that we can use the Pauli operators to describe the properties of the qubit. There are four Pauli operators, which are $\hat{\sigma}_z$, $\hat{\sigma}_x$, $\hat{\sigma}_y$, and the identity operator $\hat{I}$. We can define each of these by its action on the states in the computational basis:

$$\begin{aligned}
\hat{\sigma}_z|0\rangle &= |0\rangle, & \hat{\sigma}_z|1\rangle &= -|1\rangle, \\
\hat{\sigma}_x|0\rangle &= |1\rangle, & \hat{\sigma}_x|1\rangle &= |0\rangle, \\
\hat{\sigma}_y|0\rangle &= i|1\rangle, & \hat{\sigma}_y|1\rangle &= -i|0\rangle, \\
\hat{I}|0\rangle &= |0\rangle, & \hat{I}|1\rangle &= |1\rangle.
\end{aligned} \tag{1.31}$$

The first three Pauli operators do not mutually commute. We can write the product rule for these operators in an appealingly simple form if we introduce, by means of a scalar product, the Pauli operator for an arbitrary direction associated with the unit vector **a**:

$$\mathbf{a} \cdot \hat{\boldsymbol{\sigma}} = a_x\hat{\sigma}_x + a_y\hat{\sigma}_y + a_z\hat{\sigma}_z. \tag{1.32}$$

The product rule for two such operators is then

$$(\mathbf{a} \cdot \hat{\boldsymbol{\sigma}})(\mathbf{b} \cdot \hat{\boldsymbol{\sigma}}) = (\mathbf{a} \cdot \mathbf{b})\hat{I} + i(\mathbf{a} \times \mathbf{b}) \cdot \hat{\boldsymbol{\sigma}}, \tag{1.33}$$

as may readily be verified.

1.2.2 Information security

The use of a quantum communications channel makes three essential modifications or additions to Shannon's model. The first is that the signal sent by Alice is encoded on the quantum state of a physical system sent to Bob. This means that each signal a_i is associated with a quantum state with corresponding density operator $\hat{\rho}_i$. In addition to any intrinsic noise in the channel, we have a special role for any eavesdropper who might be listening in. This is because in order to extract any information, the eavesdropper needs to perform a measurement and, as we know, a measurement will, in general, modify the state of the observed system. Finally, Bob cannot determine the state of the system but rather has to settle for measuring one observable. Hence Bob must choose what to measure. All of these elements are essential in quantum communications and it is the combination that, in particular, makes quantum key distribution possible. Before I get to that, let us set the scene by looking at secure communications in general.

In the information age, we are all aware of the importance and difficulty of keeping information secure. The science that today underpins this effort is cryptography [82]. The history of secure communications and keeping information secure, however, is a long and interesting one [91, 92]. For important communications, we might try enciphering a message to keep it safe. The simplest such scheme, at least conceptually, is the single-key cryptosystem. The principal idea is that Alice and Bob share a secret key (a number) that Alice can use to generate the ciphertext and Bob can use to decipher it. The general scheme is depicted in Fig. 1.4.

We can write the transformation formally, involving the plaintext $\mathcal{P}$, the key $\mathcal{K}$, and the ciphertext $\mathcal{C}$. The ciphertext is a function of the plaintext and the key, to be calculated by Alice, and the plaintext may be recovered by Bob as a function of the ciphertext and the key:

$$\mathcal{C} = \mathcal{C}(\mathcal{P}, \mathcal{K}),$$
$$\mathcal{P} = \mathcal{P}(\mathcal{C}, \mathcal{K}). \tag{1.34}$$

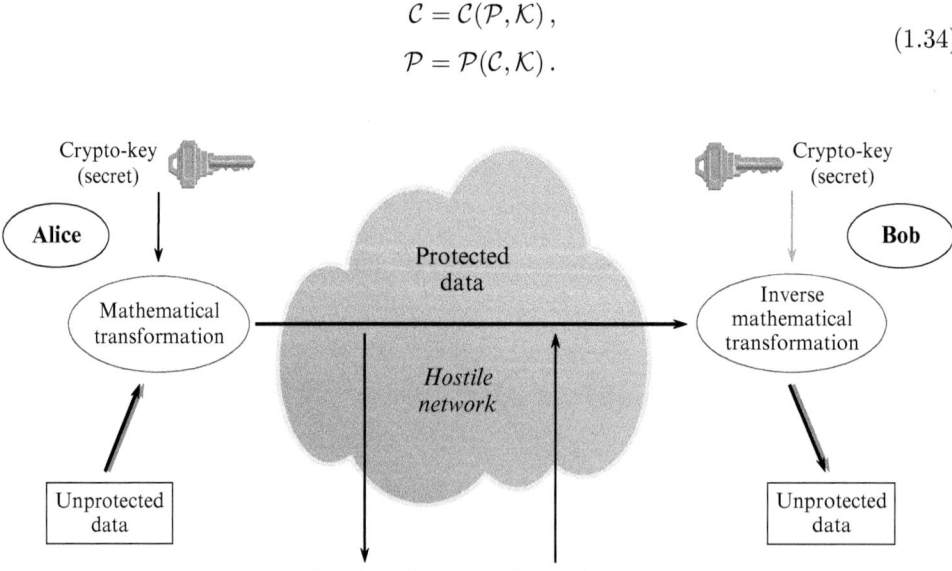

Fig. 1.4 Elements of a secret communications channel. (Reproduced, with permission, from Barnett, S. M. (2009) Quantum information. Oxford University Press, Oxford.)

For example in a substitution cipher, we replace each letter with a symbol. In principle, there are

$$26! \approx 4 \times 10^{26}$$

possible substitution ciphers, so an exhaustive search is completely impractical. A substitution cipher is easily cracked, however, using the known letter frequencies. For example the most common symbol will be E, which occurs 12.7% of the time, while Q makes up only about 0.1% of the symbols (or perhaps a bit more in these notes!). Sherlock Holmes makes use of precisely this technique in one of his cases [28].

It was Shannon who gave us an objective criterion for *perfect secrecy* [88]. Let $\{p_i\}$ be the set of possible plaintexts (messages) and $\{c_j\}$ be the set of possible cipher texts. Shannon's criterion for perfect secrecy is

$$P(c_j|p_i) = P(p_i) \qquad \forall \ i, j \,. \tag{1.35}$$

A straightforward application of Bayes' theorem shows that this is equivalent to

$$P(c_j|p_i) = P(c_j) \qquad \forall \ i, j \,. \tag{1.36}$$

This means that discovering or intercepting the ciphertext does not provide any information on the plaintext. The second condition states that any given ciphertext is equally likely to have been generated by any plaintext. A question that you might like to ponder is how many possible keys does this require?

The simplest perfectly secure cipher is the Vernam cipher, or one-time pad. It uses a key in the form of a *random* sequence of bits, $\cdots 101110 \cdots$, that is at least as long as the plaintext (which is also a sequence of bits, of course). The ciphertext is produced by bitwise modulo addition of the plaintext and the key:

$$
\begin{aligned}
0 \oplus 0 = 0, \quad 0 \oplus 1 = 1, \\
1 \oplus 0 = 1, \quad 1 \oplus 1 = 0.
\end{aligned}
\tag{1.37}
$$

A simple example is

$$
\begin{aligned}
\mathcal{P} \quad &\cdots 0011010100 \cdots, \\
\mathcal{K} \quad &\cdots 1011101000 \cdots, \\
\mathcal{C} = \mathcal{P} \oplus \mathcal{K} \quad &\cdots 1000111100 \cdots.
\end{aligned}
$$

The ciphertext $\mathcal{C}$ is *random* because the key $\mathcal{K}$ is random. Deciphering is perfumed by Bob by a second modulo addition of the key. This works because $\mathcal{K} \oplus \mathcal{K} = \cdots 0000000 \cdots$:

$$
\begin{aligned}
\mathcal{C} \quad &\cdots 1000111100 \cdots, \\
\mathcal{K} \quad &\cdots 1011101000 \cdots, \\
\mathcal{P} = \mathcal{C} \oplus \mathcal{K} \quad &\cdots 0011010100 \cdots.
\end{aligned}
$$

Clearly, the secrecy of the key is crucial, as anyone in possession of it can decipher the message.

The remaining problem, of course, is how can Alice and Bob exchange the key $\mathcal{K}$? To see how this and other secure communications are realized in practice, we need to introduce *public-key cryptography*, which is based on the fact that some mathematical operations are *easy* but the inverse operations are *very difficult* and, hopefully, effectively impossible in practice [4, 21, 68]. The classic example is multiplication and factorization.

In public-key cryptography, Bob generates *two keys*: an enciphering key e and a deciphering key d. He publishes e but keeps d secret. Alice can use e to encode her message, which should be all but impossible (she hopes!) for anyone other than Bob to decipher. This is the RSA cryptosystem [4, 21, 68].

RSA scheme

1. Bob finds two large prime numbers, p and q, and calculates $m = p \times q$. This is easy.
2. He then finds two numbers e and d such that $de = 1 \bmod (p-1)(q-1)$. There are efficient algorithms for doing this if you know p and q.
3. The *public* key is m and e. The *private* key is d.
4. Alice takes her message, which is a number x, and turns it into a ciphertext by the transformation

$$x \rightarrow x^e \bmod m \,.$$

5. By the wonders of pure mathematics (actually it is not so hard to prove this),

$$(x^e \bmod m)^d \bmod m = x \bmod m \,,$$

which is the original message. So Bob can recover the message using his secret key d.

The security of the RSA scheme is believed to be equivalent to the difficulty of factoring m into its constituent primes p and q. It is certainly true that if p and q are known, then finding d from e and m is straightforward.

How big does m have to be? Numbers of order 10^{90} can be factored in less than a day, so much larger numbers are needed. There is an RSA factoring challenge, which you might like to try. The number RSA-704, which is $\approx 7 \times 10^{212}$, was factored in 2012. There was a cash prize offered as an inducement for solving this and earlier factoring problems. Regrettably, this has now been discontinued (http://www.emc.com/emc-plus/rsa-labs/historical/the-rsa-factoring-challenge-faq.htm). If you would like to try the next challenge (for fun!), it is

RSA-896 = 412023436986659543855531365332575948179811699844327982845456

    2643387644556524842619809887042316184187926142024718886949256

    0931776375033421130982397485150944909106910269861031862704114

    8808669705649029036536588674337317208131041051908642547932826

    01391257624033946373269391

Public key cryptography is routinely used to distribute keys and for proof of identity via so-called digital signatures.

1.2.3 Quantum copying?

Quantum key distribution is a radically different approach to secure communications. It relies on the difficulty for anyone eavesdropping in determining the quantum signal generated by Alice. Before describing this in detail, we show that it is impossible to copy an *unknown* quantum state. This is the famous *no-cloning theorem* of Wootters and Zurek [104] and Dieks [33]. It works by exposing a contradiction.

Suppose that you are given a qubit in some unknown (to you) quantum state:

$$|\psi\rangle = \alpha|0\rangle + \beta|1\rangle, \tag{1.38}$$

that is, you do not know the amplitudes α and β. Clearly, you could make a measurement, but that would not tell you $|\psi\rangle$. What you would like to achieve is a transformation on this qubit and a 'blank' prepared in the state $|B\rangle$ such that

$$|\psi\rangle \otimes |B\rangle \rightarrow |\psi\rangle \otimes |\psi\rangle \qquad \forall \; |\psi\rangle. \tag{1.39}$$

Let us suppose that this works if $|\psi\rangle = |0\rangle$ or $|1\rangle$, so that

$$\begin{aligned} |0\rangle \otimes |B\rangle &\rightarrow |0\rangle \otimes |0\rangle, \\ |1\rangle \otimes |B\rangle &\rightarrow |1\rangle \otimes |1\rangle. \end{aligned} \tag{1.40}$$

The superposition principle then tells us that for a general state $|\psi\rangle$, the corresponding transformation is

$$\begin{aligned} |\psi\rangle \otimes |B\rangle &= (\alpha|0\rangle + \beta|1\rangle) \otimes |B\rangle \\ &\rightarrow \alpha|0\rangle \otimes |0\rangle + \beta|1\rangle \otimes |1\rangle \\ &\neq (\alpha|0\rangle + \beta|1\rangle) \otimes (\alpha|0\rangle + \beta|1\rangle). \end{aligned} \tag{1.41}$$

So it necessarily follows from the superposition principle that a quantum cloner cannot work perfectly for all quantum states. Having established that strictly exact cloning of an unknown state is not possible, we might very reasonably ask what is the best that is possible. There has been a great deal of work done on this topic, but probably the most important is the universal quantum-copying machine, the operation of which is the best possible with the same performance for all possible input qubit states [22, 85].

1.2.4 Optical polarization

Recall that for a plane wave, the electric field **E**, the magnetic field **H**, and the wavevector **k** are all mutually orthogonal and are oriented such that **E** × **H** points in the direction of **k** (see Fig. 1.5). This fixes **E** and **H** to be in the plane perpendicular to **k**.

Consider a complex electric field for a plane wave propagating in the z-direction:

$$\mathbf{E} = \mathbf{E}_0 e^{i(kz - \omega t)}. \tag{1.42}$$

We can characterize the polarization by the direction of the complex vector $\mathbf{E}_0$ in the x–y plane,

$$\mathbf{E}_0 = E_{0x}\mathbf{i} + E_{0y}\mathbf{j}, \tag{1.43}$$

or as a column vector, the Jones vector,

$$\mathbf{E}_0 = \begin{bmatrix} E_{0x} \\ E_{0y} \end{bmatrix}. \tag{1.44}$$

Only the relative sizes of E_{0x} and E_{0y} matter, so we can use a normalized Jones vector. The Jones vectors corresponding to horizontal/vertical, diagonal, and circular polarizations are depicted in Fig. 1.6.

For a single photon, we can associate each of the Jones vectors with a qubit state according to the mapping

$$\begin{bmatrix} \alpha \\ \beta \end{bmatrix} \to \alpha|0\rangle + \beta|1\rangle, \tag{1.45}$$

which leads to the identifications listed in Fig. 1.7. This simple idea has been used widely in experimental demonstrations of a variety of quantum information and communications protocols, most especially in quantum cryptography.

Fig. 1.5 Relative orientations of the electric and magnetic fields and the wavevector. (Reproduced, with permission, from Barnett, S. M. (2009) Quantum information. Oxford University Press, Oxford.)

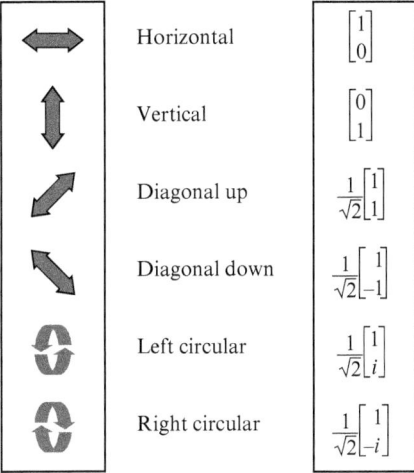

Fig. 1.6 Linear and circular polarizations, together with their Jones vectors. (Reproduced, with permission, from Barnett, S. M. (2009) Quantum information. Oxford University Press, Oxford.)

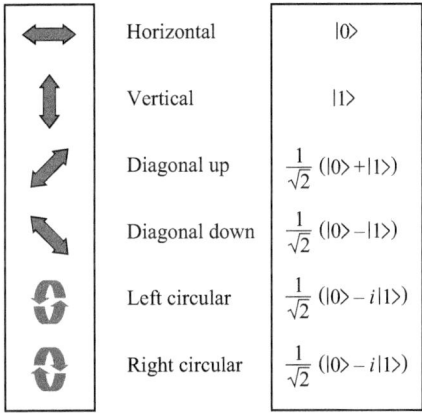

Fig. 1.7 Linear and circular polarizations for single photons, together with their associated qubit states. (Reproduced, with permission, from Barnett, S. M. (2009) Quantum information. Oxford University Press, Oxford.)

1.2.5   Quantum cryptography

Quantum cryptography, or perhaps more precisely quantum key distribution, has become an advanced experimental technique and is on the verge, perhaps, of becoming a practical technology. In these lectures, I can give only a very brief overview of the topic and, for a more complete description, I refer the reader to some of the introductory literature on the subject [4, 17, 41, 68, 81, 98].

The idea of using quantum effects for security is due to Stephen Wiesner [103], who suggested (in a paper written in about 1970 but published only much later, in 1983) the idea of unforgeable money. He supposed a high-value banknote worth, say, £10M. Each 'note' has traps for 20 single photons, each of which has been prepared in one of the polarization states $|0\rangle$, $|1\rangle$, $\frac{1}{\sqrt{2}}(|0\rangle + |1\rangle)$, and $\frac{1}{\sqrt{2}}(|0\rangle - |1\rangle)$. The note also has a serial number that identifies to the bank the states of these photon polarizations. The problem faced by a would-be forger is how to copy these unknown quantum states? The no-cloning theorem tells us, of course, that this is impossible. The situation is illustrated in Fig. 1.8. The counterfeiter needs to work out the polarization of each of the trapped photons and to do so only by performing measurements, but he or she has no idea which measurement to perform. Let us focus our attention on one of the photons that happens to be prepared in a state of vertical polarization. If the counterfeiter measures this in the horizontal/vertical basis, then the result will be a banknote with the correct horizontally polarized photon. If, however, the counterfeiter measures in the diagonal basis, then either of the two possible results will occur with equal probability and, importantly, there is no possible indication that the measurement has been performed in the incorrect basis. If the polarization is checked in the bank, then an error will be produced with a probability $\frac{1}{2}$. It follows that for a banknote produced in this way, anyone checking in the bank will see an error (indicating a forgery) with a probability of $\frac{1}{4}$ for *each* of the 20 photons. The probability that all 20 photons in the forged note pass a test in the bank is $(\frac{3}{4})^{20} \approx 0.003$. If this isn't considered low enough, then one simply has to add a few more light traps.

Quantum money is impractical, but Bennett and Brassard [10] proposed a variant on this idea: quantum key distribution. The idea is that Alice associates the bit values 0 and 1 with a linear polarization state (either horizontal/vertical or diagonal/antidiagonal):

$$H \text{ or } D \longleftrightarrow 0\,,$$
$$V \text{ or } A \longleftrightarrow 1\,. \tag{1.46}$$

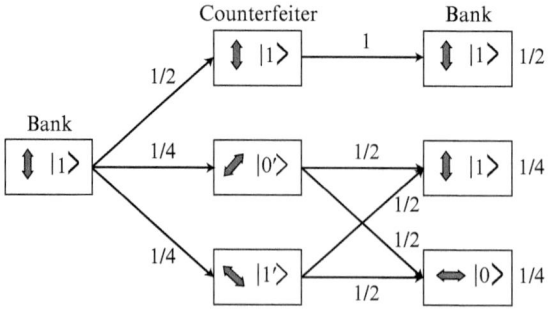

Fig. 1.8 Possible outcomes due to the intervention of a counterfeiter, together with their associated probabilities of occurrence. (Reproduced, with permission, from Barnett, S. M. (2009) *Quantum information.* Oxford University Press, Oxford.)

Suppose that Alice prepares a vertically polarized photon and that Bob measures the state in the horizontal/vertical basis. An eavesdropper, Eve, does not know that the photon was prepared in this basis and so can only make a choice of what to measure. In this way, she will introduce, for $\frac{1}{4}$ of the photons intercepted, a disagreement between the values obtained by Alice and by Bob. This 25% error rate indicates to Alice and to Bob the activity of the eavesdropper.

All that is needed is a sequence of instructions (a protocol) to ensure that Eve is trapped in this way. One such protocol is depicted in Fig. 1.9. Alice prepares a random sequence of bits and randomly encodes each onto the polarization as either a horizontally/vertically or diagonally/antidiagonally polarized photon, which she sends to Bob. For each photon, Bob randomly (and independently of Alice, of course) chooses to make a polarization measurement in one of the two bases. Roughly half the time he will guess correctly and roughly half incorrectly. Bob then uses a classical channel to tell Alice which basis he used in each time slot for which he detected a photon (but not, of course, the measured value). Alice tells Bob the slots in which he made the correct choice (time slots 2, 4, 6, 8, 9, 11, 14, 15, 16, and 19, in the case depicted in the figure). These should comprise a secret shared random bit string (0111010100 in this case), which can be used as a secret key.

It remains only to determine whether an eavesdropper has been listening in. To determine whether this has happened or not, Alice and Bob can publicly reveal a subset of their shared data and look for errors. An example of what might be expected to happen is depicted in Fig. 1.10. Alice and Bob delete those bits associated with time bins 1, 3, 7, 12, 13, 17, 18, and 20, in which the preparation and measurement were performed using different bases. In time slots 2, 4, 11, 14, 15, and 19, Eve used a different basis to Alice and Bob, and, of these, this led to a difference in the recorded bit values for Alice and Bob in time bins 4, 11, and 14. The probability that Eve has

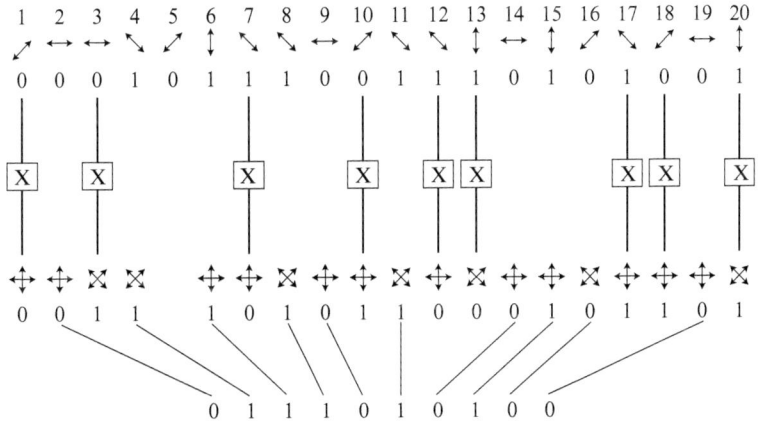

Fig. 1.9 An example of a quantum cryptographic protocol. Alice prepares single photons in one of two polarization bases and these are measured by Bob. (Reproduced, with permission, from Barnett, S. M. (2009) Quantum information. Oxford University Press, Oxford.)

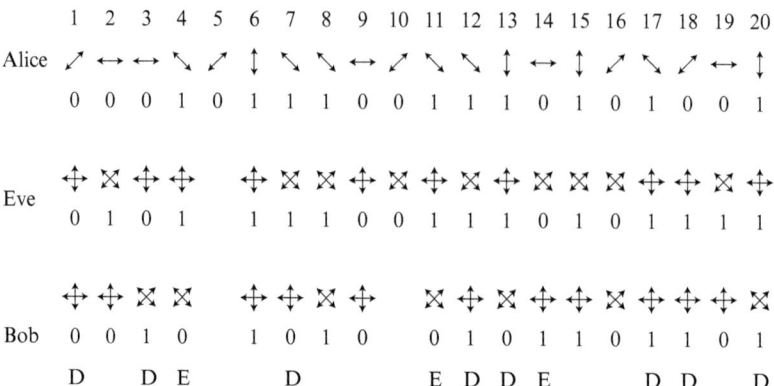

Fig. 1.10 An example of a quantum cryptographic protocol in which an eavesdropper has been active. (Reproduced, with permission, from Barnett, S. M. (2009) Quantum information. Oxford University Press, Oxford.)

been active and does not cause an error is then $\left(\frac{3}{4}\right)^N$, where N is the number of bits tested. Naturally, those bits used in the open discussion between Alice and Bob need to be discarded.

Naturally, there are many subtleties that need to be accounted for and that complicate, to a greater or lesser extent, the above simple picture. There will always be errors, and, to be safe, we need to assume that these are due to eavesdropper activity. If we were to simply discard the communication when we found an error, then communication of a key of any useful size would become impossible. We need to detect and correct errors, which is achieved by parity checks that shorten the key. The error rate is then used to place a bound on the information an eavesdropper might have, and this can then be reduced by taking as a key bit the parity of a number of bits (this is called privacy amplification). Practical systems, moreover, may not have access to single-photon sources and there will usually be other departures from the ideal. Proving and assessing the security of real-world systems has become a research topic in its own right [86].

1.3 Generalized measurements

The extraction of information from a quantum system requires us to perform measurements. Our task in this lecture is to set up a general description of this process.[3] We seek two things: (i) the probability for any given measurement result to occur and (ii) the state of the system after the measurement has been made, i.e. the post-measurement state conditioned on the measurement outcome.

---

[3] We should note that we do not include here measurements involving post-selection and so will not cover the topic of 'weak measurements' [1].

1.3.1   Ideal von Neumann measurements

Let us start with the measurement process as it is usually encountered in quantum mechanics courses. This formulation is essentially that given by von Neumann in his famous and early book on quantum mechanics [100]. We start by representing each observable A by a Hermitian operator,[4] $\hat{A}$. This operator will have a complete set of eigenvectors $|\lambda_n\rangle$ and associated eigenvalues λ_n:

$$\hat{A}|\lambda_n\rangle = \lambda_n|\lambda_n\rangle, \tag{1.47}$$

which means that we can write $\hat{A}$ in the form

$$\hat{A} = \sum_n \lambda_n|\lambda_n\rangle\langle\lambda_n|. \tag{1.48}$$

Let us assume, for the moment, that each of the eigenvalues is distinct from the others. The von Neumann description then states that if we perform a measurement of $\hat{A}$, we will find that the measurement result is one of the eigenvalues and, moreover, that the probability for finding any one of these is

$$P(\lambda_n) = |\langle\lambda_n|\psi\rangle|^2, \tag{1.49}$$

where $|\psi\rangle$ is the pre-measurement state. More generally, for a mixed state with density operator $\hat{\rho}$, we have

$$\begin{aligned} P(\lambda_n) &= \langle\lambda_n|\hat{\rho}|\lambda_n\rangle \\ &= \mathrm{Tr}(\hat{\rho}|\lambda_n\rangle\langle\lambda_n|). \end{aligned} \tag{1.50}$$

Immediately after the measurement, the von Neumann description has the system left in the eigenstate corresponding to the measurement outcome. Hence, if we make a measurement of $\hat{A}$ and find the value λ_n, then we know that the post-measurement state is $|\lambda_n\rangle$, and repeating the measurement, if it is done quickly enough, should give the same result.

There is one further point that we need to consider, which is that the eigenvalues of $\hat{A}$ might be degenerate, which means that a set of orthonormal eigenvectors, $|\lambda_n^j\rangle$, will correspond to the same measurement outcome, λ_n. To incorporate this case, it is useful to introduce a projector onto the set of states with a common eigenvalue:

$$\hat{P}_n = \sum_j |\lambda_n^j\rangle\langle\lambda_n^j|. \tag{1.51}$$

The probability that the measurement will give the result λ_n is then

$$\begin{aligned} P(\lambda_n) &= \sum_j \langle\lambda_n^j|\hat{\rho}|\lambda_n^j\rangle \\ &= \mathrm{Tr}(\hat{\rho}\hat{P}_n). \end{aligned} \tag{1.52}$$

---

[4] The distinction between Hermitian and self-adjoint operators will not concern us.

The post-measurement state will simply be that part of the pre-measurement state that was in the subspace spanned by the corresponding eigenstates or, in other words, that part of the state selected by the projector. Hence, if our measurement gives the result λ_n, then the density matrix changes as

$$\hat{\rho} \rightarrow \frac{\hat{P}_n \hat{\rho} \hat{P}_n}{\text{Tr}(\hat{\rho}\hat{P}_n)}, \tag{1.53}$$

where the denominator, which is the prior probability for the observed measurement result, ensures correct normalization of the post-measurement state.

Let us conclude this brief review of von Neumann measurements with a summary of the properties of projectors.

Properties of projectors

1. They are Hermitian: $\qquad\qquad \hat{P}_n = \hat{P}_n^\dagger$.

2. They are positive operators: $\qquad \hat{P}_n \geq \hat{1}$.

3. They are complete: $\qquad\qquad \sum_n \hat{P}_n = \hat{1}$.

4. They are orthonormal: $\qquad\quad \hat{P}_n \hat{P}_m = \hat{P}_n \delta_{nm}$.

Here $\hat{1}$ is the identity operator. We note that the first three of these properties have physical significance in that they are required in order that the probability rule, $P(\lambda_n) = \text{Tr}(\hat{\rho}\hat{P}_n)$, be true. They correspond respectively to the requirements that the projectors be observables, that they give positive probabilities, and that the probabilities for the complete set of possible outcomes must sum to unity. The final property, however, does not seem to have a similar physical significance and, indeed, we shall see that generalized measurements correspond to dropping this requirement.

1.3.2 Non-ideal measurements

Most measurements are not of the form described in the preceding subsection. Consider, for example, the operation of a photodetector. Real photodetectors have a finite efficiency and so do note resolve perfectly the number of photons. In detecting the photons, moreover, the detector absorbs the light and so leaves the field in its zero-photon (or vacuum) state. Hence neither the von Neumann forms for the detection probabilities nor those for the post-measurement state are appropriate, and something more general is required.

As a simple example, both of the problem and of how we might proceed, let us consider a measurement of a spin component for a single qubit. The ideal measurement would correspond to the pair of projectors

$$\hat{P}_0 = |0\rangle\langle 0|,$$
$$\hat{P}_1 = |1\rangle\langle 1|. \tag{1.54}$$

Suppose that our detector gives the *wrong* answer with probability p, so the two possible measurement results occur with probabilities

$$P(0) = (1 - p)\operatorname{Tr}(\hat{\rho}\hat{P}_0) + p\operatorname{Tr}(\hat{\rho}\hat{P}_1),$$
$$P(1) = (1 - p)\operatorname{Tr}(\hat{\rho}\hat{P}_1) + p\operatorname{Tr}(\hat{\rho}\hat{P}_0),$$
(1.55)

the sum of which is clearly unity, as it should be. We can write these in a form reminiscent of the von Neumann expressions,

$$P(0) = \operatorname{Tr}(\hat{\rho}\hat{\pi}_0),$$
$$P(1) = \operatorname{Tr}(\hat{\rho}\hat{\pi}_0),$$
(1.56)

by introducing the probability operators

$$\hat{\pi}_0 = (1 - p)\hat{P}_0 + p\hat{P}_1,$$
$$\hat{\pi}_1 = (1 - p)\hat{P}_1 + p\hat{P}_0.$$
(1.57)

These operators satisfy the first three properties that we listed for the projectors: they are Hermitian, positive, and complete. They are *not* orthonormal, however:

$$\hat{\pi}_0\hat{\pi}_1 = p(1 - p)\hat{\mathrm{I}}.$$
(1.58)

This is our first indication of a more general description of the measurement process, which we develop further below.

1.3.3 Probability operator measures

To calculate the portability for any given result of a generalized measurement, we need a probability rule and, in particular, a set of operators that characterize the measurement, one for each of the possible measurement outcomes. The required operators are the probability operators, the set of which is a probability operator measure or a positive operator-valued measure [4, 48, 50, 51, 79].

That the first three of the properties we listed for the projectors have physical significance, but the third does not, tells us in what way we are allowed to generalize the von Neumann description; we simply drop the fourth and final property from our list. Hence we describe any measurement by a set of probability operators $\{\hat{\pi}_m\}$ such that the probability for getting measurement outcome m is

$$P_m = \operatorname{Tr}(\hat{\rho}\hat{\pi}_m),$$
(1.59)

where the probability operators have the following *three* properties:

Properties of probability operators

1. They are Hermitian: $\hat{\pi}_n = \hat{\pi}_n^\dagger.$

2. They are positive operators: $\hat{\pi}_n \geq \hat{I}.$

3. They are complete: $\sum_n \hat{\pi}_n = \hat{I}.$

We should emphasize that there is no restriction on the number of probability operators; this can be greater or less than the dimension of the state space. For example, we shall analyse an example in which a generalized measurement on a qubit has three outcomes, even though the dimension of the state space is only 2. There is no need, moreover, for the probability operators to commute with each other.

The set of probability operators characterizing a measurement is called a probability operator measure (or POM), which you will also find called a positive operator-valued measure (POVM). The latter has become the commonly used expression, but I prefer the former.[5] The description of a measurement in terms of a POM is very useful because of two points:

1. All measurements can be described in this way. (This is at least reasonable, given the physical significance of the three properties used to define a POM.)
2. Any set of the probability operators (i.e. any POM) is realizable, at least in principle, in the laboratory.

This combination is very useful because if we seek to find the best possible measurement in any situation, we can separate out the purely mathematical optimization of the POM from the experimental question of how to achieve it. That any POM is realizable as a measurement is a consequence of Naimark's theorem. We shall not prove this here, but rather give an indication of how it works. A more complete presentation may be found in [4]. The key idea is to map the generalized measurement onto a projective, or von Neumann, measurement in an extended state space. To see how this might be done, consider a quantum system s, which we wish to measure, and let us prepare also an ancillary quantum system a, so that we know its initial state. The state of the combined system and ancilla is then $|\psi\rangle_s \otimes |A\rangle_a$. Next, we engineer an interaction between the system and the ancilla, so that a unitary transformation of our choosing occurs:

$$|\psi\rangle_s \otimes |A\rangle_a \rightarrow \hat{U}|\psi\rangle_s \otimes |A\rangle_a. \tag{1.60}$$

---

[5] POM or POVM? The term 'probability operator measure' tells us that the elements forming the measure are the probability operators. The more popular expression 'positive operator-valued measure' expresses the fact that the elements of the measure, the probability operators, are positive operators. Calling the set of operators a POM reminds us of their physical significance, while the term POVM recalls their mathematical properties.

Finally, we perform a von Neumann measurement on both the system and the ancilla, so that the probability for a given outcome will be

$$P(m, l) = \left| {}_s\langle m| \otimes {}_a\langle l|\hat{U}|\psi\rangle_s \otimes |A\rangle_a \right|^2$$
$$= {}_s\langle\psi|\hat{\pi}_{ml}|\psi\rangle_s \,,$$

(1.61)

where

$$\hat{\pi}_{ml} = {}_a\langle A|\hat{U}^\dagger|m\rangle_s \otimes |l\rangle_{aa}\langle l| \otimes {}_s\langle m|\hat{U}|A\rangle_a \,,$$

(1.62)

which is an operator acting *only* on the system state space. It is clearly Hermitian and positive. That it sums to the identity follows, moreover, from the completeness of the orthonormal states $\{|m\rangle_s\}$ and $\{|l\rangle_a\}$. It follows that the operators $\hat{\pi}_{ml}$ are probability operators.

Let us give a simple but important example of a generalized measurement, namely the simultaneous measurement of the position and momentum of a particle. This is something that we do all the time in our predominantly classical world, but, because position and momentum are incompatible observables, it *cannot* be described as a von Neumann measurement. Let us write the probability density for a joint measurement to give a position between x_m and $x_m + dx_m$ and also a momentum between p_m and $p_m + dp_m$ as

$$\mathcal{P}(x_m, p_m) = \text{Tr}\left[\hat{\rho}\hat{\pi}(x_m, p_m)\right] \,,$$

(1.63)

where the $\hat{\pi}(x_m, p_m)$ are positive operators satisfying the continuum condition

$$\int_{-\infty}^{\infty} dx_m \int_{-\infty}^{\infty} dp_m \, \hat{\pi}(x_m, p_m) = \hat{I}$$
$$\implies \int_{-\infty}^{\infty} dx_m \int_{-\infty}^{\infty} dp_m \, \mathcal{P}(x_m, p_m) = 1 \,.$$

(1.64)

A good simultaneous measurement of the position and momentum of a body will localize both quantities rather well. In order to produce a plausible measurement operator, let is introduce a Gaussian state:

$$|x_m, p_m\rangle = (2\pi\sigma^2)^{-1/4} \int_{-\infty}^{\infty} dx \exp\left[-\frac{(x - x_m)^2}{4\sigma^2} + i\frac{p_m x}{\hbar}\right]|x\rangle \,,$$

(1.65)

where $|x\rangle$ is a position eigenstate. We note that these states form an over-complete set in that

$$\frac{1}{2\pi\hbar} \int_{-\infty}^{\infty} dx_m \int_{-\infty}^{\infty} dp_m \, |x_m, p_m\rangle\langle x_m, p_m| = \hat{I} \,.$$

(1.66)

Comparing this with our normalization condition leads us to identify our probability operators as

$$\hat{\pi}(x_m, p_m) = \frac{1}{2\pi\hbar} |x_m, p_m\rangle\langle x_m, p_m| \,.$$

(1.67)

It follows, for example, that the probability distribution for the measured position is

$$P(x_m) = \int_{-\infty}^{\infty} dx \, \langle x | \hat{\rho} | x \rangle \exp\left[-\frac{(x - x_m)^2}{2\sigma^2} \right], \tag{1.68}$$

which is a convolution of the true momentum distribution, $\langle x | \hat{\rho} | x \rangle$, with a Gaussian of width associated with the state $|x_m, p_m\rangle$. A similar expression applies also for the measured momentum distribution. The variances found for the measurement results are

$$\mathrm{Var}(x_m) = \Delta x^2 + \sigma^2,$$
$$\mathrm{Var}(p_m) = \Delta p^2 + \frac{\hbar^2}{4\sigma^2}. \tag{1.69}$$

The additional contributions, over and above Δx^2 and Δp^2, are a consequence of the fact that we have performed a simultaneous measurement of two incompatible observables.

1.3.4   Optimized measurements

We can seek the best possible measurement in any given situation in two stages: first, we obtain the mathematical optimization by finding the best POM; second, we design an experiment that gives the corresponding probabilities. The mathematical optimization is simply a search over all possible sets of probability operators forming a POM. The optimal measurements to perform have been determined in a number of cases and some of these have also been realized in the laboratory [4, 6, 13, 23, 77].

Let us begin by considering a simple example motivated by our treatment of quantum communications. Suppose we have a qubit that we know to have been prepared in one of two non-orthogonal states $|\psi_1\rangle$ or $|\psi_2\rangle$:

$$\langle \psi_1 | \psi_2 \rangle \neq 0. \tag{1.70}$$

We can use our probability operators to ask whether a measurement exists that will determine which state has been prepared *with certainty*. Not surprisingly, the answer is no, but we can prove this. To proceed, we seek two probability operators, $\hat{\pi}_1$ and $\hat{\pi}_2$, such that

$$\langle \psi_1 | \hat{\pi}_1 | \psi_1 \rangle = 1 = \langle \psi_2 | \hat{\pi}_2 | \psi_2 \rangle,$$
$$\langle \psi_2 | \hat{\pi}_1 | \psi_2 \rangle = 0 = \langle \psi_1 | \hat{\pi}_2 | \psi_1 \rangle, \tag{1.71}$$
$$\hat{\pi}_1 + \hat{\pi}_2 = \hat{I}.$$

From our first condition, $\langle \psi_1 | \hat{\pi}_1 | \psi_1 \rangle = 1$, we can write

$$\hat{\pi}_1 = |\psi_1\rangle\langle\psi_1| + \hat{A}, \tag{1.72}$$

where $\hat{A}$ is a Hermitian, positive operator and $\hat{A}|\psi_1\rangle = 0$. From this, it follows that

$$\langle \psi_2 | \hat{\pi}_1 | \psi_2 \rangle = |\langle \psi_2 | \psi_1 \rangle|^2 + \langle \psi_2 | \hat{A} | \psi_2 \rangle, \tag{1.73}$$

which, being the sum of a positive quantity and a quantity greater than or equal to zero, must be greater than zero. Hence we cannot discriminate between these states perfectly. A natural question is then what is the best we can do? The answer depends, of course, on what we mean by 'best'. Here we consider only two such strategies: measurement with the minimum probability of error and unambiguous state discrimination.

1.3.4.1 *Maximum probability of being correct*

Consider the following task, motivated by our discussion of quantum communications. We are given a quantum system that we know to have been prepared in one of a set of possible states $\{\hat{\rho}_j\}$ and also the probabilities, $\{p_j\}$ with which each state was prepared. We seek to identify the state with the maximum probability of being correct or, equivalently, with the minimum value of the error probability. Hence we construct a POM with probability operators $\{\hat{\pi}_j\}$ and associate each measurement outcome, $\hat{\pi}_j$, with the corresponding state $\hat{\rho}_j$ and then minimize the quantity

$$P_{\text{error}} = 1 - \sum_j p_j \, \text{Tr}(\hat{\rho}_j \hat{\pi}_j) \,. \tag{1.74}$$

Remarkably, the necessary and sufficient conditions for reaching this minimum are known [4, 48, 49]:

$$\hat{\pi}_j \left(p_j \hat{\rho}_j - p_k \hat{\rho}_k\right) \hat{\pi}_k = 0 \,,$$

$$\sum_i p_i \hat{\rho}_i \hat{\pi}_i - p_j \hat{\rho}_j \geq 0 \,. \tag{1.75}$$

A simple derivation of these is given in [5]. These conditions do not provide a direct means to construct the minimum-error measurement and, indeed, there are many cases in which the strategy for measuring with minimum probability of error is not unique. Our approach, therefore, is try a measurement strategy, and if it satisfies these conditions, then we know it is optimal. The problem is usually a tractable one when there is some symmetry among the set of possible states, and the minimum-error strategy has been derived in a variety of cases [4, 6, 13, 23, 77]. It is interesting to note that sometimes a measurement does not help and the minimum-error strategy is simply to guess that the most likely state was prepared [52].

If we have just two possible states, $\hat{\rho}_1$ and $\hat{\rho}_2$, then the minimum-error measurement is known: we need only perform a projective (or von Neumann) measurement in which the two measurement outcomes correspond to projectors onto the positive and negative eigenvalues of $p_1\hat{\rho}_1 - p_2\hat{\rho}_2$, corresponding respectively to identifying the state as $\hat{\rho}_1$ and $\hat{\rho}_2$. This gives as the minimum-error probability the Helstrom bound:

$$P_{\text{error}}^{\text{min}} = \tfrac{1}{2} \left[1 - \text{Tr} \, |p_1\hat{\rho}_1 - p_2\hat{\rho}_2| \right] \,. \tag{1.76}$$

As an example, let us suppose that we have a single qubit that we know to have been prepared with equal probability $(p_1 = \frac{1}{2} = p_2)$ in one of the two non-orthogonal states

$$|\psi_1\rangle = \cos\theta|0\rangle + \sin\theta|1\rangle \,,$$
$$|\psi_2\rangle = \cos\theta|0\rangle - \sin\theta|1\rangle \,,$$

(1.77)

so that the overlap of the states is

$$\langle\psi_1|\psi_2\rangle = \cos 2\theta \,.$$

(1.78)

It is straightforward to confirm that the minimum-error strategy corresponds to the two probability operators

$$\hat{\pi}_1 = \tfrac{1}{2}(|0\rangle + |1\rangle)(\langle 0| + \langle 1|) = |\hat{\pi}_1\rangle\langle\hat{\pi}_1| \,,$$
$$\hat{\pi}_2 = \tfrac{1}{2}(|0\rangle - |1\rangle)(\langle 0| - \langle 1|) = |\hat{\pi}_2\rangle\langle\hat{\pi}_2| \,.$$

(1.79)

The arrangement of these states is depicted in Fig. 1.11. We see that the optimal measurement in this case corresponds to projection onto one of a pair of states $|\hat{\pi}_1\rangle$ or $|\hat{\pi}_2\rangle$ that are 'close' to the states $|\psi_1\rangle$ and $|\psi_2\rangle$. The error probability arises from the fact that $|\hat{\pi}_{1(2)}\rangle$ is not perpendicular to $|\psi_{2(1)}\rangle$. A simple experiment has been performed at this limit to discriminate between two states of photon linear polarization [2].

By no means all optimal measurements are von Neumann measurements. Indeed, a generalized measurement is normally required. As a simple illustration, consider the

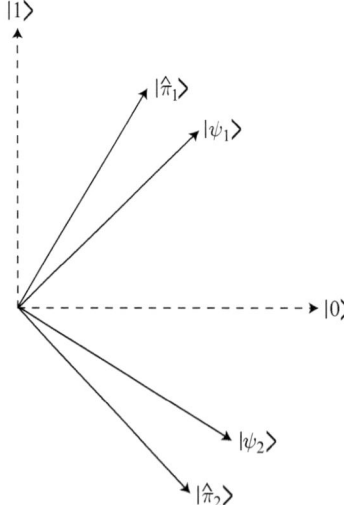

Fig. 1.11 Orientation in state space of the two non-orthogonal states to be discriminated and the directions along which it is best to measure. (Reproduced, with permission, from Barnett, S. M. (2009) Quantum information. Oxford University Press, Oxford.)

trine ensemble in which a qubit is prepared (with equal probability) in one of the three states

$$|\psi_1\rangle = |0\rangle \,,$$
$$|\psi_2\rangle = \tfrac{1}{2}\left(-|0\rangle + \sqrt{3}|1\rangle\right),$$
$$|\psi_3\rangle = \tfrac{1}{2}\left(|0\rangle + \sqrt{3}|1\rangle\right). \tag{1.80}$$

These correspond to states separated by 120° on a great circle of the Bloch sphere, as depicted in Fig. 1.12. The minimum-error strategy in this case corresponds to the three probability operators

$$\hat{\pi}_i = \tfrac{2}{3}|\psi_i\rangle\langle\psi_i| \tag{1.81}$$

and gives an error probability of $\frac{1}{3}$, so this measurement strategy determines the correct state with probability $\frac{2}{3}$. The required generalized measurement has been demonstrated using photon polarization qubits; the device is an interferometer in which the path taken through the device provides the ancillary degree of freedom [25].

1.3.4.2 *Unambiguous discrimination*

We have found the minimum-error strategy for discriminating between a pair of equiprobable qubit states, $|\psi_1\rangle$ and $|\psi_2\rangle$, but can we ever determine the state *for certain*? This might sound like a contradiction, but it is not if we allow for the possibility of an ambiguous measurement outcome. To see this, consider the von Neumann measurement in which the projectors are

$$\hat{P}_1 = |\psi_1\rangle\langle\psi_1| \,,$$
$$\hat{P}_{\bar{1}} = |\psi_1^\perp\rangle\langle\psi_1^\perp| \,, \tag{1.82}$$

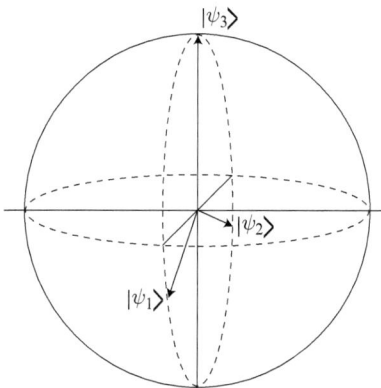

Fig. 1.12 The trine states depicted on the Bloch sphere. (Reproduced, with permission, from Barnett, S. M. (2009) Quantum information. Oxford University Press, Oxford.)

where $|\psi_1^\perp\rangle$ is the state orthogonal to $|\psi_1\rangle$. If we get the result $\bar{1}$, then we know for certain that the state was not $|\psi_1\rangle$ and hence that it *must* have been $|\psi_2\rangle$. If we get the result 1, however, then the state could have been either $|\psi_1\rangle$ or $|\psi_2\rangle$ and the outcome is ambiguous. Hence this simple measurement strategy determines the state unambiguously, but only some of the time.

It is natural to ask if we can find an optimal unambiguous discrimination strategy [4, 6, 23, 34, 54, 78]. By this, we mean a measurement strategy that identifies the state unambiguously with the highest probability or, equivalently, produces the smallest probability for an ambiguous outcome. The optimal strategy in this case is the three-element POM with probability operators

$$\hat{\pi}_1 = \frac{1}{1 + |\langle\psi_1|\psi_2\rangle|} \, |\psi_2^\perp\rangle\langle {}_2^\perp| \, ,$$

$$\hat{\pi}_2 = \frac{1}{1 + |\langle\psi_1|\psi_2\rangle|} \, |\psi_1^\perp\rangle\langle {}_1^\perp| \, , \tag{1.83}$$

$$\hat{\pi}_3 = \hat{\mathrm{I}} - \hat{\pi}_1 - \hat{\pi}_2 \, .$$

Note that the first two probability operators, $\hat{\pi}_{1(2)}$, are proportional to projectors onto the states *perpendicular* to $|\psi_{2(1)}\rangle$. This means that the measurement outcomes corresponding to these two operators *unambiguously* identify the state as $|\psi_{1(2)}\rangle$. The state will be correctly identified with probability $1 - |\langle\psi_1|\psi_2\rangle|$ and an ambiguous result occurs with probability $|\langle\psi_1|\psi_2\rangle|$. It is noteworthy that it is the modulus of the overlap of the states and not the modulus *squared* that appears in these probabilities.

To realize this unambiguous detection, we require a three-element POM, and unambiguous discrimination between two non-orthogonal photon polarization states has been demonstrated experimentally using optical fibre with polarization-dependent losses [53] and in an interferometer similar to that used for minimum-error discrimination between the trine states [24].

We have presented here only two of a variety of optimal measurements. Others that have been investigated include maximizing the mutual information and determining the state with maximum confidence. These, the relationships between them, and the experiments that have been performed are discussed further in [6].

1.3.5 Operations and the post-measurement state

We have not as yet addressed the question of how the measurement process modifies the quantum state in a generalized measurement.[6] There are two pressing reasons for proceeding beyond the von Neumann ideal in which the quantum system is left in an eigenstate corresponding to the measurement outcome. The first is that most real measurements are more destructive than this, and the second is that it gives us no idea how to describe the post-measurement state for a generalized, i.e. non-projective, measurement.

---

[6] Time did not permit me to address this question at the School, but these notes would be incomplete without at least a brief account of it.

A rigorous treatment would take us into the mathematical world of effects and operations [62]. Rather than doing this, we present only an indication of what is required. For a more complete treatment, we refer the reader to [4, 30]. We start by noting that quantum theory is linear in the state, which suggests that an allowed transformation of the density operator should be of the form

$$\hat{\rho} \rightarrow \hat{\rho}' = \sum_i \hat{A}_i \hat{\rho} \hat{B}_i \,. \tag{1.84}$$

Not every set of operators $\{\hat{A}_i, \hat{B}_i\}$ will be allowed, however, since the transformed density operator must itself be a density operator. This means that it must be Hermitian, $\hat{\rho}' = \hat{\rho}'^\dagger$, it must be positive, $\langle \psi | \hat{\rho}' | \psi \rangle$, and it must have unit trace, $\text{Tr} \, \hat{\rho}' = 1$. The first of these conditions suggests that we should set $\hat{B}_i = \hat{A}_i^\dagger$, and doing so automatically ensures that the second is fulfilled. The final one, the preservation of the unit trace, is satisfied if we set $\sum_i \hat{A}_i^\dagger \hat{A}_i = \hat{I}$.

The operator $\hat{A}_i^\dagger \hat{A}_i$ is positive and also Hermitian. The fact that we require the sum of these products to equal the identity operator, moreover, suggests the natural identification

$$\hat{\pi}_i = \hat{A}_i^\dagger \hat{A}_i \,, \tag{1.85}$$

and this allows us to complete the required description of the change of state after a measurement. If we know that a measurement has been performed but do not know the measurement outcome, then the density operator is transformed as

$$\hat{\rho} \rightarrow \sum_i \hat{A}_i \hat{\rho} \hat{A}_i^\dagger \,. \tag{1.86}$$

If we know that measurement result i was recorded, however, then the state is transformed as

$$\hat{\rho} \rightarrow \frac{\hat{A}_i \hat{\rho} \hat{A}_i^\dagger}{\text{Tr}\big(\hat{A}_i \hat{\rho} \hat{A}_i^\dagger\big)} = \frac{\hat{A}_i \hat{\rho} \hat{A}_i^\dagger}{\text{Tr}\big(\hat{\pi}_i \hat{\rho}\big)} \,. \tag{1.87}$$

In order to arrive at a unit-trace density operator in this case, we have divided by the a priori probability for the measurement result. This is directly analogous to the procedure in the von Neumann scheme, in which the transformation is

$$\hat{\rho} \rightarrow \frac{\hat{P}_n \hat{\rho} \hat{P}_n}{\text{Tr}\big(\hat{P}_n \hat{\rho}\big)} \,. \tag{1.88}$$

It should be noted that a set of Kraus operators $\{\hat{A}_i\}$ determine uniquely a corresponding set of probability operators $\{\hat{\pi}_i\}$, but the converse is *not* true; assigning the probability operators does not determine a unique transformation on the state.

1.4 Entanglement and its applications

More than any other feature, it is entanglement that gives quantum theory its distinctive character. It is the source of a number of paradoxes, the most famous of which is the much discussed EPR paradox [16, 36, 101, 102], of tests of distinctive 'quantumness', such as the violation of Bell's famous inequality [8, 9], and it underlies some of the more unexpected (and headline-grabbing) features of quantum information such as teleportation [12].

1.4.1 Entangled states and non-locality

We should start by stating what entanglement is. The simplest answer is that it is the distinctive property of entangled states, which leads us to ask 'what are entangled states?' This is, in fact, a surprisingly difficult question to answer, at least in full generality. For the purposes of these all too brief lectures, we shall avoid completely the tricky issue of entanglement for mixed states and discuss only pure states. For pure states, there is a clear definition, but it is the definition of states that are *not* entangled. A state that is not entangled is a state of two or more quantum systems that can be factorized into a product of single-system states. States that do not have this property are entangled. For two quantum systems, A and B, the combined state, $|\Psi\rangle_{AB}$, is entangled if[7]

$$|\Psi\rangle_{\mathrm{AB}} \neq |\psi\rangle_{\mathrm{A}} \otimes |\psi\rangle_{\mathrm{B}} . \tag{1.89}$$

Consider, for example, the two-qubit state

$$|\Psi\rangle = \cos\theta\,|0\rangle \otimes |1\rangle - \sin\theta\,|1\rangle \otimes |0\rangle . \tag{1.90}$$

This state will not be entangled if $\theta = n\pi/2$, but will be entangled for all other values of θ. For entangled states, the properties of the A and B systems are quantum-correlated. We can see this in the above example in that determining whether the first qubit is in the state $|0\rangle$ or the state $|1\rangle$ also determines the, previously undetermined, state of the second qubit.

The implications of the existence of entanglement are profound and many. In these lectures, we shall address only five:

1. EPR and related non-locality paradoxes.
2. The ability to perform 'magic'.
3. Quantum dense coding.
4. Teleportation.
5. Dramatic speed-up in quantum computing.

We shall treat the first four of these in this lecture and leave the last for the next and final lecture.

---

[7] For those not familiar with the notation, it is often convenient to use the symbol $\otimes$ to separate quantum states and especially operators. Thus, $\hat{\sigma}_x\hat{\sigma}_y$ denotes acting first with $\hat{\sigma}_y$ then with $\hat{\sigma}_x$ on a single qubit, but $\hat{\sigma}_x \otimes \hat{\sigma}_y$ means acting with $\hat{\sigma}_x$ on qubit 1 and with $\hat{\sigma}_y$ on qubit 2 [4].

Let us begin with the EPR paradox, in the form given by Bohm [15]. To this end, consider two qubits in the entangled state

$$|\Psi^-\rangle_{AB} = \frac{1}{\sqrt{2}} \left(|0\rangle_A \otimes |1\rangle_B - |1\rangle_A \otimes |0\rangle_B \right). \tag{1.91}$$

Let us suppose that qubit A is held by Alice and qubit B by Bob and that they are at a considerable distance from each other. If either party measures their qubit is any basis, then they find either of two possible results, each occurring with probability $\frac{1}{2}$. If they measure the same observable as each other, then they find opposite or anticorrelated results. This is a simple consequence of the eigenvalue equation

$$\mathbf{a} \cdot \hat{\boldsymbol{\sigma}} \otimes \mathbf{a} \cdot \hat{\boldsymbol{\sigma}} |\Psi^-\rangle = -|\Psi^-\rangle, \tag{1.92}$$

which holds for all unit vectors $\mathbf{a}$. The paradox is that a measurement by Alice of her particle will *instantaneously* project the state of its partner into an eigenstate of the observable selected by Alice. If we assume that Alice has a free choice of what to measure and that any influence of her measurement cannot travel arbitrarily fast, then Bob's qubit must have carried values for all possible observables, something that is clearly at odds with complementarity. We give here only a partial resolution of this in the form of the no-signalling theorem.

1.4.1.1 Bell's theorem

Correlations are also common in the classical world and, indeed, underlie classical communications. Are not the quantum correlations associated with entanglement simply the same thing? It was Bell's theorem, in the form of the violation of his celebrated inequality, that gave the definitive answer 'no'!

We present here a derivation of Bell's inequality in its most common form [9, 26]. Let us start by thinking of our qubit as a spin-half particle. A measurement of the spin along a direction in space, given by the unit vector $\mathbf{a}$, will reveal the spin to be aligned or anti-aligned with this direction. We write the first of these as +1 and the second as −1, so that in any given measurement our measurement result A will be ±1. We can do the same thing for the second qubit (entangled with the first) and write the measurement result as $B = \pm 1$. If we take the product of these two values and average over many experimental realizations, then we obtain the correlation function

$$E(\mathbf{a}, \mathbf{b}) = \langle AB \rangle, \tag{1.93}$$

which is clearly bounded in magnitude:

$$-1 \le E(\mathbf{a}, \mathbf{b}) \le 1. \tag{1.94}$$

In order to arrive at Bell's inequality, we need to suspend disbelief and wonder what a theory would look like if the indeterminacy of quantum theory arose from hidden variables or, more precisely, local hidden variables, which we denote by λ and take to be governed by a probability distribution $P(\lambda)$. To this end, we write the correlation function in the form

$$E(\mathbf{a}, \mathbf{b}) = \int d\lambda \, P(\lambda) A(\mathbf{a}, \lambda) B(\mathbf{b}, \lambda) \,. \tag{1.95}$$

Some words of explanation are in order. We say that this is a form based on a local hidden-variable theory because each measured result A and B depends only on (i) the choice of observable made at the observation site ($\mathbf{a}$ and $\mathbf{b}$, respectively) and (ii) the hidden variable λ, presumed to have been determined in the source of the particles. The measurement results do *not* depend on the choice of measurement made at the distant site. These considerations lead us to Bell's inequality as follows. First we write

$$E(\mathbf{a}, \mathbf{b}) - E(\mathbf{a}, \mathbf{b}') = \int d\lambda \, P(\lambda)[A(\mathbf{a}, \lambda)B(\mathbf{b}, \lambda) - A(\mathbf{a}, \lambda)B(\mathbf{b}', \lambda)]$$

$$= \int d\lambda \, P(\lambda)A(\mathbf{a}, \lambda)B(\mathbf{b}, \lambda)[1 \pm A(\mathbf{a}', \lambda)B(\mathbf{b}', \lambda)] \tag{1.96}$$

$$- \int d\lambda \, P(\lambda)A(\mathbf{a}, \lambda)B(\mathbf{b}', \lambda)[1 \pm A(\mathbf{a}', \lambda)B(\mathbf{b}, \lambda)] \,.$$

In the second line, we have added and subtracted the same expression. In doing so, we have assumed that it is meaningful for quantities such as $A(\mathbf{a}, \lambda)$ and $A(\mathbf{a}', \lambda)$ to coexist. This is the realism part of local realism; it states that properties exist even if we do not measure them. We recall that the product AB lies between -1 and $+1$ and hence we can obtain from this equality an inequality of the form

$$|E(\mathbf{a}, \mathbf{b}) - E(\mathbf{a}, \mathbf{b}')| \leq \int d\lambda \, P(\lambda)[1 \pm A(\mathbf{a}', \lambda)B(\mathbf{b}', \lambda)]$$

$$+ \int d\lambda \, P(\lambda)[1 \pm A(\mathbf{a}', \lambda)B(\mathbf{b}, \lambda)] \tag{1.97}$$

$$= 2 \pm [E(\mathbf{a}', \mathbf{b}') + E(\mathbf{a}', \mathbf{b})] \,,$$

which implies that

$$|E(\mathbf{a}, \mathbf{b}) - E(\mathbf{a}, \mathbf{b}')| + |E(\mathbf{a}', \mathbf{b}') + E(\mathbf{a}', \mathbf{b})| \leq 2 \,. \tag{1.98}$$

This is Bell's inequality; it states that any correlations of a local-realistic nature must satisfy this inequality. What do we find for entangled quantum states? Consider again the EPR spin state:

$$|\Psi^-\rangle_{AB} = \frac{1}{\sqrt{2}}(|0\rangle_A \otimes |1\rangle_B - |1\rangle_A \otimes |0\rangle_B) \,. \tag{1.99}$$

It is straightforward to confirm that for spin measurements on the two qubits prepared in this state, we find

$$E(\mathbf{a}, \mathbf{b}) = \langle \mathbf{a} \cdot \hat{\boldsymbol{\sigma}} \otimes \mathbf{a} \cdot \hat{\boldsymbol{\sigma}} \rangle = -\mathbf{a} \cdot \mathbf{b} \,. \tag{1.100}$$

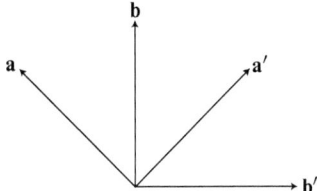

Fig. 1.13 Orientation of the optimal measurement directions for maximal violation of Bell's inequality. (Reproduced, with permission, from Barnett, S. M. (2009) Quantum information. Oxford University Press, Oxford.)

If we put this into our Bell inequality, we find

$$|E(\mathbf{a}, \mathbf{b}) - E(\mathbf{a}, \mathbf{b}')| + |E(\mathbf{a}', \mathbf{b}') + E(\mathbf{a}', \mathbf{b})| = |-\mathbf{a} \cdot (\mathbf{b} - \mathbf{b}')| + |-\mathbf{a}' \cdot (\mathbf{b} + \mathbf{b}')|. \tag{1.101}$$

To find the maximum value, we simply choose $\mathbf{a}$ and $\mathbf{a}'$ to be parallel to $\mathbf{b} - \mathbf{b}'$ and $\mathbf{b} + \mathbf{b}'$, respectively, and $\mathbf{b}$ and $\mathbf{b}'$ to be mutually perpendicular, as depicted in Fig. 1.13. With this choice, we find

$$|E(\mathbf{a}, \mathbf{b}) - E(\mathbf{a}, \mathbf{b}')| + |E(\mathbf{a}', \mathbf{b}') + E(\mathbf{a}', \mathbf{b})| = 2\sqrt{2}, \tag{1.102}$$

which clearly exceeds 2 and hence violates the inequality. It is this, perhaps more than anything else, that signified the exceptional nature of entanglement. In quantum information, it is often by violating a Bell inequality that we demonstrate most clearly the presence of entanglement: all entangled pure states violate a Bell inequality [40].

1.4.1.2 No-signalling theorem

A resolution, of sorts, of the EPR paradox is that no operation carried out on one of an entangled pair of quantum systems is detectable by an observation on its entangled partner. This means, of course, that no signal can be transmitted in this way. The no-signalling theorem was stated rigorously by Ghirardi et al. [39], but rather than following their analysis, it is instructive to make use of what we have learned already about generalized measurements [4].

Let us suppose that Alice and Bob each have one of a pair of entangled systems and that Alice attempts to communicate with Bob by making a measurement on hers. She cannot, of course, control the measurement result, but she can decide what to measure. We shall allow Alice to choose between any two measurements, which we allow to be generalized. Let Alice's measurements have the probability operators $\{\hat{\pi}_j^{A1}\}$ for measurement choice 1 and $\{\hat{\pi}_j^{A2}\}$ for choice 2. In an attempt to detect the effect of Alice's choice, Bob makes a measurement characterized by the probability operators $\{\hat{\pi}_k^{B}\}$. The joint probabilities for Alice's and Bob's measurement results are

$$P(j, k) = \langle \Psi | \hat{\pi}_j^{A1,\ A2} \otimes \hat{\pi}_k^{B} | \Psi \rangle, \tag{1.103}$$

where $|\Psi\rangle$ is the joint state of Alice's and Bob's system. If Alice's choice is to affect Bob's observation, then we need the probabilities for Bob's measurement results to depend on Alice's choice of measurement. This is clearly not the case, however, since

$$P(k) = \sum_j P(j,k) = \langle\Psi|\hat{I} \otimes \hat{\pi}_k^B|\Psi\rangle, \tag{1.104}$$

where we have used the fact that our probability operators sum to the identity operator:

$$\sum_j \hat{\pi}_j^{A1,A2} = \hat{I}. \tag{1.105}$$

The probabilities of any of Bob's possible measurement results are *independent* of Alice's choice of measurement and, indeed, are the same whether Alice performs a measurement at all. Nothing Alice does to her quantum system is detectable by Bob and hence no signalling is possible by quantum measurement of a distant system.

The no-signalling theorem is a rigorous consequence of quantum theory and can be used, in quantum information theory, to place bounds on what is and is not possible. As a simple example, we can derive a a lower bound on the probability for an inconclusive outcome in unambiguous state-discrimination using the no-signalling condition [3].

1.4.2 Quantum 'magic tricks'

Bell's inequality is only one of an impressive array of startling and testable consequences of entanglement [44, 72, 84]. Were these lectures about the mysteries of quantum theory, then we might enjoy exploring these, but our purpose is not to test entanglement and its consequences, but rather to exploit it. We shall present one example, however, of a testable consequence of entanglement as it points to a necessary distinction between logical reasoning in quantum theory and the classical world.

Hardy [43, 47] presented a simple demonstration of non-locality without an inequality. This is instructive in that it provides a warning against reasoning on the basis of what might have been measured rather than what actually was observed. The following short presentation is reproduced, essentially verbatim, from [4].

Consider a pair of qubits, one held by Alice and the other by Bob, prepared in the pure state

$$|\text{Hardy}\rangle = \frac{1}{\sqrt{3}}\left(|0\rangle_A \otimes |0\rangle_B + |1\rangle_A \otimes |0\rangle_B + |0\rangle_A \otimes |1\rangle_B\right). \tag{1.106}$$

We proceed by noting that this state can also be written in the form

$$|\text{Hardy}\rangle = \sqrt{\frac{2}{3}}|0'\rangle_A \otimes |0\rangle_B + \frac{1}{\sqrt{3}}|0\rangle_A \otimes |1\rangle_B$$

$$= \frac{1}{\sqrt{3}}|1\rangle_A \otimes |0\rangle_B + \sqrt{\frac{2}{3}}|0\rangle_A \otimes |0'\rangle_B, \tag{1.107}$$

where $|0'\rangle = \frac{1}{\sqrt{2}}(|0\rangle + |1\rangle)$ is the eigenstate of $\hat{\sigma}_x$ with eigenvalue $+1$. The following statements follow directly from the form of $|\,\text{Hardy}\rangle$:

1. If both Alice and Bob measure the observable corresponding to $\hat{\sigma}_z$, with eigenstates $|0\rangle$ and $|1\rangle$, then at least one of them will get the result $+1$, corresponding to the state $|0\rangle$.
2. If Alice measures $\hat{\sigma}_z$ and gets the value $+1$, then a measurement by Bob of $\hat{\sigma}_x$ will, with certainty, find the value $+1$, corresponding to the state $|0'\rangle$.
3. If Bob measures $\hat{\sigma}_z$ and gets the value $+1$, then a measurement by Alice of $\hat{\sigma}_x$ will, with certainty, find the value $+1$, corresponding to the state $|0'\rangle$.

Local-realistic ideas lead us to treat as simultaneously real the values ±1 of the observables corresponding to the operators $\hat{\sigma}_z$ and $\hat{\sigma}_x$. The values of these, which we denote by σ_z and σ_x, respectively, should be independent of any choice of an observation carried out on the other qubit. This leads us to express the above three properties as the following probabilities:

$$P(\sigma_z^{\text{A}} = -1, \sigma_z^{\text{B}} = -1) = 0\,,$$
$$P(\sigma_x^{\text{B}} = +1|\sigma_z^{\text{A}} = +1) = 1\,, \tag{1.108}$$
$$P(\sigma_x^{\text{A}} = +1|\sigma_z^{\text{B}} = +1) = 1\,.$$

The first of these tells us that at least one of the properties σ_z^{A} and σ_z^{B} must take the value $+1$, and the following two then tell us that at least one of σ_x^{A} and σ_z^{B} must take the value $+1$. It is a prediction of local realism, therefore, that σ_x^{A} and σ_x^{B} cannot *both* take the value -1:

$$P(\sigma_x^{\text{A}} = -1, \sigma_x^{\text{B}} = -1) = 0\,. \tag{1.109}$$

A quantum mechanical treatment, however, shows that measurement by both Alice and Bob of $\hat{\sigma}_x$ can *both* give the result $+1$ and that this occurs with probability $\frac{1}{12}$. This is clearly at odds with the local-realistic reasoning presented above.

As with all the best magic tricks, it is what the audience assumes to have happened that makes the trick appear to be impossible.

1.4.3 Ebits and shared entanglement

In quantum information, we think of entanglement not as a mystery, but rather as a *resource* to be exploited. The first question to be addressed, as with any resource, is to decide how to quantify it. This is a subtle question, but we can give, at least, a simple preliminary answer in terms of 'ebits'. If two spatially separated individuals share a pair of maximally entangled qubits, then we say that they share 1 ebit. By maximally entangled, we mean a pure state such that the reduced density operator for each of the component qubits is of the form $\hat{I}/2$. This means that we can identify the state only by bringing together the component qubits and that a measurement of just a single qubit is maximally uncertain.

It is important to ask how we can get to a situation in which two parties share an ebit. One way, of course, is to prepare an entangled state at one site and then transport (carefully) one of the entangled qubits to the other site. We might very reasonably ask, however, if it is not possible to achieve the same result by means of classical communications, by exchanging instructions between the distant sites. That this is *not* possible follows directly from what we already know. Let us suppose that Alice prepares the entangled state

$$|\Psi^-\rangle = \frac{1}{\sqrt{2}}(|0\rangle \otimes |1\rangle - |1\rangle \otimes |0\rangle). \tag{1.110}$$

To share this state with Bob, she can send the second qubit to Bob by means of a suitable quantum communications channel. The resulting quantum correlations between the qubits (at least for a sufficient number of entangled pairs) could be used to demonstrate a violation of a Bell inequality. Such a violation is not possible, of course, for any classical or classically mediated correlations. Hence, we can conclude, without need for further analysis, that classical and quantum communications are radically different in nature.

The antisymmetric state, $|\Psi^-\rangle$, is not the only maximally entangled state of two qubits. Indeed, we can create a complete basis of four orthonormal states for two qubits, each of which is similarly maximally entangled. One simple choice is the set of states now referred to as the Bell states or the Bell basis. These four states are

$$|\Psi^\pm\rangle = \frac{1}{\sqrt{2}}(|0\rangle \otimes |1\rangle \pm |1\rangle \otimes |0\rangle),$$

$$|\Phi^\pm\rangle = \frac{1}{\sqrt{2}}(|0\rangle \otimes |0\rangle \pm |1\rangle \otimes |1\rangle). \tag{1.111}$$

As these states form an orthonormal basis, we can perform a von Neumann measurement on the two qubits with each of these four corresponding to a different outcome. This 'Bell measurement' cannot be performed, however, simply by combining a von Neumann measurement on each individual qubit.

1.4.4 Quantum dense coding

A very simple application of the Bell states, perhaps the simplest, is in quantum dense coding [11]. The starting point is the observation that a (von Neumann) measurement of a qubit can provide at most one bit of information. To see how this works, we need only consider a qubit that has been prepared with equal probability in either of the two orthogonal states, $|0\rangle$ and $|1\rangle$. A simple projective measurement in this basis will reveal the state that has been prepared and provide a single bit.

By using an ebit, we can, in a sense, double this rate by communicating two bits of information for every qubit sent from Alice to Bob. Let us start with a single ebit shared by Alice and Bob, prepared in the state

$$|\Psi^-\rangle_{AB} = \frac{1}{\sqrt{2}}(|0\rangle \otimes |1\rangle - |1\rangle \otimes |0\rangle). \tag{1.112}$$

Bob selects one of four unitary transformations to perform on his qubit. These transformations, and the effect of the corresponding entangled state, are as follows:

$$\hat{U}_1 \quad \begin{matrix} |0\rangle_{\mathrm{B}} \\ |1\rangle_{\mathrm{B}} \end{matrix} \Rightarrow \begin{matrix} |0\rangle_{\mathrm{B}} \\ |1\rangle_{\mathrm{B}} \end{matrix} \qquad |\Psi^-\rangle_{\mathrm{AB}} \Rightarrow |\Psi^-\rangle_{\mathrm{AB}},$$

$$\hat{U}_2 \quad \begin{matrix} |0\rangle_{\mathrm{B}} \\ |1\rangle_{\mathrm{B}} \end{matrix} \Rightarrow \begin{matrix} |1\rangle_{\mathrm{B}} \\ |0\rangle_{\mathrm{B}} \end{matrix} \qquad |\Psi^-\rangle_{\mathrm{AB}} \Rightarrow |\Phi^-\rangle_{\mathrm{AB}},$$

$$\hat{U}_3 \quad \begin{matrix} |0\rangle_{\mathrm{B}} \\ |1\rangle_{\mathrm{B}} \end{matrix} \Rightarrow \begin{matrix} -|0\rangle_{\mathrm{B}} \\ |1\rangle_{\mathrm{B}} \end{matrix} \qquad |\Psi^-\rangle_{\mathrm{AB}} \Rightarrow |\Psi^+\rangle_{\mathrm{AB}},$$

$$\hat{U}_4 \quad \begin{matrix} |0\rangle_{\mathrm{B}} \\ |1\rangle_{\mathrm{B}} \end{matrix} \Rightarrow \begin{matrix} -|1\rangle_{\mathrm{B}} \\ |0\rangle_{\mathrm{B}} \end{matrix} \qquad |\Psi^-\rangle_{\mathrm{AB}} \Rightarrow |\Phi^+\rangle_{\mathrm{AB}}.$$

$$(1.113)$$

We see that the effect of Bob's choice is to transform the state of the entangled pair of qubits into one of the four orthogonal Bell states. Hence, by sending just his single qubit to Alice, he can send *two* bits of information (recall that $4 = 2^2$ states corresponds to 2 bits). In order to retrieve the information, Alice simply performs a Bell measurement on the two qubits comprising the ebit and extracts the two bits.[8]

It is interesting to reflect on the nature of a Bell measurement and, in particular, to try to understand it in terms of spin measurements. We know, of course, that the Pauli operators $\hat{\sigma}_x$, $\hat{\sigma}_y$, and $\hat{\sigma}_z$ do not mutually commute. It may therefore come as something of a surprise that the products of these Cartesian components of the spin, $\hat{\sigma}_x \otimes \hat{\sigma}_x$, $\hat{\sigma}_y \otimes \hat{\sigma}_y$, and $\hat{\sigma}_z \otimes \hat{\sigma}_z$, commute for *any* pair of qubits. The Bell states are the simultaneous eigenstates of these three product operators:

$$\begin{aligned}
\hat{\sigma}_x \otimes \hat{\sigma}_x |\Psi^-\rangle &= -|\Psi^-\rangle, & \hat{\sigma}_y \otimes \hat{\sigma}_y |\Psi^-\rangle &= -|\Psi^-\rangle, & \hat{\sigma}_z \otimes \hat{\sigma}_z |\Psi^-\rangle &= -|\Psi^-\rangle, \\
\hat{\sigma}_x \otimes \hat{\sigma}_x |\Psi^+\rangle &= |\Psi^+\rangle, & \hat{\sigma}_y \otimes \hat{\sigma}_y |\Psi^+\rangle &= |\Psi^+\rangle, & \hat{\sigma}_z \otimes \hat{\sigma}_z |\Psi^+\rangle &= -|\Psi^+\rangle, \\
\hat{\sigma}_x \otimes \hat{\sigma}_x |\Phi^-\rangle &= -|\Phi^-\rangle, & \hat{\sigma}_y \otimes \hat{\sigma}_y |\Phi^-\rangle &= |\Phi^-\rangle, & \hat{\sigma}_z \otimes \hat{\sigma}_z |\Phi^-\rangle &= |\Phi^-\rangle, \\
\hat{\sigma}_x \otimes \hat{\sigma}_x |\Phi^+\rangle &= |\Phi^+\rangle, & \hat{\sigma}_y \otimes \hat{\sigma}_y |\Phi^+\rangle &= -|\Phi^+\rangle, & \hat{\sigma}_z \otimes \hat{\sigma}_z |\Phi^+\rangle &= |\Phi^+\rangle.
\end{aligned}$$

$$(1.114)$$

Thus, the Bell states are those in which the products of the Cartesian components of the spin are defined and we can view a Bell measurement as a *comparison* of the three components of spin for the two qubits. You may have noticed, in fact, that it suffices to compare only two components of the spin, since the final one is then defined. The reason for this has its origins in the fact that the product of any two Pauli spin components is proportional to the third and, in particular, $\hat{\sigma}_x \hat{\sigma}_y = i\hat{\sigma}_z$. It follows that

$$(\hat{\sigma}_x \otimes \hat{\sigma}_x)(\hat{\sigma}_y \otimes \hat{\sigma}_y) = -\hat{\sigma}_z \otimes \hat{\sigma}_z \qquad (1.115)$$

---

[8] Although it sounds easy to perform a Bell measurement, achieving this in practice is, perhaps not surprisingly, rather more of a challenge [69].

and hence that the product of the eigenvalues of $\hat{\sigma}_x \otimes \hat{\sigma}_x$, $\hat{\sigma}_y \otimes \hat{\sigma}_y$, and $\hat{\sigma}_z \otimes \hat{\sigma}_z$ must be -1. This means that there can be no state of two qubits, for example, in which the x-, y-, and z-components of the spin are all the same.

1.4.5 Teleportation

Our final application is quantum teleportation. This is an important application of entanglement and came as something of a surprise when it was first proposed [12]. We should be very clear before proceeding, however, that quantum teleportation is *not* matter transportation like something out of the popular science fiction series 'Star Trek'. A better understanding may be reached, I feel, by adopting the description proposed by Haroche, which is that quantum teleportation is a 'fax machine for quantum information'. It is the quantum information, or rather the quantum state, that is transferred, and not the physical system on which the qubit state is stored.

Let us suppose that Alice wishes to send a qubit to Bob. She may know the state of the qubit or she may not. We have seen that quantum communications and classical communications are very different, and so we should not be surprised that there is, in general, no classical way to achieve this. Alice has two ways to achieve the desired result: (i) she can send the physical qubit carrying the quantum information to Bob (we might call this 'qmail') or (ii) she can teleport the quantum information to Bob using an ebit of shared entanglement (which we might call a 'qfax').

Alice can teleport the qubit state

$$|\psi\rangle_1 = \alpha|0\rangle_1 + \beta|1\rangle_1 \qquad (1.116)$$

by making use of an ebit of the form

$$|\Psi^-\rangle_{AB} = \frac{1}{\sqrt{2}}(|0\rangle_A \otimes |1\rangle_B - |1\rangle_A \otimes |0\rangle_B), \qquad (1.117)$$

shared with Bob. We can write the combined state of the three qubits as

$$\begin{aligned}
|\Psi\rangle_{1AB} &= |\psi\rangle_1 \otimes |\Psi^-\rangle_{AB} \\
&= \tfrac{1}{2}\big[|\Psi^-\rangle_{1A}(-\alpha|0\rangle_B - \beta|1\rangle_B) \\
&\quad + |\Psi^+\rangle_{1A}(-\alpha|0\rangle_B + \beta|1\rangle_B) \\
&\quad + |\Phi^-\rangle_{1A}(\alpha|1\rangle_B + \beta|0\rangle_B) \\
&\quad + |\Phi^+\rangle_{1A}(\alpha|1\rangle_B - \beta|0\rangle_B)\big].
\end{aligned} \qquad (1.118)$$

In arriving at the final expression, we have only rewritten the state, but the quantum information, in the form of the coefficients α and β, has appeared connected with Bob's qubit. We know, however, that Bob cannot yet access this information, since to do so would violate the no-signalling theorem.

If Alice performs a Bell measurement on her two qubits (qubits 1 and A), she can then tell Bob which of four possible transformations to perform on qubit B in order to change its state into $|\psi\rangle_B$. The idea is best understood in terms of an example. Let us suppose that Alice's Bell measurement gives a result corresponding to the state $|\Psi^+\rangle_{1A}$; then she knows that Bob's qubit is in the state $-\alpha|0\rangle_B + \beta|1\rangle_B$. She then has only to use a two-bit classical channel to tell Bob to perform unitary transformation number 3 so that

$$\hat{U}_3 \quad \begin{matrix} |0\rangle_B \\ |1\rangle_B \end{matrix} \Rightarrow \begin{matrix} -|0\rangle_B \\ |1\rangle_B \end{matrix} \quad -\alpha|0\rangle_B + \beta|1\rangle_B \Rightarrow \alpha|0\rangle_B + \beta|1\rangle_B , \tag{1.119}$$

and the teleportation is complete in that Bob's qubit is in the same state as that in which the original qubit started, $|\psi\rangle_1$. Each of the other three possible Bell-measurement outcomes corresponds to a different unitary transformation that Bob is required to perform.

We can understand why teleportation works by recalling that a Bell measurement is one in which we compare the Cartesian components of the two spins. In particular, the four possible measurement results tell us that

$$
\begin{aligned}
|\Psi^-\rangle &\Rightarrow \sigma_x \text{ different}, \quad \sigma_y \text{ different}, \quad \sigma_z \text{ different}, \\
|\Psi^+\rangle &\Rightarrow \sigma_x \text{ same}, \quad\quad \sigma_y \text{ same}, \quad\quad \sigma_z \text{ different}, \\
|\Phi^-\rangle &\Rightarrow \sigma_x \text{ different}, \quad \sigma_y \text{ same}, \quad\quad \sigma_z \text{ same}, \\
|\Phi^+\rangle &\Rightarrow \sigma_x \text{ same}, \quad\quad \sigma_y \text{ different}, \quad \sigma_z \text{ same}.
\end{aligned}
\tag{1.120}
$$

We know also that the shared entangled state, $|\Psi^-\rangle_{AB}$, has the three components of the spins all opposite to each other. Hence, we can reason as follows: (i) If the Bell measurement gives the state $|\Psi^-\rangle_{1A}$, then the spin components for the qubits 1 and A are anti-aligned, *but* we know also that the qubits A and B are anti-aligned and hence we are left with Bob's qubit in the initial state of qubit 1. (ii) If the Bell measurement gives the state $|\Psi^+\rangle_{1A}$, then the x- and y-components of the spins for qubits 1 and A are the same, but the z-component is different. It follows that qubit B is left in a state in which the x- and y-components of the spin are opposite to that for the initial state of qubit 1, but the z-component is the same. Instructing Bob to perform a rotation through π about the z-axis will leave his qubit in the initial state of qubit 1. Similarly, (iii) if the Bell measurement gives the state $|\Phi^-\rangle$, then Bob needs to perform a qubit rotation about the x-axis, and (iv) a Bell measurement giving the state $|\Phi^+\rangle$ means that Bob needs to perform a rotation about the y-axis.

There is much more that could be said about teleportation, but we conclude this all too brief introduction by describing a few features worthy of further thought:

1. We note that there is a sense in which Bob already has the information after Alice's measurement. This is a direct consequence of the projective nature of Alice's Bell measurement. What saves us from violating the no-signalling theorem is simply that Bob does not know where to look for it.
2. Alice's copy of the original qubit is destroyed in the teleportation process. This is inevitable and not simply a consequence of the scheme we have adopted. Were it otherwise, then we would be violating the no-cloning theorem.

3. Alice does not have to know the state that is to be teleported. We have chosen a general qubit state parametrized by the amplitudes α and β, but no process undertaken by either Alice or Bob is dependent on these amplitudes and it follows that Alice does not need to know the state to be teleported.

4. The qubit to be teleported may itself be part of an entangled state and in this way we can teleport entanglement. Thus, if Alice shares an ebit with Bob and also one with Claire, then she can teleport the entanglement to leave Bob and Claire sharing an ebit:

$$|\Psi^-\rangle_{C1} \otimes |\Psi^-\rangle_{AB} \Rightarrow |\Psi^-\rangle_{CB}. \tag{1.121}$$

This process of transferring entanglement by teleportation is commonly referred to as entanglement-swapping.

1.5 Quantum computation

No course on quantum information would be complete without a treatment of quantum computation. Yet covering such a diverse topic in a single lecture presents, if anything, an even greater challenge than the areas addressed in the preceding lectures. The material selected for this lecture can only provide a modest foretaste of a topic that spans a number of disciplines, including physics and computer science, and for a more satiating introduction I can only recommend some texts for further reading [4, 38, 59, 73, 75, 76, 95, 99].

We should start by asking why the idea of quantum computers seems to have become so prominent. There are, I think, two very good reasons. The first is in response to a very pressing need. The speed of development in computers has been truly remarkable and follows an exponential increase in performance first described by Moore in 1965 [74]. There are many versions of this law, but perhaps the simplest is that the number of transistors on chip doubles roughly every two years. The way this is achieved is by making the individual transistors ever smaller. As we make components ever smaller, however, we must inevitably run into quantum effects. Rather than fight against quantum mechanics (a battle we must lose at some size scale), perhaps it would be better to embrace the new possibilities provided by quantum information processing. The second reason is the distinctively new possibilities offered by quantum algorithms (which can run only on a quantum computer). These address problems that will always remain intractable for a computer based on classical logic and include simulations of complex quantum systems (like large molecules) and the headline-grabbing Shor algorithm for factoring, of which more later.

1.5.1 Digital electronics

We begin our discussion by describing the way in which logical bits and logic operations are implemented classically. In the simplest form, the logical bits 1 and 0 are encoded as voltages: a high voltage for 1 and ground or zero volts for 0. These values are manipulated by transistor-based devices called gates [93]. The most common of these act either on a single bit value or couple two together. There is only one single-bit

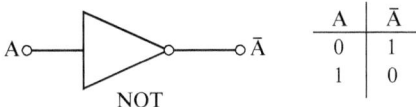

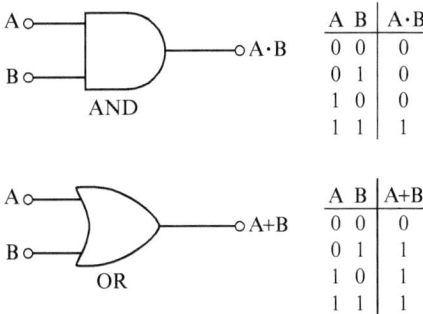

Fig. 1.14 The NOT gate and its truth table. (Reproduced, with permission, from Barnett, S. M. (2009) Quantum information. Oxford University Press, Oxford.)

Fig. 1.15 The AND and OR gates and their truth tables. (Reproduced, with permission, from Barnett, S. M. (2009) Quantum information. Oxford University Press, Oxford.)

gate, the NOT gate, which simply changes a 0 to 1 and a 1 to 0. The symbol for the NOT gate and its truth table are given in Fig. 1.14. The simplest two-bit gates are the AND and the OR gates, depicted in Fig. 1.15. Combinations of these together with a NOT gate to form NAND and NOR gates are also common. These two-bit gates, together with the NOT gate, allow us to perform any logical operation (strictly a Boolean operation) on a string of bits. Such an operation is, at a fundamental level, what we mean by a computation.

1.5.2 Quantum gates

The simplest way in which to introduce quantum elements into information processing is to replace each classical bit with a qubit and to design devices that operate upon them. We are not completely free in the operations we can devise, however, since we must respect the laws of quantum mechanics. This means, in particular, that the transformations we can perform are necessarily limited to the unitary transformations. Even at the single-qubit level, however, there is already a great deal more we can do than for a classical bit, since we are allowed to generate superpositions of the two qubit states $|0\rangle$ and $|1\rangle$. A single qubit-gate may perform any single-qubit unitary transformation, that is any rotation on the Bloch sphere. Six of the more commonly occurring one-qubit gates, together with the unitary transformations they enact, are presented in Fig 1.16.

We require in addition gates that couple qubits. Unlike their classical counterparts, these gates have two output qubits as well as two input ones; were this not the case, we would not respect unitarity. Principal among the vast array of possible two-bit

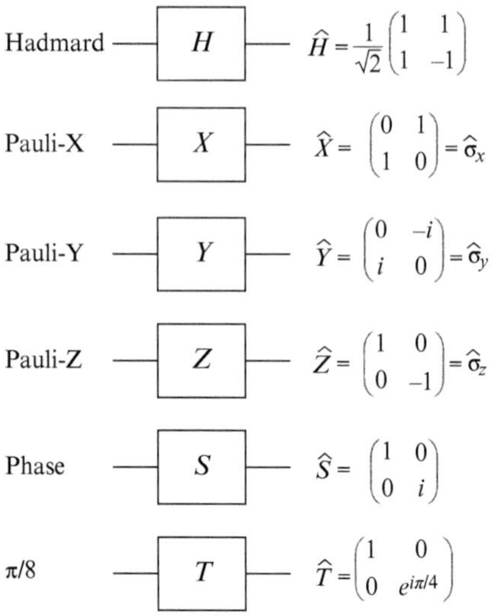

Hadmard
H
$\hat{H} = \frac{1}{\sqrt{2}}\begin{pmatrix}1 & 1\\ 1 & -1\end{pmatrix}$
Pauli-X
X
$\hat{X} = \begin{pmatrix}0 & 1\\ 1 & 0\end{pmatrix} = \hat{\sigma}_x$
Pauli-Y
Y
$\hat{Y} = \begin{pmatrix}0 & -i\\ i & 0\end{pmatrix} = \hat{\sigma}_y$
Pauli-Z
Z
$\hat{Z} = \begin{pmatrix}1 & 0\\ 0 & -1\end{pmatrix} = \hat{\sigma}_z$
Phase
S
$\hat{S} = \begin{pmatrix}1 & 0\\ 0 & i\end{pmatrix}$
π/8
T
$\hat{T} = \begin{pmatrix}1 & 0\\ 0 & e^{i\pi/4}\end{pmatrix}$

Fig. 1.16 Some of the more common one-qubit gates together with their associated unitary transformations. (Reproduced, with permission, from Barnett, S. M. (2009) Quantum information. Oxford University Press, Oxford.)

Control
Target
=
X

C	T	C	T
$\|0\rangle$	$\|0\rangle$	$\|0\rangle$	$\|0\rangle$
$\|0\rangle$	$\|1\rangle$	$\|0\rangle$	$\|1\rangle$
$\|1\rangle$	$\|0\rangle$	$\|1\rangle$	$\|1\rangle$
$\|1\rangle$	$\|1\rangle$	$\|1\rangle$	$\|0\rangle$

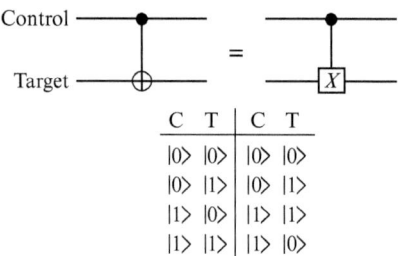

Fig. 1.17 The CNOT gate and its effect on the computational basis states. (Reproduced, with permission, from Barnett, S. M. (2009) Quantum information. Oxford University Press, Oxford.)

gates is the controlled NOT gate, or CNOT gate, depicted in Fig 1.17. The two qubits entering the gate are designated the control qubit, C, and the target qubit, T. The gate enacts the transformation $|1\rangle \to |0\rangle$, $|0\rangle \to |1\rangle$ if the control qubit is in the state $|1\rangle$ but leaves it unchanged if the if the control qubit is in the state $|0\rangle$. Despite its innocuous looking truth table, it is straightforward to show that this gate is intrinsically quantum mechanical in nature. To see this, let the control bit entering the gate be in a superposition of computational basis states. The transformation is then

$$\frac{1}{\sqrt{2}}\left(|0\rangle_C + |1\rangle_C\right)|0\rangle_T \to \frac{1}{\sqrt{2}}\left(|0\rangle_C|0\rangle_T + |1\rangle_C|1\rangle_T\right), \qquad (1.122)$$

which we recognize as one of the entangled Bell states. The fact that the CNOT gate has generated an entangled state from an unentangled one (and therefore introduced the possibility of non-locality) suffices to establish the intrinsically quantum nature of the CNOT gate.

In digital electronics, a small number of gates suffices to allow all possible Boolean operations and, remarkably, something very similar holds for quantum gates. We can perform an unitary evolution of a set of qubits by acting on them only with single-qubit gates and a suitable two-qubit gate; the CNOT gate suffices for this purpose. We do not prove this assertion here, since the required demonstration would take too long and occupy too much space. The proof is not particularly difficult, however, and may be found in [4, 75]. As an example, we can construct the three-qubit Toffoli gate, or controlled–controlled–NOT gate, using one-qubit and two-bit gates. The operation of the Toffoli gate is simplest to understand in the computational basis, $|0\rangle$, $|1\rangle$ for the three qubits:

$$|A\rangle \otimes |B\rangle \otimes |C\rangle \rightarrow |A\rangle \otimes |B\rangle \otimes |C \oplus (A \cdot B)\rangle, \qquad (1.123)$$

where A, B, and C take the values 0 or 1. In words, the gate leaves the computational basis of the first and second qubits unchanged but flips the state, $|0\rangle \leftrightarrow |1\rangle$, of the third qubit if *both* the first and second qubits are in the state $|1\rangle$. The Toffoli gate and its implementation in terms of two-qubit gates are given in Fig 1.18. This involves two CNOT gates and three controlled unitary gates, in which the designated unitary transformation,

$$\hat{W} = \frac{1-i}{2}\left(\hat{I} + i\hat{\sigma}_x\right),$$

$$(1.124)$$

$$\hat{W}^\dagger = \frac{1+i}{2}\left(\hat{I} - i\hat{\sigma}_x\right),$$

is performed on the target qubit if the control qubit is in the state $|1\rangle$ and the identity is applied if it is in the state $|0\rangle$. The controlled unitary operation may also be implemented using CNOT gates and single-qubit gates if desired [4].

If we can combine enough qubits and produce enough gates, then we can produce any desired unitary transformation. This would then constitute a quantum processor, designed to take input data in the form of an input multi-qubit state and generate output in the form of another state. The information could then be extracted by means of a measurement on the qubits. What is necessary in order to achieve this goal? The

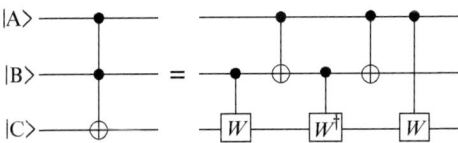

Fig. 1.18 The three-qubit Toffoli gate constructed from two-qubit gates. (Reproduced, with permission, from Barnett, S. M. (2009) Quantum information. Oxford University Press, Oxford.)

answer was provided by DiVincenzo [35] in the form of five criteria for implementing a quantum computer. These have been modified somewhat over the intervening period, but the original five serve to convey the key requirements:

1. We need a well-defined extendable qubit array that is stable.
2. The qubit array should be preparable in a suitable starting state, such as that in which all the qubits are in the state $|0\rangle$.
3. We need good isolation from the environment, i.e. long coherence times.
4. It must be possible to perform a universal set of gate operations, such as single-qubit rotations and CNOT operations for any chosen pair of qubits.
5. Finally, we need to be able to perform something close to ideal von Neumann measurements on each of the qubits.

These demands present a significant technical challenge and, although great advances have been made, I think it safe to say that, at present, no proposed implementation of a quantum processor has managed to achieve them all.

We should note that the gate model of a quantum processor described above is not the only possible one. An alternative is provided by the use of cluster states [83]. In this approach, we first prepare a highly entangled state of our qubits and then proceed by performing a sequence of single-qubit measurements followed by single-qubit unitary transformations, the forms of which depend on the preceding measurement results.

1.5.3 Principles of quantum computation

The basic idea of a quantum computation is quite simple. We start with an input string of bits, which we encode onto our initial states by preparing each in one of the states $|0\rangle$ or $|1\rangle$. We then use our collection of gates to perform on this state a unitary transformation:

$$
\begin{aligned}
101101001 &\rightarrow |1\rangle \otimes |0\rangle \otimes |1\rangle \otimes |1\rangle \otimes |0\rangle \otimes |1\rangle \otimes |0\rangle \otimes |0\rangle \otimes |1\rangle \otimes |1\rangle \\
&\rightarrow \hat{U}|1\rangle \otimes |0\rangle \otimes |1\rangle \otimes |1\rangle \otimes |0\rangle \otimes |1\rangle \otimes |0\rangle \otimes |0\rangle \otimes |1\rangle \otimes |1\rangle \,.
\end{aligned}
\tag{1.125}
$$

The computation is completed by measuring each qubit to give back a (classical) bit string that, hopefully, is the desired output:

$$
\hat{U}|1\rangle \otimes |0\rangle \otimes |1\rangle \otimes |1\rangle \otimes |0\rangle \otimes |1\rangle \otimes |0\rangle \otimes |0\rangle \otimes |1\rangle \otimes |1\rangle \rightarrow 000111010 \,.
\tag{1.126}
$$

This is not quite all there is to it. There is a problem if the function we wish to calculate gives the same value for two distinct inputs. Consider, for example, the two-bit transformation in which two bits, A and B, are transformed into A and (A AND B):

$$
00 \rightarrow 00 \,, \quad 01 \rightarrow 00 \,, \quad 10 \rightarrow 10 \,, \quad 11 \rightarrow 11 \,.
\tag{1.127}
$$

The difficulty in realizing this computation on a quantum processor in the manner indicated above is that the transformation must be unitary, and a unitary transformation necessarily maintains the overlap of the initial states; however, our computation requires

$$|0\rangle \otimes |0\rangle \rightarrow |0\rangle \otimes |0\rangle\,, \qquad |0\rangle \otimes |1\rangle \rightarrow |0\rangle \otimes |0\rangle\,,$$

$$\langle 0,0|0,1\rangle = 0\,, \qquad\qquad \langle 0,0|0,0\rangle = 1\,, \tag{1.128}$$

which clearly violates this requirement.

In order to be able to compute any function, we input into our quantum processor not one string of qubits but two. Let the input states of the two strings be $|a\rangle$ and $|b\rangle$. The first of these is the input data, with the sequence of bits encoded onto the qubits as in (1.125). The second string is to act as our output and is often prepared in the state $|b\rangle = |0\rangle^{\otimes N}$, so that every qubit is in the state $|0\rangle$. If the processor is to calculate the Boolean function $f(a)$, then we require it to realize the transformation

$$|a\rangle \otimes |b\rangle \rightarrow |a\rangle \otimes |b \oplus f(a)\rangle\,, \tag{1.129}$$

where $\oplus$ denotes bitwise modulo addition (see Fig. 1.19). It is straightforward to confirm, by explicit construction, that such a unitary transformation is always possible:

$$\hat{U}_f = \sum_a |a\rangle\langle a| \otimes \left(|f(a)\rangle\langle a| + |a\rangle\langle f(a)| + \sum_{b\neq a,f(a)} |b\rangle\langle b| \right). \tag{1.130}$$

Of course this is not the only unitary operator that performs this task, but this simple operator suffices to demonstrate that a suitable unitary operator exists.

The remarkable properties of a quantum computer derive largely from the fact that we can put into the processor not just a single number but rather a superposition of many. If each qubit in the first qubit string is prepared in the state $2^{-1/2}(|0\rangle + |1\rangle)$ then the string of N qubits is in a superposition of every number between 0 and $2^N - 1$:

$$2^{-N/2}\left(|0\rangle + |1\rangle\right)^{\otimes N} = 2^{-N/2}\sum_{a=0}^{2^N-1}|a\rangle\,. \tag{1.131}$$

The linearity of quantum mechanics means that the output state is an entangled one in which the function $f(a)$ has been calculated for *every* possible input number, as in Fig. 1.20.

A simple example may serve to illustrate the potential of a quantum processor. With this aim in mind, we present Deutsch's algorithm and its many-bit extension,

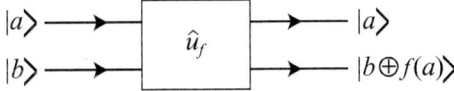

Fig. 1.19 A quantum processor designed to compute the function $f(a)$. (Reproduced, with permission, from Barnett, S. M. (2009) Quantum information. Oxford University Press, Oxford.)

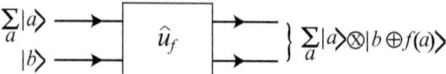

Fig. 1.20 If a superposition of possible input numbers is prepared, then the output state is an entangled one in which values of the function for each of the inputs appear. (Reproduced, with permission, from Barnett, S. M. (2009) Quantum information. Oxford University Press, Oxford.)

the Deutsch–Jozsa algorithm [27, 31, 32]. Although somewhat contrived, these serve to illustrate in a very simple way the potential for speed-up offered by a quantum computer. These are examples of a class of challenges called oracle problems.

We start with Deutsch's algorithm [31]. Let us suppose that we have a 'black box' (an oracle) and that this device evaluates a one-bit function of a one-bit input. This means that we input a single bit, with value either 0 or 1, and that this generates an output value of either 0 or 1. There are four possible such functions: two constant functions

$$f(0) = 0 \,, \qquad f(1) = 0 \,,$$

and

$$f(0) = 1 \,, \qquad f(1) = 1 \,,$$

(1.132)

and two balanced functions (balanced in the sense that both possible bit values occur in the output, one for each input)

$$f(0) = 0 \,, \qquad f(1) = 1 \,,$$

and

$$f(0) = 1 \,, \qquad f(1) = 1 \,.$$

(1.133)

If we want to know which function the oracle calculates, then we need to input both bit values and examine the corresponding outputs. Let us suppose, however, that our task is simply to determine whether the function calculated is a constant function or a balanced one. Classically, of course, this requires us to input both possible bit values and so also requires two computations. A suitable quantum processor, however, can do this in a single step. Consider the general transformation given in (1.129), but with the two qubit strings each replaced by a single qubit so that the transformation enacted by the processor is

$$|A\rangle \otimes |B\rangle \rightarrow |A\rangle \otimes |B \oplus f(A)\rangle \,. \qquad (1.134)$$

We can make use of the superposition principle to prepare *both* of our qubits in a superposition state and so produce the transformation

$$\tfrac{1}{2}(|0\rangle + |1\rangle) \otimes (|0\rangle - |1\rangle) \rightarrow \tfrac{1}{2}[(-1)^{f(0)}|0\rangle + (-1)^{f(1)}|1\rangle] \otimes (|0\rangle - |1\rangle) \,. \qquad (1.135)$$

A simple measurement carried out on the first qubit then tells us what we need to know: if it is found to be in the state $2^{-1/2}(|0\rangle + |1\rangle)$, then the function is constant, and if we find it in the orthogonal state $2^{-1/2}(|0\rangle - |1\rangle)$, then the function is balanced. We have found out what we need to know by addressing the oracle only once, rather than twice as would be required classically.

The Deutsch algorithm may seem, at least at first sight, not to be an especially convincing illustration—after all we do input two qubits into the processor. The real power appears when we consider an extension of it to the Deutsch–Jozsa algorithm. In this algorithm, our input is an n-bit number a and our output is again a single bit, 0 or 1. The function is either constant, giving either 0 or 1 for all inputs, or balanced, giving 0 for half of the inputs and 1 for the other half. Our task is to determine an algorithm that determines *with certainty* whether the function is constant or balanced. We can start putting in different strings; then, as soon as we get two different output values, we know for certain that the function computed is balanced. To know its nature for certain, however, we may have to input over half of the possible inputs, which means addressing the oracle $2^{n-1} + 1$ times. A quantum processor can achieve this task in a single shot, however. To see how this works, we let the first string consist of n qubits, each prepared in the superposition state $2^{-1/2}(|0\rangle + |1\rangle)$, with the second string being just a single qubit prepared in the state $2^{-1/2}(|0\rangle - |1\rangle)$. The oracle then performs the transformation

$$2^{-(n+1)/2}(|0\rangle + |1\rangle^{\otimes n}) \otimes (|0\rangle - |1\rangle) \;\rightarrow\; 2^{-(n+1)/2} \sum_{a=0}^{2^n - 1} (-1)^{f(a)} |a\rangle \otimes (|0\rangle - |1\rangle).$$

(1.136)

If the function is constant, then the first string of n qubits remain in their input state, as an equally weighted superposition of the states $|a\rangle$, but if the function is balanced, then the qubit string will be in a state orthogonal to this. Hence, only a single computation is required to determine whether the function is constant or balanced. This represents an improvement over the classical requirement of $2^{n-1} + 1$, which is *exponential* in n, the number of bits. Such dramatic improvements are characteristic of a number of quantum algorithms, but not all.

1.5.4 Quantum algorithms

Supposing that we have a suitable quantum processor, what might we do with it? One important thing that we might do is use it to simulate a complicated quantum process [37]. As quantum systems get larger, it becomes ever more difficult to simulate them; were this not the case, then we could emulate a large quantum computer on a classical one. We might also use a quantum computer to speed up searching in an unstructured database using Grover's algorithm, which provides a dramatic improvement (albeit not an exponential one) over classical methods [45]. The most dramatic possibility to date, however, is Shor's algorithm for determining the two prime factors of a large number. You will recall that it is the difficulty in performing this task that underlies the security of the RSA public-key cryptosystem and, with it, much of the world's secure communications [90]. It was Shor's proposal more than any other single factor that truly changed quantum computation, and with it quantum information, from a small-scale research field into a major international endeavour. There is space in these lectures to present briefly only one quantum algorithm and, because of its significance, we choose Shor's algorithm. Several others, including Grover's algorithm, and a more complete presentation of Shor's algorithm may be found in [4].

At the heart of Shor's algorithm is the remarkable ability of a quantum computer to perform, highly efficiently, a quantum Fourier transform and hence to find the period of a function. To see why this helps in factorization, we need to consider an idea from number theory. We start with three bits of input data: N, the number to be factorized, m, a small integer chosen at random, and the non-negative integers, $n = 0, 1, 2, \cdots$. We first make the series $F_N(n) = m^n \bmod N$. If we can then find the period of this function, r, such that $F_N(n + r) = F_N(n)$, then the greatest common divisor of $m^{r/2} \pm 1$ and N divides N. In other words, one of the factors of each of $m^{r/2} \pm 1$ is also a prime factor of N. There are, of course, some additional subtleties, but the number theory needed to prove this is by no means difficult, and may be found in [4]. Let us consider two examples to demonstrate the idea. First, we consider the problem of factoring 15 by means of this process. Let us select $m = 2$, so that our series is

$$2^0 \bmod 15 = 1\,,$$
$$2^1 \bmod 15 = 2\,,$$
$$2^2 \bmod 15 = 4\,,$$
$$2^3 \bmod 15 = 8\,, \tag{1.137}$$
$$2^4 \bmod 15 = 1\,,$$
$$\vdots$$

We see that the period is 4 and hence

$$m^{r/2} + 1 = 5\,,$$
$$m^{r/2} - 1 = 3\,, \tag{1.138}$$

and these are the required prime factors. We now try another example, $m = 11$, for which our series is

$$11^0 \bmod 15 = 1\,,$$
$$11^1 \bmod 15 = 11\,,$$
$$11^2 \bmod 15 = 1\,, \tag{1.139}$$
$$\vdots$$

so the period is 2. This gives

$$m^{r/2} + 1 = 12\,,$$
$$m^{r/2} - 1 = 10\,. \tag{1.140}$$

The greatest common divisor of 12 and 15 is 3, which is one of the prime factors we seek, and the greatest common divisor of 10 and 15 is 5, which is the other factor.

So why do we need a quantum computer to run this seemingly simple algorithm? The answer is that each step can be run efficiently, even for very large numbers, on a classical computer—with the single exception of finding the period of the function $F_N(n)$. This is a very difficult task and it is no exaggeration to state that it is this period-finding problem that underlies the security of the RSA cryptosystem.

Central to Shor's algorithm for factoring on a quantum computer is the quantum Fourier transform [4, 73, 75]. We do not attempt a detailed account of the quantum Fourier transform, but present instead only an outline of the steps involved in Shor's algorithm. It is, perhaps, clearest to give a list of the five major steps:

Shor's algorithm to find the two prime factors of N

1. We start by finding two integers, q and M, such that

$$q = 2^M > N^2 \tag{1.141}$$

   and prepare two registers, each containing M qubits.
2. We set each of the qubits in the first register to the state $2^{-1/2}(|0\rangle + |1\rangle)$ and each in the second register to the state $|0\rangle$, so that the state input into our quantum processor is

$$|\psi\rangle = \frac{1}{\sqrt{q}} \sum_{n=0}^{q-1} |n\rangle_1 \otimes |0\rangle_2 \,, \tag{1.142}$$

   where we have used the shorthand notation

$$
\begin{aligned}
|0\rangle &= |00\cdots000\rangle \,, \\
|1\rangle &= |00\cdots001\rangle \,, \\
|2\rangle &= |00\cdots010\rangle \,, \\
|3\rangle &= |00\cdots011\rangle \,, \\
&\;\;\vdots
\end{aligned}
\tag{1.143}
$$

3. Next we chose an integer m at random and use the quantum processor to entangle the two registers so that

$$|\psi\rangle \to \frac{1}{\sqrt{q}} \sum_{n=0}^{q-1} |n\rangle_1 \otimes |m^n \bmod N\rangle_2 \,. \tag{1.144}$$

   This can be achieved efficiently by means of a unitary transformation on a suitably programmed quantum computer.
4. Next comes the crucial quantum Fourier transform, which again can be performed efficiently as a unitary transform on a quantum computer. We use our quantum

computer to perform a quantum Fourier transform on the first register to generate the transformation

$$\frac{1}{\sqrt{q}} \sum_{n=0}^{q-1} |n\rangle_1 \otimes |m^n \bmod N\rangle_2 \ \rightarrow \ \frac{1}{q} \sum_{n=0}^{q-1} \sum_{k=0}^{q-1} |k\rangle_1 \otimes |m^n \bmod N\rangle_2 \exp\left(i2\pi\frac{kn}{q}\right).$$

(1.145)

Let us pause to think and see what has been achieved by all of this. The number $m^n \bmod N$ has period r, which means, of course, that

$$m^{n+sr} \bmod N = m^n \bmod N,$$

(1.146)

so that we have some identical states in the second register:

$$|m^n \bmod N\rangle = |m^{n+r} \bmod N\rangle = |m^{n+2r} \bmod N\rangle = \cdots.$$

(1.147)

The coefficients of these states will have phase factors

$$\exp\left(i2\pi\frac{k(n+sr)}{q}\right) = \exp\left(i2\pi\frac{kn}{q}\right) \times \exp\left(i2\pi\frac{ksr}{q}\right),$$

(1.148)

and will be in phase (or nearly in phase) for $k \approx q/r$.

5. Finally, we need only measure the first register in the computational basis, and then, because of constructive interference, we find with high probability a value k for which the amplitude is large, with $k \approx q/r$, that is, $r \approx q/k$. This allows us to greatly narrow the range of allowed values of r and hence greatly restrict the range of possible factors. All possible candidate factors can be checked efficiently and simply by dividing them into N.

This is just one example, albeit the most prominent and newsworthy, of an efficient quantum algorithm. There are others, and the field of quantum algorithm design promises to gain in importance as the first quantum processors become available.

1.5.5 Errors and decoherence

So what stops us building a quantum processor, revolutionizing computation, and breaking into RSA? The short answer is that we don't yet have any suitably scalable implementation of quantum logic elements such as qubits and gates.

Many technical challenges remain to be solved before we can build a quantum computer to challenge current devices based on classical logic. I mention here only one problem, namely that of decoherence. In a quantum computer, our information is encoded onto two-state quantum systems, our qubits, and these are made to interact with each other by the implementation of desired Hamiltonians. Yet the quantum nature of both the qubits and the Hamiltonian makes them extremely sensitive to their environment, with even weak interactions possibly having a disastrous effect. As an illustration, let us consider the effect of a single-qubit error in an implementation of Deutsch's algorithm. To keep things simple, we consider only a phase error,

$$|0\rangle \rightarrow |0\rangle \qquad |1\rangle \rightarrow -|1\rangle \,, \tag{1.149}$$

and a bit-flip error,

$$|0\rangle \rightarrow |1\rangle \qquad |1\rangle \rightarrow |0\rangle \,. \tag{1.150}$$

Recall that we obtain the desired information, namely whether the computed function is constant or balanced, by measuring the first qubit to be in the state $2^{-1/2}(|0\rangle + |1\rangle)$ or $2^{-1/2}(|0\rangle - |1\rangle)$, respectively. If a bit-flip error occurs for this qubit, then we are safe; however, if a phase error occurs, then we get the wrong answer!

Clearly, we need to work hard to suppress all sources of errors and decoherence, but this is not the end of the problem. For large-scale computations, we need a large number of qubits, and the scaling is not at all favourable. To see this, let us suppose that the probability that a single qubit has no error in a time t is

$$\exp\left(-\Gamma t\right) .$$

If we have n qubits, then the probability that *none* of these experiences an error in this time is

$$\exp\left(-n\Gamma t\right) ,$$

which is exponential in n. But worse is to come. Let us suppose that t is the typical amount of time taken to perform a single gate operation. If we need a sequence of m of these, then we find

$$\exp\left(-nm\Gamma t\right) .$$

Typically, we might need each qubit to interact with every other qubit, so that m will be of the order of n^2. We do not need to perform the gate operations one at a time, of course, and by suitably optimizing the order of operations, we might be able to have $m \approx \log n$, giving a final zero-error probability of

$$\exp(-n \log n \Gamma t) .$$

For 300 qubits, the *exponent* is about 2000 times smaller than the single-qubit and single-gate error rate, $e^{-\Gamma t}$, that we started with. The zero-error probability is the 2000th *power* of the no-error probability for a single qubit and a single gate operation.

Clearly, we need to control sources of decoherence to an extraordinary degree, but ultimately this will always be a losing battle. Is it all hopeless? By no means— the solution is essentially the same as we encountered in the first lecture. We can combat errors in a quantum computer by redundancy, just as when we combat classical errors. This is a more subtle problem than in the classical regime, however. First, there is the no-cloning theorem, which tells us that we cannot literally make multiple copies of our qubits, and there is also the problem that quantum measurements, if not performed carefully, will modify the very qubit states that we are trying to protect. Nevertheless, there are protocols for detecting and correcting qubit errors. Perhaps the most important of these is the Steane code, in which each logical qubit is encoded in terms of seven qubits [4, 94], but a description of how this works will have to wait for another time.

References

[1] Aharonov, Y., Albert, D. Z., and Vaidman, L. (1988) How the result of a measurement of a component of spin of a spin-$\frac{1}{2}$ particle can turn out to be 100. *Phys. Rev. Lett.* **60**, 1351–1354.

[2] Barnett, S. M. and Riis, E. (1997) Experimental demonstration of polarization discrimination at the Helstrom bound. *J. Mod. Opt.* **44**, 1061–1064.

[3] Barnett, S. M. and Andersson, E. (2002) Bound on measurement based on the no-signalling condition. *Phys. Rev. A* **65**, 044307.

[4] Barnett, S. M. (2009) *Quantum Information*. Oxford University Press, Oxford.

[5] Barnett, S. M. and Croke, S. (2009a) On the conditions for discrimination between quantum states with minimum error. *J. Phys. A: Math. Theor.* **42**, 062001.

[6] Barnett, S. M. and Croke, S. (2009b) Quantum state discrimination. *Adv. Opt. Photon.* **1**, 238–278.

[7] Bayes, T. (1763) An essay towards solving a problem in the doctrine of chances. *Phil. Trans. R. Soc. Lond.* **53**, 370–418.

[8] Bell, J. S. (1964) On the Einstein–Podolsky–Rosen paradox. *Physics* **1**, 195–200. Reprinted in [9, 101].

[9] Bell, J. S. (1987) *Speakable and Unspeakable in Quantum Mechanics*. Cambridge University Press, Cambridge.

[10] Bennett, C. H. and Brassard, G. (1984) Quantum cryptography: public key distribution and coin tossing. *Proceedings of IEEE International Conference on Computers, Systems and Signal Processing, Bangalore, India, 9–12 December 1984*, pp. 175–179.

[11] Bennett, C. H. and Wiesner, S. J. (1992) Communication via one- and two-particle operators on Einstein–Podolsky–Rosen states. *Phys. Rev. Lett.* **69**, 2881–2884.

[12] Bennett, C. H., Brassard, G., Crépeau, C., Jozsa, R., Peres, A., and Wootters, W. K. (1993) Teleporting an unknown quantum state via dual Einstein–Podolsky–Rosen channels. *Phys. Rev. Lett.* **70**, 1895–1899.

[13] Bergou, J. (2007) Quantum state discrimination and selected applications. *J. Phys. Conf. Ser.* **84**, 012002.

[14] Boas, M. L. (2006) *Mathematical Methods in the Physical Sciences*, 3rd edn. Wiley, Hoboken, NJ.

[15] Bohm, D. (1951) *Quantum Theory*. Prentice-Hall, Englewood Cliffs, NJ. Reprinted by Dover, New York (1989).

[16] Bohr, N. (1935) Can quantum-mechanical description of reality be considered complete? *Phys. Rev.* **47**, 696–702.

[17] Bouwmeester, D., Ekert, A., and Zeilinger, A. (eds.) (2000) *The Physics of Quantum Information.* Springer-Verlag, Berlin.

[18] Box, G. E. P. and Tiao, G. C. (1973) *Bayesian Inference in Statistical Analysis.* Wiley, New York.

[19] Bretthorst, G. L. (1988) *Bayesian Spectrum Analysis and Parameter Estimation.* Springer-Verlag, New York.

[20] Brillouin, L. (1956) *Science and Information Theory.* Academic Press, New York.

[21] Buchmann, J. A. (2001) *Introduction to Cryptography.* Springer, New York.

[22] Bužek, V. and Hillery, M. (1996) Quantum copying: beyond the no-cloning theorem. *Phys. Rev. A* **54**, 1844–1852.

[23] Chefles, A. (2000) Quantum state discrimination. *Contemp. Phys.* **41**, 401–424.

[24] Clarke, R. B. M., Chefles, A., Barnett, S. M., and Riis, E. (2001) Experimental demonstration of optimal unambiguous state discrimination. *Phys. Rev. A* **63**, 040305(R).

[25] Clarke, R. B. M., Kendon, V. M., Chefles, A., Barnett, S. M., and Riis, E. (2001) Experimental realization of optimal detection strategies for overcomplete states. *Phys. Rev. A* **64**, 012303.

[26] Clauser, J. F., Horne, M. A., Shimony, A., and Holt, R. A. (1969) Proposed experiment to test local hidden-variable theories. *Phys. Rev. Lett.* **23**, 880–884.

[27] Cleve, R., Ekert, A., Macchiavello, C., and Mosca, M. (1998) Quantum algorithms revisited *Proc. R. Soc. Lond. A* **454**, 339–354.

[28] Conan Doyle, A. (1903) The adventure of the dancing men. *Strand Magazine* **26**, December issue. Reprinted in *The Original Illustrated Sherlock Holmes*, Castle, Secaucus, NJ.

[29] Cover, T. M. and Thomas, J. A. (1991) *Elements of Information Theory.* Wiley, New York.

[30] Croke, S., Barnett, S. M. and Stenholm, S. (2008) Linear transformation of quantum states. *Ann. Phys. (NY)* **323**, 893–906.

[31] Deutsch, D. (1985) Quantum theory, the Church–Turing principle and the universal quantum computer. *Proc. R. Soc. Lond. A* **400**, 97–117.

[32] Deutsch, D. and Jozsa, R. (1992) Rapid solution of problems by quantum computation. *Proc. R. Soc. Lond. A* **439**, 553–558.

[33] Dieks, D. (1982) Communication by EPR devices. *Phys. Lett. A* **92**, 271–272.

[34] Dieks, D. (1988) Overlap and distinguishability of quantum states. *Phys. Lett. A* **126**, 303–306.

[35] DiVincenzo, D. (1997) Topics in quantum computers. In: Sohn, L. L., Kouwenhoven, L. P., and Schön, G. (eds.) *Mesoscopic Electron Transport*, pp. 657–677. Kluwer, Dordrecht [arXiv:cond-mat/9612126].

[36] Einstein, A., Podolsky, B. and Rosen, N. (1935) Can quantum-mechanical description of reality be considered complete?. *Phys. Rev.* **47**, 777–780.

[37] Feynman, R. P. (1982) Simulating physics with computers. *Int. J. Theor. Phys.* **21**, 467–488.

[38] Gay, S. and Mackie, I. (eds.) (2010) *Semantic Techniques in Quantum Computation* Cambridge University Press, Cambridge.

[39] Ghirardi, G. C., Rimini, A. and Weber, T. (1980) A general argument against superluminal transmissions through the quantum mechanical measurement process. *Lett. Nuovo Cim.* **27**, 293–298.

[40] Gisin, N. (1991) Bell's inequality holds for all non-product states. *Phys. Lett. A* **154**, 201–202. [It should be noted that this paper contains some typographical errors, the most serious of which is the title—the paper proves the exact opposite of what is stated in the title!]

[41] Gisin, N., Ribordy, G., Tittel, W., and Zbinden, H. (2002) Quantum cryptography. *Rev. Mod. Phys.* **74**, 145–195.

[42] Goldie, C. M. and Pinch, R. G. E. (1991) *Communications Theory*. Cambridge University Press, Cambridge.

[43] Goldstein, S. (1994) Nonlocality without inequalities for almost all entangled states of two particles. *Phys. Rev. Lett.* **72**, 1951–1951.

[44] Greenberger, D. M., Horne, M. A., Shimony, A., and Zeilinger, A. (1990) Bell's theorem without inequalities. *Am. J. Phys.* **58**, 1131–1143.

[45] Grover, L. (1998) Quantum computers can search rapidly by using almost any transformation. *Phys. Rev. Lett.* **80**, 4329–4332.

[46] Hamming, R. W., (1980) *Coding and Information Theory*. Prentice-Hall, London.

[47] Hardy, L. (1993) Nonlocality for two particles without inequalities for almost all entangled states. *Phys. Rev. Lett.* **71**, 1665–1668.

[48] Helstrom, C. W. (1976) *Quantum Detection and Estimation Theory*. Academic Press, New York.

[49] Holevo, A. S. (1973) Statistical decision theory for quantum systems. *J. Multivariate Anal.* **3**, 337–394.

[50] Holevo, A. S. (1982) *Probabilistic and Statistical Aspects of Quantum Theory*. North-Holland, Amsterdam.

[51] Holevo, A. S. (2001) *Statistical Structure of Quantum Theory*. Springer-Verlag, Berlin.

[52] Hunter, K. (2003) Measurement does not always aid state discrimination. *Phys. Rev. A* **68**, 012306.

[53] Huttner, B., Muller, A. Gautier, G., Zbinden, H., and Gisin, N. (1996) Unambiguous quantum measurement of nonorthogonal quantum states. *Phys. Rev. A* **54**, 3783–3789.

[54] Ivanovic, I. D. (1987) How to differentiate between non-orthogonal states. *Phys. Lett. A* **123**, 257–259.

[55] Jaynes, E. T. (1957) Information theory and statistical mechanics I. *Phys. Rev.* **106**, 620–630.

[56] Jaynes, E. T. (1957) Information theory and statistical mechanics II. *Phys. Rev.* **108**, 171–190.

[57] Jaynes, E. T. (2003) *Probability Theory: The Logic of Science*. Cambridge University Press, Cambridge.

[58] Jeffreys, H. (1939) *The Theory of Probability*. Clarendon Press, Oxford.

[59] Kaye, P., Laflamme, R., and Mosca, M. (2007) *An Introduction to Quantum Computing* Oxford University Press, Oxford.

[60] Khinchin, A. I. (1957) *Mathematical Foundations of Information Theory.* Dover, New York.

[61] Kullback, S. (1959) *Information Theory and Statistics.* Dover, New York.

[62] Kraus, K. (1983) *States, Effects and Operations.* Springer-Verlag, Berlin.

[63] Landauer, R. (1961) Irreversibility and heat generation in the computing process. *IBM J. Res. Dev.* **5**, 183–191. Reprinted in [65, 67].

[64] Lee, P. M. (1989) *Bayesian Statistics: An Introduction.* Edward Arnold, London.

[65] Leff, H. S. and Rex, A. F. (eds.) (1990) *Maxwell's Demon: Entropy, Information, Computing.* Adam Hilger, Bristol.

[66] Leff, H. S. and Rex, A. F. (1994) Entropy of measurement and erasure: Szilard's membrane model revisited. *Am. J. Phys.* **52**, 3495–3499. Reprinted in [67].

[67] Leff, H. S. and Rex, A. F. (eds.) (2003) *Maxwell's Demon 2: Entropy, Classical and Quantum Information, Computing.* Institute of Physics Publishing, Bristol.

[68] Loepp, S and Wootters, W. K. (2006) *Protecting Information: From Classical Error Correction to Quantum Cryptography.* Cambridge University Press, Cambridge.

[69] Mattle, K., Weinfurter, H., Kwiat, P. G., and Zeilinger, A. (1996) Dense coding in experimental quantum communication. *Phys. Rev. Lett.* **76**, 4656–4659.

[70] Maxwell, J. C. (1871) *Theory of Heat.* Longmans, Green, & Co., London.

[71] Mcgrayne, S. B. (2011) *The Theory That Would Not Die.* Yale University Press, New Haven, CT.

[72] Mermin, D. (1990) What's wrong with these elements of reality? *Phys. Today* June issue, 9–11

[73] Mermin, D. (2007) *Quantum Computer Science.* Cambridge University Press, Cambridge.

[74] More, G. E. (1965) Cramming More Components onto Integrated Circuits *Proc. IEEE* **86** 82–85.

[75] Nielsen, M. A. and Chuang, I. L. (2000) *Quantum Computation and Quantum Information* Cambridge University Press, Cambridge.

[76] Pachos, J. K. (2012) *Introduction to Topological Quantum Computation.* Cambridge University Press, Cambridge.

[77] Paris, M. and Řeháček, J. (eds.) (2004) *Quantum State Estimation.* Lecture Notes in Physics, No. 649. Springer-Verlag, Berlin.

[78] Peres, A. (1988) How to differentiate between two non-orthogonal states. *Phys. Lett. A* **128**, 19–19.

[79] Peres, A. (1993) *Quantum Theory: Concepts and Methods.* Kluwer, Dordrecht.

[80] Plenio, M. and Vitelli, V. (2001) The physics of forgetting: Landauer's erasure principle and information theory. *Contemp. Phys.* **42**, 25–60.

[81] Phoenix, S. J. D. and Townsend, P. D. (1995) Quantum cryptography: how to beat the code breakers using quantum mechanics. *Contemp. Phys.* **36**, 165–195.

[82] Piper, F. and Murphy, S. (2002) *Cryptography: A Very Short Introduction.* Oxford University Press, Oxford.

[83] Raussendorf, R., Browne, D. E. and Briegel, H. J. (2003) Measurement-based quantum computation on cluster states. *Phys. Rev. A* **68**, 022312.

[84] Redhead, M. (1987) *Incompleteness, Nonlocality, and Realism.* Clarendon Press, Oxford.

[85] Scarani, V., Iblisdir, S., Gisin, N., and Acín, A. (2005) Quantum cloning. *Rev. Mod. Phys.* **77**, 1225–1256.

[86] Scarani, V., Bechmann-Pasquinucci, H., Cerf, N. J., Dušek, M., Lütknehaus, N., and Peev, M. (2009) The security of practical quantum key distribution. *Rev. Mod. Phys.* **81**, 1301–1350.

[87] Shannon, C. E. (1948) A mathematical theory of communication. *Bell Syst. Tech. J.* **27**, 379–423 and 623–656.

[88] Shannon, C. E. (1949) Communication theory of secrecy systems. *Bell Syst. Tech. J.* **28**, 656–715.

[89] Shannon, C. E. and Weaver, W. (1949) *The Mathematical Theory of Communication.* University of Illinois Press, Urbana, IL.

[90] Shor, P. W. (1997), Polynomial-time algorithms for prime factorization and discrete logarithms on a quantum computer. *SIAM J. Comput.* **26**, 1484–1509 [arXiv:quant-ph/9508027v2].

[91] Singh, S. (1999) *The Code Book.* Fourth Estate, London.

[92] Singh, S. (2000) *The Science of Secrecy.* Fourth Estate, London.

[93] Smith, R. J. and Dorf, R. C. (1992) *Circuits, Devices and Systems: A First Course in Electrical Engineering,* 5th edn. Wiley, New York.

[94] Steane, A. M. Error correcting codes in quantum theory. *Phys. Rev. Lett.* **77** 793–797.

[95] Stenholm, S. and Suominen, K.-A. (2005) *Quantum Approach to Informatics.* Wiley, Hoboken, NJ.

[96] Szillard, L. (1929) Über die Entropieveminderung in einem thermodynamikschen System bei eingriffen intelligenter Wesen. *Z. Phys.* **53**, 840–856. Reprinted in translation in [65, 67, 101].

[97] Vaccaro, J. A. and Barnett, S. M. (2011) Information erasure without an energy cost. *Proc. R. Soc. Lond. A* **467**, 1770–1778.

[98] Van Assche, G. (2006) *Quantum Cryptography and Secret-Key distillation.* Cambridge University Press, Cambridge.

[99] Vedral, V. (2006) *Introduction to Quantum Information Science.* Oxford University Press, Oxford.

[100] von Neumann, J. (1955) *Mathematical Foundations of Quantum Mechanics.* Princeton University Press, Princeton, NJ.

[101] Wheeler, J. A. and Zurek, W. H. (eds.) (1983) *Quantum Theory and Measurement.* Princeton University Press, Princeton, NJ.

[102] Whitaker, A. (2006) *Einstein, Bohr and the Quantum Dilemma,* 2nd edn. Cambridge University Press, Cambridge.

[103] Wiesner, S. (1983) Conjugate coding. *SIGACT News* **15**, 78–88.

[104] Wootters, W. K. and Zurek, W. H. (1982) A single quantum cannot be cloned. *Nature* **299**, 802–803.

2

Introduction to near-field optics and plasmonics

Jean-Jacques GREFFET

Laboratoire Charles Fabry
Institut d'Optique Graduate School, CNRS
Université Paris-Sud, Palaiseau, France

Portions of this chapter are reproduced with permission from Jean-Jacques Greffet, 'Introduction to Surface Plasmon Theory', Plasmonics, Volume 167 of the series Springer Series in Optical Sciences, pp 105–148. DOI 10.1007/978-3-642-28079-5 4. Courtesy of Springer.

Jean-Jacques Greffet: *Introduction to Near-Field Optics and Plasmonics*. In: *Quantum Optics and Nanophotonics*. Edited by: Claude Fabre, Vahid Sandoghdar, Nicolas Treps, and Leticia F. Cugliandolo. Oxford University Press (2017), pp. 61–143. © Oxford University Press 2017.
DOI: 10.1093/oso/9780198768609.003.0002

Chapter Contents

The goal of this chapter is to introduce the basic concepts of near-field optics. A striking difference between near-field optics and far-field optics is the possibility of breaking the so-called diffraction limit, that is, the possibility of confining light to subwavelength spots. We first introduce the concept of evanescent waves to discuss the subwavelength confinement of light. We also review the standard concepts used to describe light, such as phase, reflection factor, and electromagnetic energy density, and point out differences with propagating plane waves. One of the key ideas put forward is that the presence of charges is required to generate highly localized fields. It is thus necessary to have a tool to compute fields in the presence of these charges. With this aim in mind, we introduce the concept of the Green tensor. This is a powerful tool to compute electromagnetic fields in inhomogeneous environments. It is also a key quantity when discussing the local density of states and therefore the control of spontaneous emission. Finally, we provide an introduction to surface plasmons, which are collective excitations of electrons. These are very useful to manipulate electromagnetic fields at the nanoscale.

2.1 Basics of near-field optics

2.1.1 Introduction

In this section, we will study in detail the fundamental properties of light in vacuum to clarify the physical origin of the limits on resolution in conventional optical systems. In particular, we will introduce evanescent waves and derive a clear definition of the near-field regime. This will also lead us to revisit diffraction theory in a wave propagation context: we show that the Huygens–Fresnel principle appears to be an approximation of the exact solution of propagation in vacuum. The second section will offer an alternative introduction to near-field optics by examining the radiation from an electric dipole. This point of view allows us to stress the link between near field, electrostatics, and magnetostatics. These two sections can also be viewed as a discussion of the concept of near field either in Fourier space (evanescent waves) or in direct space (dipole radiation). Finally, the third section will highlight some particular properties of the field: it will be seen that the familiar concepts of phase, reflection factor, field structure, and polarization need to be revisited in the near field.

2.1.2 Propagation in vacuum: angular spectrum and evanescent waves

Diffraction is usually introduced in a physical optics class using the scalar approximation and the Huygens–Fresnel principle. This approach has the great advantage of being simple. However, it introduces the disadvantage of duplicating the concepts: the connection between diffraction in physical optics and Maxwell's equations is not obvious.

Let us consider the classical problem of diffraction by an aperture. This amounts to calculating the field at a point (x, y, z) knowing the field on a plane defined by

$z = 0.$[1] Therefore, the diffraction problem is a particular class of a *propagation problem*. Hence, from a strictly electromagnetic viewpoint, we will see that the concept of diffraction is simply unnecessary. Whenever we tackle this problem using Maxwell's equations and not geometrical optics, the wave behaviour of light is automatically taken into account. In other words, it is not necessary to introduce an ad hoc principle to obtain diffraction patterns. They are nothing but the manifestation of the wave aspect of light under certain conditions. Assuming that we had only geometrical optics as a theoretical framework, the situation would be totally different and would lead to diffraction having to be considered as an unexplained phenomenon. From this point of view, the unexplained phenomenon deserves to have a name (diffraction) and a principle has to be added to the theory to be able to describe it. On the other hand, Maxwell's equations provide a framework that embodies both geometrical optics and wave propagation. Diffraction is thus merely a convenient word to refer to a particular class of propagation problems.

In order to deal with these using Maxwell's equations, we shall write the electric field in the form of a superposition of plane waves at any point in the half-space located at $z > 0$. This is the angular spectrum representation of the field. In this form, we will derive equations very close to those encountered when developing radiation theory.

2.1.2.1 *Statement of the problem*

In this section, we deal with the propagation in vacuum of monochromatic radiation having a circular frequency ω coming from a source that is not specified. We assume that we know the electric field on the plane $z = 0$. It could be produced by a laser beam that illuminates the plane $z = 0$ or by an antenna (or any other current distribution) placed at a point $z < 0$. We assume that the propagation takes place in vacuum. What we want to obtain is an explicit expression for the electric field at any point $z > 0$ (Fig. 2.1). It is clear that the diffraction of a plane wave impinging on an aperture in a screen is a particular case that falls in the class of problems that we deal with. Our formulation of the problem is general and also includes the study of any non-plane wave field, such as the spreading of a beam in vacuum, the study of the amplitude of the field in the vicinity of a focusing point, etc.

2.1.2.2 *Field propagation formula: angular spectrum*

To solve the propagation problem, we simply solve a partial differential equation with the relevant boundary conditions. For simplicity, we are going to work with a scalar amplitude. This can be considered to be a Cartesian component of a vector field, so the extension to the vector case is straightforward. The field obeys the propagation equation in vacuum and satisfies the boundary conditions at $z = 0$:

$$\nabla^2 E - \frac{1}{c^2}\frac{\partial^2 E}{\partial t^2} = 0. \tag{2.1}$$

---

[1] From a mathematical viewpoint, this entails correctly defining the boundary conditions of the problem at $z = 0$ and at $z = +\infty$. At $z = +\infty$, the condition used is the Sommerfeld radiation condition.

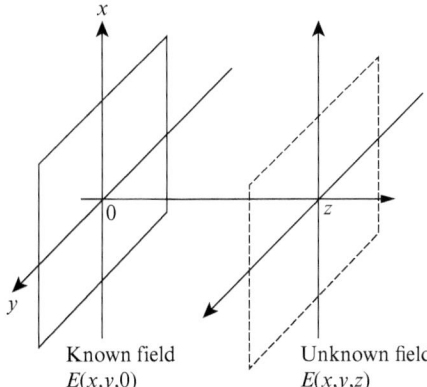

Known field
$E(x,y,0)$

Unknown field
$E(x,y,z)$

Fig. 2.1 We assume that a scalar field is known in the plane $z = 0$ and want to find the field in the plane z. This is a propagation problem. Diffraction is a particular case of the general problem.

Considering a monochromatic field at frequency ω of the form $\exp(-i\omega t)$, the propagation equation becomes a Helmholtz equation for a field that depends on space variables and ω instead of time t. For brevity, we henceforth omit the frequency dependence. Helmholtz's equation is written as follows for the scalar electric field:

$$\nabla^2 E + k_0^2 E = 0, \qquad (2.2)$$

where we have introduced $k_0 = \omega/c$, the wavevector in vacuum. With this equation and Green's theorem, it is possible to derive the electric field. This is the *Kirchhoff* or *Kirchhoff–Sommerfeld* formulation of diffraction. Here we are going to develop the field on a basis of plane waves. To do this, we note that the field $E(x, y, z)$ may be considered as a function of x and y in a fixed plane z. We now introduce the Fourier decomposition of E according to x and y only:

$$E(x, y, z) = \iint \tilde{E}(k_x, k_y, z) \exp[i(k_x x + k_y y)] \frac{dk_x}{2\pi} \frac{dk_y}{2\pi}, \qquad (2.3)$$

where k_x and k_y are real. It is always possible to introduce the Fourier transform of the field, because the field is square-integrable over the plane (x, y). This follows from the fact that the energy flux of the beam is finite.

Thus, the problem of determining $E(x, y, z)$ amounts to finding its Fourier transform. We obtain the equation satisfied by $\tilde{E}(k_x, k_y, z)$ by inserting (2.3) into (2.2). Noting that a partial derivative $\partial/\partial x$ becomes an algebraic multiplication by ik_x, we obtain

$$\iint \left\{ \frac{\partial^2 \tilde{E}(k_x, k_y, z)}{\partial z^2} + \left(\frac{\omega^2}{c^2} - k_x^2 - k_y^2 \right) \tilde{E}(k_x, k_y, z) \right\} \exp[i(k_x x + k_y y)] \frac{dk_x}{2\pi} \frac{dk_y}{2\pi} = 0.$$

$$(2.4)$$

Since the plane waves form a basis, the integrand is zero, so

$$\frac{\partial^2 \tilde{E}(k_x, k_y, z)}{\partial z^2} + \left(\frac{\omega^2}{c^2} - k_x^2 - k_y^2\right)\tilde{E}(k_x, k_y, z) = 0. \tag{2.5}$$

The general solution of this equation may be immediately written in the form of two exponentials. Let us introduce the notation:

$$\gamma = \begin{cases} \sqrt{\dfrac{\omega^2}{c^2} - k_x^2 - k_y^2} & \text{for} \quad \dfrac{\omega^2}{c^2} \geqslant k_x^2 + k_y^2, \\[4mm] \mathrm{i}\sqrt{k_x^2 + k_y^2 - \dfrac{\omega^2}{c^2}} & \text{for} \quad \dfrac{\omega^2}{c^2} < k_x^2 + k_y^2. \end{cases} \tag{2.6}$$

With this choice, the general solution can be cast in the form

$$\tilde{E}(k_x, k_y, z) = A(k_x, k_y)\exp(\mathrm{i}\gamma z) + B(k_x, k_y)\exp(-\mathrm{i}\gamma z). \tag{2.7}$$

If we consider a non-diverging field propagating in the positive z-direction, then the term B is zero. A is then found by simply writing the expression for the field in the plane $z = 0$ using (2.3) and (2.7). Thus, we obtain

$$E(x, y, 0) = \iint A(k_x, k_y)\exp[\mathrm{i}(k_x x + k_y y)]\,\frac{\mathrm{d}k_x}{2\pi}\frac{\mathrm{d}k_y}{2\pi}. \tag{2.8}$$

This expression simply shows that the complex amplitude $A(k_x, k_y)$ is the Fourier transform of the field in the plane $z = 0$:

$$A(k_x, k_y) = \tilde{E}(k_x, k_y, 0). \tag{2.9}$$

Finally, the field at any point (x, y, z) can be cast in the form

$$E(x, y, z) = \iint \tilde{E}(k_x, k_y, 0)\exp[\mathrm{i}(k_x x + k_y y + \gamma z)]\,\frac{\mathrm{d}k_x}{2\pi}\frac{\mathrm{d}k_y}{2\pi}. \tag{2.10}$$

The calculation of the Fourier transform of the electric field considered as a function of x and y yields the amplitudes of the plane waves $A(k_x, k_y) = \tilde{E}(k_x, k_y, 0)$. To give a mathematical form to this statement, the inversion formula is used:

$$\tilde{E}(k_x, k_y, 0) = \iint E(x, y, 0)\exp[-\mathrm{i}(k_x x + k_y y)]\,\mathrm{d}x\,\mathrm{d}y. \tag{2.11}$$

From (2.10), it is clear that the field is written in the form of a superposition of plane waves whose wavevectors have components (k_x, k_y, γ) (Fig. 2.2) and satisfy the following dispersion relation in vacuum:

$$k_x^2 + k_y^2 + \gamma^2 = \frac{\omega^2}{c^2}. \tag{2.12}$$

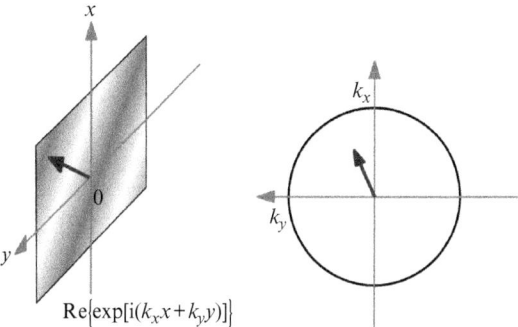

$$\mathrm{Re}\{\exp[\mathrm{i}(k_x x + k_y y)]\}$$

Fig. 2.2 Illustrating the decomposition of the field in plane waves. The components of the wavevector can also be seen as a direction and modulus.

We have seen that γ can be either real or purely imaginary. The former case corresponds to a propagating plane wave while the latter corresponds to a so-called evanescent wave. Evanescent waves do exist close to the plane $z = 0$ but decay exponentially away from this plane. Let us stress that their physical origin is very different from that of the exponentially decaying wave encountered when dealing with propagation in a metal or an absorbing system for instance. Here, the evanescent waves result from the fact that (i) the field must contain high spatial frequencies if it varies on small lengthscales, (ii) these high spatial frequencies cannot be provided by propagating waves with wavevector satisfying $k = \omega/c$.

It should be noted that (2.10) *is also valid in the near field* and that no approximation has been made so far. This exact result should not be confused with the far-field or Fraunhofer approximation introduced in the standard diffraction theory. The far-field limit can be recovered from the angular spectrum by performing an asymptotic analysis as shown in the appendix to this section (Section 2.1.7). Summarizing the last steps, it is seen that the field at any plane z is obtained by

(1) decomposing the field in the plane $z = 0$ into plane waves;
(2) propagating each plane wave by simply multiplying by the factor $\exp(\mathrm{i}\gamma z)$;
(3) adding all the resulting plane waves.

The result summarized by (2.10) is also called the field's angular spectrum. We are now going to discuss the meaning of this result.

2.1.2.3 *Propagation in the far field is a spectrum analyser*

In this subsection, we consider waves with real values of γ. Propagation is a spectrum analyser in the sense that the amplitude of a plane wave propagating in the direction (k_x, k_y, γ) (with real γ) is proportional to the Fourier transform of the field in the plane $z = 0$. When the calculation of the complex amplitude is carried out using (2.11), the amplitudes of the plane waves with components (k_x, k_y) are obtained. It is possible to give another meaning to this calculation. Taking the Fourier transform of $E(x, y, 0)$ amounts to decomposing into spatial frequencies the function $E(x, y, 0)$. For example, the component k_x can also be written in the form $2\pi f_x$, where f_x is a spatial frequency. As an example, let us consider a periodic grid with period d printed on paper with lines perpendicular to $(k_x, k_y, 0)$ (see Fig 2.2) with grey levels. The associated spatial frequency is $\sqrt{k_x^2 + k_y^2}/2\pi = 1/d$.

The first remark that can be made is that to each spatial frequency of the function $E(x, y, 0)$ there corresponds a plane wave (i.e. a propagation direction). For a wavevector with $k_y = 0$, we have $k_x^2 + \gamma^2 = \omega^2/c^2$. Note that a low spatial frequency ($k_x \ll k_z$) corresponds to a wavevector of the plane wave close to the Oz axis, while a value of k_x close to ω/c corresponds to a direction far from the Oz axis, since $k_x \gg \gamma$. Let us assume that the intensity along a propagation direction can be measured separately. To do this, it suffices to make an intensity measurement in the far field or in the focal plane of a lens.[2] The intensity is proportional to the square modulus of the field and therefore to $|\tilde{E}(k_x, k_y, 0)|^2$. In a way, *propagation plays the role of a spectrum analyser*.

Let us now return to the case of a periodic object such as the grid discussed previously. We assume that this periodic object is illuminated by a beam. This is the familiar situation of a grating diffracting light and producing several orders. From the point of view of the angular spectrum, we have a field in the plane of the object that is periodic. Therefore, its Fourier transform is discrete. The amplitude in the far field is then zero everywhere except for a discrete set of values: the orders of the grating.

2.1.3   Resolution limit: fundamentals

2.1.3.1   Resolution limit

We have just discussed the concept of a spatial frequency. We are now going to show that propagation is a low-pass filter. In other words, not all the spatial frequencies can propagate in vacuum without attenuation. This property is the basis of the resolution limit of optical instruments.

Let us first consider two representative situations for which there exist high spatial frequencies. This amounts to saying that the field varies rapidly on small lengthscales. We consider as a first example a beam blocked by a sharp edge or scattered by a small structure such as dust or a scratch. A second example is a beam illuminating a tiny aperture with sharp edges. These two examples illustrate a field that varies rapidly and therefore contains high spatial frequencies.

In order to examine the field behaviour during propagation, let us apply the propagation formula. We need to find the amplitude of each plane wave. We consider the

---

[2] This statement is rigorously supported by the asymptotic far-field expansion of the field as shown in the appendix to this section (Section 2.1.7). It can be seen that the amplitude of the field in the far field is proportional to the Fourier transform of the field in the plane $z = 0$. This is the so-called far-field or Fraunhofer approximation.

case of a rectangular aperture with sharp edges illuminated at normal incidence. The calculation of the integral (2.11) is elementary and yields the function

$$\frac{1}{k_x k_y} \sin\left(\frac{k_x a}{2}\right) \sin\left(\frac{k_y a}{2}\right).$$

This expression gives the spectrum of the object. Note that the spectrum's decay at high frequencies is slow. This is due to the presence of the field discontinuities at the edges. It is well known that the spectra of small structures include high spatial frequencies. Indeed, it is impossible to build a function with sharp edges by adding cosines of low spatial frequencies.

We have made the point that small structures always have large spatial frequencies. Now we address the following question: can these high spatial frequencies propagate in vacuum? We know that it is not possible to associate a propagation direction with a spatial frequency greater than ω/c. If we take a look at (2.6), we see that these spatial frequencies are associated with wavevectors such that γ is imaginary. This corresponds, therefore, to *waves that have an exponential decay*. They are called evanescent waves. Beyond a few wavelengths, their contribution to the field becomes negligible. Only plane waves with

$$k_x^2 + k_y^2 < \frac{\omega^2}{c^2}, \tag{2.13}$$

have a real z-component of the wavevector and thus do not decay with z. In short, therefore, only waves satisfying (2.13) may propagate (Fig. 2.3). The maximum

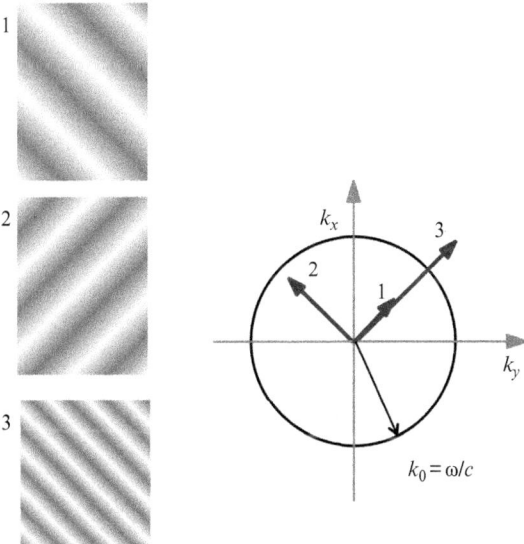

Fig. 2.3 Large wavevectors correspond to evanescent waves. Only waves with wavevectors inside the circle $k_x^2 + k_y^2 < k_0^2$ are propagating.

wavevector in the plane (x, y) corresponding to a propagating wave is $\omega/c = 2\pi/\lambda$ and the maximum spatial frequency f_x is $1/\lambda$. This has a basic consequence for teledetection, microscopy, and more generally imaging in the far field. During propagation, *all the information carried by spatial frequencies f_x higher than $1/\lambda$ is lost.* In other words, the fine details, the structures of $E(x, y, 0)$ smaller than the radiation wavelength, are lost during propagation. *In particular, any optical image cannot contain details smaller than the wavelength, since the maximum frequency is $1/\lambda$.* This is the origin of the diffraction limit.

At this point, let us make an important remark. The previous discussion dealt with the electromagnetic field: information about high spatial frequencies *of the field* is lost. So far, we have not discussed what is an object image, and therefore we have made no statement about the resolution limit for an object image. This requires us to analyse the link between a scattered field and the object that scatters the field. We will discuss this issue further at the end of this section.

2.1.3.2 *Super-resolution in the near field*

The diffraction limit may be overcome when performing a measurement of the electromagnetic field. We know that the information on the high spatial frequencies is contained in the evanescent waves. Hence, this information can be collected by measuring the field close to the object in order to detect the evanescent waves.

Let us imagine a surface on which structures smaller than the wavelength are etched. As the boundary conditions must be satisfied along the structures boundaries, the scattered fields must vary on lengthscales that depend on the geometrical form of the structures. They can be much smaller than the wavelength. Hence, the scattered field contains evanescent waves. A detector placed at a subwavelength distance from the surface can detect the evanescent waves. Therefore, it detects the high spatial frequencies and may supply images of the field with a resolution better than the wavelength used for the detection. It is useful to note at this stage that the distance z along which the radiation is propagated defines a cutoff frequency. Let us consider a large wavevector $k_x \gg \omega/c$. The decay is then given by

$$\exp(i\gamma z) \simeq \exp(-k_x z),$$

since

$$\gamma = \left(\frac{\omega^2}{c^2} - k_x^2\right)^{1/2} \simeq i k_x.$$

By observing the field at a distance z, the field associated with the wavevector $k_x = 1/z$ is attenuated by a factor $1/e$. This wavevector corresponds to a period along the x-axis $\Lambda = 2\pi/k_x = 2\pi z$. *It is useful to keep in mind that a field with period Λ along the x-axis is attenuated by a factor $1/e$ at a distance $z = \Lambda/2\pi$.* Hence, if we aim at detecting structures with Λ smaller than the wavelength λ, the distance z has to be smaller than $\lambda/2\pi$. *This condition provides a first definition of what is called the near* *field.* Let us emphasize that the standard diffraction theory based on the Huygens–Fresnel principle *does not account* for the near-field region. This will be discussed in the appendix to this section (Section 2.1.7).

2.1.3.3 *Physical origin of the evanescent waves*

The purpose of this subsection is to give some insight into the origin of the evanescent waves. When studying waves propagating in vacuum, we know that the wavevector must satisfy $k = \omega/c$. so that it seems that there is a cutoff. However, this is only valid in the absence of charges. A rapid variation of the field becomes possible in the vicinity of localized charges. This is a key point of nanophotonics: field confinement is always due to the existence of evanescent waves generated by nearby charges. These charges can be free electrons in a metal or bounded polarization charges in a dielectric. The only constraint is that there must be some material medium. Let us illustrate this idea with the simplest possible example: a scalar field produced by a monochromatic pointlike source at the origin with amplitude $\delta(\mathbf{r})S(\omega)$. The Helmholtz equation then takes the form

$$\nabla^2 E + k_0^2 E = \delta(\mathbf{r})S(\omega) \,. \tag{2.14}$$

If we introduce the spatial Fourier transform of the field, we find

$$(-k^2 + k_0^2)E(k_x, k_y, k_z) = S(\omega) \,. \tag{2.15}$$

This equation has a clear meaning. The term on the right-hand side is the spectral content of the source, both for frequency and for spatial frequencies. Since we used a pointlike source, it contains all the spatial frequencies. This term drives the field at all spatial frequencies. In other words, while the vacuum contains electromagnetic modes with $k = k_0 = \omega/c$, a pointlike source can drive the field at different spatial wavevectors imposed by the spatial spectral content of the source. This is entirely analogous to a time-harmonic oscillator driven by a pulse:

$$\frac{\mathrm{d}^2 E}{\mathrm{d}t^2} + \omega_0^2 E = S\delta(t) \,, \tag{2.16}$$

which. after Fourier transformation, becomes

$$(-\omega^2 + \omega_0^2)E(\omega) = S \,. \tag{2.17}$$

This last equation shows that the oscillator can be driven at frequencies different from ω_0 by a pulse excitation containing large temporal frequencies. In the space domain, the excitation of evanescent waves by confined current densities is entirely similar. To summarize, the existence of a pointlike source generates high spatial frequencies. Finally, we provide the explicit form of the solution of the equation 2.14, taking $S(\omega) = 1$. It is a spherical wave that can be represented as a superposition of plane waves:

$$\frac{\exp(ik_0 r)}{r} = \frac{i}{2\pi} \int \frac{1}{\gamma} \exp[i(k_x x + k_y y + \gamma|z|)] \frac{\mathrm{d}k_x}{2\pi} \frac{\mathrm{d}k_y}{2\pi} \,, \tag{2.18}$$

where γ is a function of k_x and k_y defined by (2.6) and $r = (x^2 + y^2 + z^2)^{1/2}$. This is called the Weyl expansion. Its derivation can be found in [14, 21] for instance.[3] Note that the spherical wave contains evanescent waves that are produced by the flat spatial frequency spectrum of the pointlike source. More generally, the origin of the high spatial frequencies lies in the spatial structure of induced currents in materials. The basic idea put forward in this section is that evanescent waves are intimately related to charges. Hence, solving a radiation problem is another approach to studying the near fields. This is the point of view adopted in the next subsection.

2.1.4 Radiation in the near field

We have just introduced the concept of evanescent waves. The near-field region has been defined as the region where evanescent waves contribute significantly to the field. Our discussion was thus based on a Fourier analysis of the field. Here, we show that near-field effects can be discussed without invoking evanescent waves. We start by introducing the concept of the near field by inspecting the field radiated by a dipole. We then generalize the discussion. Of particular importance here is *the possibility of using electrostatics to study the spatial structure of the electromagnetic field in the near field*.

The outline of this subsection is as follows. First, we write the expression for the electrostatic field created by a static dipole. We will show that this field may be written by means of a tensor. Then we will discuss the field radiated by an oscillatory dipole. We will show that the latter reduces to the former for distances such that $k_0 r \ll 1$, i.e. in the near field. We then generalize the discussion of the applicability of electrostatics for computing electric fields in the near field.

2.1.4.1 *Dipolar electrostatic field*

We denote by **p** the dipole moment that is the source of an electrostatic field in vacuum. It can be shown that the field is given by the equation:

$$\mathbf{E} = \frac{-1}{4\pi\epsilon_0 r^3}[\mathbf{p} - 3(\mathbf{u}_r \cdot \mathbf{p})\mathbf{u}_r],$$

where $\mathbf{u}_r = \mathbf{r}/r$ and $\mathbf{r} = (x, y, z)$. Two well-known aspects of the electrostatic field deserve to be highlighted here. On the one hand, the electrostatic field is larger along the axis of the dipole moment while we know, in contrast, that the radiated field at a large distance is zero in this direction. On the other hand, this static field decays as $1/r^3$. It should be noted at this stage that the field is not parallel to the dipole moment. This is why the relation between the dipole and the electric field cannot be given simply by a function. We need instead a linear relation connecting two vectors. In a particular basis, this is given by a matrix. It turns out to be practical to give an

---

[3] It may be disturbing to have a representation of a wave with spherical symmetry displaying evanescent waves decaying only along z. Here, the choice of the z-axis is arbitrary—any axis can be chosen, so there may be evanescent waves decaying along other directions.

explicit form introducing the field $\mathbf{X}$ created by a dipole moment of unit amplitude carried by the unit vector $\mathbf{e}_x$, the field $\mathbf{Y}$ created by a dipole moment of unit amplitude carried by the unit vector $\mathbf{e}_y$, and the field $\mathbf{Z}$ created by a dipole moment of unit amplitude carried by the unit vector $\mathbf{e}_z$. It is clear that the total field created by any dipole moment may be cast in the form

$$\mathbf{E} = p_x \mathbf{X} + p_y \mathbf{Y} + p_z \mathbf{Z}\,.$$

This form may be further condensed into the product of a matrix whose columns are the vectors $\mathbf{X},\mathbf{Y},\mathbf{Z}$ multiplied by the vector $\mathbf{p}$. Since this structure does not depend on the basis chosen, the relation between the two vectors is given by a tensor. Note that this tensor relationship may be written in the dyadic form

$$\mathbf{E} = \frac{-1}{4\pi\epsilon_0 r^3}\left(\overset{\leftrightarrow}{\mathbf{I}} - 3\mathbf{u}_r\mathbf{u}_r\right)\mathbf{p}\,, \tag{2.19}$$

where $\overset{\leftrightarrow}{\mathbf{I}}$ stands for the unit tensor and $\mathbf{u}_r\mathbf{u}_r$ is the dyadic notation for a tensor defined by $(\mathbf{u}_r\mathbf{u}_r)\mathbf{X} = \mathbf{u}_r(\mathbf{u}_r \cdot \mathbf{X})$. To summarize, the tensor structure is necessary to account for this elementary property: the radiated field is not parallel to the dipole moment.

2.1.4.2 *Electric field in the near field: field radiated by a dipole*

Let us turn now to the field radiated by an oscillating dipole with circular frequency ω and dipole moment $\mathbf{p}$. The radiated field may be cast in the form

$$\mathbf{E} = \frac{k_0^2}{4\pi\epsilon_0}\frac{\exp(ik_0 r)}{r}\left[\overset{\leftrightarrow}{\mathbf{I}} - \mathbf{u}_r\mathbf{u}_r + \left(\frac{i}{k_0 r} - \frac{1}{k_0^2 r^2}\right)\left(\overset{\leftrightarrow}{\mathbf{I}} - 3\mathbf{u}_r\mathbf{u}_r\right)\right]\mathbf{p}\,. \tag{2.20}$$

Like the electrostatic field, the radiated field is not parallel to the dipole moment, so we need again a tensor. In addition, we note that when the electrostatic limit is considered, the previous result (2.19) is recovered.

The electrostatic limit corresponds to the limit where the speed of light goes to infinity, so retardation effects become negligible. Taking this limit amounts to considering that $k_0 = \omega/c$ tends to zero. As $k_0 = 2\pi/\lambda$, this is also equivalent to considering an infinite wavelength. It can be seen from (2.20) that in this limit, the remaining terms decay as $1/r^3$.

It is interesting to note that we obtain exactly the same result by keeping c unchanged and taking the limit where r goes to zero. The $1/r^3$ terms dominate, so we again obtain the electrostatic field. This equation may be used to obtain a second definition of the near-field region: it is the region where the $1/r^3$ and $1/r^2$ terms are no longer negligible in the radiated field. According to (2.20), this region is defined by $r \leqslant \lambda/2\pi$. Thus, we recover the condition that appeared in our discussion of evanescent waves; we also note here that the lengthscale that characterizes the near field is not λ, but rather $\lambda/2\pi$.

This behaviour is illustrated in Fig. 2.4, where we show the field scattered by a dipolar scatterer. We have plotted the square of the electric field in a plane at three

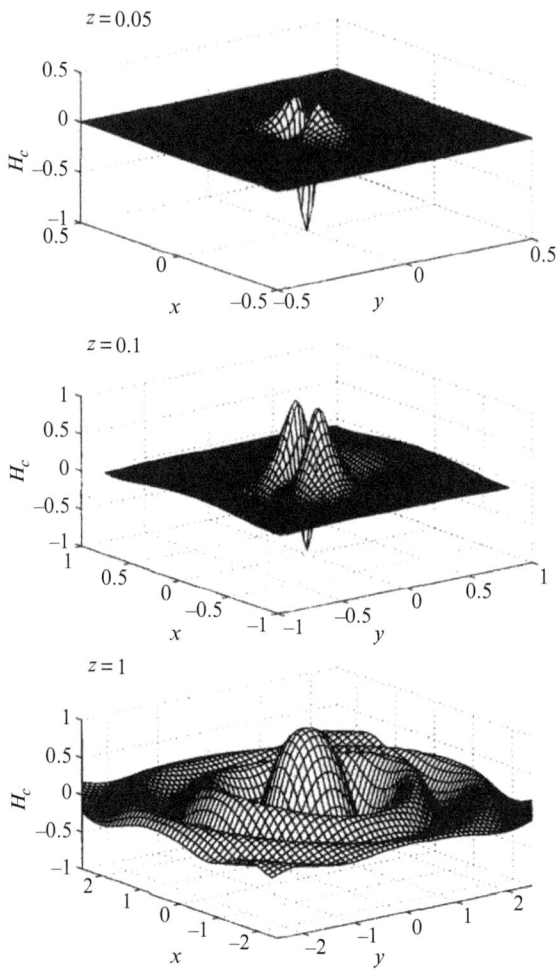

Fig. 2.4 Interference pattern of the incident field and the field scattered by a dipolar isotropic scatterer at different distances from the dipole. The incident field is polarized along the x-axis. (Reprinted from Greffet, J.J. & Carminati, R. (1997). Image formation in near-field optics. Progress in Surface Science, 56(3), 133–237. doi:10.1016/s0079-6816(98)00004-5, with permission from Elsevier.)

different distances from the scatterer, $z = 0.05$, 0.1, and 1 in wavelength units. The incident field is polarized along the x-axis. The oscillations are due to interference between the incident plane wave and the scattered field. There are many remarkable features in this figure. Let us first note the different lengthscales along the x- and y-axes, demonstrating that the field is very confined in the plane $z = 0.05\lambda$. It can be seen that for a distance $z = \lambda/10$, the width of the field distribution is already of the order of the wavelength, illustrating the previous discussion pointing out that the limit of the near-field region is given by $\lambda/2\pi$. The second remarkable feature is the polarization structure. Whereas the intensity distribution depends strongly on the polarization in the near field, it can be seen that at a distance $z = \lambda$, the distribution of the intensity becomes almost isotropic.

From the previous discussion, we can draw a very important conclusion. The electric field has a strongly polarized structure in the near field. As it is obvious from the analytic expression and from the figure, the scalar approximation is simply meaningless in the near field.

We have seen that in the near field, the spatial structure of the electric field is given by the electrostatic field. This has been found for a particular case: the dipole field. However, this is a general result, as will be shown in the next subsection.

2.1.4.3   *Electric field in the near field: arbitrary charge distribution*

Here, we generalize the previous discussion to any charge distribution. We start by considering the retarded scalar potential for an arbitrary time-dependent charge distribution $\rho(\mathbf{r}, t)$ in vacuum. The scalar potential produced by a charge density $\rho(\mathbf{r}, t)$ in a finite volume V is given by

$$\Phi(\mathbf{r}, t) = \frac{1}{4\pi\varepsilon_0} \int_V \frac{\rho(\mathbf{r}', t - |\mathbf{r} - \mathbf{r}'|/c)}{|\mathbf{r} - \mathbf{r}'|} \, \mathrm{d}x' \, \mathrm{d}y' \, \mathrm{d}z' \, .$$

It can be seen that the retardation term $|\mathbf{r} - \mathbf{r}'|/c$ becomes negligible when the velocity of light tends to infinity. More precisely, we consider systems with a characteristic size L such that L/c is negligible when compared with a typical timescale of the charge distribution (the period of the electric field for instance). In this limit, the retarded potential becomes

$$\Phi(\mathbf{r}, t) = \frac{1}{4\pi\varepsilon_0} \int_V \frac{\rho(\mathbf{r}', t)}{|\mathbf{r} - \mathbf{r}'|} \, \mathrm{d}x' \, \mathrm{d}y' \, \mathrm{d}z' \, .$$

Hence, in this limit, the spatial structure of the electric field is given by $\mathbf{E}(\mathbf{r}, t) = -\nabla_{\mathbf{r}}\Phi(\mathbf{r}, t)$. For a monochromatic potential, the complex amplitude can be cast in the form

$$\mathbf{r}\Phi(\mathbf{r}) = \frac{1}{4\pi\varepsilon_0} \int_V \rho(\mathbf{r}') \frac{\exp(i\omega|\mathbf{r} - \mathbf{r}'|/c)}{|\mathbf{r} - \mathbf{r}'|} \, \mathrm{d}x' \, \mathrm{d}y' \, \mathrm{d}z' \, .$$

The retardation term becomes $\exp(i\omega|\mathbf{r} - \mathbf{r}'|/c)$. This is the phase of a spherical wave. If the phase can be neglected, the potential is equal to the electrostatic potential. This approximation is valid in three cases: (i) the circular frequency is zero; (ii) the velocity of light tends to infinity; (iii) $|\mathbf{r} - \mathbf{r}'| \ll \lambda/2\pi$. The third condition defines the near-field region. It is the space region where retardation effects can be neglected, so it has an extension smaller than $\lambda/2\pi$. As we have seen, in this region, the spatial structure of the electric field is given by the electrostatic form.

Let us give some numerical examples of $\lambda/2\pi$ in different frequency regimes. In the visible, the near-field region corresponds to a lengthscale of 100 nm, so near-field optics coincide with optics at the nanoscale. For microwaves at $1\,\text{GHz}$ (mobile phones are operated at $1.8\,\text{GHz}$), $\lambda = 30\,\text{cm}$, so $\lambda/2\pi \approx 5\,\text{cm}$. For radio waves at $200\,\text{kHz}$, $\lambda = 1.5\,\text{km}$, so $\lambda/2\pi \approx 250\,\text{m}$, and therefore any building having an antenna operating at this frequency is within its near field.

2.1.5 Revisiting simple concepts in the near field

2.1.5.1 Phase in the near field

At first glance, introducing phases in the near field seems contradictory to the statement that the near field is correctly described in the electrostatic limit. Indeed, in electrostatics, the fields are real and the phase is not a relevant concept. Note, however, that the dielectric constant is a relevant concept in electrostatics. Now, when dealing with fields in the optical regime, we need to describe the responses of materials using their dielectric properties at optical frequencies. Hence, dielectric constants are complex numbers in the frequency domain. For instance, a metallic nanoparticle will be polarized by an incident field. The induced dipole moment depends on the phase of the polarizability. It will be described later in this chapter how the electrons in a metallic nanoparticle have a collective motion that can be excited resonantly by the incident field. Depending on the frequency detuning between the driving incident field and the resonance frequency, the phase can change from 0 to π. Clearly, the field scattered by the particle has a phase that depends on the induced dipole phase. This example shows that the phase of the fields is strongly related to the material properties. Other examples involve the familiar lumped elements of electrical circuits, which introduce phases in the responses of subwavelength objects. We have emphasized above the importance of the polarization in the structure of the electric field in the near field. Here, we emphasize that the phases are different for different components of the electric field. This result is obvious upon inspection of the structure of the field radiated by a dipole. Indeed, it can be seen from (2.20) that the terms varying as $1/r^2$ are in quadrature with the terms varying as $1/r$ or $1/r^3$.

2.1.5.2 Reflection factor in the near field

We now briefly consider the concept of reflection in the near field. We start by discussing the intuitive picture of a collimated beam impinging on a surface. We note that such a beam has a width that is larger than the wavelength, so the reflection factor cannot be defined for a subwavelength spot. If we examine the reflection factor more closely, we note that it is usually defined as the response of a planar interface to an incident plane wave with a well-defined incident wavevector. From that perspective, reflection appears to be perfectly localized in k-space and poorly defined in direct space. The reflection factor of a beam can be defined and appears to be a non-local operator in the general case [50].[4]

    While Fresnel reflection factors are usually defined for real angles and therefore only for wavevectors with a modulus smaller than ω/c, it turns out that Fresnel reflection

---

[4] It turns out to be given by the Wigner transform of the Fresnel reflection factor.

factors can also be defined for incident evanescent waves with large wavevectors k_x. Care has to be taken when defining an incident evanescent wave. Such a wave decays exponentially as it approaches the interface and diverges at infinity, which seems to be unphysical. However, this evanescent wave can be used to represent the field within a *finite volume above the interface*. Consider for instance a pointlike dipole located at $\mathbf{r} = (0, 0, z_0)$ above an interface defined by $z = 0$. The field produced by a dipole contains evanescent waves, as discussed above. They have the form $\exp(i\gamma|z - z_0|)$, so they decay away from the source. Since they are only defined between 0 and z_0, there is no divergence issue. In other words, the angular spectrum of the field produced by a dipole above an interface contains evanescent waves.

We now ask the question: what are the reflection and transmission factors for an evanescent wave ? It turns out that the same procedure can be applied to either propagating or evanescent waves when using a real wavevector k_x to describe the incident wave. With this notation (as opposed to using angles), there are no differences between the two cases $k_x > k_0$ and $k_x < k_0$. It follows that the usual forms of the Fresnel reflection factors are valid. We can in particular seek the limiting values when $k_x \gg k_0$. It is easy to check that for the two polarizations, we get the following results:

$$
\begin{aligned}
r_\mathrm{p} &= \frac{\epsilon_1 \gamma_2 - \epsilon_2 \gamma_1}{\epsilon_1 \gamma_2 + \epsilon_2 \gamma_1} \approx \frac{\epsilon_1 - \epsilon_2}{\epsilon_1 + \epsilon_2}, \\
r_\mathrm{s} &= \frac{\gamma_1 - \gamma_2}{\gamma_1 + \gamma_2} \approx \frac{\epsilon_2 - \epsilon_1}{4} \frac{k_0^2}{k_x^2}.
\end{aligned}
\tag{2.21}
$$

In other words, the s-polarized reflectivity tends to zero, whereas the p-polarized reflection factor tends to the factor that appears when computing the image charge of a pointlike charge above an interface. Hence, we recover the electrostatic and magneto-static limit. It is important to note that this near-field reflectivity can be greater than 1. While this would contradict energy conservation in the far field, here the reflection does not have any energy conservation constraints. Indeed, we are dealing with evanescent waves, so these do not contribute to the energy flux. We know that the image charge is an effective concept to account for the surface charges that screen the electric field at an interface. A reflection factor greater than 1 simply indicates that the localized source has resonantly excited surface charge modes of the interface, so the field amplitude can be larger than the incident excitation amplitude, as for any resonance. The possibility of resonant surface waves is one of the most important aspects of nanophotonics. This resonance is associated with a microscopic resonance of the polarization of the material. Section 2.3 is devoted to the study of these excitations called surface polaritons.

2.1.5.3 *Electric field and magnetic field in the near field*

In this section, we analyse the structure of the electromagnetic field in the near field. Here, we stress that the well-known relations between the electric and the magnetic fields for a plane wave are no longer valid. In particular, we will see that the fields are not necessarily perpendicular in the geometrical sense. We will also show that the ratio between electric and magnetic field in vacuum is no longer given by the familiar plane wave ratio $|B| = |E|/c$. In particular, the magnetic field may be the leading term in the energy density, depending on the polarization. It is therefore very important to carefully analyse the fields before discarding the magnetic contribution when working in the near field.

We start by considering an evanescent wave with a wavevector given by $\mathbf{k} = k_x \mathbf{e}_x - \gamma \mathbf{e}_z$. Here, we define a reference plane containing the propagation axis Ox and the decay axis Oz. This plane allows us to define two linear polarizations: we define the s-polarized field when the electric field is perpendicular to this plane and the p-polarized field when the magnetic field is perpendicular to the plane. Hence, the electric field is given by $\mathbf{E} = (0, E, 0)$ for an s-polarized field. The magnetic field follows from the Maxwell–Faraday equation $\nabla \times \mathbf{E} = i\omega \mathbf{B}$ and is given by $\mathbf{B} = (-\gamma, 0, k_x) E/\omega$. It can be checked that this is a transverse solution of Maxwell's equations, since $\nabla \cdot \mathbf{E} = i\mathbf{k} \cdot \mathbf{E} = 0$. Note, however, that for an evanescent wave, $\gamma = i|\gamma|$ is a purely imaginary complex number, so the equality $\mathbf{k} \cdot \mathbf{E} = 0$ does not have the usual interpretation in terms of perpendicular vectors.

Let us now compare the moduli of the magnetic and electric fields: $|\mathbf{B}|^2 = \mathbf{B} \cdot \mathbf{B}^* = (|\gamma|^2 + k_x^2)|\mathbf{E}|^2 \omega^2$. For a propagating wave, γ is real, so $|\gamma|^2 + k_x^2 = \gamma^2 + k_x^2 = \omega^2/c^2$. Hence, the equality yields the usual result

$$|\mathbf{B}| = \frac{|\mathbf{E}|}{c},$$

whereas for an evanescent wave, we have $|\gamma|^2 = -\gamma^2$, so

$$|\mathbf{B}| = \frac{|\mathbf{E}|}{c} \sqrt{2 \frac{k_x^2}{k_0^2} - 1}.$$

It follows that for large wavevectors such that $k_x \gg k_0$, we have

$$\frac{cB}{E} \approx \sqrt{2} \frac{|k_x|}{k_0}.$$

Clearly, the magnetic field can be much larger than E/c. It is therefore necessary to think twice before discarding magnetic effects in the near field. A similar analysis shows that for p-polarized evanescent waves, the electric field dominates. In that case, we find

$$\frac{E}{cB} \approx \sqrt{2} \frac{|k_x|}{k_0}.$$

In summary, the ratio between electric and magnetic fields depends strongly on the polarization in the near field. The relation $E/cB = 1$ is no longer valid. This remark can be reformulated in a somewhat different manner. It is well known that electric and magnetic fields are uncoupled in the non-retarded (static) limit. Here, we see that the fields tend to be uncoupled in the near-field regime, when retardation can be neglected. Furthermore, the above analysis shows that p-polarization claims the electric field, whereas s-polarization claims the magnetic field.

2.1.5.4 Polarization

When discussing polarization, it is often assumed that the electric field is perpendicular to a privileged direction given by the propagation direction of a collimated beam, so that only two components need to be specified to characterize the polarization. Obviously, there is no such privileged direction in the near field, so one needs to consider all field components. Let us also point out that when dealing with an evanescent waves, the condition $\nabla \cdot \mathbf{E} = 0$ yields $k_x E_x + \gamma E_z$, as for a propagating wave. However, as γ is purely imaginary, this equation does not mean that the field is perpendicular to a real vector $(k_x, 0, \gamma)$. Instead, it shows that

$$E_x = -\frac{i|\gamma|}{k_x} E_z \, ,$$

indicating that the electric field is elliptically polarized in the (x, z) plane.

2.1.6 Super-resolution imaging

It has been shown above that propagation is a low-pass filter. We now discuss to what extent this is a limitation on super-resolution imaging. We first emphasize that this limit applies only in the framework of linear optics, where a Fourier analysis is meaningful. We start by discussing the so-called diffraction limit in the framework of the angular spectrum. We then point out that the loss of high spatial frequencies *of the field* does not necessarily mean that it is impossible to obtain images with resolution better than half a wavelength, even though standard linear optics is used. We do not cover in this subsection the recent techniques achieving super-resolution from far-field measurements based on nonlinear effects such as saturation effects (stimulated emission depletion [39]) or blinking sources [18, 96] spatially separated so that the resolution is improved because the error on positioning of the centre of a point spread function (PSF) is significantly smaller than the radius of the PSF.

2.1.6.1 Analytic continuation of the field in Fourier space

The analysis of propagation of light has shown that waves with large wavevector k_x cannot propagate over distances larger than $1/k_x$. As the typical focal length of a microscope objective is of the order of several millimetres, and the typical decay length of an evanescent wave is smaller than a wavelength, it is clear that all information carried by evanescent waves is lost. Hence, as discussed in the previous subsections, the resolution *on the field* is limited by the wavelength. *This statement is a true for a physicist and wrong for a mathematician.* From a mathematical point of view, it can be shown that if an *analytic function* is known on a bounded part of its domain, then an analytic continuation can be performed so that all information can be recovered. Hence, knowing the Fourier transform of the electric field in the disk $k_x^2 + k_y^2 < \omega^2/c^2$, one should be able to recover the values of the field for larger wavevectors. With this result in mind, it seems that there is no resolution limit. This topic has been studied by several authors [85, 112]. In practice, the signal-to-noise ratio severely limits reconstruction from data. Indeed, it is now well understood that this problem belongs to a class of ill-posed inverse problems.

2.1.6.2 Smallest light spot that can be produced with an objective

Let us go a step further and examine what is the smallest spot size that can be produced using a microscope objective. We assume that the objective subtends a maximum angle α_M, so that the largest wavevector component is given by

$$k_{x,M} = n\frac{2\pi}{\lambda}\sin\alpha_M,$$

where n is the refractive index of the medium. By applying the uncertainty relation $\Delta x\, k_{x,M} \approx 2\pi$, which follows from the Fourier structure of the field, we find

$$\Delta x \approx \frac{\lambda}{NA},$$

where we have introduced the numerical aperture $NA = n\sin\alpha_M$. Defining the resolution as the half-width $\Delta x/2$, we recover the familiar Abbe resolution limit. This result is usually derived in the framework of classical diffraction theory and assuming a circular diaphragm. The intensity distribution is then given by the so-called Airy function.

2.1.6.3 Imaging fields and imaging structures

Let us stress that we have only pointed out that subwavelength information *on the field* is lost. Yet, the link between an object to be imaged and the field scattered/reflected/transmitted by this object is neither trivial nor linear. Hence, to draw a conclusion regarding the resolution limit requires a model of this relation. Indeed, we have not proved that subwavelength information *on the sample* itself is lost. Since the emitted or scattered field depends on an integral involving both the optical properties of the sample and the field, the *low spatial frequencies of the field may convey subwavelength information on the sample*.

Let us reformulate this point. We have used the Fourier analysis to stress that information was lost. Fourier analysis is only valid for linear systems. Here, the link between the field at $z = 0$ and the field at $z > 0$ is linear. It is thus correct to state that information on the subwavelength features of the field was exponentially damped. However, *the imaging issue is not a field reconstruction problem*. Let us explain the difference between field reconstruction and imaging. Imaging usually assumes that the electromagnetic field at a surface conveys information on the local reflectivity of that surface. In other words, an image is a map of the reflectivity of the object. It is often taken for granted that the local object properties and the local field are related: this is a very strong assumption that is valid only for lengthscales much larger than the wavelength. In that low-resolution context, the field is locally related to the object (i.e. the product of the incident field by the local reflectivity) and therefore there is no difference between reconstructing a field distribution and reconstructing a reflectivity map and thus an object image. Of course, the concept of Fresnel reflectivity is well defined for an infinite plane surface, so it is no longer relevant when dealing with subwavelength objects, as we have already pointed out. Hence, *we need a more fundamental definition of the nature of imaging*. From a theoretical point of view, the light scattered by an object is entirely determined by the knowledge of the dielectric permittivity of the object, $\varepsilon(\mathbf{r})$, and the illumination conditions.[5] This defines a well-posed scattering problem. Hence, a complete description of the object, *from the point of view of its interaction with light*, is the distribution of permittivity in space. Hence, imaging amounts to reconstruct a *permittivity map $\varepsilon(\mathbf{r})$* from a finite subset of *scattered field measurements*. Here, we stress that *the link between the scattered field and the distribution of permittivity in a medium is not linear*: it amounts to solving a non-trivial inverse problem. This point will be further discussed later in this chapter, when we introduce a theoretical framework to deal with scattering. Yet, the key point is that the knowledge of the field on a limited set of wavevectors does contain information on the high spatial frequencies of the object. This discussion should not be confused with the studies on super-resolution based on the properties of analytic functions briefly mentioned above.

To summarize this discussion, the main idea that we have put forward is that the loss of information on the high spatial frequencies of the *electromagnetic field* does not mean that the information on *the details of the permittivity $\varepsilon(\mathbf{r})$* is lost. In other words, the fundamental resolution limit does apply to imaging inasmuch as imaging is simply based on assuming that the field distribution is an image of an object. Strictly speaking, this limit does not apply to the imaging problem. It is thus possible to produce images with subwavelength resolution by *solving the inverse problem numerically using only far-field data*. This issue is further discussed in the next section. Some examples of subwavelength reconstruction with far-field data can be found in [28, 49, 101].

2.1.7 Appendix: the Huygens–Fresnel principle revisited

The purpose of this appendix is to show the equivalence between the rigorous solution outlined at the start of this section and the Huygens–Fresnel principle. To some extent, we are essentially trying to establish a link between two different representations of the field: either a sum of plane waves or a sum of spherical waves. The problem is therefore likely to be solved if we know how to decompose a spherical wave on a basis of plane waves.

We start by reproducing the plane-wave expansion of a spherical wave:

$$\frac{\exp(ik_0 r)}{r} = \frac{2i}{\pi} \int \frac{1}{\gamma} \exp[i(k_x x + k_y y + \gamma|z|)] \frac{dk_x}{2\pi} \frac{dk_y}{2\pi}, \qquad (2.22)$$

---

[5] Note that magnetic permeability is μ_0 at optical frequencies.

where γ is a function of k_x and k_y defined by (2.6) and $r = (x^2 + y^2 + z^2)^{1/2}$. Subsequently, we shall need the Fourier transform of the sole numerator $\exp[i\gamma(k_x, k_y)|z|]$. Taking the z-derivative of (2.22) for $z > 0$, we obtain

$$\frac{\partial}{\partial z}\left[\frac{\exp(ik_0 r)}{r}\right] = -2\pi \int \exp[i\gamma(k_x, k_y)|z|]\,\exp[i(k_x x + k_y y)]\,\frac{dk_x}{2\pi}\frac{dk_y}{2\pi}. \tag{2.23}$$

2.1.7.1   *The Huygens–Fresnel principle*

Let us return to the starting point, namely the expression of the field given by (2.10). Note that the field at the point (x, y, z) is given by the Fourier transform of the product of two functions: $\tilde{E}(k_x, k_y, 0)$ and $\exp[i\gamma(k_x, k_y)|z|]$. We know that we may write it in the form of a convolution product of the two Fourier transforms. Hence, by using the result (2.23), we obtain

$$E(x, y, z) = \frac{-1}{2\pi}\iint E(x', y', 0)\frac{\partial}{\partial z}\left[\frac{\exp(ik_0 R)}{R}\right]dx'\,dy', \tag{2.24}$$

where $R^2 = (x - x')^2 + (y - y')^2 + z^2$. This expression is identical to the result obtained using Green's theorem, but is different from the historical Huygens–Fresnel principle. Note that the exact result includes a near-field term varying as $1/R^2$. The standard diffraction formula is recovered by doing two approximations: (i) neglecting the near-field term, (ii) using a paraxial approximation, replacing the z-derivative by $ik\cos\theta$, where θ is the angle between the observation direction and the Oz axis. Before the generalized use of computers, a field expression in the form of a Fourier transform was of little help in most cases. This explains why the presentation of diffraction as an expansion in plane waves was rarely used in textbooks on physical optics and electromagnetism. However, it is a simple matter to perform a Fourier transform numerically nowadays, so (2.10) is often more useful than the Huygens–Fresnel principle. In particular, it is exact in the Fresnel region and also *in the near field*.

2.1.7.2   *Far-field or Fraunhofer approximation. Stationary phase approximation*

Despite the presence of a Fourier transform in (2.10) obtained in Section 2.1.2.2, the angular spectrum should not be confused with the far-field approximation (or Fraunhofer approximation). Equation (2.10) is an exact form of the field valid at any point. Starting from (2.24), we may derive an asymptotic form of the field diffracted by an aperture with a characteristic length L. This expression is valid at large distances.[6] The same asymptotic result may be derived directly from (2.10) using an asymptotic

---

[6] Note in passing that the concept of far field is defined here by the condition $r \gg L^2/\lambda$, where L is a typical transverse extension of the field in the plane $z = 0$. Moreover, it is identical to the far-field concept defined in radiation theory. Note that the concept of near field is not defined as the opposite of far field. The near field is defined as the region where evanescent waves have a non-negligible amplitude. Thus, the near field is confined to distances less than typically $\lambda/10$.

form of the integral valid for large distances. This can be obtained using the so-called stationary phase approximation. For integrals with the structure

$$U(x, y, z) = \iint a(m, p) \exp[ik(mx + py + qz)] \, dm \, dp, \tag{2.25}$$

it can be shown that

$$U(x, y, z) \approx -\frac{2i\pi}{k_0} \frac{z}{r} \, a\left(\frac{x}{r}, \frac{y}{r}\right) \frac{\exp(ik_0 r)}{r}. \tag{2.26}$$

This form is often useful when calculating fields scattered at large distances. Here, we apply this form to the asymptotic evaluation of the field $E(x, y, z)$ given by (2.10). We see that, in the far field, the amplitude $E(x, y, z)$ is proportional to the Fourier transform $\tilde{E}(k_x = k_0 x/r, k_y = k_0 y/r, 0)$.

2.2 Computing near fields with integral equations. Green tensor and local density of electromagnetic states (LDOS)

In the previous section, we saw that the electromagnetic field may vary on lengthscales smaller than the wavelength in the vicinity of matter. It is thus important to be able to compute the field in this situation. When dealing with fields scattered in the far field, the exact form of the field at the scatterer position is often estimated using simple approximations. In contrast, in the near field, we are interested in the detailed form of the electromagnetic fields close to edges, tips, surfaces, etc. There are many theoretical ways to compute the electromagnetic field. A useful and powerful tool is the integral formulation of the field equations. A nice feature of this formulation is that it allows some physical insight to be gained into the scattering mechanisms. In simple words, an integral formulation amounts to writing the electric field at any point as the sum of the fields radiated by all the currents (either in the sources or induced in the matter). Thus, the basic quantity is the field radiated by an elementary volume. Since a small volume can be modelled by a dipole, it turns out that the key quantity is the field radiated by a dipole. It proves convenient to introduce the tensor describing the electric field radiated by a pointlike dipole; this is known as Green tensor. In the next subsection, we derive an integral equation for the electric field. We also derive the exact expression for the Green tensor and discusses its singularity as well as its physical meaning. For the sake of completeness, we also introduce the Green tensor as the solution of a partial differential equation whose source term is a pointlike dipole. Then, we apply these concepts to the study of the scattering of light by particles. In particular, we introduce the first Born approximation, and we are then able to discuss the possibility of achieving super-resolution by using far-field data and solving the inverse problem.

   The second part of this section deals with the analysis of the link between the Green tensor and the local density of electromagnetic states. This quantity plays a key role in the study of spontaneous emission in the presence of nanostructures in the weak-coupling regime.

2.2.1 Green tensor and integral formulation of electromagnetism

The most usual formulation of the scattering of electromagnetic fields is based on partial differential equations. The propagation equation in a linear isotropic inhomogeneous medium characterized by a permittivity $\epsilon_r(\mathbf{r})$ can be cast in the form

$$\nabla \times \nabla \times \mathbf{E} - \epsilon_r(\mathbf{r})k_0^2\mathbf{E} = 0 \,. \tag{2.27}$$

Here ϵ_r can be a function of both the considered point $\mathbf{r}$ and the frequency for inhomogeneous and dispersive media. Note also that its imaginary part accounts for losses. Once boundary conditions have been specified, the problem is well posed. This formulation is very practical for simple shapes such as planes or spheres. However, the problem becomes untractable for complex geometries, since general solutions are then difficult to find. Instead, it is possible to reformulate the problem in terms of an integral equation that has a transparent physical meaning. The basic idea is to consider the field propagating in vacuum and account for the presence of material media by introducing additional source terms. To proceed, let us add the term $k_0^2(\epsilon_r - 1)\mathbf{E}$ to both sides of (2.27). We find

$$\nabla \times \nabla \times \mathbf{E} - k_0^2\mathbf{E} = k_0^2(\epsilon_r - 1)\mathbf{E} = i\omega\mu_0\mathbf{j} \,, \tag{2.28}$$

where $\mathbf{j} = -i\omega\epsilon_0(\epsilon_r-1)\mathbf{E}$. As announced, this is a propagation equation in vacuum with an (unknown) source term. In other words, we have transformed a reflection/scattering problem into a radiation problem. Yet, we still need to find the sources. In this subsection, we will start from the retarded potentials and derive an integral expression of the field in terms of the currents either in the source or induced in the matter. A more direct and abstract derivation will be given in the next subsection. An advantage of the first approach is to point out the presence of a singularity and to reveal its physical meaning. To proceed, we start from the general expression for the field in terms of the potential. The vector potential is

$$\mathbf{A}(\mathbf{r}) = \frac{\mu_0}{4\pi} \int \mathbf{j}(\mathbf{r}') \frac{\exp(ik_0R)}{R} \, \mathrm{d}^3\mathbf{r}' \,, \tag{2.29}$$

where $R = |\mathbf{r} - \mathbf{r}'|$. This solution assumes that the field decays at infinity. In this solution, the currents can be considered to be either currents produced by a source and denoted external currents $\mathbf{j}_{\mathrm{ext}}$ or the induced currents in the material media $\mathbf{j} = -i\omega\epsilon_0(\epsilon_r - 1)\mathbf{E}$. The former are the source of the incident field; the latter are the source of the field scattered by the object. The vector potential can be cast in the form

$$\mathbf{A} = \mathbf{A}_{\mathrm{inc}} + \mathbf{A}_{\mathrm{ind}} = \mathbf{A}_{\mathrm{inc}} - \frac{i\omega}{c^2} \int (\epsilon_r - 1)\mathbf{E}(\mathbf{r}')g(R) \, \mathrm{d}^3\mathbf{r}' \,, \tag{2.30}$$

where $g(R) = \exp(ik_0R)/4\pi R$. We have denoted by $\mathbf{A}_{\mathrm{inc}}$ the potential vector generated by the external currents. This corresponds to the field incident on the medium. Using the general definition of the field, we obtain

$$\mathbf{E}(\mathbf{r}) = -\frac{\partial \mathbf{A}}{\partial t} - \nabla V \,. \tag{2.31}$$

The scalar potential can be derived from the expression for the potential vector using the Lorentz gauge:

$$\nabla \cdot \mathbf{A} - \frac{\mathrm{i}\omega V}{c^2} = 0 \,. \tag{2.32}$$

Inserting this expression for the scalar potential V into (2.31) yields

$$\mathbf{E} = \mathrm{i}\omega \left[\mathbf{A} + \frac{1}{k_0^2} \nabla (\nabla \cdot \mathbf{A}) \right] \,. \tag{2.33}$$

Using the explicit expression for the vector potential, we finally obtain

$$\mathbf{E} = \mathbf{E}_{\mathrm{inc}} + k_0^2 \left(\overset{\leftrightarrow}{\mathbf{I}} + \frac{1}{k_0^2} \nabla\nabla \right) \int (\epsilon_r - 1) \mathbf{E}(\mathbf{r}') g(R) \, \mathrm{d}^3\mathbf{r}' \,. \tag{2.34}$$

This formula yields the electric field both in the sources and outside the sources. When used in the sources (either external or induced sources like the currents in a scattering object), some care has to be taken because of the divergence of the integrand. This particular point will be discussed below.

2.2.1.1 *Singularity of the Green tensor and integral equation*

Taking the derivative $\nabla\nabla$ of the integral does not introduce any difficulty, provided that the field is evaluated at a point that does not belong to the distribution of currents. Indeed, in this case, there is no singularity of the integrand. In contrast, if the point belongs to the integral, there is a logarithmic divergence of the latter. This is due to the second derivative, which yields a term that diverges as $1/r^3$. When combined with the r^2 dependence of the volume element, there is a $1/r$ term that produces a logarithmic divergence of the integral. In order to avoid this difficulty, it is possible to introduce an exclusion volume δV around the point $\mathbf{r}$ at which the field is evaluated. In what follows, we shall use a spherical exclusion volume and we let its radius tend to zero. The field appears as the sum of two contributions:

$$\lim_{\delta V \to 0} \nabla\nabla \int_{V - \delta V} \mathbf{P}(\mathbf{r}') g(\mathbf{r} - \mathbf{r}') \, \mathrm{d}^3\mathbf{r}' + \lim_{\delta V \to 0} \nabla\nabla \int_{\delta V} \mathbf{P}(\mathbf{r}') g(\mathbf{r} - \mathbf{r}') \, \mathrm{d}^3\mathbf{r}', \tag{2.35}$$

where $\mathbf{P}$ denotes the polarization vector $\epsilon_0(\epsilon - 1)\mathbf{E}$. It is possible to exchange the order of differentiation and integration for the first term, since the singularity has been removed. The evaluation of the second term is more involved. A derivation can be found in [21, 114]. However, the result is almost obvious from a physical analysis of the problem. Indeed, it can be seen that the contribution of the small exclusion volume to the electric field is simply the field produced by a small sphere that has been uniformly polarized. As we discussed previously, the electrostatic approximation is valid under these circumstances. Thus, the contribution of the exclusion volume is given exactly by $-\mathbf{P}/3\epsilon_0$. We finally obtain

$$\lim_{\delta V \to 0} \int_{V - \delta V} \nabla\nabla \mathbf{P}(\mathbf{r}') g(\mathbf{r} - \mathbf{r}') \, \mathrm{d}^3\mathbf{r}' - \frac{\mathbf{P}}{3\epsilon_0} \,. \tag{2.36}$$

It is worth noting that introducing an exclusion volume and letting it tend to zero is the definition of the principal value of the integral. Inserting this result into (2.34), we obtain a compact integral equation for the electric field:

$$\mathbf{E}(\mathbf{r}) = \mathbf{E}_{\text{inc}}(\mathbf{r}) + k_0^2 \int [\epsilon_r(\mathbf{r}') - 1] \overleftrightarrow{\mathbf{G}}(\mathbf{r}, \mathbf{r}') \mathbf{E}(\mathbf{r}') \, d^3\mathbf{r}', \qquad (2.37)$$

where the Green tensor $\overleftrightarrow{\mathbf{G}}$ is given by

$$\overleftrightarrow{\mathbf{G}} = \text{PV}\left(\overleftrightarrow{\mathbf{I}} + \frac{1}{k_0^2} \nabla\nabla \right) \frac{\exp(ik_0 R)}{4\pi R} - \frac{\overleftrightarrow{\mathbf{I}}}{3k_0^2} \delta(\mathbf{r} - \mathbf{r}'). \qquad (2.38)$$

Here, 'PV' indicates that the integral is evaluated in the sense of a principal value.[7] It can be checked that the Green dyadic coincides with the field radiated by a dipole, except that we have characterized the singularity at the origin and identified its physical meaning. The expression that we have derived for the electric field is an integral equation for the electric field: the unknown electric field appears in the integrand. The physical meaning of this equation is transparent. It can be seen that the field at any point is the sum of two contributions: the incident field and the field radiated by the induced currents in the scattering object. Indeed, each volume element $d^3\mathbf{r}'$ of the scattering object has a dipole moment $\delta\mathbf{p} = \epsilon_0[\epsilon_r(\mathbf{r}') - 1]\mathbf{E}(\mathbf{r}') \, d^3\mathbf{r}'$ and radiates a field $k_0^2 \overleftrightarrow{\mathbf{G}}(\mathbf{r}, \mathbf{r}')[\epsilon_r(\mathbf{r}') - 1]\mathbf{E}(\mathbf{r}') \, d^3\mathbf{r}'$. Hence, we see that the Green tensor is simply the operator that gives the field produced by an electric dipole. Since every point is illuminated by the fields produced by currents excited at all the other points of the medium, this is a multiple-scattering problem. Its solution generally involves taking into account all the interactions.

Let us stress here that the Green tensor is the Green tensor in vacuum. This may be surprising at first glance, since a volume element in a metal will emit a wave that will not be able to propagate through the metal. The decay of the wave emitted from a dipole in a bulk metal will emerge as the result of the coherent sum of all the scattered fields produced by all volume elements.

2.2.1.2 *Green tensor as the solution of a PDE*

A more direct introduction of the Green tensor is based on the partial differential equation (PDE) satisfied by the Green dyadic. This approach closely follows the introduction of the Green function in the context of the solution of the equation for the

---

[7] It is important to note that the shape of the exclusion volume is important. If the exclusion volume is chosen to be a cube or a sphere, then the electrostatic contribution to the electric field of the exclusion volume is $-\mathbf{P}/3\epsilon_0$. If the shape changes, then the value of the contribution also changes. However, the sum of the principal value and the singularity contributions does not depend on the choice of the shape of the exclusion volume.

potentials. To illustrate the procedure, we start by solving the equation for the scalar potential. We then move to the vector case. The equation for the scalar potential is

$$\nabla^2 V + k_0^2 V = -\frac{\rho}{\epsilon_0}. \tag{2.39}$$

The Green function is defined as the solution of the equation

$$\nabla^2 G + k_0^2 G = -\delta(\mathbf{r} - \mathbf{r}') \tag{2.40}$$

plus the radiation boundary conditions. The solution of this equation is given by

$$G(\mathbf{r} - \mathbf{r}') = \frac{\exp(ik_0|\mathbf{r} - \mathbf{r}'|)}{4\pi|\mathbf{r} - \mathbf{r}'|}. \tag{2.41}$$

The solution of (2.39) is then simply obtained by superposing all the contributions of the different points of the source by taking advantage of the linearity of the equation. This yields the explicit result

$$V(\mathbf{r}) = \frac{1}{4\pi\epsilon_0} \int d^3\mathbf{r}' \, \rho(\mathbf{r}') \frac{\exp(ik_0|\mathbf{r} - \mathbf{r}'|)}{|\mathbf{r} - \mathbf{r}'|}. \tag{2.42}$$

PDE satisfied by the Green dyadic   To derive the integral form of the electric field, we start from the PDE for the electric field. It is given by

$$\nabla \times \nabla \times \mathbf{E} - k_0^2 \mathbf{E} = i\omega\mu_0 \mathbf{j}. \tag{2.43}$$

From this equation, we can derive the equation for the Green function. However, it can be seen that we cannot solve for a scalar function, because we are seeking a linear object that relates the current density (the source) to the electric field. If the equation related the x-component of the current density to the x-component of the electric field, and so on, as it is the case for the potential vector, then a scalar function could be introduced. However, the operator $\nabla \times \nabla\times$ introduces relations between different components of the electric field to each component of the current density. A simpler way of expressing this fact is to point out that the electric field is not parallel to the current density. Thus, the object needed is a tensor. This comes as no surprise, since we have the same structure for the electrostatic case. Therefore, we write the equation for the Green tensor in the form

$$\nabla \times \nabla \times \overset{\leftrightarrow}{\mathbf{G}} - k_0^2 \overset{\leftrightarrow}{\mathbf{G}} = \overset{\leftrightarrow}{\mathbf{I}} \delta(\mathbf{r} - \mathbf{r}'). \tag{2.44}$$

The source term is a pointlike source. As compared with the scalar case, there is a unit tensor $\overset{\leftrightarrow}{\mathbf{I}}$. This indicates that the source can be a dipole with three different possible orientations. This equation can be solved using techniques similar to the solution for the scalar case. When dealing properly with the singularity, one finds the result given in (2.19). The general solution of the problem is then obtained in a straightforward manner by a linear superposition as before:

$$\mathbf{E}(\mathbf{r}) = \int \overset{\leftrightarrow}{\mathbf{G}}(\mathbf{r}, \mathbf{r}') i\omega\mu_0 \mathbf{j}(\mathbf{r}') \, d^3\mathbf{r}'. \tag{2.45}$$

2.2.1.3 Alternative derivation of the integral equation

It is now a simple matter to establish the integral form of the electric field. We first note that there are two components in the current density $\mathbf{j}$. One is the current density in the sources $\mathbf{j}_{\mathrm{inc}}$. This produces the incident field $\mathbf{E}_{\mathrm{inc}}(\mathbf{r})$. The second contribution is due to the source term $i\omega\mu_0\mathbf{j}_{\mathrm{ind}}$, where $\mathbf{j}_{\mathrm{ind}}$ is the induced current density given by $-i\omega\epsilon_0(\epsilon_r - 1)\mathbf{E}$. Taking the convolution product of the Green tensor with the source term yields immediately

$$\mathbf{E}(\mathbf{r}) = \mathbf{E}_{\mathrm{inc}}(\mathbf{r}) + k_0^2 \int [\epsilon_r(\mathbf{r}') - 1]\overleftrightarrow{\mathbf{G}}(\mathbf{r},\mathbf{r}')\mathbf{E}(\mathbf{r}')\,\mathrm{d}^3\mathbf{r}'. \tag{2.46}$$

This approach is much more efficient in deriving the integral equation. Its drawback is that it introduces a singularity in the Green tensor without revealing its physical meaning. A very good discussion of the Green singularity from a mathematical perspective can be found in [107, 114, 117, 121]. Equation (2.46) is an integral equation for the field in the region where the dielectric contrast $\epsilon_r(\mathbf{r}') - 1$ is non-zero. On solving this integral equation, the field is known within the scattering region. The field can then be evaluated anywhere by performing the integral appearing in (2.46).

2.2.1.4 Far-field approximation

It is interesting to derive an explicit far-field form of the scattered field. To proceed, we keep only the terms varying like $1/r$, which is a valid approximation if $R \gg \lambda/2\pi$. This amounts to taking the derivative of the exponential but not the derivative of $1/r$. Thus,

$$\partial_j g(R) \approx ik_0 \frac{R_j}{R} g(R),$$

where $\mathbf{R} = \mathbf{r} - \mathbf{r}'$. Next, we assume that the scattering region has a maximum size L and we evaluate the field in the so-called far-field region $r \gg L^2/\lambda$, so that $|\mathbf{r} - \mathbf{r}'| \approx r - \mathbf{u}\cdot\mathbf{r}'$ and $\mathbf{r}/r \approx \mathbf{R}/R$, resulting in

$$\frac{R_j}{R} \approx \frac{r_j}{r} = u_j.$$

It follows that

$$G_{ij} = \left(\delta_{ij} + \frac{\partial_i\partial_j}{k_0^2}\right)g(R) \approx (\delta_{ij} - u_i u_j)\frac{\exp(ik_0 r)}{4\pi r}\exp(-ik_0\mathbf{u}\cdot\mathbf{r}'). \tag{2.47}$$

The tensor $\delta_{ij} - u_i u_j$ is the tensor that takes the transverse part of the field. Apart from this, the rest is essentially the usual far-field form of the potential vector. We finally find the explicit form of the scattered electric field:

$$\mathbf{E}_s(\mathbf{r}) \approx k_0^2 \frac{\exp(ik_0 r)}{4\pi r}\left(\overleftrightarrow{\mathbf{I}} - \mathbf{u}\mathbf{u}\right)\int [\epsilon_r(\mathbf{r}') - 1]\exp(-ik_0\mathbf{u}\cdot\mathbf{r}')\,\mathbf{E}(\mathbf{r}')\,\mathrm{d}^3\mathbf{r}'. \tag{2.48}$$

This formula contains all the information connecting the scattered field and the permittivity of the object. The result is remarkably simple: it shows that the scattered field amplitude is given by the Fourier transform of the polarization density in the scatterer. Hence, the angular width of the scattered field is given by λ/L, where L is a typical transverse length of the scatterer. Thus, scattering by subwavelength particles is expected to be quasi-isotropic, whereas, in order to get scattering in the forward direction, it is necessary to have large scatterers.

2.2.2 Scattering by a particle

2.2.2.1 *Scattering by a particle with low dielectric contrast: first Born approximation*

Here, we show how the integral formulation of the electric field can be used to obtain in a very straightforward way an expression for the field scattered by an object of arbitrary shape. The problem that we address here is the computation of the field scattered by a particle illuminated by a plane wave. The electric field can be derived everywhere using its integral form, provided that the field within the scatterer is known.

The first and more difficult problem to solve is finding the induced currents in the scatterer. To proceed, one can use the integral equation 2.46. The first term is the incident field, while the second term represents the field produced by the induced currents. There are a number of cases where the scattered field is much smaller than the incident field, so that it is possible to simply assume that the field within the scatterer can be approximated by the incident field. This is known as first Born approximation. It is difficult to give an accurate criterion to characterize the accuracy of this approximation. The general idea is that the scatterer should be small and that the dielectric contrast should be small. A qualitative criterion is given by $|\mathbf{E} - \mathbf{E}_{\text{inc}}|/|\mathbf{E}_{\text{inc}}| \ll 1$ for any point within the scatterer. One can also derive sufficient conditions by deriving an upper bound on the integral that yields the electric field. In general, it is necessary to account for the explicit structure of the scatterer and the form of the incident field. For the scattering of plane waves by more or less spherical objects, it can be shown that the set of conditions $|\epsilon_r - 1| \ll 1$ and $L|n|/\lambda \ll 1$ are sufficient. The first expresses the fact that the transmission factor is close to unity, the second expresses the fact that the phase difference during propagation through the particle can be neglected. In the case of an opaque material, this means that the size of the scatterer should be much smaller than the skin depth. The scattered field $\mathbf{E}_s$ can thus be written as

$$\mathbf{E}_s(\mathbf{r}) = k_0^2 \int [\epsilon_r(\mathbf{r}') - 1] \overleftrightarrow{\mathbf{G}}(\mathbf{r}, \mathbf{r}') \mathbf{E}_{\text{inc}}(\mathbf{r}') \, \mathrm{d}^3\mathbf{r}', \qquad (2.49)$$

under the first Born approximation. In the case of an incident plane wave, the amplitude can be written as $\mathbf{E}_{\text{inc}} \exp(i\mathbf{k}_{\text{inc}}) \cdot \mathbf{r}$. It follows that the far-field scattered amplitude becomes

$$\mathbf{E}_s(\mathbf{r}) \approx k_0^2 \frac{\exp(ik_0 r)}{4\pi r} \left(\overleftrightarrow{\mathbf{I}} - \mathbf{u}\mathbf{u}\right) \int [\epsilon_r(\mathbf{r}') - 1] \exp(i\mathbf{q} \cdot \mathbf{r}') \, \mathbf{E}_{\text{inc}}(\mathbf{r}') \, \mathrm{d}^3\mathbf{r}', \qquad (2.50)$$

where $\mathbf{q} = \mathbf{k}_{\text{inc}} - k_0 \mathbf{u} = k_0(\mathbf{u}_{\text{inc}} - \mathbf{u})$ is the difference between the incident and scattered wavevectors.

2.2.2.2 *Scattering by a small spherical particle with arbitrary dielectric contrast*

Here, we consider the case of a spherical particle with arbitrary dielectric contrast and radius $a \ll \lambda$. According to (2.37) and (2.38), the field can be written as

$$\mathbf{E}(\mathbf{r}) = \mathbf{E}_{\text{inc}}(\mathbf{r}) + k_0^2 \int [\epsilon_r(\mathbf{r}') - 1]\overset{\leftrightarrow}{\mathbf{G}}_{\text{NS}}(\mathbf{r}, \mathbf{r}')\mathbf{E}(\mathbf{r}')\,\mathrm{d}^3\mathbf{r}' - \frac{[\epsilon_r(\mathbf{r}) - 1]\mathbf{E}(\mathbf{r})}{3}. \quad (2.51)$$

We have explicitly separated the contribution of the singularity from the contribution of the non-singular Green tensor denoted by $\overset{\leftrightarrow}{\mathbf{G}}_{\text{NS}}$. When the particle radius a becomes very small, the contribution of the principal value goes to zero as a^3. The contribution of the singularity is thus the leading term. Keeping only this term in (2.51) yields the field in the sphere:

$$\mathbf{E}(\mathbf{r}) + \frac{[\epsilon_r(\mathbf{r}) - 1]\mathbf{E}(\mathbf{r})}{3} = \frac{\epsilon_r(\mathbf{r}) + 2}{3}\mathbf{E}(\mathbf{r}) = \mathbf{E}_{\text{inc}}(\mathbf{r}). \quad (2.52)$$

This result is very interesting: it yields the relationship between the field inside the sphere and the field that illuminates the sphere. In other words, it accounts for the dielectric screening. It could have been found using an electrostatic approximation. Note that if the particle shape is different, then the result is also different, as pointed out previously. Since the field inside a spherical particle is given by $\{3/[\epsilon_r(\mathbf{r}) + 2]\}\mathbf{E}_{\text{inc}}(\mathbf{r})$, the induced dipole moment of the particle is

$$\mathbf{p} = 4\pi a^3 \frac{\epsilon_r(\mathbf{r}) - 1}{\epsilon_r(\mathbf{r}) + 2}\mathbf{E}_{\text{inc}}(\mathbf{r}). \quad (2.53)$$

The field scattered by the particle can then be derived using the standard expression of the field radiated by a dipole. Before moving to a more general approach to solve the integral equation, let us point out that the approximation made above has neglected higher-order terms in the expansion in powers of a. These terms are needed for energy conservation.

2.2.2.3 *Scattering by an arbitrary particle: the discrete dipole approximation*

In this subsection, we briefly show how it is possible to solve numerically the problem of scattering by a particle of arbitrary shape and permittivity. The basic idea amounts to considering the particle as a collection of small volume elements that can be modelled by dipoles. This is possible provided that the volume elements are taken to be much smaller than the wavelength. To proceed, we start form the integral equation in the form given by (2.51) and cast in a slightly different form:

$$\frac{\epsilon_r(\mathbf{r}) + 2}{3}\mathbf{E}(\mathbf{r}) = \mathbf{E}_{\text{inc}}(\mathbf{r}) + k_0^2 \int [\epsilon_r(\mathbf{r}') - 1]\overset{\leftrightarrow}{\mathbf{G}}_{\text{NS}}(\mathbf{r}, \mathbf{r}')\mathbf{E}(\mathbf{r}')\,\mathrm{d}^3\mathbf{r}'. \quad (2.54)$$

The left-hand side has a simple physical meaning: if one considers a small sphere, then the field inside this sphere is given by $\mathbf{E}(\mathbf{r})$ and the field outside is given by $\frac{1}{3}[\epsilon_r(\mathbf{r})+2]\mathbf{E}(\mathbf{r})$. Let us denote the illuminating field of a small volume element by $\mathbf{E}_{\text{ill}}$. We can write the following integral equation for this field:

$$\mathbf{E}_{\text{ill}}(\mathbf{r}) = \mathbf{E}_{\text{inc}}(\mathbf{r}) + k_0^2 \int \frac{\epsilon_r(\mathbf{r}') - 1}{\epsilon_r(\mathbf{r}') + 2} \overset{\leftrightarrow}{\mathbf{G}}_{\text{NS}}(\mathbf{r}, \mathbf{r}')\mathbf{E}(\mathbf{r}')\,\mathrm{d}^3\mathbf{r}'. \tag{2.55}$$

By extracting the singularity of the Green tensor, we have shifted from the internal field to the illuminating field. The next step in solving the problem is to introduce a mesh splitting the integral into several integrals over small volumes δV_n. Assuming that the electric field is constant over each volume, one can convert the integral equation into a linear system that can be solved using standard numerical routines. This yields the field for each cell. Once this has been solved, the computation of the field at any point is a simple integral.

2.2.3   Super-resolution in the far field: Born approximation, inverse problem, and structured illumination

2.2.3.1   *Spatial resolution with Born approximation*

We now address the issue of the resolution limit when dealing with low dielectric contrast such that the Born approximation is valid. Equation (2.50) clearly shows that the far field is *linearly related* to the structure of the object. This paves the way to a reconstruction of the structure of the object directly from knowledge of the field. Note, however, that the far field does not contain information on the high spatial frequencies of the object. Indeed, the wavevector modulus $|\mathbf{q}|$ is bounded by $2\pi/\lambda$, so $\mathbf{k}$ is bounded by $4\pi/\lambda$. Hence, by using several incident angles, it is possible to recover this information, but there is a cutoff at $4\pi/\lambda$. We can conclude that *in the Born approximation regime, namely in the single-scattering approximation*, the information content on the object suffers from a fundamental limit. This is only twice as large as the resolution limit on the diffracted field.

2.2.3.2   *Super-resolution in the far field: inverse problem*

It is important to realize that if the dielectric contrast is large enough, then there is multiple scattering. This can be clearly seen by using a symbolic notation for (2.37) and solving iteratively:

$$\begin{aligned}
\mathbf{E} &= \mathbf{E}_{\text{inc}} + VG\,\mathbf{E} \\
&= \mathbf{E}_{\text{inc}} + VG\,\mathbf{E}_{\text{inc}} + VGVG\,\mathbf{E} \\
&= \mathbf{E}_{\text{inc}} + VG\,\mathbf{E}_{\text{inc}} + VGVG\,\mathbf{E}_{\text{inc}} + VGVGVG\,\mathbf{E}.
\end{aligned} \tag{2.56}$$

This iterative solution allows us to properly exhibit single- and double-scattering terms. It also shows clearly that the field is no longer related linearly to the dielectric constant function as soon as multiple-scattering terms are included. As a consequence, the amplitude of the scattered field at low spatial frequencies contains information on the high spatial frequencies of the dielectric constant. Recovering this information amounts to collecting the field amplitudes scattered in the far field and attempting to reconstruct the scattering object. This is a typical inverse problem. The key remark to be made here is that the inverse problem does not suffer from any a priori fundamental resolution limit. Solving the inverse problem is a difficult task. It requires some a priori knowledge on the object. Yet, it can be done and remarkable resolution can be obtained [49]. Let us be more specific. We assume that the dielectric constant can be represented by a function $\epsilon_r(\mathbf{r}) - 1 = \chi\theta(\mathbf{r})$, where the characteristic function θ takes the value 1 if $\mathbf{r}$ is in the scatterer and zero otherwise. χ is the dielectric contrast $\epsilon_r - 1$, assuming that the scatterer is homogeneous. With these definitions, it can be seen that the far-field amplitude is given by the Fourier transform of the product of two functions: the characteristic function $\theta(\mathbf{r})$ and the field $\mathbf{E}(\mathbf{r})$. It follows that the field detected along a direction $\mathbf{q}$ has an amplitude proportional to

$$\int \chi(\mathbf{k})\mathbf{E}(\mathbf{q} - \mathbf{k})\,d\mathbf{k}. \tag{2.57}$$

Hence, the amplitude in direction $\mathbf{q}$ contains information on arbitrary large values of $\mathbf{k}$, provided that the field also contains high spatial frequencies. This remark is the basis for super-resolution imaging using only far-field data.

Obviously, finding $\chi(\mathbf{r})$ is not an easy task. One needs to solve the inverse problem, namely to recover the function $\chi(\mathbf{r})$ from knowledge of the scattered field. This requires that as much information as possible be gathered by measuring both the amplitude and phase of the scattered field for as many directions as possible. Then one needs to solve the inverse problem. A priori knowledge of the scatterer is always required to improve the quality of the solution.

2.2.3.3 Structured illumination

The second route that can be followed is to generate a structured illumination containing on purpose high spatial frequencies. Let us assume that the dielectric contrast is low so that the Born approximation is valid. Then we can replace the electric field in the scatterer by the incident electric field, so the scattered amplitude is given by

$$\int \chi(\mathbf{k})\mathbf{E}_{\text{inc}}(\mathbf{q} - \mathbf{k})\,d\mathbf{k}. \tag{2.58}$$

As $\mathbf{q}$ has a modulus given by ω/c, we can gain access to the characteristic function χ for large values of $\mathbf{k}$ only if the incident field has a spatial spectrum that contains high spatial frequencies. This is the principle of structured illumination: to generate an illuminating field with very rapid spatial variations. In practice, one can deposit the sample to be imaged on a grating with a *subwavelength* period d. When illuminating this grating by a plane wave with wavevector $k_{x,\text{inc}}$, the grating generates high-order diffracted fields with wavevectors $k_{x,n} = k_{x,\text{inc}} + n2\pi/d$ that can be very large. These high-order terms are evanescent waves if $k_{x,n} > \omega/c$. It can be seen from (2.58) that such a field will be able to translate in k-space a high frequency of the characteristic function. This approach is not strictly speaking a pure far-field technique, since a structured illumination containing very high spatial frequencies is necessarily an illumination with evanescent waves and therefore a near-field illumination. However, the use of a textured substrate will allow pure far-field illumination and detection [53, 101].

2.2.4   Local density of states

So far, we have discussed how the scattered field is related to the permittivity of the object to be imaged. We have shown that the Green tensor is a very useful tool in that context. The purpose of this subsection is to briefly introduce the concept of the local density of states (LDOS). In particular, we will show how the LDOS is related to the Green tensor. We will discuss briefly applications to light emission assisted by surface waves. We will also show that the influence of surface plasmons on the LDOS plays a key role in the Casimir force [25], a pure quantum electrodynamical phenomenon.

2.2.4.1   Elementary introduction to the density of electromagnetic states

Before starting the discussion, we make a remark on semantics regarding the meanings of state, mode, and eigenfunction. The word 'state' is usually used in the context of quantum mechanics or statistical physics, whereas the word 'mode' is often used in the context of wave theory. Both deal with eigenfunctions of linear operators so that the terms are often interchanged. The term 'density of states' (DOS) is used for $g(\omega)$ such that $g(\omega)\,\mathrm{d}\omega$ is the number of states (modes) with frequency in the interval $[\omega, \omega + \mathrm{d}\omega]$. To begin with, we briefly recall how to derive the density of electromagnetic states or, in other words, how to count the number of different solutions (plane waves) of Maxwell's equations in vacuum. It is useful to introduce a virtual cubic box of size L and look for fields satisfying periodic boundary conditions. Indeed, this allows us to discretize the solutions and thereby count them. From the periodic condition, it follows that the wavevector components are of the form $k_x = n2\pi/L, k_y = m2\pi/L, k_z = l2\pi/L$, where n, m, l are integers. In k-space, the volume occupied by a state is therefore $(2\pi/L)^3$, so the number of states in the volume element $\mathrm{d}k_x\,\mathrm{d}k_y\,\mathrm{d}k_z$ is $2L^3\,\mathrm{d}k_x\,\mathrm{d}k_y\,\mathrm{d}k_z/(2\pi)^3$, where the factor 2 accounts for the two possible polarizations of each state. The density of states in k-space per unit volume is thus given by $1/4\pi^3$. We can now easily find the number of states with a given frequency $\omega = ck$. They occupy the volume $4\pi k^2\,\mathrm{d}k$ in k-space. Using the dispersion relation $k = \omega/c$, we find the number of states per unit volume in the range $[\omega, \omega + \mathrm{d}\omega]$:

$$g_{\mathrm{v}}(\omega)\,\mathrm{d}\omega = \frac{\omega^2}{\pi^2 c^3}\,\mathrm{d}\omega\,, \tag{2.59}$$

where we have introduced the DOS per unit volume $g_{\mathrm{v}}(\omega)$ in vacuum. We now illustrate the importance of this concept using three examples. Let us first count how many states $N(\omega)$ are available between 0 and ω in a volume V:

$$N(\omega) = V \int_0^\omega g_{\mathrm{v}}(\omega')\,\mathrm{d}\omega' = V\frac{\omega^3}{3\pi^2 c^3} = \frac{8\pi}{3}\frac{V}{\lambda^3}\,. \tag{2.60}$$

The simple rule to remember is that the number of states with frequency smaller than ω is roughly given by the volume divided by $(\lambda/2)^3$. The second example is the form of the energy of the blackbody radiation. Each mode has a quantum of energy $\hbar\omega$, and the mean excitation number is given by Bose–Einstein statistics $n_{\mathrm{BE}}(\omega) = 1/[\exp(\hbar\omega/k_{\mathrm{B}}T)-1]$. The product of these two terms by the DOS yields the blackbody density of energy at temperature T:

$$u(\omega, T) = \frac{\hbar\omega^3}{\pi^2 c^3} \frac{1}{\exp(\hbar\omega/k_{\mathrm{B}}T) - 1} . \tag{2.61}$$

We now show how the LDOS plays a key role in the lifetime of the excited state in a two-level system. From Fermi's golden rule, it is known that the lifetime is proportional to the number of final states. When studying the rate of radiative relaxation, the radiative decay rate is therefore proportional to the number of electromagnetic states at the corresponding frequency. This can be seen by comparing the stimulated and spontaneous emission rates given by the Einstein coefficients. Their ratio is given by

$$\frac{A_{21}}{B_{21}\hbar\omega} = \frac{\omega^2}{\pi^2 c^3} , \tag{2.62}$$

which is nothing but the vacuum DOS. For stimulated emission, only the mode of the incident photon has to be considered, whereas for spontaneous emission, one has to sum over all possible electromagnetic states. Hence, the spontaneous emission coefficient is proportional to the DOS. Let us finally point out a slight difference in the definition of the LDOS depending on the application: evaluating the equilibrium energy or evaluating spontaneous emission. A two-level system that is coupled to the electromagnetic field through an electric dipole moment can couple only to the component of the electric field parallel to the electric dipole. The relevant form of the LDOS is thus called the projected LDOS, since only one component of the field matters. In vacuum, this is simply a factor 3 difference, since the field is isotropic. In more complex situations, the DOS can be different for different polarizations. It is well known for instance that the lifetime of a molecule close to an interface depends on the orientation of its dipole moment.

To summarize, the concept of the DOS plays a key role when looking at the radiative decay of a two-level system and when looking at the thermodynamic properties of electromagnetic radiation. In what follows, we shall analyse why it is necessary to introduce a *local* DOS and how it can be related to the Green tensor.

2.2.4.2 *Electron and phonon DOS*

Let us first compare the DOS for light with that for electrons or phonons. The DOS in k-space is also given by $1/4\pi^3$ for electrons. Here, we have accounted for the degeneracy due to the spin half of the electron. The total number of states in a crystal of volume V with N atoms is given by $2N$ for an s-band. Similarly, the total number of phonon states is given by 3N, because this is simply the total number of degrees of freedom of the atoms. Hence, the number of states per unit volume is roughly N/V, which is the inverse of the volume a^3 of a unit cell of the crystal. If we now compare the number of states available for visible light in vacuum $((2/\lambda)^3 \approx 10^{19})$ and for condensed matter excitations $(1/a^3 \approx 10^{30})$, we find a difference of 11 orders of magnitude.

This very crude estimate shows what is the key of the efficiency of plasmons or optical phonons in increasing the energy density or in reducing the lifetime of quantum emitters. The DOS of polaritons benefits from the large number of states of condensed matter excitations (electrons or phonons). However, many of these modes do not couple to light, so a careful analysis is required to obtain an accurate description of the DOS. This is the purpose of the following subsections.

2.2.4.3 *Elementary introduction to the concept of the LDOS*

Here, we discuss why it is necessary to introduce an LDOS. Let us first introduce this concept in the context of electronic states in quantum mechanics. It is simpler to introduce the LDOS in this context because the electron wavefunction is scalar and because the probability interpretation of the wavefunction is familiar. We first recall the standard density of electronic states for a homogeneous system. The number of states $dN(E)$ having an energy between E and δE is given by the DOS, $g(E)$:

$$dN(E) = g(E)\,\delta E\,.$$

If the spectrum is discrete, then the DOS can be written in the form

$$g(E) = \sum_n \delta(E - E_n)\,.$$

It is readily seen that this expression does indeed yield the number of states whose energy is in the range $[E, E + \delta E]$:

$$\int_E^{E+\delta E} \left[\sum_n \delta(E' - E_n) \right] dE' = dN(E)\,.$$

Note that this definition is analogous to the definition of the density of charges:

$$\rho(\mathbf{r}) = \sum_i q_i \delta(\mathbf{r} - \mathbf{r}')\,.$$

Let us stress that this concept is not specific to quantum mechanics. It can be introduced for other waves (e.g. electromagnetic or acoustic) in a cavity. The number of modes having an eigenvalue in a given frequency interval can always be defined for a particular cavity. The natural definition is then

$$g(\omega) = \sum_n \delta(\omega - \omega_n)\,.$$

We now introduce the concept of the LDOS in the case of electronic states. The motivation is to account for the fact that not all points are equivalent. For example, when a cobalt atom is adsorbed on a metal surface, the density of electrons is perturbed around this atom. The interference between an incident wave and a scattered wave produces oscillations of the LDOS. A similar effect appears at any interface between a metal and vacuum. The interference between incident and reflected waves produces oscillations known as Friedel oscillations of the LDOS. For these two examples, the eigenmodes $\Psi_n(\mathbf{r})$ are no longer plane waves as in a translationally invariant crystal. Instead, they exhibit maxima and minima due to interferences. The concept of a *local* DOS recognizes this spatial dependence. In order to introduce the definition of the LDOS, we recall that the probability density of finding a particle in a given volume element is proportional to $|\Psi(\mathbf{r})|^2$. Hence, we can introduce a weighting factor as follows to introduce the LDOS:

$$\rho(\mathbf{r},\omega) = \sum_n |\Psi_n(\mathbf{r})|^2 \delta(\omega - \omega_n)\,. \tag{2.63}$$

In other words, the contribution of a given mode Ψ_n to the LDOS is proportional to $|\Psi_n(\mathbf{r})|^2$. It is clear that on integration over space, the previous form is easily recovered owing to the normalization condition. We now move to the definition of an LDOS for electromagnetic waves. For the sake of simplicity, we adopt here a scalar approximation. The derivation is not significantly modified when dealing with vector waves.

2.2.4.4   *Derivation of the LDOS*

The goal of this section is to establish an explicit form of the LDOS in terms of the Green function. We start by defining the Green function as the solution of the equation

$$\nabla^2 G(\mathbf{r},\mathbf{r}',\omega) + \frac{\omega^2}{c^2} G(\mathbf{r},\mathbf{r}',\omega) = -\delta(\mathbf{r}-\mathbf{r}')\,, \tag{2.64}$$

with the appropriate boundary conditions. We now consider the complete orthogonal set of modes $\{\Psi_n\}$ of the Laplacian operator:

$$\nabla^2 \Psi_n = -\frac{\omega_n^2}{c^2}\Psi_n\,,$$

with eigenvalues $-\omega_n^2/c^2$. We expand the Green function in this basis:

$$G(\mathbf{r},\mathbf{r}',\omega) = \sum_n c_n(\mathbf{r}',\omega)\Psi_n(\mathbf{r}).$$

Inserting this expansion into (2.64), we obtain

$$\sum_n c_n(\mathbf{r}',\omega)\left(\frac{\omega^2}{c^2} - \frac{\omega_n^2}{c^2}\right)\Psi_n(\mathbf{r}) = -\delta(\mathbf{r}-\mathbf{r}')\,. \tag{2.65}$$

Since the $\{\Psi_n\}$ form an orthonormal basis, we have

$$\int \Psi_n \Psi_m^* \, \mathrm{d}^3\mathbf{r} = \delta_{n,m}\,.$$

On multiplying (2.65) by Ψ_n^* and integrating, we obtain

$$c_m(\mathbf{r}', \omega) = -\frac{\Psi_m^*(\mathbf{r}')}{\dfrac{\omega^2}{c^2} - \dfrac{\omega_m^2}{c^2}} \,.$$

Finally, the Green function can be cast in the form

$$G(\mathbf{r}, \mathbf{r}', \omega) = -\sum_n c^2 \frac{\Psi_n(\mathbf{r})\Psi_n^*(\mathbf{r}')}{\omega^2 - \omega_n^2} \,.$$

We can now obtain the expression for the LDOS. The key point is to note that[8]

$$\lim_{\eta \to 0} \operatorname{Im}\left[\frac{1}{(\omega + i\eta)^2 - \omega_n^2}\right] = \frac{\pi}{2\omega_n}[\delta(\omega - \omega_n) - \delta(\omega + \omega_n)]\,.$$

Using this result, we find

$$\sum_n |\Psi_n(\mathbf{r})|^2 \frac{\delta(\omega - \omega_n)}{2\omega_n} = \lim_{\eta \to 0} \frac{1}{\pi c^2} \operatorname{Im}[G(\mathbf{r}, \mathbf{r}, \omega + i\eta)]\,,$$

which is valid for positive values of ω. Finally, we get

$$\rho(\mathbf{r}, \omega) = \sum_n |\Psi_n(\mathbf{r})|^2 \delta(\omega - \omega_n)$$

$$= \lim_{\eta \to 0} \frac{2\omega}{\pi c^2} \operatorname{Im}[G(\mathbf{r}, \mathbf{r}, \omega + i\eta)]\,.$$

2.2.5 Local density of electromagnetic states in the presence of losses

The previous derivation was presented because it is based on the explicit form of the LDOS given by (2.63). It was based on the existence of an orthogonal set of eigenmodes. This approach is more difficult to implement in the presence of losses. In this subsection, we follow a different route to sketch a derivation of the LDOS in the presence of losses and accounting for the vector nature of electromagnetic fields. The basic idea is to use the fact that the energy density at thermodynamic equilibrium is related to the DOS. We consider a system made of arbitrary lossy objects in vacuum. We are interested in deriving the LDOS in vacuum in the presence of these objects.

---

[8] We use

$$\left(\frac{\omega^2}{c^2} - \frac{\omega_m^2}{c^2}\right)^{-1} = \frac{1}{2\omega_m}\left(\frac{1}{\omega - \omega_m} - \frac{1}{\omega + \omega_m}\right)$$

and

$$\lim_{\eta \to 0} \frac{1}{\omega - \omega_m + i\eta} = \operatorname{PV}\left(\frac{1}{\omega - \omega_m}\right) - i\pi\delta(\omega - \omega_m)\,.$$

We start by noting that the energy density at thermal equilibrium in vacuum without any object (i.e. blackbody radiation) can be written in the form

$$U(\omega) = \rho_v(\omega)\hbar\omega \left[\frac{1}{2} + \frac{1}{\exp(\hbar\omega/k_{\mathrm{B}}T) - 1}\right], \tag{2.66}$$

where $\rho_v(\omega)$ is the vacuum DOS. Now, in the presence of lossy objects such as an interface separating a lossy half-space from vacuum, the energy density is no longer uniform. Hence we can write it as

$$U(\mathbf{r}, \omega) = \rho(\mathbf{r}, \omega)\hbar\omega \left[\frac{1}{2} + \frac{1}{\exp(\hbar\omega/k_{\mathrm{B}}T) - 1}\right], \tag{2.67}$$

by definition of the local density of electromagnetic states $\rho(\mathbf{r}, \omega)$. On the other hand, the energy density can be expressed in terms of the electric and magnetic fields:

$$\langle U \rangle = \frac{\epsilon_0}{2} \left\langle |\mathbf{E}(\mathbf{r}, t)|^2 \right\rangle + \frac{\mu_0}{2} \left\langle |\mathbf{H}(\mathbf{r}, t)|^2 \right\rangle. \tag{2.68}$$

As we are considering a point that is in vacuum, albeit in the presence of objects made of lossy materials, the usual form of the energy in vacuum is valid. In thermodynamic equilibrium, the field correlation functions are stationary and depend only on $t - t'$. It follows that they can be expressed in terms of cross-spectral densities:

$$\langle E_k(\mathbf{r}, t) E_l(\mathbf{r}', t') \rangle = \mathrm{Re}\left\{\int_0^\infty \frac{\mathrm{d}\omega}{2\pi} \, \mathcal{E}_{kl}^{(N)}(\mathbf{r}, \mathbf{r}', \omega) \exp[-\mathrm{i}\omega(t - t')]\right\}, \tag{2.69}$$

$$\langle H_k(\mathbf{r}, t) H_l(\mathbf{r}', t') \rangle = \mathrm{Re}\left\{\int_0^\infty \frac{\mathrm{d}\omega}{2\pi} \, \mathcal{H}_{kl}^{(N)}(\mathbf{r}, \mathbf{r}', \omega) \exp[-\mathrm{i}\omega(t - t')]\right\}. \tag{2.70}$$

The fluctuation–dissipation theorem [2, 64] establishes a connection between these cross-spectral densities and the Green tensor. It follows [64] that the energy can be cast in the form

$$U(\mathbf{r}, \omega) = \frac{\hbar\omega}{\exp(\hbar\omega/k_B T) - 1} \frac{\omega}{\pi c^2} \mathrm{Im} \mathrm{Tr}[\overset{\leftrightarrow}{\mathbf{G}}^{EE}(\mathbf{r}, \mathbf{r}, \omega) + \overset{\leftrightarrow}{\mathbf{G}}^{HH}(\mathbf{r}, \mathbf{r}, \omega)]. \tag{2.71}$$

A detailed definition of the Green tensor can be found in [63, 64]. Here, we simply point out that these tensors connect either the electric field at point $\mathbf{r}$ to an electric dipole moment at point $\mathbf{r}'$ or a magnetic field at point $\mathbf{r}$ to a magnetic dipole moment at point $\mathbf{r}'$. Note that these quantities yield the field produced by the dipoles in the presence of the objects.

A comparison of (2.67) and (2.71) shows that the LDOS is the sum of an electric contribution ρ^E and a magnetic contribution ρ^H:

$$\rho(\mathbf{r}, \omega) = \frac{\omega}{\pi c^2} \mathrm{Im} \mathrm{Tr}[\overset{\leftrightarrow}{\mathbf{G}}^{EE}(\mathbf{r}, \mathbf{r}, \omega) + \overset{\leftrightarrow}{\mathbf{G}}^{HH}(\mathbf{r}, \mathbf{r}, \omega)] = \rho^E(\mathbf{r}, \omega) + \rho^H(\mathbf{r}, \omega). \tag{2.72}$$

In what follows, we shall discuss a few examples to illustrate the modification of the LDOS due to the presence of a lossy half-space.

2.2.5.1 Vacuum

In the vacuum, the Green tensors $\overleftrightarrow{\mathbf{G}}^{EE}$ and $\overleftrightarrow{\mathbf{G}}^{HH}$ obey the same equation and have the same boundary conditions. Therefore, their contributions to the electromagnetic energy density are equal:

$$\mathrm{Im}\big[\overleftrightarrow{\mathbf{G}}^{EE}(\mathbf{r},\mathbf{r},\omega)\big] = \mathrm{Im}\big[\overleftrightarrow{\mathbf{G}}^{HH}(\mathbf{r},\mathbf{r},\omega)\big] = \frac{\omega}{6\pi c}\overleftrightarrow{\mathbf{I}}. \tag{2.73}$$

The LDOS is thus obtained by multiplying the electric field contribution by 2. After taking the trace, the usual result for vacuum is retrieved:

$$\rho_v(\mathbf{r},\omega) = \rho_v(\omega) = \frac{\omega^2}{\pi^2 c^3}. \tag{2.74}$$

As expected, we note that the LDOS is homogeneous and isotropic.

2.2.5.2 Plane interface

We now consider a plane interface separating vacuum (medium 1 in the upper half-space) from a semi-infinite material (medium 2, in the lower half-space) characterized by its complex dielectric constant $\epsilon_2(\omega)$. The material is assumed to be homogeneous, linear, isotropic and non-magnetic. The expression for the LDOS at a given frequency and at a given height z above the interface in vacuum is obtained by inserting the expressions for the electric and magnetic Green tensors for this geometry [102] into (2.72). Note that in the presence of an interface, the magnetic and electric Green tensors are no longer the same. Indeed, the boundary conditions at the interface are different for the electric and magnetic fields.

Let us consider some specific examples for real materials like metals and dielectrics. We first calculate $\rho(z,\omega)$ for aluminium at different heights. Aluminium is a metal whose dielectric constant is well described by a Drude model for near-ultraviolet, visible and near-infrared frequencies [88]:

$$\epsilon(\omega) = 1 - \frac{\omega_\mathrm{p}^2}{\omega(\omega + i\Gamma)}, \tag{2.75}$$

with $\omega_\mathrm{p} = 1.747\times10^{16}\,\mathrm{rad\,s}^{-1}$ and $\Gamma = 7.596\times10^{13}\,\mathrm{rad\,s}^{-1}$. We have plotted in Fig. 2.5 the LDOS $\rho(\mathbf{r},\omega)$ in vacuum in the near-ultraviolet to near-infrared frequency domain at four different distances from the interface. We first note that the LDOS increases drastically when the distance from the material is reduced. As discussed in the previous paragraph, at larger distances from the material, one retrieves the vacuum DOS. Note that at a given distance, it is always possible to find a sufficiently high frequency for which the corresponding wavelength is small compared with the distance so that a far-field situation is retrieved. This can clearly be seen when looking at the curve for $z = 1\,\mu\mathrm{m}$, which coincides with the vacuum LDOS. When the distance to the material is decreased, additional modes are present: these are the evanescent modes, which are confined close to the interface and which cannot be seen in the far field. Moreover, aluminium exhibits a resonance around $\omega = \omega_p/\sqrt{2}$. We anticipate the next section by noting that below this frequency, the material supports surface-plasmon polaritons, so there are additional modes contributing to the LDOS in the near field, close to the interface. The enhancement is particularly important at a resonant frequency that corresponds to $\mathrm{Re}[\epsilon_2(\omega)] = -1$. Also note that in the low-frequency regime, the LDOS increases. Finally, Fig. 2.5 shows that it is possible to have a LDOS smaller than that of vacuum at some particular distances and frequencies. Figure 2.6 shows the propagating and evanescent wave contributions to the LDOS above an aluminium sample at a distance of 10 nm. The propagating contribution is very similar to that of the vacuum LDOS. As expected, the evanescent contribution dominates at low frequency. It also dominates close to the surface-plasmon polariton resonance, where pure near-field contributions dominate.

Let us now turn to the comparison of $\rho(z,\omega)$ with the usual definition often encountered in the literature [29, 120], which corresponds to $\rho^E(z,\omega)$. In Fig. 2.7, we plot ρ, ρ^E, and ρ^H above an aluminium surface at a distance $z = 10$ nm. In this figure, it is possible to identify three different domains for the LDOS behaviour. We note again that in the far-field situation (corresponding here to high frequencies, i.e. $\lambda/2\pi \ll z$), the LDOS reduces to the vacuum situation. In this case, $\rho(z,\omega) = 2\rho^E(z,\omega) = 2\rho^H(z,\omega)$. Around the resonance, the LDOS is dominated by the electric contribution ρ^E. Conversely, at low frequencies, $\rho^H(z,\omega)$ dominates. Thus, Fig. 2.7 shows that we have to be very careful when using the approximation $\rho(z,\omega) = \rho^E(z,\omega)$. Above aluminium and at a distance $z = 10$ nm, it is only valid in a narrow range between $\omega = 10^{16}\,\mathrm{rad\,s^{-1}}$ and $\omega = 1.5 \times 10^{16}\,\mathrm{rad\,s^{-1}}$, i.e. around the frequency at which the surface wave exists.

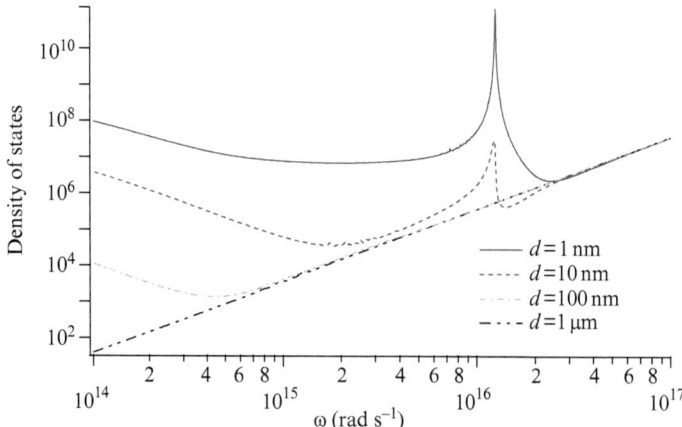

Fig. 2.5 LDOS versus frequency at different heights above a semi-infinite sample of aluminium. (Reprinted with permission from Joulain, K., Carminati, R., Mulet, J.P. & Greffet J.J. (2003). Definition and measurement of the local density of electromagnetic states close to an interface. Phys. Rev. B, 68(24). doi:10.1103/physrevb.68.245405. Copyright 2003 by the American Physical Society.)

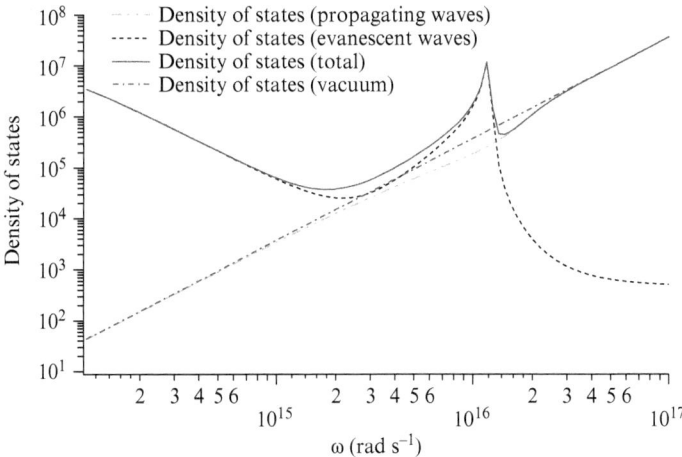

Fig. 2.6 DOS contributions due to the propagating and evanescent waves compared with the total DOS and the vacuum DOS. These quantities are calculated above an aluminium sample at a distance of 10 nm. (Reprinted with permission from Joulain, K., Carminati, R., Mulet J.P. & Greffet J.J (2003). Definition and measurement of the local density of electromagnetic states close to an interface. Phys. Rev. B, 68(24). doi:10.1103/physrevb.68.245405. Copyright 2003 by the American Physical Society.)

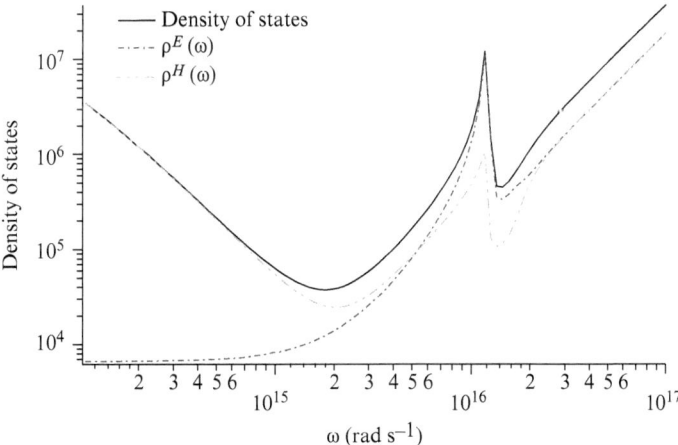

Fig. 2.7 LDOS at a distance $z = 10$ nm above a semi-infinite aluminium sample. Comparison with $\rho^E(z, \omega)$ and $\rho^H(z, \omega)$. (Reprinted with permission from Joulain, K., Carminati, R., Mulet, J.P. & Greffet J.J. (2003). Definition and measurement of the local density of electromagnetic states close to an interface. Phys. Rev. B, 68(24). doi:10.1103/physrevb.68.245405. Copyright 2003 by the American Physical Society.)

2.2.5.3 *Asymptotic form of the LDOS in the near-field*

In order to obtain more physical insight, we have calculated the asymptotic LDOS behaviour in the three regimes mentioned above. As we have already seen, the far-field regime ($\lambda/2\pi \ll z$) corresponds to the vacuum case. To study the near-field situation ($\lambda/2\pi \gg z$), we focus on the evanescent contribution due to the large wavevectors K, as suggested by the results shown in Fig. 2.6. In this (quasistatic) limit, the Fresnel reflection factors reduce to

$$\lim_{K \to \infty} r_{12}^{\mathrm{s}} = \frac{\epsilon_2 - 1}{4(K/k_0)^2} , \tag{2.76}$$

$$\lim_{K \to \infty} r_{12}^{\mathrm{p}} = \frac{\epsilon_2 - 1}{\epsilon_2 + 1} . \tag{2.77}$$

Asymptotically, the expressions for $\rho^E(z, \omega)$ and $\rho^H(z, \omega)$ are [63]

$$\rho^E(z, \omega) = \frac{\rho_{\mathrm{v}}}{|\epsilon_2 + 1|^2} \frac{\epsilon_2''}{4k_0^3 z^3} , \tag{2.78}$$

$$\rho^H(z, \omega) = \rho_{\mathrm{v}} \left(\frac{\epsilon_2''}{8k_0 z} + \frac{\epsilon_2''}{2|\epsilon_2 + 1|^2 k_0 z} \right) . \tag{2.79}$$

At a distance $z = 10\,\mathrm{nm}$ above an aluminium surface, these asymptotic expressions match almost perfectly with the evanescent contributions ($K > k_0$) of ρ^E and ρ^H. These expressions also show that for a given frequency, one can always find a distance z from the interface below for which the dominant contribution to the LDOS will be that due to the imaginary part of the electric field Green function, which varies like $(k_0 z)^{-3}$. But for aluminium at a distance $z = 10\,\mathrm{nm}$, this is not the case for all frequencies. As we have already mentioned, this is only true around the resonance. For example, at low frequencies, and for $z = 10\,\mathrm{nm}$, the LDOS is actually dominated by the magnetic contribution $\rho_{\mathrm{v}} \epsilon_2'' / (8k_0 z)$. This is due to the fact that the magnetic field is continuous at the interface, while the electric field has a very large dielectric screening.

2.2.5.4 *Spatial oscillations of the LDOS*

Let us now focus on the LDOS variations at a given frequency ω versus the distance z from the interface. There are essentially three regimes. First, for distances much larger than the wavelength, the LDOS is given by the vacuum expression ρ_{v}. The second regime is observed close to the interface, where oscillations are observed. Indeed, at a given frequency, each plane wave incident on the interface can interfere with its reflected counterpart. This generates an interference pattern with a fringe spacing that depends on the angle and the frequency. On adding the contributions of all the plane waves over angles, the oscillating structure disappears except close to the interface. This leads to oscillations around distances of the order of the wavelength. This phenomenon is the electromagnetic analog of the Friedel oscillations that can be

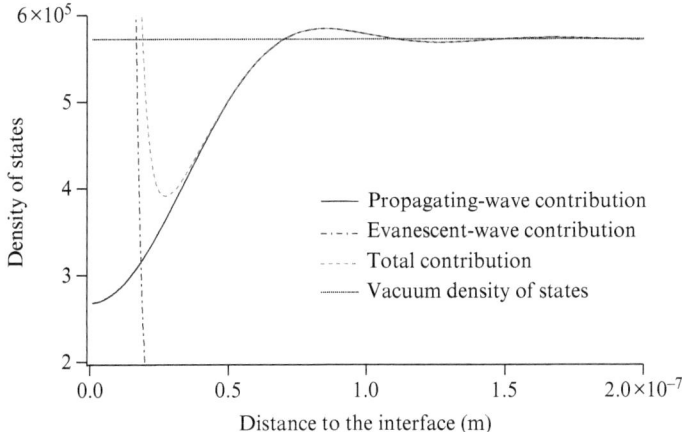

Fig. 2.8 LDOS versus the distance z from an aluminium–vacuum interface at the aluminium resonant frequency. (Reprinted with permission from Joulain, K., Carminati, R., Mulet J.P. & Greffet J.J. (2003). Definition and measurement of the local density of electromagnetic states close to an interface. Phys. Rev. B, 68(24). doi:10.1103/physrevb.68.245405. Copyright 2003 by the American Physical Society.)

observed in the electronic DOS near interfaces [12]. For a highly reflecting material, the real parts of the reflection coefficients are negative, so the LDOS decreases on approaching the surface. These two regimes are clearly observed for aluminium in Fig. 2.8. The third regime is observed at small distances, as can be seen in Fig. 2.8. It is due to the contribution of surface waves. Its behaviour is thus dependent on frequency. We consider the particular case of the frequency corresponding to $\mathrm{Re}[\epsilon_2(\omega)] + 1 = 0$. It can be seen that the evanescent contribution dominates and ultimately the LDOS always increases as $1/z^3$, following the behaviour found in (2.78). This is the usual quasistatic contribution that is always found at short distances [58]. At a frequency slightly lower than the resonance frequency, surface waves are still excited on the surface but for a single wavevector. These modes increase the LDOS according to an exponential law.

2.3 Introduction to surface plasmons

The term 'surface plasmon' is used both for polarization oscillations of metallic nanoparticles and for surface waves propagating along a metallic interface. Surface waves are bound to an interface and decay exponentially away from the interface. This section will cover mostly the latter case. From the point of view of electrodynamics, surface plasmons are a particular case of a surface wave, a topic that was extensively covered in the early days of radio-wave propagation around the Earth [14, 23, 44, 68]. From the point of view of optics, surface plasmons are modes of an interface. They were extensively studied in the 1970s and 1980s. Several excellent monographs are available [3, 20, 94], and more recent achievements are summarized in [16, 98, 123]. Finally, from the point of view of solid state physics, a plasmon is a collective excitation of electrons. Excellent introductions can be found in well-known textbooks [12, 69] and in more advanced texts [70, 89, 124]. The goal of this section is to provide an introduction to the three different points of view. We will first show how the plasma frequency can be seen as the natural oscillation frequency of electrons in a thin film. This (solid state) point of view will then be generalized to bulk plasmons in an electron gas using a hydrodynamic model. This analysis will serve the purpose of explaining the concept of the polariton: an electromagnetic wave coupled to a polarization excitation in the material. We will then adopt the macroscopic electrodynamics point of view and derive the dispersion relation of a surface wave. In this approach, the material properties are accounted for by using a dielectric constant without any microscopic model. We will discuss the similarities and differences between different types of surface waves (lateral waves, Zenneck modes, Sommerfeld modes, and quasicylindrical waves), without invoking any specific model of the dielectric constant. Hence, this discussion will be equally valid for radio waves or optical waves, for metals or dielectrics. We will then focus on the case of surface plasmon polaritons. With this aim in mind, it is often convenient to use the Drude model, but it is also essential to be aware of its limitations. We will then introduce surface phonon polaritons and radio surface waves. We will outline the key properties that make surface plasmons unique, not least with the aim of identifying those of their fundamental properties that may help in deciding when they may be useful for optics applications. Finally, we then turn to the subtle issue of the dispersion relation of surface plasmons on lossy materials.

2.3.1 Surface and particle electron oscillation modes: introductory examples

To introduce the concept of plasma oscillation, we consider a thin metallic film. The metal is described by a simple model: we assume that there are n free and independent electrons per unit volume. The crystal is modelled by a uniform positively charged background. This is the so-called jellium model. The purpose of this subsection is to illustrate the essence of a plasmon: it is an oscillatory collective mode of the electrons. To proceed, we assume that classical mechanics can be used.

Let us now assume that a positive static homogeneous electric field $E_{\text{ext}}\hat{\mathbf{x}}$ is applied normally to the film along the x-axis (see Fig. 2.9). A force $-eE_{\text{ext}}\hat{\mathbf{x}}$ is exerted on the electrons, so they will be displaced by x (with $x < 0$). A negative static surface charge nex will appear on the left interface and a positive surface charge on the right. These surface charges produce a static field that cancels the external field in the metal. Let us now assume that the external electrostatic field is turned off at time $t = 0$. The electrons in the film will be accelerated by the electric field generated by the surface charges. When they return to their initial positions, they have acquired a momentum, so they keep moving along the positive x-axis and therefore generate an electric field of opposite sign. This process will be repeated and will produce an oscillation. It is

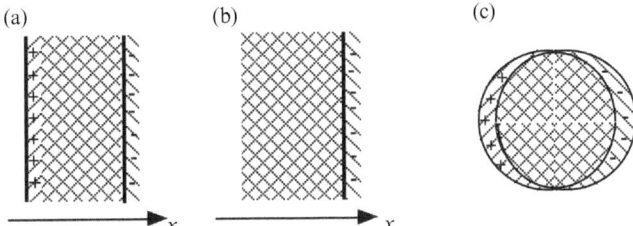

Fig. 2.9 Oscillation of the electron gas for (a) a thin film, (b) vacuum–metal interface, and (c) a nanosphere.

easy to describe quantitatively this phenomenon by using Newton's law applied to a single electron:

$$m\frac{\mathrm{d}^2 x}{\mathrm{d}t^2} = -eE_x \,, \tag{2.80}$$

where we have neglected the magnetic force. Using Gauss's theorem, it can be shown that the field generated by a sheet carrying a surface charge nex is $(nex/2\epsilon_0)\mathbf{u}$, where $\mathbf{u}$ is an outward unit vector. It follows that the field generated by the surface charge nex for a displacement x is found to be $E_x = 2nex/2\epsilon_0$, where the factor 2 accounts for the presence of two interfaces. It follows from Newton's equation that the movement of one electron is given by

$$\frac{\mathrm{d}^2 x}{\mathrm{d}t^2} + \frac{ne^2}{m\epsilon_0}x = 0 \,. \tag{2.81}$$

This straightforward argument allows us to introduce in a simple way the plasma frequency ω_{p}:

$$\omega_{\mathrm{p}}^2 = \frac{ne^2}{m\epsilon_0} \,, \tag{2.82}$$

which appears to be the frequency of the collective oscillation of the electrons in the bulk of the film. To summarize, the oscillation is produced by an electric field due to all the electrons. This is why it is called a collective oscillation. With this simple argument, we have captured the essence of the plasmon: *it is the natural collective oscillation of the electrons characterized by the plasma frequency.*

We now consider the case of a single interface. In other words, we consider that the thickness of the film goes to infinity, so the force is due only to the charge density of one interface. It follows that the electric field is due to only one interface instead of two and takes the value $nex/2\epsilon_0$. The oscillation frequency is therefore $\omega_{\mathrm{p}}/\sqrt{2}$.

We finally consider the case of a nanosphere. For a sphere much smaller than the wavelength, retardation effects can be neglected, so we can use the electrostatic form of the field generated by a uniform polarization field P_x of the particle, $E_x = -P_x/3\epsilon_0$ [61]. Inserting this form of the electric field into (2.80) yields

$$m\frac{\mathrm{d}^2 x}{\mathrm{d}t^2} = -e\frac{-P_x}{3\epsilon_0} \,. \tag{2.83}$$

The polarization P_x is due to the displacement of the electrons, so we have $P_x = -nex$. On inserting this expression into (2.83), we find

$$\frac{\mathrm{d}^2 x}{\mathrm{d}t^2} + \frac{ne^2}{3m\epsilon_0} x = 0\,. \tag{2.84}$$

so the resonance frequency of the plasmon in a nanosphere is given by $\omega_{\mathrm{p}}/\sqrt{3}$.

In Fig. 2.10, we show the results of an experiment reporting the energy lost by electrons impinging on a surface. After interacting with either bulk plasmons or surface plasmons, they can lose an energy given by $n\hbar\omega_{\mathrm{p}} + m\hbar\omega_{\mathrm{p}}/\sqrt{2}$ where n, m are integers. The values measured are 15.3 and 10.3 eV. The ratio is 1.49, which is slightly greater than the predicted 1.41. This experiment illustrates the quantization of the exchange of

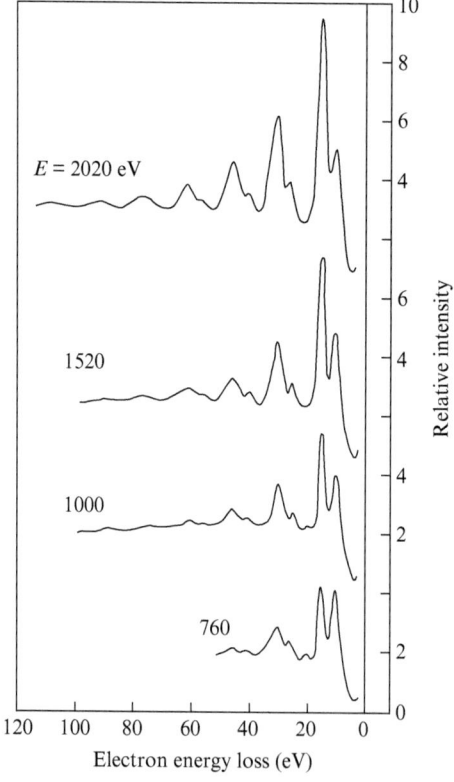

Fig. 2.10 Electron energy loss spectroscopy showing quantization of the energy of plasmons and surface plasmons. Electrons accelerated with different energies (2020, 1520, and 1000 eV) are incident at 45° and are scattered by a thin aluminium foil (of thickness estimated to be in the range 5–10 nm). The curve displays the number of reflected electrons at 45° versus the lost energy. (Redrawn from Powell, C. J., & Swan, J. B. (1959). Origin of the Characteristic Electron Energy Losses in Aluminum. Phys. Rev., 115(4), 869–875. doi:10.1103/physrev.115.869.)

energy between electrons and plasmons. The experiment can be simply interpreted by considering that the electrons emits bulk plasmons or surface plasmons whose energies are quantized.

In this section, we will focus on surface modes that can propagate along a flat interface while decaying exponentially on both sides of the interface. Here we simply make a comment on terminology. It turns out that both nanosphere modes and modes propagating along an interface are called surface modes, although they are different.

2.3.2 Bulk plasmon

2.3.2.1 *Hydrodynamic model: the concept of polariton*

We now consider a more general analysis of the concept of plasmon. We do not consider a specific geometry. Instead, we look for a general equation describing a charge density wave in an infinite homogeneous electron gas. Our primary objective here is to illustrate the concept of polariton in the particular case of a plasmon. The key idea that will be introduced is that when an electromagnetic wave propagates in a material medium, the field polarizes the medium and therefore excites mechanical movement of charges. It follows that field and charges are coupled. This coupled excitation is called a polariton. In the case of a metal, the field can couple to a longitudinal charge density wave that can be viewed as an acoustic wave in the electron gas. The resulting polariton is called a plasmon polariton. In an ionic crystal like NaCl, an electromagnetic field can excite mechanical motion of the ions (a phonon) and therefore generate a polarization oscillation. This is called a phonon polariton. Finally, the coupling between the field and an electron–hole pair (an exciton) is called an exciton polariton. The purpose of this subsection is to provide a simple explicit model of this coupling in the case of a metal within the jellium model introduced above. We will use a hydrodynamic model to derive the equation for the charge density wave. To begin, we write Euler's equation, the mass conservation equation, and the Gauss–Maxwell equation:

$$nm\left(\frac{\partial \mathbf{v}}{\partial t} + \mathbf{v}\cdot\nabla\mathbf{v}\right) = -ne\mathbf{E} - \nabla P_{\mathrm{e}}\,,$$

$$\nabla\cdot(n\mathbf{v}) = -\frac{\partial n}{\partial t}\,, \tag{2.85}$$

$$\nabla\cdot\mathbf{E} = \frac{n-n_0}{\epsilon_0}(-e)\,,$$

where P_{e} is the electronic pressure, $\mathbf{v}$ is the electron velocity, and $n(x,t)$ is the number of electrons per unit volume. Finally, we introduce the compressibility of the electron gas, $\partial P_{\mathrm{e}}/\partial n = m\beta^2$, where β^2 is of the order of v_{F}^2 and v_{F} is the Fermi velocity [12]. A detailed discussion of the value of β is given by Halevi [54]. For high frequencies $\omega\tau \gg 1$, where τ is a typical relaxation time, we have $\beta^2 = 3/5v_{\mathrm{F}}^2$. When looking for a small-amplitude perturbation $n_1(x,t) = n(x,t) - n_0$ and $P_{\mathrm{e}1} = P_{\mathrm{e}} - P_{\mathrm{e}0}$, where x_0 indicates the equilibrium value of x, the nonlinear term $\mathbf{v}\cdot\nabla\mathbf{v}$ can be neglected. Let us comment on this set of equations. For a neutral gas, the electric force $-ne\mathbf{E}$ in Euler's equation would be suppressed. One would then find the usual propagation equation for acoustic waves. Here, after linearizing, we find a set of two coupled linear equations:

$$\nabla^2 n_1 - \frac{5}{3v_{\mathrm{F}}^2} \frac{\partial^2 n_1}{\partial t^2} = -\frac{5n_0 e}{3mv_{\mathrm{F}}^2} \nabla \cdot \mathbf{E},$$

$$\nabla \cdot \mathbf{E} = \frac{n_1}{\epsilon_0}(-e). \tag{2.86}$$

This system clearly exhibits the coupling between the acoustic wave and the electric field. It can be seen that the electron density satisfies a propagation equation with a source term given by the divergence of the electric field. Similarly, the equation describing the longitudinal component of the electric field is driven by the electron density modulation $n - n_0$. The resulting coupled oscillation is called a polariton. The key idea here is that the acoustic and electromagnetic fields are no longer modes of the system. The mode of the system is a coupled mode called a polariton. It can be viewed as an object that is half a photon and half a phonon.

It is now a simple matter to eliminate the electric field and find the propagation equation for the electron density wave that accounts for both the pressure force and the electric force. When searching for a solution of the form $\exp(\mathrm{i}kx - \mathrm{i}\omega t)$, we find the dispersion relation

$$\omega^2 = \omega_{\mathrm{p}}^2 + \frac{3}{5} v_{\mathrm{F}}^2 k^2. \tag{2.87}$$

It turns out that the electric force yields the ω_{p}^2 contribution, which is much larger than the pressure contribution $v_{\mathrm{F}}^2 k^2$ for wavevectors in the optical regime (i.e. $k \ll \omega/v_{\mathrm{F}}$). It follows that in the optical regime, the dependence of ω on the wavevector can be neglected.

To summarize this subsection, it has been shown that the plasmon appears as an acoustic wave in an electron gas. As the particles are charged, an additional electric force has to be accounted for. It turns out that this electric force yields the dominant contribution, so the waves are essentially non-dispersive. Note also that with this approach, it is clear that the electric field is parallel to the wavevector since it is due to the charge density gradient.

2.3.2.2 *Bulk plasmon: electromagnetic model*

When studying the propagation of waves in vacuum, we always focus on transverse waves, since longitudinal solutions do not exist. This is no longer the case in a material medium. Plasmons are longitudinal solutions of Maxwell's equations. In the previous subsection, we studied the propagation of coupled mechanical and electromagnetic waves using a hydrodynamic model of a metal. We found that the electromagnetic solution has an electric field that is parallel to the wavevector. From a more general perspective, this solution is a longitudinal solution, namely a solution that satisfies $\nabla \times \mathbf{E} = 0$. Such a solution is therefore fully described by the equation $\nabla \cdot \mathbf{D} = 0$. In this subsection, we investigate the existence of a longitudinal solution of Maxwell's equations without invoking a specific model of the medium. If we assume that the medium is linear, homogeneous, and isotropic, we can introduce a dielectric constant. The most general linear form includes a dependence on the frequency and the wavevector $\epsilon(\mathbf{k}, \omega)$:

$$\mathbf{D}(\mathbf{k}, \omega) = \epsilon(\mathbf{k}, \omega)\mathbf{E}(\mathbf{k}, \omega). \qquad (2.88)$$

The dependence of $\epsilon(\mathbf{k}, \omega)$ on ω is called dispersion and the dependence on $\mathbf{k}$ is called spatial dispersion. This dependence on the wavevector leads to a non-local relation between the electric field and the vector $\mathbf{D}$ in direct space, so spatial dispersion and a non-local dielectric constant are two aspects of the same property:

$$\mathbf{D}(\mathbf{r}, \omega) = \int \frac{d\mathbf{k}}{8\pi^3} \epsilon(\mathbf{k}, \omega)\mathbf{E}(\mathbf{k}, \omega)\exp(i\mathbf{k} \cdot \mathbf{r}) = \int \epsilon(\mathbf{r} - \mathbf{r}', \omega)\mathbf{E}(\mathbf{r}', \omega)\,d\mathbf{r}'. \qquad (2.89)$$

The equation $\nabla \cdot \mathbf{D} = 0$ can be cast in the form

$$
\begin{aligned}
\nabla \cdot \mathbf{D} &= \nabla \cdot \int \frac{d\mathbf{k}}{8\pi^3}\frac{d\omega}{2\pi} \epsilon(\mathbf{k}, \omega)\mathbf{E}(\mathbf{k}, \omega)\exp(i\mathbf{k} \cdot \mathbf{r} - i\omega t) \\
&= \int \frac{d\mathbf{k}}{8\pi^3}\frac{d\omega}{2\pi} \epsilon(\mathbf{k}, \omega)[i\mathbf{k} \cdot \mathbf{E}(\mathbf{k}, \omega)]\exp(i\mathbf{k} \cdot \mathbf{r} - i\omega t) = 0.
\end{aligned}
\qquad (2.90)
$$

If we seek a non-zero longitudinal electric field, then $\mathbf{k} \cdot \mathbf{E}(\mathbf{k}, \omega) \neq 0$, so $\epsilon(\mathbf{k}, \omega) = 0$. A local medium has a dielectric constant that does not depend on $\mathbf{k}$, so the dispersion relation of the longitudinal solution is given by $\epsilon(\omega) = 0$.

For the particular case of a non-lossy Drude model, $\epsilon(\omega) = \epsilon_0(1 - \omega_p^2/\omega^2)$, so we find $\omega = \omega_p$, in agreement with the local approximation of (2.87). The discrepancy with the previous subsection illustrates the fact that the Drude model is an approximation that does not account for the $\mathbf{k}$ dependence of the dielectric constant. This is usually an excellent approximation, as we have discussed above. However, it is necessary to be aware that for the Drude model to be valid, we must have $k \ll \omega_p v_F$. Models accounting for the k-dependence of the dielectric constant (i.e. non-local models) are discussed in [12, 45, 89, 124].

2.3.3 Surface electromagnetic wave

So far we have introduced the concept of polariton and the particular case of a bulk plasmon polariton. Let us emphasize that we have only discussed waves propagating in a bulk medium. Moreover, we have studied longitudinal electromagnetic modes. We now consider waves propagating along an interface, which are transverse. The aim of this subsection is to seek a solution confined close to the interface. More precisely, we look for a solution that decays exponentially away from the interface. At this stage, we do not make any particular assumptions regarding the specific properties of the media. Hence, the surface wave dispersion relation that we will find can be applied to any material (e.g. a metal or a dielectric) and any frequency range (e.g. radio, infrared, or visible). The only assumption made in what follows is that the media are local and isotropic. Hence, each is characterized by a complex frequency-dependent dielectric constant ϵ_r and a complex frequency-dependent permeability μ_r. We indicate

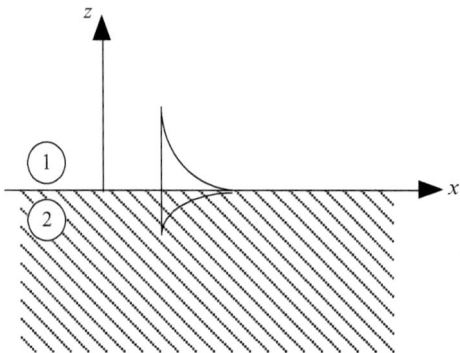

Fig. 2.11 Schematic representation of the exponential decay along z of the amplitude in media 1 (dielectric) and 2 (metal) of a surface wave propagating along the x-axis.

the upper medium ($z > 0$) by the index 1 and the lower medium ($z < 0$) by the index 2 as shown in Fig. 2.11. We denote the wavevector by $\mathbf{k}$, its Cartesian components by (k_x, k_y, γ), and its modulus by k.

2.3.3.1 *Dispersion relation for the non-magnetic case*

The electric field obeys Helmholtz's equation in both media:

$$\nabla^2 \mathbf{E}_i + \mu_i \epsilon_i \frac{\omega^2}{c^2} \mathbf{E}_i = 0 \,, \tag{2.91}$$

with $i = 1, 2$. For a p-polarized solution (also called TM, for 'transverse magnetic'), we seek a solution of the form

$$
\begin{aligned}
z > 0 : \quad & E_{x1} = E_0 \exp(\mathrm{i}k_x x + \mathrm{i}\gamma_1 z) \,, \\
z < 0 : \quad & E_{x2} = E_0 \exp(\mathrm{i}k_x x - \mathrm{i}\gamma_2 z)
\end{aligned}
\tag{2.92}
$$

that satisfies the continuity condition along the interface. Here,

$$\gamma_1 = \left(\frac{\mu_1 \epsilon_1 \omega^2}{c^2} - k_x^2 \right)^{1/2} , \tag{2.93}$$

with $\mathrm{Im}(\gamma_1) > 0$, and

$$\gamma_2 = \left(\frac{\mu_2 \epsilon_2 \omega^2}{c^2} - k_x^2 \right)^{1/2} \tag{2.94}$$

with $\mathrm{Im}(\gamma_2) > 0$, so the waves decay exponentially far from the interface. We look for transverse waves, so, by definition, $\nabla \cdot \mathbf{E} = 0$. In Fourier space, this relation becomes $\mathbf{k} \cdot \mathbf{E} = 0$, where $\mathbf{k} = (k_x, 0, \gamma)$. We stress that this equation does not have the usual geometrical meaning of two perpendicular real vectors, because $\mathbf{k}$ is complex. In other words, transverse (i.e. $\nabla \cdot \mathbf{E} = 0$) should not be confused with perpendicular. It follows that

$$z > 0: \quad E_{z1} = -\frac{kE_0}{\gamma_1} \exp(\mathrm{i}k_x x + \mathrm{i}\gamma_1 z),$$

$$z < 0: \quad E_{z2} = \frac{kE_0}{\gamma_2} \exp(\mathrm{i}k_x x - \mathrm{i}\gamma_2 z). \tag{2.95}$$

If we now enforce the continuity condition on the z-component of $\epsilon\mathbf{E}$ at the interface, we obtain

$$\epsilon_1 \gamma_2 = -\epsilon_2 \gamma_1. \tag{2.96}$$

These equations are the dispersion relations of the surface wave. To obtain a more explicit form, we take the squares of both terms. Note that we lose the sign at this point, so we will need to check that the final solution satisfies the original dispersion relation. For TM polarization, the solution for k_x is denoted by $k_{\mathrm{SP}}(\omega)$ and is given by

$$k_{\mathrm{SP}}^2(\omega) = \frac{\omega^2}{c^2} \frac{\epsilon_1 \epsilon_2}{\epsilon_1 + \epsilon_2}. \tag{2.97}$$

When using this solution, we should remember that it has two branches. One corresponds to the solution of $\epsilon_1 \gamma_2 = -\epsilon_2 \gamma_1$ and the other to the solution of $\epsilon_1 \gamma_2 = \epsilon_2 \gamma_1$. Only the first is of interest to us. At this point, it is not obvious that this dispersion relation provides a surface wave. We need to make sure that the wave decays exponentially away from the interface. Let us assume that *both media are non-lossy*, so ϵ_1 and ϵ_2 are real numbers. Let us further assume that medium 1 is a non-lossy dielectric, so its refractive index n_1 is a real positive number. To ensure that we obtain a surface wave, we need to check that $k_{\mathrm{SP}} > n_1\omega/c$ so that the wave decays exponentially in medium 1. This entails that

$$\frac{\epsilon_2}{\epsilon_1 + \epsilon_2} > 1. \tag{2.98}$$

This condition cannot be satisfied if $\epsilon_2 > 0$. Hence, we see that ϵ_2 is real and negative. It follows that the wave is necessarily exponentially decaying in medium 2, since $\gamma_2 = (\epsilon_2 \omega^2/c^2 - k_x^2)^{1/2}$ is purely imaginary. Finally, in order to have $\epsilon_1/(\epsilon_1 + \epsilon_2) > 0$, we need to satisfy the condition

$$\epsilon_2 < -\epsilon_1. \tag{2.99}$$

A similar calculation for the magnetic case in s-polarization yields

$$\mu_1 \gamma_2 = -\mu_2 \gamma_1. \tag{2.100}$$

2.3.3.2 *Polarization of the surface wave*

We have seen that the electric field of a surface wave propagating along the x-axis has two components along the x- and z-axes. Moreover, the z-component of the electric field is complex. Hence, the electric field has an elliptic polarization in the (x, z) plane. This peculiar polarization can be understood from the following remark. The existence of a z-component of the electric field entails the presence of a surface charge $P_{z2} - P_{z1}$ along the interface. Hence, the surface wave can be viewed as a surface charge density wave propagating along the x-axis as depicted in Fig. 2.12. Since in the vacuum above the interface, the field lines must be continuous, there must be an x-component of the field to close the field lines (see Fig. 2.12).

It is worth emphasizing a difference between the current density and the surface charge associated with the surface plasmon. Although the current density $\mathbf{j} = -i\omega\mathbf{P} = -i\omega\epsilon_0(\epsilon_2 - 1)\mathbf{E}$ penetrates the metal over the skin depth, the surface charge does not penetrate the metal, since $\nabla \cdot \mathbf{P} = \epsilon_0(\epsilon_2 - \epsilon_1)\nabla \cdot \mathbf{E} = 0$ below the interface. The contribution to the surface charge is a pure surface term given by $P_{z2} - P_{z1}$. For a metal–vacuum interface, it is simply given by P_{z2}. From a physical point of view, a surface charge must have some finite extension along the z-axis. One has to account for non-local effects to introduce the relevant lengthscale. This is the Thomas–Fermi lengthscale, which is of the order of 0.1 nm.

2.3.3.3 *Lengthscales of a surface wave*

There are three different lengths characteristic of a surface wave. It can be seen from (2.97) that the wavevector is complex if there are losses. The imaginary part of k_{SP} accounts for the decay of the surface wave on propagation along the interface. A characteristic decay length can be defined by $\delta_x = 1/\mathrm{Im}(k_{\mathrm{SP}})$. There are two other characteristic lengths accounting for the exponential decay of the surface wave away from the interface. They are given by $\delta_{zi} = 1/\mathrm{Im}(\gamma_i)$ in media $i = 1, 2$ and are found by inserting (2.97) into (2.93) and (2.94):

$$\frac{1}{\delta_{zi}} = \mathrm{Im}(\gamma_i) = \frac{\omega}{c}\, \mathrm{Im}\left[\left(\frac{\epsilon_i^2}{\epsilon_1 + \epsilon_2}\right)^{1/2}\right]. \tag{2.101}$$

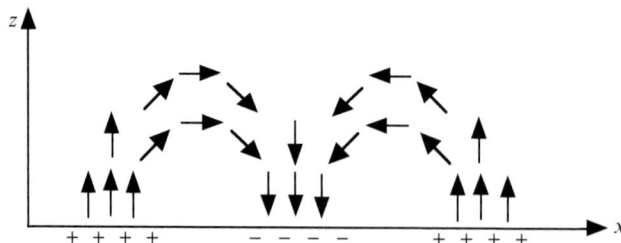

Fig. 2.12 Polarization of the surface plasmon polariton. The figure illustrates the surface charge density wave. It follows that the electric field has a normal component at the interface that oscillates. It can be seen that continuity of the field in the vacuum requires curvature of the field lines. The electric field is thus elliptically polarized in the (x, z) plane.

Table 2.1 Decay length for a surface plasmon propagating along a gold–vacuum interface.

λ_i (μm)	0.633	1	10	36
δ_x (μm)	9.8	91.6	38 880	504 243
δ_{z1} (μm)	0.165	0.51	57.3	702.67
δ_{z2} (μm)	0.014	0.012	0.011	0.013

Typical orders of magnitude are given in Table 2.1. It can clearly be seen that the surface plasmon has a decay length along the x-axis that is of the order of a few micrometres in the visible range but is considerably larger in the infrared. Regarding the spatial extension of the wave in the metal (medium 2), it can be seen that the decay length is almost constant. It is mainly given by the skin depth in the metal and is of the order of 12 nm. In contrast, the extension of the surface wave given by δ_{z1} in medium 1 changes dramatically. It varies between 165 nm in the visible (633 nm) to 700 μm in the infrared at 36 μm. Hence, it can be seen that the surface wave is not confined close to the interface in the infrared. Since most of the energy of the mode is in the vacuum, Joule losses are negligible, so the decay length upon propagation is drastically reduced. This is a very important remark: surface plasmon losses are reduced when most of the energy of the mode is in the non-lossy dielectric and not in the metal. It is often useful to design hybrid metallo-dielectric structures to reduce losses.

We finally note that the confinement in a dielectric is due to the fact that $k_x^2 > \epsilon_1 \omega^2 / c^2$. We note in particular that for very large wavevectors k_x, $\delta_{zi} = 1/\mathrm{Im}(\gamma_i) \approx 1/k_x$, so large vectors are strongly confined.

2.3.3.4 *Link with resonances of the reflection factor*

An alternative approach to find the dispersion relation consists in looking for the poles of the Fresnel reflection factor. The reason for looking at the Fresnel factor is simple. Since we can write $E_r^{\mathrm{s,p}} = r_{\mathrm{s,p}} E_{\mathrm{inc}}^{\mathrm{s,p}}$, it can be seen that the reflection factor $r_{\mathrm{s,p}}$ can be considered to be a linear response factor to the incident field $E_{\mathrm{inc}}^{\mathrm{s,p}}$ viewed as an external excitation. As for any linear system, a resonant response is the signature of the excitation of a mode of the system. Hence, putting the denominator of $r_{\mathrm{s,p}}$ equal to zero yields the pole, which is the signature of a mode of the interface. It can be checked that $\epsilon_1 \gamma_2 + \epsilon_2 \gamma_1$ is indeed the denominator of the Fresnel reflection factor for p-polarization for a non-magnetic material. We can now generalize the approach to a magnetic material for both polarizations. The Fresnel reflection factors can be cast in the form

$$r_{\mathrm{s}} = \frac{\mu_2 \gamma_1 - \mu_1 \gamma_2}{\mu_2 \gamma_1 + \mu_1 \gamma_2},$$

$$r_{\mathrm{p}} = \frac{\mu_1 \epsilon_2 \gamma_1 - \mu_2 \epsilon_1' \gamma_2}{\mu_1 \epsilon_2 \gamma_1 + \mu_2 \epsilon_1 \gamma_2}. \tag{2.102}$$

It follows that the corresponding dispersion relation can be written as

$$\mu_2\gamma_1 + \mu_1\gamma_2 = 0\,,$$

$$\mu_1\epsilon_2\gamma_1 + \mu_2\epsilon_1\gamma_2\,.$$

(2.103)

It can be seen that a surface wave can be obtained in the case of a magnetic material in s-polarization if the permeabilities μ_i have opposite signs. It is also of interest to note that the zeros and poles of the reflection factor are given by very similar equations. We will return to this point in Section 2.3.5.1. This approach is of particular interest when dealing with more complex systems such as multilayers. It accounts for guided modes, interface modes, and the coupling between these modes.

A technical remark might be useful here. The reader may be familiar with a presentation of the Fresnel reflection factor using the incident angle θ_i as a variable instead of the parallel component of the wavevector, k_x. For the case of a propagating incident wave in a lossless dielectric medium with refractive index n_1, it is essentially a matter of taste whether to use k_x or $n_1(\omega/c)\sin\theta_i$. If we seek the zero of the denominator, we need to use a real value of k_x that is larger than $n_1\omega/c$ in order to generate an evanescent wave. However, if we use the notation $n_1(\omega/c)\sin\theta_i$, a complex angle is needed to obtain a number larger than $n_1\omega/c$. Such a complex angle is not a very convenient concept. We now move to the question of the excitation of a surface wave. It is necessary to generate large wavevectors.

2.3.3.5   *Generation of a surface wave*

One possibility to generate large wavevectors is to use a pointlike source. According to Weyl's expansion, the spherical wave generated by a point source contains evanescent waves. There are other techniques to generate evanescent waves, i.e. to generate large wavevectors:

1. One can use a metal film with a thickness smaller than the skin depth separating two dielectric media with different dielectric constant n_1 and $n_2 > n_1$. Here, the key idea is to take advantage of a large refractive index to increase the modulus of the wavevector. By illuminating from the side of the high-refractive-index medium, it is possible to excite through the film with a plane wave with wavevector $k_x = n_2\omega/c\sin\theta_i$ and excites the surface wave on the other side by taking advantage of the fact that the incidence angle can be chosen so that $k_x > n_1\omega/c$.
2. A grating with period d can be used so that the nth order of the grating has a wavevector $k_{x,n} = n_1\omega/c\sin\theta_i + n2\pi/d$ that can be equal to k_{SP}.

2.3.4   **Surface plasmon polariton**

In this subsection, we will consider the specific case of surface waves propagating at the interface between a metal and a dielectric. These surface waves are called surface plasmon polaritons. Some authors [3, 123] call surface plasmons the electrostatic limit (or large-wavevector limit) of the surface plasmon polariton as introduced in

Section 2.3.3. However, most authors use the term 'surface plasmon' as a generic term without making this distinction.

2.3.4.1 Dielectric constant of a metal

Drude model   We start the discussion by introducing the Drude model of the dielectric constant for a metal described by an electron gas. The relative dielectric constant can be cast in the form

$$\epsilon_{\mathrm{r}}(\omega) = 1 - \frac{\omega_{\mathrm{p}}^2}{\omega^2 + \mathrm{i}\Gamma(\omega)\omega}. \tag{2.104}$$

This relation clearly shows that there is a strong frequency dependence (i.e. dispersion) of the dielectric constant. Since the Fourier transform of a product is a convolution product, the relation $\mathbf{D}(\mathbf{r}, \omega) = \epsilon_0 \epsilon_{\mathrm{r}}(\omega)\mathbf{E}(\mathbf{r}, \omega)$ becomes, in the time domain,

$$\mathbf{D}(\mathbf{r}, t) = \int_{-\infty}^{t} \epsilon_0 \epsilon_{\mathrm{r}}(t - t')\mathbf{E}(\mathbf{r}, t')\,\mathrm{d}t'. \tag{2.105}$$

Two timescales are included in the model. On the one hand, the plasma frequency is the mode frequency of the charge density oscillation. The plasma frequency lies in the near-ultraviolet for most metals. A second timescale appears in this formula, namely the relaxation time $\tau(\omega) = 1/\Gamma(\omega)$. When dealing with transport in the DC regime, this relaxation time describes collision processes of electrons with phonons and defects in the crystal. It plays a major role in the electrical conductivity $ne^2\tau/m$. It is well known that by decreasing the temperature, the DC conductivity increases. When dealing with absorption of an electromagnetic wave in the optical regime, the microscopic process at the origin of absorption is different. It is an intraband excitation of the electrons. In other words, the absorption corresponds to the transition of a single electron in the conduction band with energy E to a state in the same conduction band with energy $E + \hbar\omega$. Hence, the parameter Γ describes microscopic processes that are completely different at zero frequency and in the optical regime. In particular, the temperature dependence of Γ in the optical regime is very different from what is observed in the DC regime. In what follows, we will discuss in more detail the microscopic physical processes responsible for absorption in a metal. However, before doing so, we will first introduce a few orders of magnitude for different timescales in the physics of absorption by metals.

Typical decay time for fields and electrons   We start by giving the order of magnitude of the collision time for an electron in a metal. Here, we are discussing the transport of electricity in the diffusive regime in a bulk metal. The electron collision time is typically the mean duration between two collisions. It is of the order of 10–20 fs for noble metals. On multiplying this time by the Fermi velocity, one finds the mean free path.

We now discuss the decay time of a bulk plasmon. We will show that it is related to the Γ effective parameter of the Drude model. From a macroscopic point of view, we can search the complex eigenfrequency of the bulk plasmon mode. Its imaginary part contains the information on the decay time. For instance, for a bulk plasmon, the complex frequency is the solution of $\varepsilon(\omega) = 0$, so $\omega^2 + i\omega\Gamma - \omega_p^2 = 0$. The root of this equation is $\sqrt{\omega_p^2 - \Gamma^2/4} - i\Gamma/2$. Hence, the relaxation time of the plasmon appears to be given by Γ, as expected. The typical decay time for an electromagnetic field in a metal is also of the order of 10 fs, although the microscopic process is different. The temperature dependence is very different.

We now discuss the relaxation time for an electron occupying an excited state in the conduction band. We first note that this is an out-of-equilibrium situation, so the dielectric constant does not contain any information about this process. To study these effects, a convenient experiment consists in using a pump–probe technique. A femtosecond laser shines on a thin metallic film. Absorbed electrons are promoted to the conduction band. A second femtosecond laser is then used to measure the reflectivity as a function of the time delay between the two pulses. This technique allows measurement of the dynamics of the excited electrons in the conduction band. Such experiments show that the excited electrons exchange energy through electron–electron interactions and thermalize on a typical timescale of the order of 300–500 fs. This rapid thermalization allows the definition of an electronic temperature for the excited electrons in the conduction band above the Fermi level. This first step is followed by a relaxation process whereby the electrons transfer their energy to the phonons of the lattice. This electron–phonon decay time is of the order of 1 ps [116].

Absorption processes in bulk metals   The absorption of an electromagnetic wave in the DC regime or at radiofrequencies is due to to electron collisions. In this low-frequency regime, the classical picture used to describe transport is valid: (i) the electromagnetic field transfers energy to the electron, (ii) the electron loses energy through collisions. The microscopic mechanism responsible for collisions is mostly an electron–phonon interaction, although electron–electron interactions play a minor role.

We now turn to the absorption of energy of an electromagnetic wave at visible or infrared frequencies (including a plasmon). This process corresponds to electron–hole pair generation. There are different interaction mechanisms: intraband absorption assisted by phonons, interband absorption, and Landau damping.

Intraband absorption   For low values of the wavevector $k < \omega/v_F$ and low frequencies, the leading absorption mechanism is intraband absorption. This is the only term that is included in the Drude model. As momentum conservation cannot be satisfied because the photon momentum is negligible compared with the change in electron momentum, this process needs to be a three-body interaction. Hence, this process is possible only if a phonon is either emitted or absorbed. The temperature dependence of this process differs from that of the collision time. One of the practical conclusions of the situation described in this paragraph is that metal losses at optical frequencies cannot be significantly reduced by reducing the temperature. The reader will find more information on electron losses in [7, 36, 52, 67, 103, 113].

Interband absorption   The second absorption mechanism for an electromagnetic wave in a metal is related to interband absorption. For noble metals, electrons in d-shells can be promoted to states above the Fermi level and with the same wavevector (so-called vertical transition) if the frequency is above a threshold. This mechanism is well known [12] and accounts for the colours of copper and gold. It can be included in the model by adding a specific term of the dielectric constant. It cannot be accounted for by the standard Drude form.

Landau damping   The Landau damping mechanism is observed only for large wavevectors such that $k > \omega/v_F$. It can be understood using either a wave or a particle picture. The wave picture is due to Landau and has been introduced to explain the absorption in plasmas. When the electron velocity v_F is equal to the phase velocity ω/k of the field, the electron velocity and the field are always in phase, so energy transfer is very efficient. Another point of view is to consider that the absorption of a plasmon allows the generation of an electron–hole pair. This is possible because a plasmon has a wavevector much larger than that of a photon. Let us consider an interaction between an electron with initial momentum $\hbar q$ and initial energy $\hbar^2 q^2/2m$ with a surface plasmon with energy $\hbar\omega$ and momentum $\hbar k$. The electron is close to the Fermi surface, so $\hbar q = m v_F$. The interaction must conserve energy and momentum, so we have

$$\hbar\omega = \frac{\hbar^2}{2m}[(\mathbf{q} + d\mathbf{q})^2 - \mathbf{q}^2] \approx \frac{\hbar^2 q\, dq}{m} \approx \hbar v_F\, dq\,,$$

$$\hbar\mathbf{k} = \hbar\, d\mathbf{q}\,,$$

(2.106)

where we have given a rough estimate of $\hbar\omega$. Eliminating dq between the two equations, it can be seen that for $k \approx \omega/v_F$, the interaction satisfies energy and momentum conservation. For usual electromagnetic excitations, this process is forbidden because the electromagnetic wavevector $k \approx \omega/c$ is too small. However, surface plasmons may have large wavevectors and therefore this process can take place. We note here that this process can also take place when a dipole is close to a metal interface at a distance d smaller than v_F/ω, since its near field contains large wavevectors. In summary, for values of k larger than ω/v_F, the plasmon is damped because it can relax by generating an electron–hole pair. Clearly, this process introduces a cutoff spatial frequency for the surface plasmons. More information on the non-local description of the optical properties of solids can be found in [12, 45, 89, 124]. For noble metals, the typical Landau damping lengthscale v_F/ω is of the order of 0.5 nm.

Beyond the Drude model   Although the Drude model can be a very useful tool, it is important to keep in mind that its accuracy is much better in the infrared than in the visible range. The reason is that the Drude model accounts for the intraband absorption due to free electrons but not for interband absorption. The latter introduce serious deviations from the Drude model. This can be accounted for by developing fits of the measured dielectric constant as reported in several references [41, 55, 77, 115]. It is of course essential to use these realistic models when studying plasmons using time-domain calculations. Figure 2.13 illustrates the large difference in the optical part of the spectrum between a Drude model obtained by fitting experimental data and a more detailed fit of the data reported by Johnson and Christy [62] for gold. The

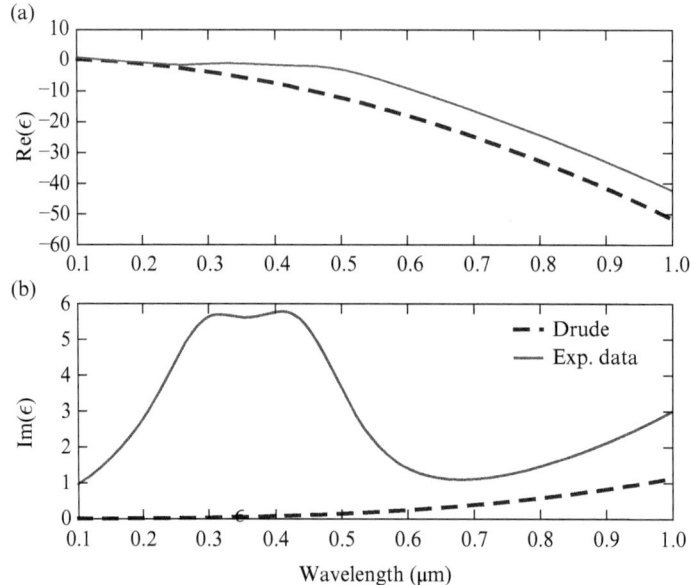

Fig. 2.13 Comparison of experimental data and a Drude fit of the dielectric constant for gold: (a) real and (b) imaginary parts of the complex dielectric constant.

absorption band observed in the imaginary part of the dielectric constant between 0.3 and 0.4 μm is due to absorption by d-band electrons.

Another limitation of the Drude model is that it does not account for non-locality, also called spatial dispersion. Spatial dispersion means that the dielectric constant depends on the wavevector. In direct space, the polarization at a point **r** depends on the value of the electric field not only at this point but also in its vicinity—hence the name non-locality [76]. The dielectric constant has two lengthscales, corresponding to two different phenomena. The first effect is the screening of the field at an interface. Classical local electromagnetism assumes that the normal component is divided by ϵ_r at the interface between vacuum and a metal. The microscopic phenomenon responsible for this effect is screening of the field by the electrons. This requires a certain length for it to take place. This screening length is the so-called Thomas–Fermi length for metals. Its typical value is 0.1 nm for noble metals. For electrolytes, the corresponding screening length is called the Debye–Hückel length. This phenomenon corresponds to longitudinal fields.

We now consider the second lengthscale that appears in non-local models of the optical response of metals. This lengthscale is the length travelled by an electron at a velocity v_F during an optical cycle. When this lengthscale is much smaller than the wavelength, the optical properties are not affected. In contrast, for wavevectors $k > \omega_p/v_F$, non-local corrections are expected. We have already seen that this condition corresponds also to the onset of the Landau damping absorption mechanism, so surface plasmons virtually disappear for wavevectors larger than v_F/ω. A discussion of non-local effects can be found in [43, 45].

2.3.5   Dispersion relation of an SPP

2.3.5.1   *Dispersion relation of an SPP for a non-lossy Drude metal*

Here, we discuss the dispersion relation of a surface plasmon based on on the Drude model. As already stated, this is a crude model for noble metals when the frequency approaches the plasma frequency. Nevertheless, we shall use it for the sake of simplicity to discuss a few key issues. Although losses play a very important role, we start by neglecting them in order to base our introductory discussion on the simplest analytical formulas. However, we emphasize that the results obtained are only a rough approximation of the actual properties. We consider that the upper medium is a dielectric with a real dielectric constant ϵ_1 and the lower medium is a metal described by a non-lossy ($\Gamma = 0$) Drude model. Inserting the Drude form of the dielectric constant into the dispersion relation given by (2.97), we obtain

$$k_{\mathrm{SP}} = \frac{n_1 \omega}{c} \left[\frac{\omega^2 - \omega_{\mathrm{p}}^2}{(1 + \epsilon_1)\omega^2 - \omega_{\mathrm{p}}^2} \right]^{1/2}. \tag{2.107}$$

It can be seen from this formula that the dispersion relation has an asymptote for a frequency $\omega_{\mathrm{p}}/\sqrt{1 + \epsilon_1}$ (see Fig. 2.14). When plotting this equation, a second branch is obtained for frequencies larger than ω_{p}. This branch is not a surface wave. Indeed, for $\omega > \omega_{\mathrm{p}}$, the metal dielectric constant is a positive real number, so the metal is a dielectric from the optical point of view. In this regime, the waves can propagate, although the refractive index is smaller than 1, indicating that the phase velocity is larger than c. The meaning of this branch of the dispersion relation is clear if one

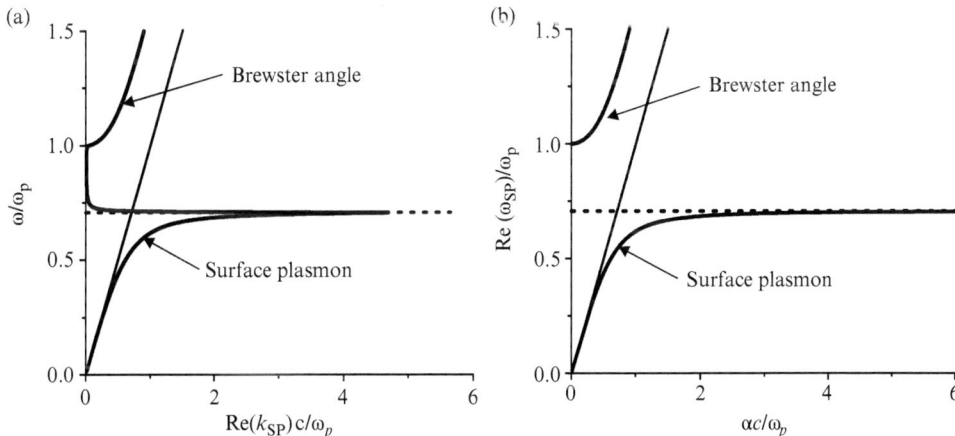

Fig. 2.14 Dispersion relation of a surface plasmon propagating along an interface separating a lossy metal described by the Drude model from vacuum. The implicit dispersion relation can be solved, searching for a real frequency and a complex wavevector or vice versa. Two different dispersion relations are obtained. (a) Frequency versus real part of complex k_{SP}. (b) Real part of complex frequency versus real wavevector k_x.

remembers that we neglected the sign when solving the dispersion relation $\epsilon_1\gamma_2 + \epsilon_2\gamma_1 = 0$. As ω becomes larger than ω_p, the sign of the dielectric constant changes. In this range of frequencies, the equation (2.97) is simply the solution of $\epsilon_1\gamma_2 - \epsilon_2\gamma_1 = 0$ or, in other words, the zero of the reflection factor. This is connected to the Brewster angle θ_B by the relation $k_x = n_2(\omega/c)\sin\theta_B$.

2.3.5.2 *SPP on a lossy Drude metal. Is it a surface plasmon?*

In this discussion, we now do not neglect losses. From (2.97), it can be seen that for a real value of the frequency ω, we find a complex value of the wavevector. Alternatively, it is possible to search for a solution of the equation with complex ω and real k_x. Both possibilities are equally valid. When plotting the real part of ω as a function of the real part of k_x, we find different dispersion relations, as illustrated in Fig. 2.14, depending on the choice. This raises the question of the interpretation of the physical content of each dispersion relation. We shall come back to this subtle issue in Section 2.3.10. We now compare the case of low frequencies with the case of optical frequencies. The question raised here is the nature of the surface wave for different frequencies. We start by analysing the Drude model in the low- and high-frequency regimes. It is easily seen that we can approximate the dielectric constant by

$$
\begin{aligned}
\omega \gg \Gamma(\omega): \quad & \epsilon(\omega) \approx 1 - \frac{\omega_p^2}{\omega^2}, \\
\omega \ll \Gamma(\omega): \quad & \epsilon(\omega) \approx 1 + \mathrm{i}\frac{\omega_p^2}{\omega\Gamma} = 1 + \frac{\mathrm{i}\sigma}{\omega\epsilon_0},
\end{aligned}
\tag{2.108}
$$

where $\sigma = ne^2/m\Gamma$ is the DC conductivity. This form is enlightening since it shows that the optical properties of the metal are dominated by the plasmon response in the optical regime but by the drag force in the low-frequency regime. In the former case, the dielectric constant is almost a negative real number, whereas in the latter case, it is almost a pure imaginary number. It follows that the plasmonic (oscillatory) character of the surface wave is only meaningful in the regime $\omega \gg \Gamma$. Indeed, if we rewrite the equations of motion of the electrons including the friction term as in (2.81), we find

$$
-\omega^2 x = \mathrm{i}\omega\Gamma x - \frac{ne^2}{m\epsilon_0}x.
\tag{2.109}
$$

For large and small frequencies, we have different approximate expressions:

$$
\begin{aligned}
\omega \gg \Gamma(\omega): \quad & -\omega^2 x + \frac{ne^2}{m\epsilon_0}x = 0, \\
\omega \ll \Gamma(\omega): \quad & -\mathrm{i}\omega\Gamma x + \frac{ne^2}{m\epsilon_0}x = 0.
\end{aligned}
\tag{2.110}
$$

It can clearly be seen that the oscillatory regime that is the essence of a plasmon corresponds only to the case $\omega \gg \Gamma$. Instead, for small frequencies, the electronic response of the medium is dominated by the viscous term. Since the typical value of Γ

is 10^{14} Hz, we note that a surface wave is hardly a surface plasmon for frequencies in the infrared or smaller. On the other hand, the plasmonic behaviour (i.e. oscillatory behaviour) dominates the metal response in the case of femtosecond pulses.

In summary, from a macroscopic electrodynamics point of view, there is one surface wave defined for all frequencies that may be called a surface plasmon. However, from a microscopic point of view, the underlying behaviour of the electrons is very different in the low- and high-frequency regimes. The essence of the plasmon is a collective oscillation that takes place only in the regime $\omega \gg \Gamma(\omega)$. Beyond this semantic remark, this distinction is important, because the detailed form of the dispersion relation is different for low and high frequencies, as we examine in more detail below.

2.3.5.3 *Dispersion relation of surface waves in the radiofrequency range*

Let us analyse in more detail the nature of the surface waves propagating along a conductive surface at low frequencies. We can give a more explicit form of the dispersion relation (2.97) in this regime. At radiofrequencies, the dielectric constant of a metal can be cast in the form $\epsilon = 1 + \mathrm{i}\sigma/\omega\epsilon_0 \approx \mathrm{i}\sigma/\omega\epsilon_0$, so the modulus of the dielectric constant is much larger than 1. A Taylor expansion of the wavevector of the surface wave can thus be written in the form

$$k_{\mathrm{SP}} = \frac{\omega}{c} \left(\frac{1}{1 + 1/\epsilon} \right)^{1/2} \approx \frac{\omega}{c} \left(1 + \frac{\mathrm{i}\omega\epsilon_0}{2\sigma} \right). \tag{2.111}$$

It can be seen that in the limit of a perfect conductor, σ becomes infinite, so the wavevector becomes $k = \omega/c$. This entails that the wave is almost unconfined close to the interface. Here, we recover the concept of surface wave used in the context of radio waves propagating along perfectly conducting wires or impinging on perfectly conducting structures. Note in particular that, in this limit, there is no longer any damping, since the wavevector becomes real. Let us now check that (2.111) is a valid solution of the equation $\epsilon_1\gamma_2 + \epsilon_2\gamma_1$. In the radio regime, the analysis is very different from the case of a surface plasmon in the optical regime. We consider the case of an interface separating a metal (medium 2) from vacuum (medium 1) and we use the notation $\epsilon = |\epsilon| \exp(\mathrm{i}\phi_\epsilon)$. We have

$$\gamma_2 \approx \frac{\omega}{c}\epsilon^{1/2} \approx \frac{\omega}{c}|\epsilon|^{1/2} \exp(\mathrm{i}\phi_\epsilon/2),$$

$$\gamma_1 \approx \frac{\omega}{c} \left(\frac{1}{\epsilon} \right)^{1/2} \approx -\frac{\omega}{c} \left(\frac{1}{|\epsilon|} \right)^{1/2} \exp(-\mathrm{i}\phi_\epsilon/2). \tag{2.112}$$

The choice of square root is imposed by the condition $\mathrm{Im}(\gamma) > 0$, so we had to include a minus sign for γ_1. It is clear then that the condition $\epsilon\gamma_1 + \gamma_2 = 0$ is satisfied.

To summarize, inserting the Drude model into the dispersion relation of a surface wave yields two limits: a surface plasmon for $\omega > \Gamma(\omega)$ and a surface wave with $k = \omega/c$ for $\omega \ll \Gamma$. The latter is the so-called Sommerfeld mode. Note that when dealing with terahertz waves, the nature of the wave is closer to a radio surface wave than to a surface plasmon.

2.3.5.4 *Electrostatic limit*

Here, we extend the previous discussion to the non-retarded limit. Let us consider a dipole source oscillating at a frequency $\omega = 2\pi c/\lambda$ at a distance d above an air–metal interface such that $d \ll \lambda$. As the distance is much smaller than the wavelength, the interaction between the source and the interface can be described within the electrostatic approximation. Here, we mean that the spatial structure of the field in the near field of a small object can be computed by solving an electrostatic problem. It follows that the interface can be modelled by introducing an image charge. The standard electrostatic formalism [61] allows us to introduce an electrostatic reflection factor given by $(\epsilon - 1)/(\epsilon + 1)$. Using a non-lossy Drude model, this yields a resonance for $\epsilon + 1 = 0$ and hence for $\omega = \omega_p/\sqrt{2}$. This frequency can also be derived by seeking a mode of the Laplace equation for the scalar potential for a system with one interface separating two homogeneous media [123]. It can be seen that the factor presents a resonance and becomes much larger than 1. This point has already been discussed in Section 2.3.1. It corresponds to a resonance of the surface charge at the interface.

2.3.6 Surface phonon polaritons

In the case of a metal and a frequency in the range $[\Gamma, \omega_p]$, we have seen that the dielectric constant is negative. In this regime, the surface wave has the character of a surface plasmon. There are other situations such that the dielectric constant becomes negative. In agreement with the Kramers–Kronig relations, they always correspond to frequencies close to resonant excitation of the medium. In the infrared, crystals can absorb light owing to coupling to the optical phonons. There is a frequency range called the reststrahlen band where the dielectric constant is negative. A simple model allows us to show that the dielectric constant can be cast in the form

$$\epsilon(\omega) = \epsilon_\infty \frac{\omega_L^2 - \omega^2 - i\Gamma\omega}{\omega_T^2 - \omega^2 - i\Gamma\omega}, \qquad (2.113)$$

where ω_L is the longitudinal frequency and ω_T is the transverse optical frequency. These frequencies are due to the presence of optical phonons. As in the case of electrons, a longitudinal solution exists at ω_L. It corresponds to a charge density wave. Here, it is a polarization charge density. A detailed discussion can be found in the books by Ziman [124] and by Ashcroft and Mermin [12] for example.

There are some differences with the plasmon case. The dielectric constant is negative only in the range $[\omega_T, \omega_L]$. This range corresponds to a wavelength range of the order of $1\,\mu m$ typically. The central frequency is typically between 10 and $40\,\mu m$. Hence, the surface phonon polariton can exist only in the mid-infrared or near-terahertz frequency ranges. A very important similarity of the surface phonon polariton with the surface plasmon is the existence of a horizontal asymptote in the dispersion relation. This indicates the presence of a peak in the LDOS close to the interface, as will be discussed later [63, 99]. Figure 2.15 is an example of a dispersion relation of a surface phonon polariton. It corresponds to the case of a wave propagating at the interface between GaAs and vacuum.

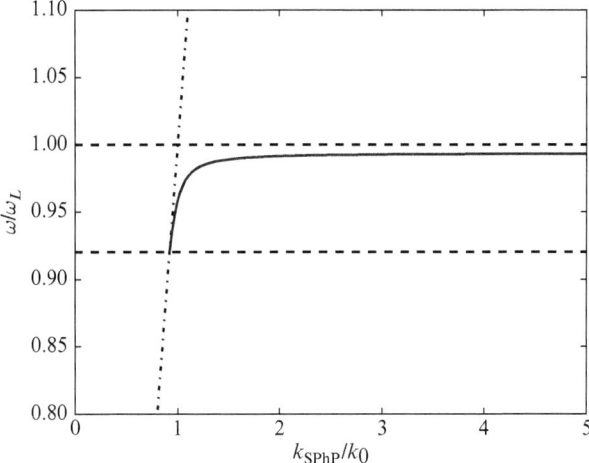

Fig. 2.15 Dispersion relation of a surface phonon polariton propagating along a GaAs–vacuum interface with dielectric constant given by a non-lossy Lorentz model $\epsilon(\omega) = \epsilon_\infty(\omega_L^2 - \omega^2)/(\omega_T^2 - \omega^2)$. The wavevector axis has been normalized by ω_L/c and the frequency axis by ω_L. Here, $\omega_L = 292.1\,\mathrm{cm}^{-1}$, $\omega_T = 267.8\,\mathrm{cm}^{-1}$, and $\epsilon_\infty = 11$.

2.3.7 A potpourri of surfaces waves: Sommerfeld or Zenneck modes, quasicylindrical or lateral waves

In the previous subsections, we have introduced surface waves as solutions of Maxwell's equations in the presence of interfaces. We have seen that there are several cases where a solution can be found for both s- and p-polarizations. Surface waves are often given different names depending on their frequency range (e.g. visible or radio waves) or their type (e.g. electromagnetic, seismic, or acoustic). It turns out that the topic of surface waves is much broader than the topic of surface plasmons. Electromagnetic surface waves were studied in the context of radio-wave propagation well before surface plasmons were discovered. Surface waves have also been studied in a variety of other contexts. The purpose of this subsection is not to give a detailed discussion, but rather to introduce and explain the terminology.

2.3.7.1 Historical perspective

The concept of surface wave was introduced by Zenneck at the beginning of the twentieth century. His motivation was to explain the mechanism of long-range propagation of radio waves, with the basic idea being that a surface wave decays as $1/r$ instead of $1/r^2$ as in three dimensions. (It finally turned out that the correct explanation for long-range propagation is the presence of the ionosphere, with the space between the Earth and the ionosphere acting as a waveguide.) The concept of surface wave was further studied by Sommerfeld, who derived a rigorous solution for the field generated by a dipole above a flat interface separating two homogeneous media. In Sommerfeld's derivation, the surface mode is defined as the contribution of the pole of the reflection factor to the field produced by a dipole above an interface. This definition agrees with our remark in Section 2.3.3.4 linking the surface wave with the pole of the reflection factor. The much debated existence of a surface wave in the radio literature is due to the fact that the field has a complex structure. It should be stressed that there is perfect consensus on the integral form of the field radiated by a dipole above an interface. However, such an integral is not useful for practical applications. Several authors have therefore derived approximate analytical expressions valid in different cases. It is only when it comes to the interpretation of the different contributions that there is a debate. The reader will find a detailed account of these works in the books by Brekhovskikh [23], Banos [14], Felsen and Marcuvitz [44], and King et al. [68], and in the review paper by Collin [30]. To make a long story short, let us summarize the situation as follows. The field radiated by a dipole can be decomposed into a sum of plane waves. In the presence of an interface, each plane wave is transmitted and reflected. The total field is hence the field radiated by the source plus a sum of plane waves weighted by the corresponding Fresnel factors. When performing the integral over all reflected plane waves, one can extract the contribution due to the poles of the reflection factors. This contribution yields the surface waves. There is a second contribution that appears when analytical techniques are used to evaluate asymptotically the integral in the complex plane: this is the contribution along the branch cuts in the complex plane. It is often called a lateral wave. In summary, the field can be viewed either as the sum of an infinite number of plane waves or as the sum of a surface wave and a lateral wave. We shall now give a very brief account of the latter.

2.3.7.2 Lateral wave

The second noteworthy contribution is called a lateral wave in the radio-wave and seismology community. It has been investigated only recently [73] in the optics community and has been called a quasicylindrical wave [32, 74] or a Norton wave [122]. An extensive discussion of this wave in the context of propagation of radio waves along the Earth has been given by King [68]. In this section, we briefly describe a physical picture of the lateral wave. Let us consider a source located at $z = -d$ in a material medium with refractive index n. The lateral wave is the wave corresponding to propagation from the source to the interface followed by a refraction at the interface and propagation by a plane wave parallel to the interface in the medium $z > 0$ (Fig. 2.16). When the medium at $z < 0$ is absorbing and the medium $z > 0$ is not absorbing, this is the most efficient mechanism for propagation over large distances, since most of the energy is in the non-absorbing medium. When excited by a line, this two-dimensional wave decays as $1/L^{3/2}$, where L is the distance of propagation along the interface. Instead, the surface wave has an exponential decay.

The lateral wave is very well known in seismology. An excellent account of surface waves in elastic media can be found in the textbook by Aki and Richards [4]. In the context of radio waves, it has received much attention in the 1960s and 1970s. The reader will find a detailed account in the works quoted above [14, 23, 30, 44, 68]. Of particular interest in optics are [32, 73, 74, 122], where it is shown that these waves

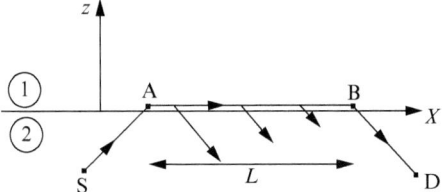

Fig. 2.16 Illustration of a lateral wave. The wave propagates from the source to the interface at the critical angle. After refraction, it propagates along the interface as a plane wave and reenters the medium continuously. This contribution becomes dominant when the medium is lossy so that direct propagation in the medium is rapidly damped.

cannot be neglected in many cases when studying propagation and scattering along metallic surfaces, including the resonant transmission phenomenon [81].

In what follows, we focus on key properties of surface plasmons related to (i) the LDOS, (ii) spatial confinement, and (iii) the fast temporal response. With regard to these three aspects, surface plasmons differ greatly from lateral waves. These properties are intimately linked to the underlying electronic character of the surface plasmon polariton.

2.3.8 Key properties of surface plasmon polaritons

Plasmonics has become a very active area of research owing to its large number of applications. However, all these applications rely on a small number of key properties. Here, we shall discuss these key properties in simple and general terms. We will first analyse the importance of having a dispersion relation with very large wavevectors. We will discuss the implications in terms of field confinement and the LDOS. Finally, we will discuss the spectral width of surface plasmon excitations and its significance in terms of ultrafast response.

2.3.8.1 Confinement of the field

Many applications of surface plasmons rely on the possibility of producing highly confined fields and/or producing or observing light at a lengthscale smaller than its wavelength in vacuum. Here, we shall review the basic properties of surface plasmons underlying these applications. Let us first consider the case of surface plasmons propagating along flat interfaces. It is necessary to distinguish between confinement of the field along the normal to the interface and confinement in the plane of the interface. We have already given some orders of magnitude of the field confinement away from the interface in Table 2.1. We now discuss the potential of surface waves for in-plane confinement.

Lateral confinement Losses impose a limit to the extension of surface waves along an interface. The decay length is given by $1/\text{Im}(k_{\text{SP}})$. This value depends significantly on the losses of the material. Typically, for noble metals and visible frequencies, it is a few micrometres. It should be noted that this length is considerably reduced for frequencies close to the asymptote of the dispersion relation. The main reason for this is connected to our previous discussion. A photon-like surface plasmon is poorly localized close to the interface and has most of its energy in the dielectric above the metal, where there are no losses. In contrast, a plasmon-like surface plasmon has a large part of its energy in the metal, so it is very sensitive to losses. It is a general rule that modes with most of the field energy in the dielectric have a longer propagation length. In practice, another mechanism can reduce the propagation length, namely radiative leakage of energy (often called radiative losses) as a result either of scattering by random roughness or of diffraction by a periodic structure such as a grating.

We have discussed the longer lengthscale of a surface plasmon along the interface. We now address the issue of the shortest lengthscale. It was seen in Fig. 2.14(b) that the maximum wavevector given by the plasmon dispersion relation may be much larger than ω/c. This seems to pave the way to a strong confinement of the field. There have been some attempts to take advantage of this strong localization of the field. It has been proposed by Pendry that this property could be used to realize a superlens [91] with a simple thin film supporting surface plasmons. Experimental implementations have been reported by two groups in the visible using silver [42] and in the infrared using silicon carbide [109]. Another proposal for super-resolution based on the use of surface plasmons was made in [104]. The key idea was to take advantage of the large wavevectors that are seen in the dispersion relation (with the choice of a real wavevector and a complex frequency). A debate followed this proposal [38, 105]. However, in practice, it is not possible to take full advantage of this property of surface plasmons because losses play an important role. We shall show in Section 2.3.10 that the relevant dispersion relation that must be used for discussing confinement of the field is the one shown in Fig. 2.14(a). It can be seen that the wavevector modulus is limited because of so-called backbending of the dispersion relation. This entails that the surface plasmons always have an intrinsic limit in terms of resolution. This is a rather severe limitation, since, in many cases, the maximum value of the wavevector is hardly larger than $2\omega/c$. Let us stress that, so far, the resolution obtained in different experimental results [42, 109] is indeed limited by losses.

However, surface plasmons are often used to produce highly localized spots that go well beyond these limitations. In all practical examples, the origin of the localization of the fields lies in the spatial structure of the material. In most cases, one uses nanometre-scale particles or nanostructures such as nanowires, nanoholes, or indentations in a metallic substrate. Examples include the first implementation of near-field optical microscopy [78, 92], the strong confinement obtained using nanoholes [79], and the use of tips as nanosources [82, 119]. One might then ask what is the role of plasmons in that case? A simple answer can be obtained by analysing the fields produced by a subwavelength spherical particle of dielectric constant ϵ and radius a. If $a \ll \lambda$, then the non-retarded approximation is valid, so the spatial structure of the field can be found using an electrostatic approximation. The scattered field is given by the field of a dipole for a distance $r > a$. The field decays as $1/r^3$ for $r > a$, so the confinement is only limited by a. This confinement is due to the geometry and not to the plasmon; it is independent of the material properties. However, the amplitude of the field depends on the material properties at the particular frequency considered. For instance, when dealing with a spherical nanoparticle with radius a, its polarizability α_{p} can be written as

$$\alpha_{\mathrm{p}} = 4\pi a^3 \, \frac{\epsilon(\omega) - 1}{\epsilon(\omega) + 2} \,. \tag{2.114}$$

It can be clearly seen from this equation that the only limitation to confinement is the fact that the amplitude of the dipole tends to zero with a. However, if the frequency is such that $\epsilon(\omega) + 2$ is almost zero, a resonance is excited, as will be discussed in more detail in Section 2.3.9. In the Drude model, this condition is satisfied for $\omega = \omega_{\mathrm{p}}/\sqrt{3}$. Thus, it can be seen that a surface plasmon resonance of a particle is useful for light confinement indirectly: its role is to compensate for the small value of the dipole moment of a small object.

In the above example, we have seen that the form of the electric field close to a particle can be factorized into two terms. The dependence on space variables is given by the electrostatic form of the dipolar field, while the dependence on frequency is given by the polarizability. Only the latter exhibits resonant behaviour, which is the signature of the plasmon resonance.

This concept of confinement by geometry coupled to resonant enhancement was put forward by Li, Stockman, and Bergman who proposed to use a chain of nanoparticles to realize an efficient nanolens [80]. This is the basic property at work in the heat-assisted magnetic recording (HAMR) technique proposed to increase the capabilities ofs magnetic storage [26].

Finally, we mention another possibility for confining the field. Metal–dielectric–metal (MDM) structures can support surface modes that are plasmonic even if the dielectric is only of a few nanometres thick so that these modes are highly localized in the gap [66]. Remarkable guiding properties have been demonstrated using channel surface polaritons, which essentially rely on this type of structure [22]. Other applications of confinement in MDM structures includes light emission in the weak-coupling regime [40] and the achievement of strong light–matter coupling [111]. These are intimately related to the LDOS, as we now describe.

2.3.8.2 *Surface plasmon contribution to the LDOS*

The lifetime of an atom can typically be reduced by orders of magnitude when it is located close to an interface. This was first demonstrated experimentally by Drexhage [37]. Excellent discussions can be found in [15, 27, 45]. Similarly, the thermodynamic equilibrium (blackbody) electromagnetic energy density can be increased by orders of magnitude close to an interface [95, 99]. Both phenomena depend on the density of electromagnetic states. The changes by orders of magnitude are a clear signature of a change in the physical phenomena that determines the LDOS. In both cases, the contribution of surface waves plays a key role. Here, we shall show how surface waves may drastically modify the LDOS. We will discuss briefly applications to light emission assisted by surface waves. We will also show that the influence of surface plasmons on the LDOS plays a key role in the Casimir force, a pure quantum electrodynamics phenomenon.

2.3.8.3 Increasing the DOS: surface waves, slow light, and microcavities

It is known that low-velocity systems can be used to increase the DOS. The mechanism is depicted in Fig. 2.17. As the dispersion relation becomes flat close to the band edge, the number of states (represented by dots) with a frequency in the interval $\Delta\omega$ increases. This behaviour is known as a van Hove singularity. A major advantage of photonic crystals compared with plasmonic systems is that there are almost no losses in dielectric media. Since the density of states diverges (the group velocity becomes zero), these systems seem to be the perfect solution to engineer the optical properties of quantum emitters. However, it is important to emphasize that the number of states available when using a waveguide is always finite. A plasmonic system can provide an LDOS that is orders of magnitude larger than what can be achieved with a slow waveguide. In order to understand this apparent paradox, let us first recall the derivation of the DOS for a one-dimensional system. We consider a waveguide branch characterized by a dispersion relation $k_z(\omega)$ for a mode propagating along the z-axis in a periodic waveguide with period a. In order to count the number of modes, we again consider that the system has a finite length L and we introduce the so-called Born–von Kármán (or periodic) boundary conditions stipulating that the system is periodic with period L along z. It follows that $k_z = p\,2\pi/L$. The number of modes in the interval $\mathrm{d}\omega$ corresponding to an interval $\mathrm{d}k_z$ is given by

$$g(\omega)\,\mathrm{d}\omega = \frac{L}{2\pi}\,\mathrm{d}k_z = \frac{L}{2\pi}\frac{\mathrm{d}k_z}{\mathrm{d}\omega}\,\mathrm{d}\omega. \tag{2.115}$$

It can be seen that the DOS diverges as the group velocity goes to zero. However, one should keep in mind that *this divergence is integrable*, so the number of states in a finite interval $[\omega_1,\omega_2]$ always remains finite.

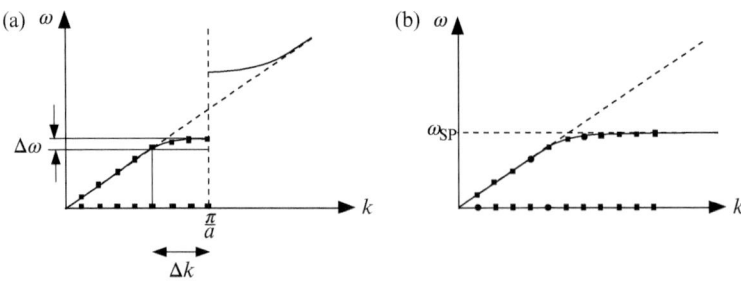

Fig. 2.17 Increasing the DOS. States are characterized by k and ω. They correspond to points located on the dispersion relation. In k-space, the DOS per unit length is constant and takes the value $1/2\pi$. It can be seen in (a) that in ω space, the DOS increases close to the gap edge, while (b) shows that the increase close to the asymptote is considerably larger, since the asymptote is not limited along the k-axis.

This is clear from Fig. 2.17, where we represent schematically the dispersion relation of a guided wave close to a band edge and the dispersion relation of a surface plasmon. It can be seen that the modes are simply redistributed close to the band edge, so this only concerns a finite number of modes, although mathematically the DOS diverges. An upper limit on the number of modes involved is clearly the size of the Brillouin zone $2\pi/a$ divided by the interval between two modes $2\pi/L$, which gives L/a. The period of a photonic crystal is of the order of the wavelength, so we obtain an estimate of an upper bound on the number of modes in a photonic crystal given by L/λ. This is orders of magnitude less than the number of modes available with surface waves at resonance.

We now discuss briefly another possibility for increasing the density of states originally proposed by Purcell. The idea is to use a cavity with a single mode. The number of states per unit volume is thus $1/V$, where V is the cavity volume. Taking into account the finite value of the quality factor of the cavity, we obtain for a Lorentzian resonance with width Γ the following DOS:

$$g(\omega) = \frac{1}{V}\frac{\Gamma}{2\pi}\frac{1}{(\omega - \omega_0)^2 + \Gamma^2/4}. \tag{2.116}$$

At resonance, the DOS is thus given by

$$g(\omega_0) = \frac{2}{\pi V \Gamma}. \tag{2.117}$$

The Purcell factor is the LDOS in the cavity normalized by the DOS in vacuum:[9]

$$F_{\mathrm{p}} = \frac{2}{\pi V \Gamma}\frac{3\pi^2 c^3}{\omega^2} = \frac{3}{4\pi^2}Q\frac{\lambda^3}{V}, \tag{2.118}$$

where $Q = \omega_0/\Gamma$. This derivation is based on a poorly defined volume. A more accurate description should account for the polarization of the mode field, as well as its space dependence. Indeed, the field in a cavity is not uniform, so the coupling between a mode and an emitter will strongly depend on the exact location of the emitter. A more detailed analysis can be found in [47, 51, 56].

2.3.8.4   *LDOS due to surface waves*

We begin the discussion of the role of surface waves on the LDOS with a qualitative discussion based on the dispersion relation. We then present a more quantitative analysis. Figure 2.17 suggests that the number of states provided by a surface plasmon at resonance is infinite, since the dispersion relation appears to be flat and unbounded. This is not correct and is a consequence of the model of the dielectric constant not accounting for non-locality. A non-local dielectric constant introduces a cutoff [45] in the dispersion relation given by $v_{\mathrm{F}}\omega_{\mathrm{SP}}/\sqrt{2}$, where v_{F} is the Fermi velocity as already

---

[9] In the context of Fermi's golden rule, a factor of one-third is introduced to account for the fact that a given dipole can couple to only one component of the electric field.

discussed. We can now easily compare the DOS due to surface plasmons with the vacuum DOS. A rough estimate of the number of surface plasmons per unit area is given by dividing the area of a disk of radius $\pi k_{\text{SP,max}}^2$ by the area per state, $1/4\pi^2$:

$$\frac{\pi k_{\text{SP,max}}^2}{4\pi^2} \approx \frac{\omega_{\text{SP}}^2}{4\pi v_{\text{F}}^2}, \tag{2.119}$$

which is clearly much larger than what we found for dielectrics, where the order of magnitude is $1/\lambda^2 \approx \omega^2/c^2$. We remark that c/v_{F} is typically of the order of 300.

We now turn to a quantitative analysis of the LDOS due to surface waves. An explicit form can be derived from the imaginary part of the Green tensor. The reader will find a detailed analysis in [63]. Of particular interest is the asymptotic expression for the LDOS at a distance z from the interface such that $z \ll \lambda$:

$$\frac{g(z,\omega)}{g_{\text{v}}(\omega)} = \frac{\text{Im}(\epsilon)}{|\epsilon+1|^2} \frac{1}{4(k_0 z)^3}, \tag{2.120}$$

where $g_{\text{v}}(\omega)$ is the vacuum DOS. The surface plasmon resonant contribution is clearly given by the term $1/|\epsilon+1|^2$. Figure 2.18 illustrates the contribution of this term to the LDOS at a distance of 10 nm from a gold surface and a GaAs surface.

In both cases, the presence of the surface wave results in a peak in the LDOS enhancement. It clearly appears that the enhancement is orders of magnitude larger for surface phonon polaritons than for surface plasmons.

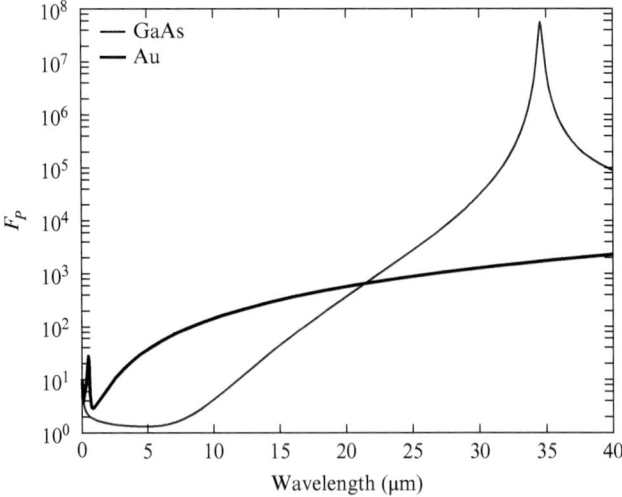

Fig. 2.18 LDOS at a distance of 10 nm above an interface separating vacuum from gold or GaAs.

2.3.8.5 LDOS and energy transfer at the nanoscale

Another remarkable consequence of the increase in the DOS due to surface waves is the increase in the energy density at thermodynamic equilibrium. Since the modes are thermally excited at equilibrium, there is a large energy density close to the interface. Figure 2.19 illustrates the evolution of the energy spectral density at different distances from an interface separating SiC from vacuum. Two features appear clearly in this figure: (i) the energy density normalized by the maximum energy density of a blackbody is increased by several orders of magnitude, (ii) the spectrum becomes quasimonochromatic. The existence of thermally excited surface waves has been observed experimentally by de Wilde et al. [35], who were able to obtain near-field images of samples without external illumination. Note that in the near field, the spectrum becomes quasimonochromatic, indicating that the field is partially temporally coherent. The coherence time is essentially the decay time of the surface phonon polariton, as discussed in [64, 99].

As a consequence of this increase in energy density close to the interface, the heat transfer between two half-spaces separated by a distance smaller than the wavelength increases. This heat transfer mechanism can be viewed as being mediated by

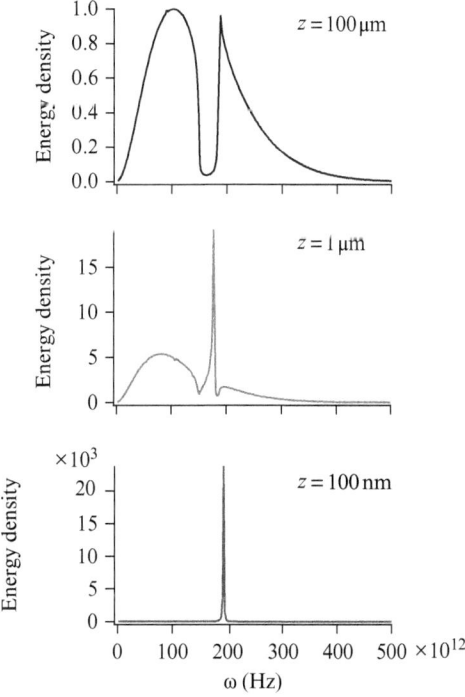

Fig. 2.19 Electromagnetic energy density at equilibrium at 300 K as a function of distance from an interface between vacuum and SiC. The energy density is normalized by the maximum blackbody value at 300 K.

surface phonon polaritons. Since the number of surface waves increases dramatically at small distance, the heat flux through vacuum gap can be enhanced by orders of magnitude. This effect due to surface waves was predicted in [83, 84] and has now been measured [95, 100]. The heat transfer through the gap due to this surface phonon interaction can be viewed as a phonon tunnelling phenomenon. It can also be described in a form similar to the Landauer conductance. It has been shown [19] that each mode characterized by (k_x, k_y, ω) yields a contribution to the radiative heat conductance proportional to the thermal quantum of conductance $\pi^2 k_B^2 T/3h$ and a transmission factor (here h is Planck's constant).

2.3.8.6 LDOS and light emission assisted by surface waves

We have already cited the pioneering experiment by Drexhage showing that the lifetime of an emitter can be drastically reduced close to an interface. When the distance is below 5 nm, the energy goes into heat in the substrate and surface plasmons do not make a significant contribution to this mechanism. However, if the distance is larger than approximately 10 nm, most of the energy goes into the surface plasmon for appropriate frequencies. In the case of a flat surface, this energy is then converted into heat. However, if, for example, a grating is ruled on the surface, the energy can be radiated. In that case, the surface works as an antenna: the energy of the source is efficiently coupled into the surface (owing to the large LDOS) and then efficiently radiated by the surface (owing to the grating). This idea has been put forward in the context of light-emitting diodes [48, 71]. The influence of the surface plasmon resonance on single-molecule fluorescence assisted by a resonant particle has been investigated theoretically [13, 110]. Quantum well light emission assisted by surface plasmons has been demonstrated more recently [86]. A remarkable demonstration of optical nanoantennas at the level of a single emitter has been reported using metallic nanospheres. An emitter located at a distance of the order of 10 nm excites efficiently the surface plasmon of the particle. By appropriately choosing the particle radius, it is possible to increase the ratio of power emitted to power absorbed in the particle so that the nanosphere becomes an efficient nanoantenna. Two experiments have clearly demonstrated how metallic nanospheres can be used as efficient nanoantennas to increase molecular fluorescence [8, 72]. More recently, several metallic structures have been proposed as antennas to control angular emission and also increase the emission rate [17, 31, 40, 108]. Nanopatch antennas have been reported with a Purcell factor of the order of 1000 while maintaining an efficiency of 0.5 [5].

2.3.8.7 LDOS and the Casimir force

Another consequence of the contribution of surface plasmons to the LDOS is the Casimir force between two metallic parallel plates. The Casimir force is a pure quantum electrodynamics effect that manifests itself at a macroscopic scale. Casimir [25] predicted that there is an attractive force between two parallel perfectly conducting surfaces at 0 K separated by a gap of width d. Since then, his remarkable prediction has been measured experimentally with great accuracy [34, 65, 75]. However, when comparing these measurements with theory, the assumption of a perfect conductor can no longer be used [46]. A careful analysis shows that surface plasmons are responsible for the forces actually observed [57, 59, 60]. We give here a brief qualitative discussion of this effect. We refer the reader to [57] for a further discussion. The gap behaves as a waveguide with a set of modes. From the quantum electrodynamics point of view, each mode (k, ω) has a zero-point energy at $0\,\mathrm{K}$ given by $\hbar\omega$. It follows that the total energy is given by $\sum_n \hbar\omega_n$, where the sum is over all modes of the gap. In the original derivation by Casimir, modes of a planar waveguide with perfectly conducting surfaces were used. When accounting properly for the optical properties of metals, it turns out that the density of states DOS is dominated by the surface plasmon contribution [57].

2.3.9 Broad spectrum and fast response

When comparing plasmonic resonators with dielectric resonators, the presence of losses in the metal is the main difference. Here, we establish the connection between the quality factor of the plasmonic resonance of a nanosphere and the imaginary part of the dielectric constant. We then discuss the implications of the presence of losses. Close to the sphere's resonance frequency, we can expand the dielectric constant. We use the notation $\epsilon(\omega) = \epsilon_\mathrm{R}(\omega) + i\epsilon_\mathrm{I}(\omega)$ and assume that $\epsilon_\mathrm{R}(\omega_0) = -2$. We have

$$
\begin{aligned}
\epsilon(\omega) &\approx \epsilon_\mathrm{R}(\omega_0) + \frac{\mathrm{d}\epsilon_\mathrm{R}}{\mathrm{d}\omega}(\omega - \omega_0) + i\epsilon_\mathrm{I}(\omega_0) \\
&\approx -2 + \left(\frac{\mathrm{d}\epsilon_\mathrm{R}}{\mathrm{d}\omega}\right)_{\omega_0} \left(\omega - \omega_0 + i\frac{\Gamma}{2}\right),
\end{aligned}
\tag{2.121}
$$

where $\Gamma = 2\epsilon_\mathrm{I}(\omega_0)/(\mathrm{d}\epsilon_\mathrm{R}/\mathrm{d}\omega)_{\omega_0}$. The polarizability can be approximated by a Lorentzian profile:

$$
\begin{aligned}
\alpha(\omega) &= 4\pi a^3 \frac{\epsilon(\omega) - 1}{\epsilon(\omega) + 2} \\
&\approx \frac{12\pi a^3}{(\mathrm{d}\epsilon_\mathrm{R}/\mathrm{d}\omega)_{\omega_0}} \frac{1 - i\epsilon_\mathrm{I}(\omega_0)/3}{\omega_0 - \omega - i\Gamma}.
\end{aligned}
\tag{2.122}
$$

It follows that the quality factor is given by

$$
Q = \frac{\omega_0}{\Gamma} = \frac{\omega_0}{2\epsilon_\mathrm{I}} \frac{\mathrm{d}\epsilon_\mathrm{R}}{\mathrm{d}\omega}.
\tag{2.123}
$$

We have seen already that metallic losses in the optical frequency regime are due to electron–electron and electron–phonon interactions. These processes have a weak dependence on temperature, so they cannot be suppressed. Hence, these losses are a specific property of a plasmonic resonator. They result in two properties of surface plasmons. The quality factor of the resonance depends on the imaginary part of the dielectric constant at the resonance frequency. A typical quality factor for plasmons is between 10 and 100. Accordingly, the relaxation time of the system in the visible is of the order of a few femtoseconds. A small quality factor can be viewed as a drawback in terms of the Purcell effect for instance. On the other hand, a resonator with a large bandwidth can be very interesting. On the one hand, when using the resonator as an antenna, it can be operated with many different emitters [5, 17, 40]. On the other hand, a broad resonator can be used to perform coherent control of pulses on extremely short timescales. This particular property is the basis of a large number of recent contributions [1, 106].

Finally, we should emphasize that the relaxation time of a plasmonic resonance also depends on the geometry of the structure. To illustrate this, we discuss the example of the so-called long-range and short-range surface plasmons on symmetric thin metallic films in a dielectric first investigated by Sarid [97]. In order to understand why the geometry influences the relaxation time of a plasmonic mode, it is sufficient to realize that the field of the long-range surface plasmon energy is mostly in the non-lossy dielectric, whereas the short-range surface plasmon is more confined in the metal, where losses take place. This has also been widely discussed in the context of shell structures [93]. In both cases, the coupling between two modes depends on the thickness of the metal film and allows one to control the resonance frequencies. In contrast, if retardation effects are negligible, i.e. for objects much smaller than the wavelength in vacuum, the quality factor depends only on the dielectric properties of the metal [118].

2.3.10 Surface plasmon polaritons on lossy materials: dispersion relations

As already mentioned, the dispersion relation given by (2.97) cannot be solved using a real frequency and a real wavevector when the dielectric constant is complex. However, a solution can be found using either a complex k_x and a real ω or vice versa. These two choices lead to different shapes of the dispersion relation, as seen in Fig. 2.14. One dispersion relation has an asymptote for very large values of k_x, while the other has limited values of k_x and exhibits backbending. We have discussed the key properties of surface plasmons with an emphasis on the lateral confinement (necessary, for example, for a superlens) and on the large LDOS (necessary, for example, for a nanoantenna). Since the existence of a horizontal asymptote plays a key role both for the transverse confinement of the field (large-k surface plasmons are needed) and for the LDOS (a flat dispersion curve), it is important to establish a prescription for which choice should be made when looking for a dispersion relation. Here, we discuss the origin and physical content of these two dispersion relations, summarizing [10], where further details can be found.

2.3.10.1 First interpretation

The existence of two different forms of the dispersion relation $\omega(k)$ was first pointed out by Alexander [6]. It was first thought that the dispersion relation with backbending was an unphysical mathematical curiosity. However, Arakawa [9] remarked that when the positions of the dips in a reflectivity experiment in which the angle of incidence is varied at fixed frequency are plotted in the (k, ω) plane, the dispersion relation exhibits a backbending branch, whereas when the spectrum is obtained at a fixed angle, there is no backbending. This approach provides a practical prescription for the comparison of attenuated total reflection (ATR) experiments with theory. If the spectrum is taken at a fixed angle, then one should choose the solution of the dispersion relation with complex frequency and real angle, and vice versa for a fixed frequency. It is possible to remember this choice by noting that the spectrum taken at a fixed angle has a peak whose width is given by the imaginary part of the frequency. This points to a real angle and a complex frequency. Nevertheless, this remark is not an explanation. Furthermore, this discussion does not provide a general prescription that can be used to discuss all possible issues. To illustrate this point, we consider two questions regarding important properties of surface plasmons: confinement of the fields and large DOS. What can we learn from the dispersion relation regarding these questions? The dispersion relation with backbending predicts a cutoff spatial frequency k_{co}: it follows that the LDOS has an upper bound and that the maximum confinement of the field is also limited by $1/k_{\mathrm{co}}$. In contrast, the dispersion relation without backbending predicts a divergence of the LDOS at the frequency corresponding to the asymptote of the dispersion relation. It also predicts no limit to the possible resolution. Hence, we have conflicting answers regarding key questions. It is thus clear that a general discussion on the applicability of the different dispersion relations is needed.

2.3.10.2 *Representation of the fields*

In order to analyse the meaning of the dispersion relation and to clarify this issue, it is necessary to investigate the meaning of the choice of a real or complex wavevector. In what follows, we emphasize that the relevant quantity is not a field with real or complex wavevector but rather a field that depends on time and position. A standard and convenient representation is the Fourier transform of the field with respect to both time and position. In a Fourier representation, the three wavevector components and the frequency are real, so there is no room for complex quantities.

When looking at the field excited by any distribution of sources in the presence of an interface supporting surface waves, it is possible to cast the field in the form of a linear superposition of plane waves reflected or transmitted by the interface, with an amplitude given by the Fresnel factors. Following Sommerfeld's prescription, the contribution of the pole of the Fresnel factors to the integral is the surface wave. The contribution of the pole can be evaluated analytically using the residue theorem. This first analytic integration can be done either over frequencies or over the wavevector. Since the pole is complex, the analytic integration yields either a complex wavevector or a complex frequency, depending on the choice. When following this programme, we find two different representations of the surface plasmon field that are equally valid because the final value of the integral does not depend on the way it is evaluated. One representation uses surface modes with real frequency and complex wavevector, whereas the other uses complex frequencies and real wavevectors. We skip all the details here and give the result of the integration reported in [10].

2.3.10.3 Field representation with a real wavevector

The field can be cast in the form of a linear superposition of modes with real wavevector $\mathbf{K}$ and complex frequency $\omega_{\rm SP}$. We denote by $\mathbf{K}$ the projection of the wavevector parallel to the interface, $\mathbf{K} = (k_x, k_y, 0)$, and then

$$\mathbf{E}_{\rm SP} = 2\,{\rm Re}\left[\int \frac{{\rm d}^2\mathbf{K}}{(2\pi)^2}\,E(\mathbf{K},t)\left(\hat{\mathbf{K}} - \frac{K}{\gamma_m}\mathbf{n}_{\rm m}\right)\exp[{\rm i}(\mathbf{K}\cdot\mathbf{r} + \gamma_{\rm m}|z| - \omega_{\rm SP}t)]\right], \quad (2.124)$$

where $\mathbf{n}_{\rm m} = -\hat{\mathbf{z}}$ if $z < 0$ and $\hat{\mathbf{z}}$ if $z > 0$, and $\hat{\mathbf{K}} = \mathbf{K}/K$. The surface plasmon field takes a form that looks as a mode superposition, *except that the amplitude $E(\mathbf{K},t)$ depends on the time t*. Indeed, when describing a stationary field using modes that have an exponential decay, the amplitude is necessarily time-dependent. In order to obtain a superposition of modes with fixed amplitudes, it is necessary to assume that all sources are extinguished after time $t = 0$ so that we observe the field after it has been excited. In that case, the decay of the mode is well described by the imaginary part of $\omega_{\rm SP}$. Equation (2.124) is thus well suited for fields excited by pulses. Note that the polarization of each mode is specified by the complex vector $\hat{\mathbf{K}} - (K/\gamma_{\rm m})\mathbf{n}_{\rm m}$, whose component along the z-axis depends on the medium from which the field is evaluated.

2.3.10.4 Field representation with a real frequency

A different representation of the field can be derived using modes characterized by a real frequency ω and a real k_y. The x-component of the wavevector is complex and is given by

$$K_{x,\rm SP} = (k_{\rm SP}^2 - k_y^2)^{1/2}. \quad (2.125)$$

The z-component of the wavevector is given by the usual form $\gamma = (\epsilon_{\rm m}\omega^2/c^2 - k_{\rm SP}^2)^{1/2}$. With this notation, the field can be cast in the form

$$\mathbf{E} = \int \frac{{\rm d}\omega}{2\pi}\int \frac{{\rm d}k_y}{2\pi}\left\{E_>(k_y,\omega,x)\left(\hat{\mathbf{K}}^+ - \frac{k_{\rm SP}}{\gamma_{\rm m}}\mathbf{n}_{\rm m}\right)\exp[{\rm i}(K_{x,\rm SP}x + k_y y + \gamma_{\rm m}|z| - \omega t)]\right.$$

$$\left. + E_<(k_y,\omega,x)\left(\hat{\mathbf{K}}^- - \frac{k_{\rm SP}}{\gamma_{\rm m}}\mathbf{n}_{\rm m}\right)\exp[{\rm i}(-K_{x,\rm SP}x + k_y y + \gamma_{\rm m}|z| - \omega t)]\right\}, \quad (2.126)$$

where $\hat{\mathbf{K}}^+ = (K_{x,SP}\hat{\mathbf{x}} + k_y\hat{\mathbf{y}})/k_{\rm SP}$ and $\hat{\mathbf{K}}^- = (-K_{x,SP}\hat{\mathbf{x}} + k_y\hat{\mathbf{y}})/k_{\rm SP}$. Note that the *mode amplitudes depend on x*. A proper mode representation should use only fixed amplitudes. This is possible if all the sources lie in the $x < 0$ region and the region of interest is the $x > 0$ region. It can be shown in that case that the surface plasmon field can be cast in the form

$$\mathbf{E} = \int \frac{{\rm d}\omega}{2\pi}\int \frac{{\rm d}k_y}{2\pi}\left(\hat{\mathbf{K}} - \frac{k_{\rm SP}}{\gamma_{\rm m}}\mathbf{n}_{\rm m}\right)E_>(k_y,\omega)\exp[{\rm i}(\mathbf{K}\cdot\mathbf{r} + \gamma_{\rm m}|z| - \omega t)], \quad (2.127)$$

where $\mathbf{K} = K_{x,\mathrm{SP}}\hat{\mathbf{x}} + k_y\hat{\mathbf{y}}$ is complex and $\hat{\mathbf{K}} = \mathbf{K}/k_{\mathrm{SP}}$. We conclude that *stationary monochromatic fields excited by sources confined in a bounded domain are well described outside this domain by a representation that uses complex wavevectors and real frequencies.*

2.3.10.5 *Complex ω or complex k? a simple prescription*

To summarize, we have shown that the surface plasmon field can be represented using modes that have either a complex frequency or a complex wavevector. However, these mode amplitudes may still depend on either time or space in the more general case. However, there are two cases where these modes with a complex wavevector or complex frequency can be used with amplitudes that are constant. The first case is the representation of a field excited by a pulse for times after the end of the excitation. Then a representation with modes having a complex frequency is appropriate. The second case is a field excited by a stationary but localized source. Then a representation using complex wavevectors is appropriate. To each situation there corresponds a specific dispersion relation. This simple analysis yields a simple prescription to choose the proper dispersion relation. Note that in the case of pulses limited in space, both representations can be used.

2.3.10.6 *Implications for the LDOS*

Let us now discuss the LDOS. We have already pointed out the connection between the dispersion relation and the LDOS in Section 2.3.8.6. In particular, we have seen that the DOS diverges when the group velocity goes to zero. A quick look at Fig. 2.14 shows that different dispersion relations seem to predict different LDOS. While Fig. 2.14(b) predicts a very large peak at $\omega_{\mathrm{SP}}/\sqrt{2}$ due to the asymptote (zero group velocity) and no states above this frequency, Fig. 2.14(a) predicts a smaller peak and a non-zero LDOS between $\omega_{\mathrm{SP}}/\sqrt{2}$ and ω_{SP}. There must be a unique answer, since the LDOS determines the energy density at equilibrium and the lifetime of emitters, which are well-defined physical quantities. We actually know that the LDOS is given by the trace of the imaginary part of the Green tensor. This quantity diverges close to the surface, as discussed in Section 2.2.5.3. Again, we see that a prescription is needed to choose the right dispersion relation.

A standard procedure to derive the DOS in reciprocal space is based on periodic boundary conditions. Assuming a surface of side L, the wavevector takes the form

$$\mathbf{K} = n_x\frac{2\pi}{L}\hat{\mathbf{x}} + n_y\frac{2\pi}{L}\hat{\mathbf{y}}. \tag{2.128}$$

When performing this analysis, both k_x and k_y are real. Thus, *the relevant representation uses real wavevectors and complex frequencies. The corresponding dispersion relation has no backbending* and therefore exhibits a singularity. Of course, this divergence is non-physical. It is related to the modelling of the medium using a continuous description of the metal without accounting for the non-locality and the Landau damping that introduce a cutoff at $k = \omega/v_{\mathrm{F}}$.

2.3.10.7 *Implications for super-resolution and strong confinement*

Let us first discuss the issue of resolution when imaging with a surface plasmon driven at frequency ω by an external source. If the dispersion curve with asymptotic behaviour is chosen, there seems to be no diffraction limit (if we neglect the cutoff due to Landau damping) and only the amplitude decay of surface plasmon due to Ohmic losses in the metal limits the resolution. The effect of the backbending of surface plasmon dispersion discussed in [38] limits the surface plasmon wavelength $2\pi/\mathrm{Re}(k_{\mathrm{SP}})$ and therefore, the resolution. Clearly, the two dispersion relations do not lead to the same conclusion and a prescription to choose one or the other is needed. Let us consider a situation where a surface plasmon is excited locally by a stationary monochromatic field. From Section 2.3.10.2, we know that it is valid to use a representation with fixed amplitudes using modes with complex wavevectors and real frequencies. This implies that the dispersion relation with real frequency (with backbending) is relevant. Hence, there is a cutoff spatial frequency. Indeed, as k_x may be complex, the propagation term $\exp(\mathrm{i}k_x x)$ introduces damping. In the case of a lossy medium, damping may be due to losses. However, even for a non-lossy medium (where k_{SP} is real), $k_x = (k_{\mathrm{SP}}^2 - k_y^2)^{1/2}$ can be imaginary. This occurs when k_y exceeds the value k_{SP}. This situation is the two-dimensional analogue of the evanescent waves with wavevector k larger than ω/c that cannot propagate in vacuum. Clearly, k_{SP} is a cutoff frequency and the propagation term $\exp(\mathrm{i}k_x x)$ works as a low-pass filter that prevents propagation of fields associated with spatial frequencies larger than k_{SP}. When dealing with lossy media, it is the real part of k_{SP} that specifies the cutoff spatial frequency. It can be seen from Fig. 2.14 that this real part is limited by the backbending of the dispersion relation. However, this analysis suggest that by using a time-dependent pulse, the resolution may be improved. This has been studied and confirmed in [11].

In summary, when discussing *imaging with stationary monochromatic surface plasmons, the relevant representation is based on modes with a complex wavevector and a real frequency* given by (2.127). This corresponds to a *dispersion relation with backbending*. It follows that *the resolution is limited* by the cutoff spatial frequency given by the maximum value of $\mathrm{Re}(k_{\mathrm{SP}})$.

References

[1] Aeschlimann, M., Bauer, M., Bayer, D., Brixner, T., García de Abajo, F. J., Pfeiffer, W., Rohmer, M., Spindler, C., and Steeb, F. (2007). Nature **446**, 301.

[2] Agarwal, G. S. (1975). Phys. Rev. A **11**, 253.

[3] Agranovich, V. M. Mills D. L. (eds.) (1982). *Surface Polaritons*. North-Holland, Amsterdam.

[4] Aki, K. and Richards, P. G. (2002). *Quantitative Seismology*, University Science Books, Sausalito, CA.

[5] Akselrod, G. M., Argyropoulos, C., Hoang, T. B., Ciracì, C., Fang, C., Huang, J., Smith, D. R., and Mikkelsen, M. H. (2014). Nat. Photonics **8**, 835.

[6] Alexander, R. W., Kovener, G. S., and Bell, R. J. (1974). Phys. Rev. Lett. **32**, 154.

[7] Allen, P.B. (1971). Phys. Rev. B **3**, 305.

[8] Anger, P., Bharadwaj, P., and Novotny, L. (2006). Phys. Rev. Lett. **96**, 113002.

[9] Arakawa, E. T., Williams, M. W., Hamm, R.N., and Ritchie, R.H. (1973). Phys. Rev. Lett. **31**, 1127.

[10] Archambault, A., Teperik, T., Marquier, F., and Greffet, J.-J. (2009). Phys. Rev. B **79**, 195414.

[11] Archambault, A., Besbes, M., and Greffet, J.-J. (2012). Phys. Rev. Lett. **109**, 097405.

[12] Ashcroft, N. W. and Mermin, N. D. (1976). *Solid State Physics*. Holt-Saunders, New York.

[13] Azoulay, J., Débarre, A., Richard, A., and Tchénio, P. (2000). Europhys. Lett. **51**, 374

[14] Banos, A. (1994). *Dipole Radiation in the Presence of a Conducting Half-Space*. IEEE Press, Piscataway, NJ.

[15] Barnes, W. L. (1998). J. Mod. Optics **45**, 661.

[16] Barnes, W. L., Dereux, A., and Ebbesen, T. W. (2003). Nature **424**, 824.

[17] Belacel, C., Habert, B., Bigourdan F., Marquier, F., Hugonin, J.-P., Michaelis de Vasconcellos, S., Lafosse, X., Coolen, L., Schwob, C., Javaux, C., Dubertret, B., Greffet, J.-J., Senellart, P., and Maitre, A. (2013). Nano Lett. **13**, 1516.

[18] Betzig, E., Patterson, G. H., Sougrat, R., Lindwasser, O. W., Olenych, S., Bonifacino, J. S., Davidson, M. W., Lippincott-Schwartz, J., and Hess, H. F., (2006). Science **313**, 1642.

[19] Biehs, S. A., Rousseau, E., and Greffet, J.-J. (2010). Phys. Rev. Lett. **105**, 234301.

[20] Boardman, A. D. (ed.) (1982). *Electromagnetic Surface Modes*. Wiley, New York.

[21] Born, M. and Wolf, E. (1999). *Principles of Optics*, 7th edn. Cambridge University Press, Cambridge.

[22] Bozhevolnyi, S. I., Vokov, V. S., Devaux, E., Laluet, J. Y., and Ebbesen, T. (2006). Nature **440**, 508.

[23] Brekhovskikh, L. M. (1960). *Waves in Layered Media*, Academic Press, New York.

[24] Carminati, R. and Greffet, J.-J. (1999). Phys. Rev. Lett. **82**, 1660.

[25] Casimir, H. B. G. (1948). Proc. Koninkl. Ned. Akad. Wetenschap. **51**, 793.

[26] Challener, W. A., Peng, C., Itagi, A. V., Karns, D., Peng, W., Peng, Y., Yang, X., Zhu, X., Gokemeijer, N. J., Hsia, Y.-T., Ju, G., Rottmayer, R. E., Seigler, M. A., and Gage, E. C. (2009). Nat. Photonics **3**, 220.

[27] Chance, R. R., Prock, A., and Silbey, R. (1978). Adv. Chem. Phys. **37**, 1.

[28] Chen, F. C. and Chew, W. C. (1998). Appl. Phys. Lett. **72**, 3080.

[29] Chicanne, C., David, T., Quidant, R., Weeber, J. C., Lacroute, Y., Bourillot, E., Dereux, A., Colas des Francs, G., and Girard, C. (2002). Phys. Rev. Lett. **88**, 097402.

[30] Collin, R. E. (2004). IEEE Trans. Antennas Propagation Magazine **46**, 64.

[31] Curto, A., Volpe, G., Taminiau, T. H., Kreuzer, M. P., Quidant, R., and van Hulst, N. F. (2010). Science **329**, 930.

[32] Dai, W. and Soukoulis, C. M. (2009). Phys. Rev. B **80**, 155407.

[33] Dawson, P., de Fornel, F., and Goudonnet, J. P. (1994). Phys. Rev. Lett. **72**, 2927.

[34] Decca, R. S., Lopez, D., Fishbach, E., and Krause, D. E. (2003). Phys. Rev. Lett. **91**, 050402.

[35] de Wilde, Y., Formanek, F., Carminati, R., Gralak, B., Lemoine, P.-A., Joulain, K., Mulet, J.-P., Chen, Y., and Greffet, J.-J. (2006). Nature **444**, 740.

[36] Del Fatti, N. (1999). PhD Dissertation, Ecole Polytechnique.

[37] Drexhage, K. H. (1974). In *Progress in Optics*, ed. E. Wolf. North-Holland, Amsterdam.

[38] Drezet, A., Hohenau, A., and Krenn, J. R. (2007). Phys. Rev. Lett. **98**, 209703.

[39] Dyba, M. and Hell, S. (2002). Phys. Rev. Lett. **88**, 163901.

[40] Esteban, R., Teperik, T., and Greffet, J.-J. (2010). Phys. Rev. Lett. **104**, 026802.

[41] Etchegoin, P. G., Le Ru, E.C., and Meyer, M. (2006). J. Chem. Phys. **125**, 164705.

[42] Fang, N., Lee, H., Sun, C., and Zhang, X. (2005). Science **308**, 534.

[43] Feibelman, P. J. (1982). Prog. Surf. Sci. **12**, 287.

[44] Felsen, L. and Marcuvitz, N. (1996). *Radiation and Scattering of Waves*. Oxford University Press, New York.

[45] Ford G. W. and Weber, W. H. (1984). Phys. Rep. **113**, 195.

[46] Genet, C., Lambrecht, A., and Reynaud, S. (2000). Phys. Rev. A **62**, 012110.

[47] Gérard, J. M. and Gayral, B. (1999). J. Lightwave Technol. **17**, 2089.

[48] Gifford, D. K. and Hall, D. G. (2002). Appl. Phys. Lett. **81**, 4315.

[49] Giovannini, U. (2013). Phys. Rev. Lett. **111**, 053902.

[50] Greffet, J.-J. and Nieto-Vesperinas, M. (1998) J. Opt. Soc. Am. A **15**, 2735

[51] Greffet, J.-J., Laroche, M., and Marquier, F. (2010). Phys. Rev. Lett. **105**, 117701.

[52] Gurzhi, R. N., Ya Azbel', M., and Lin, H. P. (1963). Sov. Phys. Solid State **5**, 554.

[53] Gustafsson, M. (2005). Proc. Natl Acad. Sci. USA **102**, 13081.

[54] Halevi, P. (1995). Phys. Rev. B **51**, 7497.

[55] Hao, F. and Nordlander, P. (2007). Chem. Phys. Lett. **446**, 115.

[56] Haroche, S. (1992). In: *Fundamental Systems in Quantum Optics. Proceedings of the 1990 Les Houches Summer School, Session LIII*, ed. J. Dalibard. Elsevier, Amsterdam.

[57] Henkel, C., Joulain, K., Mulet, J. P., and Greffet, J.-J. (2004). Phys. Rev. A **69**, 023808.

[58] Henkel, C., Joulain, K., Carminati, R., and Greffet, J.-J. (2000). Opt. Commun. **86**, 57.

[59] Intravaia, F. and Lambrecht, A. (2005). Phys. Rev. Lett. **94**, 110404.

[60] Intravaia, F., Henkel, C., and Lambrecht, A. (2007). Phys. Rev. A **76**, 033820.

[61] Jackson, J. D. (1999). *Classical Electrodynamics*, 3rd edn. Wiley, New York.

[62] Johnson, P. B. and Christy, R. W. (1972). Phys. Rev. B **6**, 4370.

[63] Joulain, K., Carminati, R., Mulet, J. P., and Greffet, J.-J. (2003). Phys. Rev. B **68**, 245405.

[64] Joulain, K., Mulet, J. P., Marquier, F., Carminati, R., and Greffet, J.-J. (2005). Surf. Sci. Rep. **57**, 59.

[65] Jourdan, G., Lambrecht, A., Comin, F., and Chevrier, J. (2009). Europhys. Lett. **85**, 31001.

[66] Jun, Y. C., Kekatpure, R.D., White, J. S., and Brongersma, M. L. (2008). Phys. Rev. B **78**, 153111.

[67] Kaveh, M. and Wiser, N. (1984). Adv. Phys. **33**, 257.

[68] King, R. W. P., Owens, M., and Wu, T. T. (1992). *Lateral Electromagnetic Waves*. Springer-Verlag, New York.

[69] Kittel, C. (2005). *Introduction to Solid State Physics*, 8th edn. Wiley, New York.

[70] Kittel, C. (1987). *Quantum Theory of Solids*, 2nd edn. Wiley, New York.

[71] Kock, A., Gornik, E., Hauser, M., and Beistingl, W. (1990). Appl. Phys. Lett. **57**, 2327.

[72] Kuhn, S., Hakanson, U., Rogobete, L., and Sandoghdar, V. (2006). Phys. Rev. Lett. **97**, 017402.

[73] Lalanne, P. and Hugonin, J. P. (2006). Nat. Phys. **2**, 551.

[74] Lalanne, P., Hugonin, J. P., Liu, H. T., and Wang, B. (2009). Surf. Sci. Rep. **64**, 453.

[75] Lamoreaux, S. K. (1997). Phys. Rev. Lett. **78**, 5.

[76] Landau, L. D., Pitaevskii, L. P., and Lifshitz, E. M. (2004). *Electrodynamics of Continuous Media*. Elsevier, Amsterdam.

[77] Laroche, T. and Girard, C. (2006). Appl. Phys. Lett. **89**, 233119.

[78] Lewis, A., Isaacson, M., Harootunian, A., and Murray, A. (1984). Ultramicroscopy **13**, 227.

[79] Lezec, H. J., Degiron, A., Devaux, E., Linke, R. A., Martin-Moreno, L., Garcia-Vidal, F. J., and Ebbesen, T. W. (2002). Science **297**, 820.

[80] Li, K., Stockman, M. I., and Bergman, D. J. (2003). Phys. Rev. Lett. **91**, 227402.

[81] Liu, H. and Lalanne, P. (2008). Nature **452**, 728.

[82] Madrazo, A., Carminati, R., Nieto-Vesperinas, M., and Greffet, J.-J. (1998). J. Opt. Soc. Am. A **15**, 109.

[83] Mulet, J. P., Joulain, K., Carminati, R., and Greffet, J.-J. (2001). Appl. Phys. Lett. **78**, 2931.

[84] Mulet, J. P., Joulain, K., Carminati, R., and Greffet, J.-J. (2002). Nanoscale Microscale Thermophys. Eng. **6**, 209.

[85] Nieto-Vesperinas, M. (2006). *Scattering and Diffraction in Physical Optics*. World Scientific, New York.

[86] Okamoto, K., Niki, I., Shvartser, A., Narukawa, Y., Mukai, T., and Scherer, A. (2004). Nat Mater **3**, 601.

[87] Ordal, M. A., Bell, R. J., Alexander, R. W., Long, L. L., and Querry, M. R. (1985). Appl. Opt. **24**, 4493.

[88] Palik, E. D. (ed.) (1997). *Handbook of Optical Constants of Solids*. Elsevier, Amsterdam.

[89] Pines, D. (1964). *Elementary Excitations in Solids: Lectures on Phonons, Electrons and Plasmons*. W. A. Benjamin, London.

[90] Powell, C. J. and Swan, J. B. (1959). Phys. Rev. **115**, 869.

[91] Pendry, J. B. (2000). Phys. Rev. Lett. **85**, 3966.

[92] Pohl, D.W., Denk, W., and Lanz, M. (1984). Appl. Phys. Lett. **44**, 651.

[93] Prodan, E., Radloff, C., Halas, N. J., and Nordlander, P. (2003). Science **302**, 419.

[94] Raether, H. (1988). *Surface Plasmons*. Springer-Verlag, Berlin.

[95] Rousseau, E., Siria, A., Jourdan, G., Volz, S., Comin, F., Chevrier, J., and Greffet, J.-J. (2006). Nat. Photonics **3**, 514.

[96] Rust, M., Bates, M., and Zhuang, X. (2006). Nat. Methods **3**, 793.

[97] Sarid, D. (1981). Phys. Rev. Lett. **47**, 1927.

[98] Schuller, J. A., Barnard, E. S., Cai W., Jun, Y. C., White, J. S., and Brongersma, M. L. (2010). Nat. Mater. **9**, 193.

[99] Shchegrov, A. V., Joulain, K., Carminati, R., and Greffet, J.-J. (2000). Phys. Rev. Lett. **85**, 1548.

[100] Shen, S., Narayanaswamy, A., and Chen, G. (2009). Nano Lett. **9**, 2909.

[101] Sentenac, A., Chaumet, P. C., and Belkebir, K. (2006). Phys. Rev. Lett. **97**, 243901.

[102] Sipe, J. E. (1987). J. Opt. Soc. Am. A **4**, 481.

[103] Smith, J. B. and Ehrenreich, H. (1982). Phys. Rev. B **25**, 923.

[104] Smolyaninov, I. I., Davis, C. C., and Zayats, A. V. (2005). New J. Phys. **7**, 175.

[105] Smolyaninov, I. I., Davis, C. C., Elliot, J., and Zayats, A. V. (2007). Phys. Rev. Lett. **98**, 209704.

[106] Stockman, M. I., Faleev, S. V., and Bergman, D. J. (2002). Phys. Rev. Lett. **88**, 067402.

[107] Tai, C. T. (1993). *Dyadic Green Functions in Electromagnetic theory*. IEEE Press, New York.

[108] Taminiau, T. H., Stefani, F. D., Segerink, F. B., and van Hulst, N. F. (2008). Nat. Photonics **2**, 234.

[109] Taubner, T., Korobkin, D., Urzhumov, Y., Shvets, G., and Hillenbrand, R. (2006). Science **313**, 1595.

[110] Thomas, M., Greffet, J.-J., Carminati, R., and Arias-Gonzalez, J. R. (2004). Appl. Phys. Lett. **85**, 3863.

[111] Todorov, Y., Andrews, A. M., Colombelli, R., De Liberato, S., Ciuti, C., Klang, P., Strasser, G., and Sirtori, C. (2010). Phys. Rev. Lett. **105**, 196402.

[112] Toraldo di Francia, G. (1969). J. Opt. Soc. Am. **69**, 799.

[113] Tsai, C.-Y., Tsai, C.-Y., Chen, C.-H., Sung, T.-L., Wu, T.-Y., and Shih, F.-P. (1998). IEEE J. Quantum Electron. **34**, 552.

[114] van Bladel, J. (1991). *Singular Electromagnetic Fields and Sources*. Clarendon Press, Oxford.

[115] Vial, A., Grimault, A. S., Macias, D., Barchiesi, D., an Lamy de la Chapelle, M. (2005). Phys. Rev. B **71**, 085416.

[116] Voisin, C., Del Fatti, N., Christofilos, D., and Vallée, F. (2001). J. Phys. Chem. B **105**, 2264.

[117] Wang J. J. H. (1991). *Generalized Moment Methods in Electromagnetics*. Wiley, New York.

[118] Wang, F. and Shen, Y. R. (2006). Phys. Rev. Lett. **97**, 206806.

[119] Wessel, J. (1985). J. Opt. Soc. Am. B **2**, 1538.

[120] Wijnands, F., Pendry, J. B., Garcia-Vidal, F. J., Bell, P. M., Roberts, P. J., and Martin Moreno, L. (1997). Opt. Quantum Electron. **29**, 199.

[121] Yaghjian, A. (1980). Proc. IEEE **68**, 248.

[122] Yu Nikitin, A., Rodrigo, S. G., Garcia-Vidal, F. J. and Martin-Moreno, L. (2009). New J. Phys. **11**, 123020.

[123] Zayats, A. V., Smolyaninov, I. I., and Maradudin, A. A. (2005). Phys. Rep.**408**, 131.

[124] Ziman, J. M. (1964). *Principles of the Theory of Solids*. Cambridge University Press, Cambridge.

3

Transformation optics

Ulf LEONHARDT

Weizmann Institute of Science
Rehovot, Israel

Ulf Leonhardt: *Transformation Optics*. In: *Quantum Optics and Nanophotonics*. Edited by:
Claude Fabre, Vahid Sandoghdar, Nicolas Treps, and Leticia F. Cugliandolo. Oxford University
Press (2017), pp. 145–168. © Oxford University Press 2017.
DOI: 10.1093/oso/9780198768609.003.0003

Chapter Contents

Transformation optics applies ideas from Einstein's general theory of relativity to optical and electrical engineering with the aim of designing devices that can do the (almost) impossible: invisibility cloaking, perfect imaging, levitation, and the creation of analogues of the event horizon. These lecture notes give an introduction to this field requiring minimal prerequisites.

3.1 Introduction

According to Einstein's general theory of relativity, the geometry of space–time is curved by the momentum and energy of macroscopic objects. This curvature is what we perceive as gravity, because it influences the motion of particles such as Newton's apple falling from a tree in the space–time geometry curved by the Earth or the motion of the planets circling around in the space–time geometry curved by the Sun. Gravity also influences the propagation of waves, the most striking demonstration of which is gravitational lensing, where light from distant stars or galaxies is deflected and focused in the space–time geometry created by other stars or galaxies. Gravity is universal, because the geometry of space and time sets the scene for everything, particle and wave alike.

Analogues of gravity occur when the geometry of space–time appears to be altered by other means than momentum and energy. The most natural example of analogue gravity is the propagation of light in a medium. The medium—a piece of glass, the water in a vase, or any other transparent substance—distorts images in much the same way as stars and galaxies distort light. We may say, with some justification to be given later in these lecture notes, that media apparently alter the geometry of space–time for light. This geometry differs from the natural space–time geometry of gravity: the medium establishes a virtual geometry different from the natural geometry of physical space. The virtual geometry is created by a completely different physical process than the geometry of physical space, and also it is not universal but restricted to certain physical phenomena. In the case of light in a medium, the virtual geometry differs from the real geometry only for light (and, typically, only for light within a certain frequency range).

In these lecture notes, we show how and when virtual geometries arise for light, or indeed for electromagnetic waves in general. For this, we combine ideas from two of the most beautiful theories of physics, namely Maxwell's electromagnetism and Einstein's general relativity, such that they become transformable into each other (Fig. 3.1). This connection between general relativity and electromagnetism in a medium is not new, dating back to ideas of Gordon [11] published in 1923 and of Tamm [43, 44] at about the same time, and further back to Newton, who allegedly toyed with the idea that gravity is mediated by a medium before settling for Newtonian gravity [32], and also to the enigmatic genius of Fermat [25]. What is new is the application of these ideas to electrical and optical engineering, as design concepts for novel devices that do the (almost) impossible, for example invisibility cloaking and perfect imaging. This new research area with old and deep roots, called transformation optics [5, 12, 15, 16, 19, 31], has been regarded as one of the most fascinating research insights of the last

Fig. 3.1 Einstein, 'Einwell', Maxwell. Transformation optics combines ideas from Einstein's general relativity and Maxwell's electromagnetism. In particular, it uses transformations of space—like the transformation between Einstein and Maxwell shown by these pictures.

decade [41]. In these lecture notes, we derive the foundations of this area and explain a few of the key applications. We also show how transformation optics is related to the physics of the event horizon.

3.2 Maxwell's electromagnetism

3.2.1 Maxwell's equations

Let us begin at the beginning, with Maxwell's equations in empty, flat space in Cartesian coordinates:

$$\boldsymbol{\nabla} \cdot \boldsymbol{E} = 0 \qquad \text{GAUSS'S LAW,}$$

$$\boldsymbol{\nabla} \times \boldsymbol{B} = \frac{1}{c^2} \frac{\partial \boldsymbol{E}}{\partial t} \qquad \text{AMPÈRE'S LAW WITH}$$
$$\text{MAXWELL'S DISPLACEMENT CURRENT,} \qquad (3.1)$$

$$\boldsymbol{\nabla} \times \boldsymbol{E} = -\frac{\partial \boldsymbol{B}}{\partial t} \qquad \text{FARADAY'S LAW OF INDUCTION,}$$

$$\boldsymbol{\nabla} \cdot \boldsymbol{B} = 0 \qquad \text{ABSENCE OF MAGNETIC MONOPOLES.}$$

As usual, c denotes the speed of light in vacuum. Throughout these lecture notes, we use SI units for the electromagnetic fields. Now suppose we change the spatial coordinates, for example by using spherical coordinates $\{r, \theta, \phi\}$ instead of the Cartesian $\{x, y, z\}$. The differentials of the new coordinates appear in a different way in the line element ds than the Cartesian differentials, for example as

$$\mathrm{d}s^2 = \mathrm{d}x^2 + \mathrm{d}y^2 + \mathrm{d}z^2 = \mathrm{d}r^2 + r^2\,\mathrm{d}\theta^2 + r^2\sin^2\theta\,\mathrm{d}\phi^2 \qquad (3.2)$$

in spherical coordinates. In general, curved coordinates x^i contribute to the line element as

$$ds^2 = g_{ij}\, dx^i dx^j \,, \tag{3.3}$$

where we sum over repeated indices (running from 1 to 3). The g_{ij} usually depend on the x^i, as the line element (3.2) of the spherical coordinates shows. They constitute the metric tensor with determinant g and matrix inverse g^{ij}:

$$g \equiv \det(g_{ij}) \,, \quad g^{ij} \equiv (g_{ij})^{-1} \,. \tag{3.4}$$

Differential geometry [19] tells us how to express the divergences and curls in Maxwell's equations (3.1) in terms of curved coordinates:

$$\frac{1}{\sqrt{g}}\, \partial_i \sqrt{g}\, g^{ij} E_j = 0 \qquad \text{GAUSS'S LAW,}$$

$$\epsilon^{ijk} \partial_j B_k = \frac{1}{c^2} \frac{\partial g^{ij} E_j}{\partial t} \qquad \begin{array}{l}\text{AMPÈRE'S LAW WITH} \\ \text{MAXWELL'S DISPLACEMENT CURRENT,}\end{array} \tag{3.5}$$

$$\epsilon^{ijk} \partial_j E_k = -\frac{\partial g^{ij} B_j}{\partial t} \qquad \text{FARADAY'S LAW OF INDUCTION,}$$

$$\frac{1}{\sqrt{g}}\, \partial_i \sqrt{g}\, g^{ij} B_j = 0 \qquad \text{ABSENCE OF MAGNETIC MONOPOLES,}$$

where the Levi-Civita tensor ϵ^{ijk}, appearing in the curls, is given in terms of the completely antisymmetric symbols $[ijk]$ as [19]:

$$\epsilon^{ijk} = \pm \frac{1}{\sqrt{g}} [ijk] \,. \tag{3.6}$$

The $\pm$ sign depends on the handedness of the coordinate system: '+' in right-handed systems and '−' in left-handed systems.

Maxwell's equations (3.5) are valid not only in a flat space expressed in curved coordinates, but also in genuine curved spaces.[1] The reason is the following: Maxwell's equations (3.5) are first-order partial differential equations containing maximally first derivatives of the metric tensor g_{ij}. Now, a theorem from differential geometry [19, 39] says that, for any given point, we can always construct a local Cartesian coordinate system where $g_{ij} = \delta_{ij}$ and $\partial_k g_{ij} = 0$ at that point, regardless how curved the geometry is. If the space is curved, these local Cartesian systems do not form a single, global Cartesian frame, but rather represent a patchwork of frames that are not consistent. The inconsistency is caused by the spatial curvature. However, there is always a local coordinate transformation from each local frame to the global frame of the curved

---

[1] We consider purely spatial geometries first and then, in Section 3.5, we generalize our theory to space–time geometries.

manifold. As Maxwell's equations depend maximally on first derivatives of g_{ij}, we can assume them in the form (3.1) in each local Cartesian frame and then transform to the general form (3.5) in the global frame. Therefore, the form (3.5) describes electromagnetism in curved space as well.

3.2.2 The medium of a geometry

Consider now a case similar to gravitational lensing,[2] where we assume a given spatial geometry. In the following, we show how this geometry appears as a medium. For this, we express Maxwell's equations (3.5) for the quantities $\boldsymbol{E}$ (electric field strength), $\boldsymbol{D}$ (dielectric displacement), $\boldsymbol{H}$ (magnetic field), and $\boldsymbol{B}$ (magnetic induction) familiar from macroscopic electromagnetism in a medium:

$$\nabla \cdot \boldsymbol{D} = 0 \qquad \text{Gauss's law,}$$

$$\nabla \times \boldsymbol{H} = \frac{\partial \boldsymbol{D}}{\partial t} \qquad \begin{array}{l} \text{Ampère's law with} \\ \text{Maxwell's displacement current,} \end{array} \qquad (3.7)$$

$$\nabla \times \boldsymbol{E} = -\frac{\partial \boldsymbol{B}}{\partial t} \qquad \text{Faraday's law of induction,}$$

$$\nabla \cdot \boldsymbol{B} = 0 \qquad \text{Absence of magnetic monopoles.}$$

Considering Gauss's law, we can write it as $\nabla \cdot \boldsymbol{D} = \partial_i D^i = 0$, with

$$D^i = \varepsilon_0 \, \varepsilon^{ij} E_j \qquad (3.8)$$

and $\varepsilon^{ij} \propto \sqrt{g}\, g^{ij}$. Here ε_0 denotes the electric permittivity of the vacuum. Let us see whether this definition of the dielectric displacement $\boldsymbol{D}$ is consistent with the other place $\boldsymbol{D}$ occurs in Maxwell's equations (3.7), namely Ampère's law with Maxwell's displacement current:

$$\epsilon^{ijk} \partial_j B_k = \pm \frac{1}{\sqrt{g}} \, [ijk] \, \partial_j B_k = \frac{1}{c^2} \frac{\partial g^{ij} E_j}{\partial t} \,. \qquad (3.9)$$

If we write

$$H_k = \varepsilon_0 c^2 B_k \qquad (3.10)$$

and

$$\varepsilon^{ij} = \pm \sqrt{g}\, g^{ij} \,, \qquad (3.11)$$

we obtain Ampère's law for $\boldsymbol{D}$ given by the definition (3.8). Using the same arguments for the remaining two of Maxwell's equations, we get

$$B^i = \mu_0 \mu^{ij} H_j \,, \qquad (3.12)$$

---

[2] In gravitational lensing, gravity alters primarily the measure of time, but, owing to the conformal invariance of electromagnetism (see Section 3.5.2), this is equivalent to altering the measure of space.

with the magnetic permeability of the vacuum

$$\mu_0 = \frac{1}{\varepsilon_0 c^2} \tag{3.13}$$

and the relative magnetic permeability

$$\mu^{ij} = \pm\sqrt{g}\, g^{ij}\,. \tag{3.14}$$

In the formalism developed here [19], $\boldsymbol{E}$ and $\boldsymbol{H}$ are the fundamental fields, whereas $\boldsymbol{D}$ and $\boldsymbol{B}$ are derived from E_i and H_i by raising the index with g^{ij} and multiplying by $\sqrt{g}$. Mathematically, $\boldsymbol{E}$ and $\boldsymbol{H}$ are one-forms, whereas $\boldsymbol{D}$ and $\boldsymbol{B}$ are vector densities with respect to the spatial geometry [19]. The fact that ε^{ij} and μ^{ij} are matrices that depend on two indices indicates that the medium representing the geometry g_{ij} is anisotropic in general. We also see that a spatial geometry appears as a medium with

$$\varepsilon^{ij} = \mu^{ij}\,. \tag{3.15}$$

In electrical engineering, such media are called impedance-matched (to the vacuum).

3.2.3 The geometry of a medium

The converse is also true: impedance-matched media, i.e. media satisfying the condition (3.15), can be understood as spatial geometries, from the following argument. We calculate the determinant $\det\varepsilon$ of the matrix ε^{ij} from the relation (3.11) and obtain

$$\det\varepsilon = \pm\sqrt{g}\,. \tag{3.16}$$

Consequently, we can write g^{ij} in terms of ε^{ij} and, by virtue of impedance matching (3.15), of μ^{ij} as well, as

$$g^{ij} = \frac{\varepsilon^{ij}}{\det\varepsilon} = \frac{\mu^{ij}}{\det\mu}\,. \tag{3.17}$$

Spatial geometries appear as impedance-matched media, and impedance-matched media make virtual geometries. Impedance matching establishes an exact virtual geometry where electromagnetic fields are identical in all aspects to such fields in a real geometry. Without impedance matching, the geometric picture is not exact. In particular, the propagation of light in non-impedance-matched anisotropic media depends on the polarization, a phenomenon called birefringence. However, non-impedance-matched media can still be used for establishing geometries in planar media for specific polarizations (see e.g. the experiments described in [6, 22, 23, 38, 47]), because in such cases not all components of the ε^{ij} and μ^{ij} tensors are needed. Furthermore, in optically isotropic media (where $\varepsilon^{ij} = \varepsilon\,\delta_{ij}$ and $\mu^{ij} = \mu\,\delta_{ij}$), the geometric picture gives an excellent approximation for electromagnetic waves within the validity range of geometrical optics [19].

3.3   Spatial transformations

In general, the virtual geometry of light in media is curved. To give a simple example, a lens focuses parallel light rays in the focal point; parallels are thus no longer parallel, but meet, which violates Euclid's parallel axiom of flat space. Curved space is commonplace in optics. It is much harder to create a virtual geometry that is flat. What would it do? As any flat geometry can be reduced to Cartesian coordinates by a coordinate transformation, the material would just perform a transformation of space where each point of physical space appears to be at a position in virtual space that may deviate from the real one. If the new coordinates agree with the old ones outside a device that is made of the medium, then we would not see the difference between propagation in the medium and propagation in empty, flat space. In short, the device would be completely invisible.

3.3.1   Invisibility cloaking

Such invisible devices could be used to make other things invisible, too: they can be turned into invisibility devices as follows. Suppose the device performs the following transformation (Fig. 3.2): an extended region in physical space (Fig. 3.2(b)) is condensed into a single point in virtual space (Fig. 3.2(a)). Anything inside this region has thus become as small as a single point, invisibly small. Everything inside is hidden and, as the device itself is invisible, the very act of hiding is hidden as well. The transformation in Fig. 3.2 makes a perfect cloaking device [31].

Such a cloaking device has been demonstrated for microwaves [38] or, to be more precise, an approximation without impedance matching was made [38]. However, this device only operated for one polarization, because it used a medium that was not impedance-matched. Moreover, it could only operate for one frequency only, for a fundamental reason [16] that applies to all purely transformation-based cloaking devices [31]. This reason is easy to understand: consider a light ray that just straddles the invisible point in virtual space. As the device performs a spatial transformation from virtual to physical space, the light propagation in both spaces is synchronized. Therefore, the light must go around the invisible region in precisely the time it takes to pass a single point in virtual space: zero time. Consequently, the speed of light must tend to infinity at the inner surface of the cloaking device [4, 16]. By speed of light we mean the phase velocity here, whereas the group velocity turns out to approach zero [4]. Relativistic causality does not prohibit an infinite phase velocity in a medium, but it allows it only for a single frequency—a single colour—and thus in a purely stationary regime where nothing changes and no new information is transferred. Any change would cause distortions, which defeats the point of a cloaking device;[3] one might as well use a hologram of the background.

Perfect invisibility is impossible, but this does not prevent invisibility that is good enough. One might be content with deforming a surface by conformal [15] or quasi-conformal transformations [13], which do not make an object disappear altogether,

---

[3] It might be reassuring to know that perfect deception is impossible; the truth will always appear in the end.

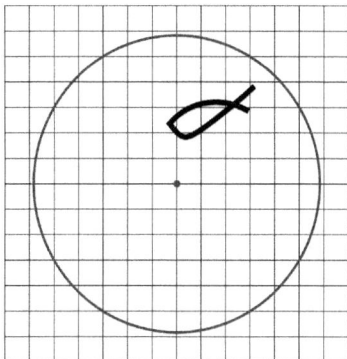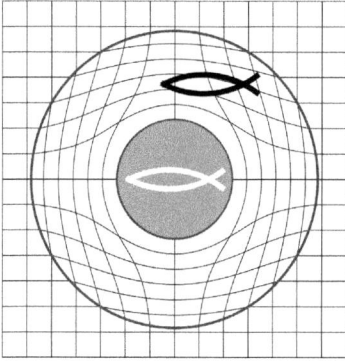

Fig. 3.2 Transformation of space. Optical materials appear to change the perception of space; objects (fish) in physical space (right) appear at positions (left) different from where they actually are. Suppose the medium performs a coordinate transformation from physical space (right) to virtual space (left) and vice versa. Virtual space is empty and so light propagates along straight lines. In virtual space, we may draw a coordinate system as a rectangular grid of light rays (the grid in (left)). In physical space, the light rays are curved; the coordinate grid of virtual space is transformed into a curved coordinate system in physical space (the grid in (right)). As the coordinate transformation changes space only within a circle, this circle marks the boundary of the optical material used to transform space. We see that the image of the black fish is distorted in virtual space, because the coordinate transformation illustrated here is not uniform. Moreover, the white fish has completely disappeared, because it was swimming within a region of physical space (grey) that, in virtual space, is contracted to a single, invisible point. Such an optical material makes an invisibility device.

but make it optically flat; the fugu in Fig. 3.3 is turned into a flatfish. One could then use conventional camouflage (which flatfish are masters of) to disguise the flat object. Or one might use non-Euclidean cloaking devices where virtual space is not flat but curved in an appropriate way [15, 17, 33]. In this case, the speed of light is finite in the device (and can be made slower than c [33]), but the price to be paid is a time delay of the light making the detour in the cloaking device. Furthermore, relativistic causality is of little concern to the cloaking of sound waves, because the normal speed of sound is several orders of magnitude slower than c. Near-perfect acoustic cloaking over a broad band of frequencies is possible and has been demonstrated [48].

3.3.2 Transformation media

What does it take to build a cloaking device and similar transformation devices? Let us work out the electromagnetic properties of devices required for transforming space. We use x^i to denote the coordinates of physical space and $x^{i'}$ for virtual space (the prime at the index does not mean that we simply use a different index variable but shall indicate the different coordinates of virtual space). As the device performs a transformation from physical to virtual space and vice versa, the $x^{i'}$ of virtual space

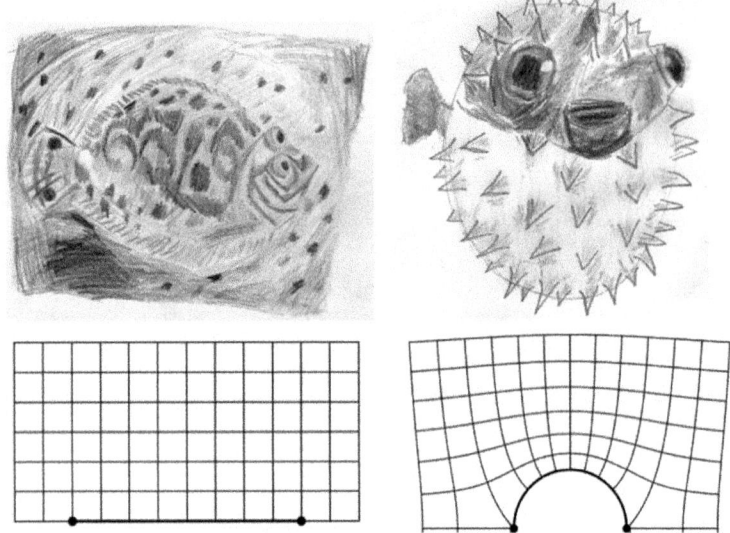

Fig. 3.3 From fugu to flatfish. A coordinate transformation may turn a voluminous object in physical space (fugu) into a flat object (flatfish). (Drawing: Maria Leonhardt.)

are thus functions $x^{i'}(x^i)$ of the physical-space coordinates x^i—the device performs a coordinate transformation. Suppose that virtual space is flat and empty, and that we describe it in Cartesian coordinates. In this case we obtain for the line element

$$\mathrm{d}s^2 = \delta_{i'j'}\,\mathrm{d}x^{i'}\,\mathrm{d}x^{j'} = \delta_{i'j'}\,\frac{\partial x^{i'}}{\partial x^i}\,\frac{\partial x^{j'}}{\partial x^j}\,\mathrm{d}x^i\mathrm{d}x^j\,,\tag{3.18}$$

and thus, according to the definition (3.3),

$$g_{ij} = \Lambda^{i'}{}_i\,\delta_{i'j'}\,\Lambda^{j'}{}_j\,,\qquad\text{with}\quad \Lambda^{i'}{}_i \equiv \frac{\partial x^{i'}}{\partial x^i}\,.\tag{3.19}$$

From this expression we get for the matrix inverse of g_{ij}:

$$g^{ij} = \Lambda^i{}_{i'}\,\delta^{i'j'}\,\Lambda^j{}_{j'}\,,\qquad\text{with}\quad \Lambda^i{}_{i'} \equiv \frac{\partial x^i}{\partial x^{i'}}\,,\tag{3.20}$$

or, in matrix notation,

$$\boldsymbol{g}^{-1} = \boldsymbol{\Lambda}\boldsymbol{\Lambda}^{\mathrm{T}}\,,\tag{3.21}$$

where $\boldsymbol{g}$ denotes the matrix g_{ij} and

$$\boldsymbol{\Lambda} \equiv \left(\frac{\partial x^i}{\partial x^{i'}}\right)\,.\tag{3.22}$$

Now we can calculate the matrices $\boldsymbol{\varepsilon}$ and $\boldsymbol{\mu}$ of the ε^{ij} and μ^{ij} according to the recipes (3.11) and (3.14). There we need the determinant g of g_{ij}, which is $g = (\det \boldsymbol{\Lambda})^{-2}$ according to the formula (3.21). We thus obtain

$$\boldsymbol{\varepsilon} = \boldsymbol{\mu} = \frac{\boldsymbol{\Lambda}\boldsymbol{\Lambda}^{\mathrm{T}}}{\det \boldsymbol{\Lambda}} \,. \tag{3.23}$$

Equation (3.23) formulates a simple recipe for calculating the required electromagnetic properties of a spatial transformation device. This recipe is valid in Cartesian coordinates, where both virtual and physical space are described by Cartesian grids, but it can also be easily extended to curved coordinates [16, 19].

3.3.3 Perfect imaging with negative refraction

Apart from invisibility cloaking, another prominent application of transformation optics is perfect imaging [16]. Ordinary imaging systems made of ordinary lenses and mirrors suffer from a fundamental limit to their resolution known as the diffraction limit [3]. Even with the strongest microscope, we cannot see atoms, molecules, or nanostructures; for this, we need electron or atomic-force microscopy. The wavelength of light sets the limit. However, there are ways around the diffraction limit, one of which we discuss in this section, namely perfect imaging with negative refraction, since it turns out [16] to be an application of transformation optics.

Imagine a device that performs the following transformation in Cartesian coordinates:

$$x = x(x'), \quad y = y', \quad z = z', \tag{3.24}$$

where $x(x')$ is folded as shown in Fig. 3.4. We see in Fig. 3.4 that in the fold of the function $x(x')$, each point x' in virtual space has three faithful images in physical space, so the electromagnetic field at these three points is the same as that at the point x'. Electromagnetic fields at each of the three points in physical space are therefore perfectly imaged at the other two: the device is a perfect lens. The transformation (3.24) has the transformation matrix $\boldsymbol{\Lambda} = \mathrm{diag}(\mathrm{d}x/\mathrm{d}x', 1, 1)$ and we find from the recipe (3.23) that

$$\boldsymbol{\varepsilon} = \boldsymbol{\mu} = \mathrm{diag}\left(\frac{\mathrm{d}x}{\mathrm{d}x'}, \frac{\mathrm{d}x'}{\mathrm{d}x}, \frac{\mathrm{d}x'}{\mathrm{d}x}\right). \tag{3.25}$$

Inside the device, i.e. inside the fold in the transformation of Fig. 3.4, the derivative $\mathrm{d}x'/\mathrm{d}x$ becomes negative and the coordinate system changes handedness. The electromagnetic left-handedness of such a material appears through a transformation to a left-handed coordinate system; the material is called a left-handed material and also a material with negative refraction. When the negative slope in the transformation is $\mathrm{d}x'/\mathrm{d}x = -1$, (3.25) gives a perfect lens made of isotropic material with $\varepsilon = \mu = -1$ (otherwise the material (3.25) is anisotropic). As Fig. 3.4 shows, the imaging range is equal to the thickness of the lens in this case.

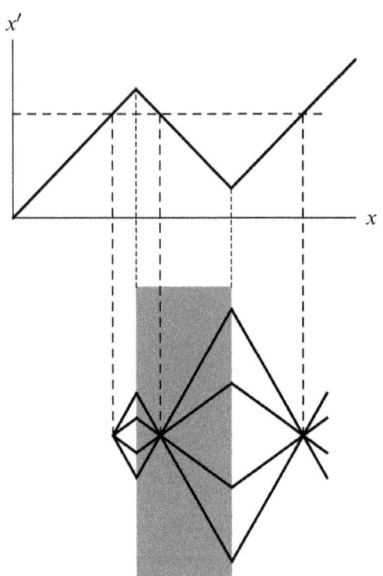

Fig. 3.4 Perfect lens. Negatively refracting perfect lenses employ transformation media. (top) A suitable coordinate transformation from the physical x-axis to the electromagnetic x'. (bottom) The corresponding device. The inverse transformation from x' to x is either triple- or single-valued. The triple-valued segment on the physical x-axis corresponds to the focal region of the lens: any source point has two images, one inside the lens and one on the other side. Since the device facilitates an exact coordinate transformation, the images are perfect.

Perfect lensing was first analysed through the imaging of evanescent waves in a slab of negatively refracting material [30]. These are waves that may carry images finer than the optical resolution limit. Various aspects of this idea have been subject to a considerable theoretical debate [29], but experiments have confirmed negative refraction (see e.g. [46]). Sub-resolution imaging was observed for a 'poor-man's perfect lens' [9] where the lens is effectively implemented by a sub-wavelength sheet of silver. Our pictorial argument leads to a simple intuitive explanation of why such lenses are indeed perfect. It also reveals some of the practical limitations of perfect imaging by negative refraction.

We have seen that if the imaging device performs the spatial transformation (3.24) illustrated in Fig. 3.4, the electromagnetic field is identical in three separate regions of space. How is this possible? The electromagnetic field cannot instantly hop from one region to another—this is forbidden by relativistic causality, but it can settle to identical field structures over time in a stationary regime. But this implies that the stationary response of the electromagnetic material must be very different from the instantaneous response: the material must be dispersive. Dispersion is always accompanied by dissipation, and dissipation turns out [42] to severely reduce the resolution of imaging by negative refraction in left-handed materials. Therefore only 'poor man's perfect lenses' have worked [9], where the imaging distance is a mere fraction of the wavelength. Nevertheless, the tantalizing ideas of negative refraction have inspired the entire research area of metamaterials and transformation optics [41].

3.4   Curved space

Spatial transformations are simple and intuitive in theory, but often difficult to implement in practice. In particular, for the most interesting transformations such as invisibility cloaking and perfect imaging, fundamental problems prevent their practical realization in a meaningful way, as we have seen. The alternative to spatial transformations is the implementation of a curved virtual space, which is the standard case in isotropic media anyway (even non-impedance-matched isotopic media appear to electromagnetic waves as geometries, as long as the approximation of geometrical optics is valid [19]). The theory of light in curved space is harder, but the experiments are much easier than in implementations of flat space. In some cases, curved spaces have extraordinary optical properties that makes their practical use in optical devices highly desirable. Let us discuss such a case that also represents the simplest curved space in three dimensions: the surface of the four-dimensional hypersphere. There we can use a lot of the intuition we have about the surface of the ordinary three-dimensional sphere to predict interesting physical phenomena without calculation, just by drawing pictures.

3.4.1   Einstein's universe and Maxwell's fish eye

Our case is closely related to a famous cosmological model due to Einstein [8] that happens to be wrong for the universe, but, when turned into an electromagnetic device, may be very useful in down-to-Earth applications. Einstein assumed that the universe is static and for this case derived an exact solution of the equations of general relativity that describe how matter curves space–time. Astronomical observations have shown, however, that the universe is not static but expanding, but this should not deter us from turning Einstein's solution into a practical device. In Einstein's static universe, light propagates as if it were confined to the three-dimensional surface of a four-dimensional hypersphere in $\{X, Y, Z, W\}$ space with

$$X^2 + Y^2 + Z^2 + W^2 = a^2 \,, \tag{3.26}$$

where the constant a describes the radius of the hypersphere. Suppose that the hypersurface (3.26) is the virtual space of an optical device. Furthermore, the Cartesian coordinates $\{x, y, z\}$ of physical space will be connected to the virtual hypersphere by stereographic projection (Fig. 3.5):

$$x = \frac{X}{1 - W/a} \,, \qquad y = \frac{Y}{1 - W/a} \,, \qquad z = \frac{Z}{1 - W/a} \,. \tag{3.27}$$

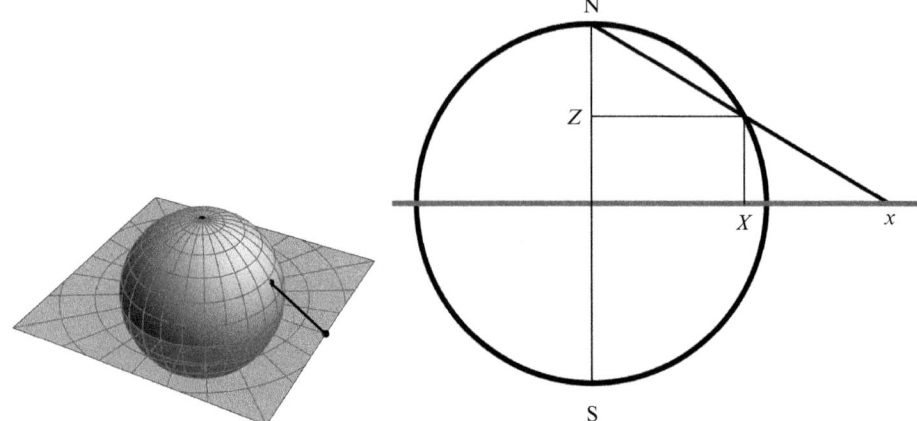

Fig. 3.5 Stereographic projection. Points on the sphere (or hypersphere) are projected to the plane (or hyperplane) as follows. A line is drawn through the north pole N and the point on the sphere. Where this line intersects the plane cut through the equator, there lies the projected point. The picture shows a cut through the sphere and plane where a point on the sphere is characterized by the coordinates $\{X, Z\}$ and the point on the plane by x.

One easily verifies that the virtual-space coordinates are given by the following inverse stereographic projection:

$$X = \frac{2x}{1 + r^2/a^2}, \quad Y = \frac{2y}{1 + r^2/a^2}, \quad Z = \frac{2z}{1 + r^2/a^2}, \quad W = a\,\frac{r^2 - a^2}{r^2 + a^2}, \qquad (3.28)$$

where r denotes the radius in physical space, with

$$r^2 = x^2 + y^2 + z^2. \qquad (3.29)$$

In order to deduce the effective geometry in physical space and hence the medium required to implement it, we express the line element in virtual space in terms of the differentials in physical space with the help of the inverse stereographic projection (3.28). We obtain

$$ds^2 = dX^2 + dY^2 + dZ^2 + dW^2 = n^2 \left(dx^2 + dy^2 + dz^2\right), \qquad (3.30)$$

with the radius-dependent prefactor

$$n = \frac{2}{1 + r^2/a^2}. \qquad (3.31)$$

From the line element (3.30), we read off the metric tensor g_{ij}, its determinant g, and its matrix inverse g^{ij} as

$$g_{ij} = n^2 \delta_{ij}, \qquad g = n^6, \qquad g^{ij} = n^{-2}\delta_{ij}. \qquad (3.32)$$

According to the relations (3.11) and (3.14), this spatial geometry corresponds to a medium with

$$\varepsilon = \mu = n\mathbb{1} \,. \tag{3.33}$$

As ε and μ are proportional to the unit matrix $\mathbb{1}$, the medium is optically isotropic, and it has the refractive-index profile (3.31). Isotropic media are usually the easiest to implement, which is the reason why we have chosen the stereographic projection (3.27) and not any other mapping from virtual to physical space.

The index profile (3.31) of Einstein's static universe [8] was written down by Maxwell as a student at Trinity College, Cambridge [26]. Maxwell was not aware of its relation to the stereographic projection of a sphere—that was discovered in optics by Luneburg [21] much later—Maxwell was simply fascinated by the extraordinary optical properties of a device with the profile (3.31). It reminded him of the eye of a fish and therefore such a device is called Maxwell's fish eye.

3.4.2 Perfect imaging with positive refraction

Maxwell found two properties of the fish eye particularly fascinating: (1) light goes in circles and (2) all light rays from any point meet at a corresponding image point. These properties turn out to be simple mathematical consequences of the light propagation on the virtual hypersphere and the stereographic projection to physical space. Let us, instead of the four-dimensional hypersphere, imagine an ordinary three-dimensional sphere—the hypersphere is not greatly different (Fig. 3.6). The light rays on the sphere are the geodesics, the great circles. Now, the stereographic projection always maps circles on the sphere to circles in physical space [19] (some degenerate into lines, i.e.

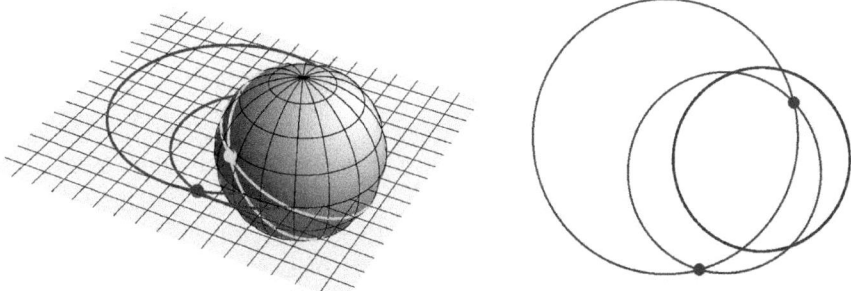

Fig. 3.6 Light propagation in Einstein's universe and Maxwell's fish eye. The virtual space (left) is a sphere or hypersphere—Einstein's universe. Light follows the geodesics, here the great circles. All light rays emitted from one point must come together again at the antipodal point. The physical space (right) is Maxwell's fish eye with the refractive-index profile (3.31). As the stereographic projection (Fig. 3.5) maps circles into circles, light goes in circles in physical space. Moreover, all light rays emitted at an arbitrary given point must focus at a corresponding image point. Maxwell's fish eye makes a perfect imaging device.

circles with infinite radius). From this follows property (1) of Maxwell's fish eye—light goes in circles. Property (2) is the easiest to understand. Consider the great circles of light rays emitted from a point P on the virtual sphere. All great circles from a given point P must intersect at the antipodal point. To see this, just rotate P to the north pole of the sphere. In this case, the great circles are the lines of longitude, and they all meet at the two poles. Therefore, the geodesics from the north pole intersect at the south pole. If we rotate the point P back to its original position, the rotated south pole turns into the antipodal point. Now, as the stereographic projection maps the surface of the hypersphere to physical space, the trajectories of light rays are simply the projections of the great circles. Therefore, they must all meet in physical space as well, at the stereographic projection of the antipodal point. All light rays emitted from any point in Maxwell's fish eye meet at the corresponding image point. Devices where all light rays from any point within an object region intersect at the points of the image region are called absolute optical instruments [3].

Maxwell's fish eye is an absolute optical instrument and it has other curious optical properties, too—for example, light goes in circles—but it has never been built in its original form (3.31), for two good reasons that are connected to each other. First, the profile of Maxwell's fish eye fills the entire physical space. Second, for $r \to \infty$, the refractive index (3.31) tends to zero, i.e. the speed of light becomes infinite at infinity. The two reasons are connected because Maxwell's fish eye represents the geometry of a finite space, the virtual hypersphere, in an infinitely extended space. There, the speed of light must go to infinity to keep propagation times finite. However, there is a remedy [18, 20] that solves both problems in one stroke. Imagine we place a mirror around the equator of the virtual sphere. The mirror would create the illusion that the light propagates in the entire virtual sphere, whereas in reality it is confined to one of the hemispheres, say the southern hemisphere. In physical space, the southern hemisphere corresponds to the region with $r \leq a$, where the refractive index ranges only from 1 at $r = a$ to 2 in the centre and the device is now finite. Therefore, Maxwell's fish eye with a mirror can be built, and this has recently been done [10, 23] in two dimensions. In three-dimensional space, the mirror should be a spherical shell at $r = a$ that encloses the index profile (3.31) of the fish eye.

As far as ray optics is concerned, absolute optical instruments like Maxwell's fish eye create perfect images, because all light rays from all object points faithfully arrive at the corresponding image points. However, the resolution of optical instruments is normally restricted by the wave nature of light [3] and cannot be made much finer than the wavelength. Is perfect imaging possible with Maxwell's fish eye? It is wise to consider this problem in virtual space, on the sphere (representing the virtual hypersphere for the three-dimensional fish eye). Any source can be regarded as a collection of point sources, so it suffices to investigate the wave produced by a single point source of arbitrary position on the sphere. A wave propagates from the point of emission around the sphere and focuses at the antipodal point; this corresponds to emission from a point in the plane of the actual device and focusing at the image point in physical space. The wave propagating around the virtual sphere would come to the antipodal point and focus there. Because of the symmetry of the sphere, the initially outgoing wave from the source points turns into an ingoing wave at the image point, i.e. a time-reversed outgoing wave. However, the time reversal is only complete if one essential element is present at the image point: a reversed source, i.e. a drain. The drain at the image point is something natural in imaging, where one wishes to detect an image, for example by photochemical reactions or in a CCD array. The drain represents a detector. Without the detector, the image is not infinitely sharp, but is limited by the wavelength. The perfect image may appear—but only if one looks.

The crucial point concerning perfect imaging is that, given a choice of detectors in the image area, the light localizes at the correct ones. The fact that the correct light localization naturally happens in Maxwell's fish eye is also understandable if we imagine the absorption in an array of detectors as the time reverse of the emission by a collection of point sources. Given perfect time symmetry, the light must settle down at the image points that correspond to the actual source points and avoid those corresponding to potential source points that did not emit. In this way, a sharp image is formed, with a resolution given by the cross section of the detectors and not by the wave nature of light. The cross section can be made small for non-resonant detectors. Perfect imaging is possible with positive refraction [18, 20]. This idea has stirred up controversy [45], but computer simulations [27] and microwave experiments [28] indicate that it does indeed work.

3.5   Space–time media

So far, we have considered only spatial geometries for light—we have shown that they appear as impedance–matched media and that impedance-matched media appear as spatial geometries. We have discussed applications of spatial transformations and curved virtual space such as invisibility cloaking and perfect imaging. Let us now include time and extend our theory to space–time geometries.

3.5.1   Space–time geometries

We distinguish space–time coordinates x^α by Greek indices running from 0 to 3, where the 0th coordinate refers to time, whereas purely spatial coordinates are indicated by Latin indices. For example, $x^\alpha = \{ct, x, y, z\}$ are the Galilean coordinates of Minkowski space–time. The lines in space–time are worldlines—they describe the trajectories of particles in space and time. The line element $\mathrm{d}s/c$, with

$$\mathrm{d}s^2 = -c^2\,\mathrm{d}t^2 + \mathrm{d}x^2 + \mathrm{d}y^2 + \mathrm{d}z^2 \tag{3.34}$$

in Minkowski space–time, measures the proper time experienced by a particle on its way. The line element $\mathrm{d}s$ characterizes the space–time geometry. The square of $\mathrm{d}s$ is a quadratic form of the increments of the space–time coordinates,

$$\mathrm{d}s^2 = g_{\alpha\beta}\,\mathrm{d}x^\alpha \mathrm{d}x^\beta\,. \tag{3.35}$$

As in the case of spatial geometries, we denote the determinant of $g_{\alpha\beta}$ by g and the matrix inverse by $g^{\alpha\beta}$; note that g is usually negative. For example, in Minkowski space–time, $g_{\alpha\beta} = \eta_{\alpha\beta}$, with

$$\eta_{\alpha\beta} = \mathrm{diag}(-1,\,1,\,1,\,1) = \eta^{\alpha\beta} \tag{3.36}$$

and $g^{\alpha\beta} = \eta^{\alpha\beta}$ and $g = -1$. We know that purely spatial geometries act as impedance-matched media on electromagnetic fields; how do genuine space–time geometries appear?

3.5.2 Magneto-electric media

Plebanski [35] deduced a description of electromagnetism in space–time geometries that closely resembles the familiar form of constitutive equations. Here we will simply state Plebanski's result; readers interested in its derivation are referred to the Appendices of [16] and [19]. Plebanski's constitutive equations are

$$\boldsymbol{D} = \varepsilon_0 \boldsymbol{\varepsilon} \boldsymbol{E} + \frac{\boldsymbol{w}}{c} \times \boldsymbol{H}\,, \qquad \boldsymbol{B} = \mu_0 \boldsymbol{\mu} \boldsymbol{H} - \frac{\boldsymbol{w}}{c} \times \boldsymbol{E}\,, \tag{3.37}$$

with the electromagnetic properties

$$\varepsilon^{ij} = \mu^{ij} = \mp \frac{\sqrt{-g}}{g_{00}} g^{ij}\,, \qquad w_i = \frac{g_{0i}}{g_{00}}\,. \tag{3.38}$$

The constitutive equations (3.37) and (3.38) show that space–time geometries with $g_{0i} \neq 0$ mix electric and magnetic fields; they appear as magneto-electric media (also known as bi-anisotropic media [40]). The magneto-electric coupling vector $\boldsymbol{w}$ has the physical dimensions of a velocity. We will show in the next subsection that $\boldsymbol{w}$ is related to the local velocity of a moving medium. In a moving medium, the material responds to the electric and magnetic fields in locally comoving frames where the medium appears to be at rest. There it is described by the constitutive equations (3.8) and (3.12). The Lorentz transformations to such locally comoving frames mix electric and magnetic fields, which gives Plebanski's constitutive equations.

The dielectric tensors $\boldsymbol{\varepsilon}$ and $\boldsymbol{\mu}$ of (3.38) closely resemble the tensors (3.11) and (3.14) of purely spatial geometries, except that g and g_{00} are negative. Furthermore, only the ratio of $\sqrt{-g}\,g^{ij}$ and g_{00} matters to electromagnetic fields, and so does the ratio of g_{0i} and g_{00} in the magneto-electric coupling, which reflects an important property of light (and electromagnetic radiation in general) known as conformal invariance. This originates from the fact that the proper time of a light ray is zero—light does not experience time. Therefore, light does not recognize the magnitude of the line element (3.35), only the relative contributions of the increments $\mathrm{d}x^{\alpha}$ to $\mathrm{d}s^2$. This implies that we can multiply the line element (3.35) by an arbitrary non-vanishing function of the space–time coordinates without any effect on the propagation of light. Thus, suppose we make the following change:

$$g_{\alpha\beta} \to \Omega^2 g_{\alpha\beta}\,. \tag{3.39}$$

As $g^{\alpha\beta} \to \Omega^{-2} g^{\alpha\beta}$ and $g \to \Omega^8 g$, Plebanski's constitutive equations (3.38) are invariant; hence electromagnetism is conformally invariant.

3.5.3 Moving media

Let us discuss an instructive example of a medium that corresponds to a space–time geometry, namely a moving isotropic impedance-matched medium. We denote the local velocities of the medium by $\boldsymbol{u}$, where $\boldsymbol{u}$ may vary. For a given space–time point x^α, we can always erect a locally comoving frame in which $\boldsymbol{u}$ vanishes at that point (where the medium is locally at rest). As the medium is impedance-matched, we can describe it in the locally comoving frame by a space–time geometry with

$$g_{\alpha\beta} = \mathrm{diag}\left(-1,\, n^2,\, n^2,\, n^2\right). \tag{3.40}$$

In view of the conformal invariance of electromagnetism, we can replace this $g_{\alpha\beta}$ by

$$g_{\alpha\beta} = \mathrm{diag}\left(-n^{-2},\, 1,\, 1,\, 1\right). \tag{3.41}$$

Using the definition (3.36) of the metric tensor of Minkowski space–time and introducing

$$u_\alpha = (-1, 0, 0, 0), \tag{3.42}$$

we can write $g_{\alpha\beta}$ as

$$g_{\alpha\beta} = \eta_{\alpha\beta} + \left(1 - n^{-2}\right) u_\alpha u_\beta. \tag{3.43}$$

Now, the Lorentz transformation from the locally comoving frame back to the laboratory frame maintains the space–time geometry (3.34) of Minkowski space and hence leaves $\eta_{\alpha\beta}$ invariant. Additionally, we write u_α in terms of quantities with a simple geometrical meaning in space–time:

$$u_\alpha = \eta_{\alpha\beta}\, u^\beta, \qquad u^\alpha = \frac{\mathrm{d}x^\alpha}{\mathrm{d}s}, \tag{3.44}$$

where $\mathrm{d}s$ refers to the Minkowski line element (3.34). The u^α form the local four-velocity of the medium. In the comoving frame, $x^\alpha = (ct, 0, 0, 0)$ and $\mathrm{d}s = c\,\mathrm{d}t$, and so the four-velocity agrees with the expression (3.42). As $\mathrm{d}s$ is a Lorentz invariant, the four-velocity behaves like x^α under a Lorentz transformations to the laboratory frame. Applying the formula (3.35) and $\boldsymbol{u} = \mathrm{d}\boldsymbol{x}/\mathrm{d}t$, we obtain the explicit expressions

$$u^\alpha = \frac{(1, \boldsymbol{u}/c)}{\sqrt{1 - u^2/c^2}}, \qquad u_\alpha = \frac{(-1, \boldsymbol{u}/c)}{\sqrt{1 - u^2/c^2}}. \tag{3.45}$$

In this way, we can easily express the tensor (3.43) in the laboratory frame. This $g_{\alpha\beta}$ describes the space–time geometry established for light by the moving medium. It was discovered by Gordon [11] in 1923 and independently rediscovered several times [14, 36, 37]. The matrix inverse $g^{\alpha\beta}$ of $g_{\alpha\beta}$ is

$$g^{\alpha\beta} = \eta^{\alpha\beta} + \left(1 - n^2\right) u^\alpha u^\beta, \tag{3.46}$$

as one easily verifies by calculating the matrix product $g_{\alpha\gamma}g^{\gamma\beta}$, which gives δ_α^β, the unit matrix. With these expressions, we can calculate the dielectric properties of moving media. They are particularly instructive in the limit of low velocities. In this case, we obtain from Plebanski's constitutive equations (3.38):

$$\varepsilon = \mu \approx n\,\mathbb{1}\,, \qquad w \approx (n^2 - 1)u\,.\tag{3.47}$$

We see that the magneto-electric coupling vector w is proportional to the velocity. The proportionally factor is the susceptibility $n^2 - 1$, which vanishes in empty space, where $n = 1$. In the case of large velocities, w and also $\varepsilon = \mu$ depend in a more complicated way on the velocity of the moving medium [19].

3.5.4 Space–time transformations

Suppose the medium moves in one direction only, say the z direction (but possibly with varying velocity) and that n may also only vary in z. We will show that the light propagation in z direction is equivalent to a transformation in space and time; the one-dimensionally moving medium appears as a space–time transformation medium [16]. Our starting point is Gordon's space–time geometry (3.46). If u has only a z-component u and all other components vanish, we have

$$g^{\alpha\beta} = \begin{pmatrix} \dfrac{u^2 - c^2 n^2}{c^2 - u^2} & 0 & 0 & \dfrac{(1 - n^2)cu}{c^2 - u^2} \\ 0 & 1 & 0 & 0 \\ 0 & 0 & 1 & 0 \\ \dfrac{(1 - n^2)cu}{c^2 - u^2} & 0 & 0 & \dfrac{c^2 - n^2 u^2}{c^2 - u^2} \end{pmatrix}.\tag{3.48}$$

We introduce the new coordinates t' and z' defined by

$$t' \mp \frac{z'}{c} = t - \int \frac{\mathrm{d}z}{v_\pm}\,.\tag{3.49}$$

Here $v_\pm$ denote the relativistic addition of the velocity of light in the medium in the positive and negative directions, $\pm c/n$, and the velocity of the medium, u:

$$v_\pm = \frac{u \pm c/n}{1 \pm u/(cn)}\,.\tag{3.50}$$

Gordon's $g^{\alpha\beta}$ tensor appears in the new coordinates as

$$g^{\alpha'\beta'} = \Lambda^{\alpha'}{}_\alpha\, g^{\alpha\beta}\, \Lambda^{\beta'}{}_\beta\tag{3.51}$$

with the transformation matrix

$$\Lambda^{\alpha'}{}_\alpha = \begin{pmatrix} 1 & 0 & 0 & \dfrac{(n^2 - 1)\,c\,u}{c^2 - n^2 u^2} \\ 0 & 1 & 0 & 0 \\ 0 & 0 & 1 & 0 \\ 0 & 0 & 0 & \dfrac{n(c^2 - u^2)}{c^2 - n^2 u^2} \end{pmatrix}.\tag{3.52}$$

The result is the diagonal matrix

$$g^{\alpha'\beta'} = \text{diag}\left(-\frac{n^2(c^2 - u^2)}{c^2 - n^2u^2}, 1, 1, \frac{n^2(c^2 - u^2)}{c^2 - n^2u^2}\right), \tag{3.53}$$

with the inverse

$$g_{\alpha'\beta'} = \text{diag}\left(-\frac{c^2 - n^2u^2}{n^2(c^2 - u^2)}, 1, 1, \frac{c^2 - n^2u^2}{n^2(c^2 - u^2)}\right). \tag{3.54}$$

The determinant of the metric is

$$g' = -\frac{(c^2 - n^2u^2)^2}{n^4(c^2 - u^2)^2}. \tag{3.55}$$

The metric $g_{\alpha'\beta'}$ describes the geometry in virtual space–time. To find out how this geometry appears as a medium, we use Plebanski's constitutive equations (3.38) in virtual space–time, with primed instead of unprimed tensors. Since $g_{\alpha'\beta'}$ is diagonal, the magneto-electric coupling vector $\boldsymbol{w}'$ vanishes: in virtual space–time, the medium is at rest. For the dielectric tensors, we obtain

$$\boldsymbol{\varepsilon}' = \boldsymbol{\mu}' = \text{diag}\left(1, 1, \varepsilon'_{zz}\right), \tag{3.56}$$

with some ε'_{zz} that we do not need to specify here. Since electromagnetic waves propagating in the z-direction are polarized in the (x, y) plane, their electromagnetic fields only experience the x- and y-components of the dielectric tensors. Consequently, for one-dimensional wave propagation, virtual space–time is empty—waves are free here. In virtual space–time, left- and right-moving wavepackets are functions of either $l' + z'/c$ or $l' - z'/c$, in physical space–time, they are modulated wavepackets according to the transformation (3.49). Instead of the velocity of light in vacuum, they experience the relativistic velocity addition (3.50) of the speed of light in the medium and the velocity of the medium.

Suppose that at some place, say $z = 0$, the velocity of the moving medium reaches the speed of light in the medium, as illustrated in the aquatic analogue in Fig. 3.7:

$$|u(0)| = \frac{c}{n(0)}. \tag{3.57}$$

Without loss of generality, we assume that $u < 0$ around $z = 0$ (the medium moves from right to left) and consider electromagnetic waves propagating against the flow as wavepackets with velocity v_+ that, according to the addition (3.50) of velocities, vanishes at $z = 0$. Lineariszing v_+ at $z = 0$,

$$v_+ = \alpha z, \tag{3.58}$$

we see that the integral in the virtual coordinates (3.49) develops a logarithmic singularity. Virtual space–time appears as two branches, one corresponding to $z < 0$ and the other to $z > 0$, both being independent Minkowski spaces. It follows from this that right-propagating light confined to either the region where $z < 0$ or the region

The 'Point of no return' is the analogue of the event horizon for fish in a river. To its left, the water flows faster than a fish can swim. So, if a fish happens to drift beyond this line, it can never get back up-stream; it is doomed to be crushed in Singularity Falls.

Fig. 3.7 Aquatic analogue of the event horizon. (Reproduced with friendly permission of Yan Nascimbene.)

where $z > 0$ will remain there; right-propagating light cannot cross $z = 0$, the horizon, where the velocity of the medium reaches the speed of light. The horizon is the point of no return; on one side of the horizon light can escape, but beyond the horizon, the light is dragged away by the moving medium, as the medium exceeds the speed of light there.

How to create a horizon for light in practice? It seems rather hopeless for a realistic medium to reach the speed of light, because although normal media reduce the velocity of light, they do not reduce c by much; light is still rather fast. However [34], one can mimic a moving medium with an intense pulse of light in a suitable material with nonlinear optical response [2]. In such a case, the refractive index of the material, n, may get an additional contribution that is proportional to the intensity of the pulse (this is called the Kerr effect). The pulse enhances the refractive index as if an additional piece of material were added to the medium. As this effective piece of material is made by light, it naturally moves at the speed of light in the material. Consider two light fields now: the pulse and the probe light that experiences the pulse as a contribution to the refractive index. One can control the speed of both the pulse and the probe by using the frequency and polarization dependence of the refractive index [34] and by the lateral profiles [1] of the light fields; different frequencies, polarizations, and lateral profiles propagate at different speeds. The place where the velocity of the probe is sufficiently reduced by the pulse to reach the speed of the pulse makes a horizon [1, 34].

Acknowledgements

I am most grateful for the discussions I have had with many distinguished scientists about geometry, light, and a wee bit of magic. In particular, I thank my former group at

St Andrews and wish them well: Simon Horsley, Susanne Kehr, Thomas Philbin, Sahar Sahebdivan, and William Simpson. My work is supported by the European Research Council, the Israel Science Foundation, and the Weizmann Institute of Science.

References

[1] F. Belgiorno, S. L. Cacciatori, M. Clerici, V. Gorini, G. Ortenzi, L. Rizzi, E. Rubino, V. G. Sala, and D. Faccio. Phys. Rev. Lett. **105**, 203901 (2010).

[2] R.W. Boyd. *Nonlinear Optics* (Academic Press, San Diego, 1992).

[3] M. Born and E. Wolf. *Principles of Optics*, 7th edn (Cambridge University Press, Cambridge, 1999).

[4] H. Chen and C. T. Chan. J. Appl. Phys. **104**, 033113 (2008).

[5] H. Chen, C. T. Chan, and P. Sheng. Nat. Mater. **9**, 387 (2010).

[6] X. Chen, Y. Luo, J. Zhang, K. Jiang, J. B. Pendry, and S. Zhang. Nat. Commun. **2**, 176 (2011).

[7] I. E. Dzyaloshinskii, E. M. Lifshitz, and L. P. Pitaevskii. Adv. Phys. **10**, 165 (1961).

[8] A. Einstein. Sitzungsberichte der Königlich Preußischen Akademie der Wissenschaften (Berlin), 142 (1917).

[9] N. Fang, H. Lee, C. Sun, and X. Zhang. Science **308**, 534 (2005).

[10] L. H. Gabrielli and M. Lipson. J. Opt. **13**, 024010 (2011).

[11] W. Gordon. Ann. Phys. (Leipzig) **72**, 421 (1923) [in German].

[12] A. Greenleaf, M. Lassas, and G. Uhlmann. Math. Res. Lett. **10**, 685 (2003).

[13] J. Li and J. B. Pendry. Phys. Rev. Lett. **101**, 203901 (2008).

[14] U. Leonhardt and P. Piwnicki. Phys. Rev. A **60**, 4301 (1999).

[15] U. Leonhardt. Science **312**, 1777 (2006).

[16] U. Leonhardt and T. G. Philbin. New J. Phys. **8**, 247 (2006).

[17] U. Leonhardt and T. Tyc. Science **323**, 110 (2009).

[18] U. Leonhardt. New J. Phys. **11**, 093040 (2009).

[19] U. Leonhardt and T. G. Philbin. *Geometry and Light: The Science of Invisibility* (Dover, Mineola, NY, 2010).

[20] U. Leonhardt and T. G. Philbin. Phys. Rev. A **81**, 011804 (2010); **84**, 049902 (2011).

[21] R. K. Luneburg. *Mathematical Theory of Optics* (University of California Press, Berkeley and Los Angeles, 1964).

[22] Y. G. Ma, C. K. Ong, T. Tyc, and U. Leonhardt. Nat. Mater. **8**, 639 (2009).

[23] Y. G. Ma, S. Sahebdivan, C. K. Ong, T. Tyc, and U. Leonhardt. New J. Phys. **13**, 033016 (2011).

[24] Y. G. Ma, S. Sahebdivan, C. K. Ong, T. Tyc, and U. Leonhardt. New J. Phys. **14**, 025001 (2012).

[25] M. S. Mahoney. *The Mathematical Career of Pierre de Fermat, 1601–1665* (Princeton University Press, Princeton, NJ, 1994).

[26] J. C. Maxwell. Cambridge Dublin Math. J. **8**, 188 (1854).

[27] J. C. Miñano, R. Marqués, J. C. González, P. Benítez, V. Delgado, D. Grabovičkič, and M. Freire. New J. Phys. **13**, 125009 (2011).

[28] J. C. Miñano, J. Sánchez-Dehesa, J. C. González, P. Benítez, D. Grabovičkič, J. Carbonell, and H. Ahmadpanahi. New J. Phys. **16** 033015 (2014).

[29] J. R. Minkel. Phys. Rev. Focus **9**, 23 (2002).

[30] J. B. Pendry. Phys. Rev. Lett. **85**, 3966 (2000).

[31] J. B. Pendry, D. Schurig, and D. R. Smith. Science **312**, 1780 (2006).

[32] R. Penrose (private communication).

[33] J. Perczel, T. Tyc, and U. Leonhardt. New J. Phys. **13**, 083007 (2011).

[34] T. G. Philbin, C. Kuklewicz, S. Robertson, S. Hill, F. König, and U. Leonhardt. Science **319**, 1367 (2008).

[35] J. Plebanski. Phys. Rev. **118**, 1396 (1960).

[36] P. M. Quan. C. R. Acad. Sci. (Paris) **242**, 465 (1957).

[37] P. M. Quan. Arch. Rat. Mech. Anal. **1**, 54 (1957/58).

[38] D. Schurig, J. J. Mock, B. J. Justice, S. A. Cummer, J. B. Pendry, A. F. Starr, and D. R. Smith. Science **314**, 977 (2006).

[39] B. F. Schutz. *A First Course in General Relativity* (Cambridge University Press, Cambridge, 2009).

[40] A. Serdyukov, I. Semchenko, S. Tretyakov, and A. Sihvola. *Electromagnetics of Bi-anisotropic Materials* (Gordon and Breach, Amsterdam, 2001).

[41] R. Service and A. Cho. Science **330**, 1622 (2010).

[42] M. I. Stockman. Phys. Rev. Lett. **98**, 177404 (2007).

[43] I. Y. Tamm, J. Russ. Phys.-Chem. Soc. **56**, 2–3, 248 (1924) [in Russian].

[44] I. Y. Tamm. J. Russ. Phys.-Chem. Soc. **56**, 3–4, 1 (1925) [in Russian].

[45] T. Tyc and X. Zhang. Nature **480**, 42 (2011).

[46] J. Valentine, S. Zhang, T. Zentgraf, E. Ulin-Avila, D. A. Genov, G. Bartal, and X. Zhang. Nature **455**, 376 (2008).

[47] B. Zhang, Y. Luo, X. Liu, and G. Barbastathis. Phys. Rev. Lett. **106**, 033901 (2011).

[48] S. Zhang, C. Xia, and N. Fang. Phys. Rev. Lett. **106**, 024301 (2011).

4

Temporal and spectral properties of quantum light

Birgit STILLER, Ulrich SEYFARTH
and Gerd LEUCHS

Max-Planck-Institut für die Physik des Lichts
Erlangen, Germany

Birgit Stiller, Ulrich Seyfarth and Gerd Leuchs: *Temporal and spectral properties of quantum light.*
In: *Quantum Optics and Nanophotonics.* Edited by: Claude Fabre, Vahid Sandoghdar, Nicolas
Treps, and Leticia F. Cugliandolo. Oxford University Press (2017), pp. 169–227. © Oxford
University Press 2017. DOI: 10.1093/oso/9780198768609.003.0004

Chapter Contents

Colour figures.   For those figures in this chapter that use colour, please see the version of these lecture notes at arXiv:1411.3765v2 [quant-ph]. These figures are indicated by '[Colour online]' at the start of the caption.

4.1   Introduction

The modes of the electromagnetic field are solutions of Maxwell's equations taking into account the material boundary conditions. The field modes of classical optics—properly normalized—are also the mode functions of quantum optics. Quantum physics adds that the excitation within each mode is quantized in close analogy to the harmonic oscillator. A complete set of mode functions forms a basis with which any new modes can be reconstructed. In full generality, each electromagnetic mode function in the four-dimensional space–time is mathematically equivalent to a harmonic oscillator.

The quantization of the electromagnetic field defines the excitation per mode and the correlation between modes. In classical optics, there can be oscillations and stochastic fluctuations of amplitude, phase, polarization, etc. In quantum optics, there are in addition uncertain quantum field components, quantum correlations, and quantized energies. This range of topics is discussed in the book by Mandel and Wolf [44]. Here, we present selected topics ranging from classical to quantum optics.

The spectrum of the light field is determined by its temporal evolution. In quantum optics, this corresponds to the time-dependent part of the mode function. The different properties of the light field that can be measured are the frequency and intensity, as well as phase differences. In particular, one can measure the spectral densities that are associated with these parameters. These different spectral densities are related in a non-trivial way. Therefore, we start in the next section with the classical optics description of a light field and its spectral densities, their measurement, and their interpretation. In the third section, the quantum properties of a single light mode are reviewed, as well as ways to measure these quantum properties. Gaussian states of a light mode are emphasized, i.e. states for which the Wigner function has a two-dimensional Gaussian shape. The fourth section is concerned with more than one mode, presenting a unifying approach to quadratic Hamiltonians, including phase conjugation, which is related to time reversal.

4.2   Temporal and spectral properties of classical light

4.2.1   Classical light fields

A spatial mode of a given frequency ω as defined by Maxwell's equations can be written as

$$\mathbf{E}(\mathbf{r}, t) = \mathbf{u}(\mathbf{r})\mathcal{E}(t)e^{i\omega t} + \text{c.c.}, \tag{4.1}$$

where $\mathbf{u}(\mathbf{r})$ represents the normalized mode function and $\mathcal{E}(t) = A(t)e^{i\phi(t)}$ is equivalent to the slowly varying optical field (Fig. 4.1).

Assuming that the light field is stationary, the outcome of the measurement does not depend on *when* the measurement is made. It is difficult to measure all aspects of

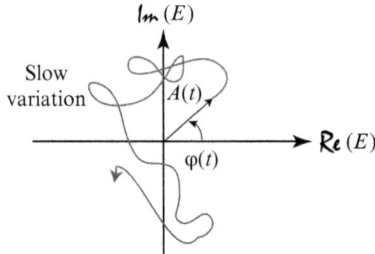

Fig. 4.1 [Colour online] Electric field diagram ($\equiv$ phase space).

the optical field directly. Different types of measurements, respectively measurement devices, are helpful:

- spectrometer: spectrum $\tilde{I}(\omega)$;
- Fabry–Perot interferometer: $(\omega - \omega_0)(t) = \Delta\omega(t)$;
- homodyne measurement: $\Delta\varphi(t)$;
- direct detection: $I(t) = (c\epsilon\epsilon_0)E^*(t)E(t)$, where $A(t) = \sqrt{I(t)}$.

The possible measurements on a monochromatic field are summarized in Fig. 4.2. The quantity $\tilde{I}(\omega)$ measured in a spectrometer can be interpreted as a result of interference. For one particular frequency ω at the output, one finds

$$\tilde{I}(\omega) \sim \tilde{E}^*(\omega)\tilde{E}(\omega). \tag{4.2}$$

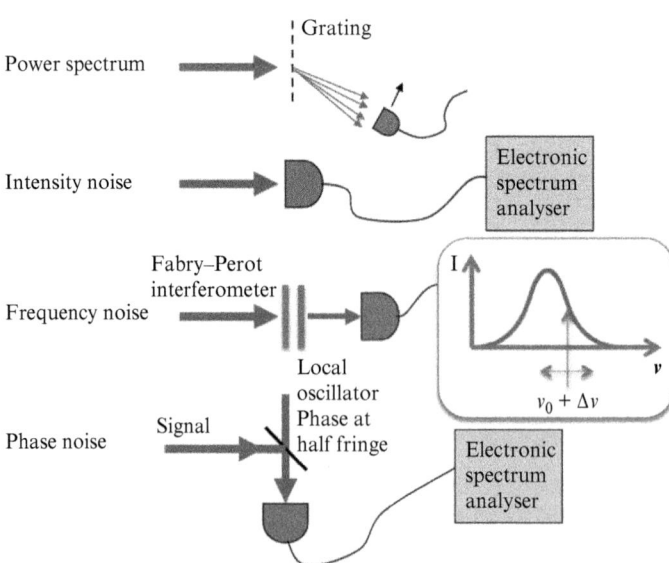

Fig. 4.2 [Colour online] Sketches of experimental setups for measuring different parameters of a light field. Note that the measurement shown for phase noise only works if the phase noise is small enough.

It is worth noting that $\tilde{E}^*(\omega)\tilde{E}(\omega)$ is not the Fourier transform of $I(t) = |E(t)|^2$.
$\tilde{I}(\omega)$ is closely related to the spectral density of the electric field:

$$S_E(\omega) \sim \int_{-\infty}^{\infty} E^*(t)e^{-i\omega t}\,\mathrm{d}t \int_{-\infty}^{\infty} E(t')e^{i\omega t'}\,\mathrm{d}t'. \tag{4.3}$$

However, in an experiment, the Fourier transform will always be based on a time series of finite length. Thus, we rewrite

$$S_E(\omega) = \lim_{T \to \infty} \frac{1}{2T} \int_{-T}^{T} E^*(t)e^{-i\omega t}\,\mathrm{d}t \int_{-T}^{T} E(t')e^{i\omega t'}\,\mathrm{d}t'$$

$$= \lim_{T \to \infty} \frac{1}{2T} \iint_{2T} E^*(t)E(t')e^{i\omega(t'-t)}\,\mathrm{d}t\,\mathrm{d}t'$$

$$= \int_{-\infty}^{\infty} \left\{ \lim_{T \to \infty} \frac{1}{2T} \int_{-T}^{T} \mathrm{d}t\, \tfrac{1}{2}[E^*(t)E(t+\tau) + E^*(t+\tau)E(t)] \right\} e^{i\omega\tau}\,\mathrm{d}\tau. \tag{4.4}$$

Note that $S_E(\omega)$ is real, as is $\tilde{I}(\omega)$. The last step is achieved by enforcing the symmetry between t and t', i.e. by averaging over the substitutions $t' \to t+\tau$ and $t \to t'+\tau$. By using $\langle E^*(t)E(t+\tau)\rangle \equiv \lim_{T\to\infty}(1/2T)\int_{-T}^{T}\mathrm{d}t\,E^*(t)E(t+\tau)$, we can write

$$S_E(\omega) = \mathcal{R}e\left\{ \int_{-\infty}^{\infty} \langle E^*(t)E(t+\tau)\rangle e^{i\omega\tau}\,\mathrm{d}\tau \right\}, \tag{4.5}$$

which is the *Wiener–Khinchin theorem*. The quantity $\langle E^*(t)E(t+\tau)\rangle$ can be measured by a Michelson interferometer. The spectrum $S_E(\omega)$ is obtained by Fourier transformation of the field correlation function $\langle E^*(t)E(t+\tau)\rangle$, which itself is obtained by moving one of the interferometer mirrors. Such a spectrometer is thus called a *Fourier transform spectrometer*.

4.2.2   The spectrum of intensity noise

Taking the Fourier transform of $I(t)$ directly gives a different quantity. It corresponds to the experiment shown in Fig. 4.3, with

$$S_I(f) = \int_{-\infty}^{\infty} \underbrace{\langle I(t)I(t+\tau)\rangle}_{\equiv G^{(2)}(\tau)} \cos(2\pi f\tau)\,\mathrm{d}\tau, \tag{4.6}$$

where the parameter f indicates a low frequency in the radiofrequency range (see also [17]). The correlation function of a real-valued function is an even function of time. Hence, $\exp(i2\pi f\tau)$ can be replaced by $\cos(2\pi f\tau)$. The spectral density $S_I(f)$ is given by $[\tilde{I}(f)]^2$, much like $S_E(\omega)$ is given by $|\tilde{E}(\omega)|^2$. $S_I(f)$ is closely related to the amplitude spectral density defined correspondingly:

$$S_A(f) = \int_{-\infty}^{\infty} \langle A(t)A(t+\tau)\rangle \cos(2\pi f\tau)\,\mathrm{d}\tau. \tag{4.7}$$

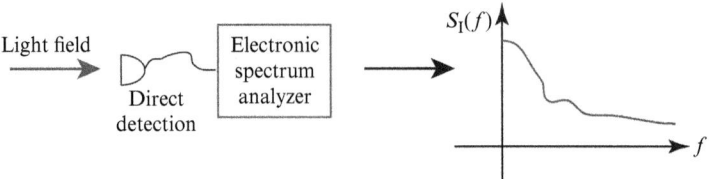

Fig. 4.3 [Colour online] Direct detection scheme.

Yet another quantity is the phase spectral density

$$S_\varphi(f) = \int_{-\infty}^{\infty} \langle \varphi(t)\varphi(t+\tau)\rangle \cos(2\pi f\tau)\,\mathrm{d}\tau\,. \tag{4.8}$$

The spectral density of the frequency noise is closely related by $\delta\nu = (1/2\pi)\,\mathrm{d}\varphi/\mathrm{d}t$ as

$$S_\nu(f) = f^2 S_\varphi(f)\,, \tag{4.9}$$

and by using $\Omega = 2\pi f$ we find

$$S_\nu(\Omega) = \left(\frac{\Omega}{2\pi}\right)^2 S_\varphi(\Omega)\,. \tag{4.10}$$

A relation between $S_A(f)$ and $S_I(f)$ is obtained in a straightforward manner, as

$$\begin{aligned}
\langle I(t)I(t+\tau)\rangle &= \langle A(t)A(t)A(t+\tau)A(t+\tau)\rangle \\
&= \langle A(t)A(t)\rangle\langle A(t+\tau)A(t+\tau)\rangle + 2\langle A(t)A(t+\tau)\rangle^2 \\
&= \langle I\rangle^2 + 2\langle A(t)A(t+\tau)\rangle^2\,,
\end{aligned} \tag{4.11}$$

where the second step assumes Gaussian statistics relating the fourth moment to products of second moments ($\langle\sigma_1\sigma_2\sigma_3\sigma_4\rangle = \langle\sigma_1\sigma_2\rangle\langle\sigma_3\sigma_4\rangle + \langle\sigma_1\sigma_3\rangle\langle\sigma_2\sigma_4\rangle + \langle\sigma_1\sigma_4\rangle\langle\sigma_2\sigma_3\rangle$). Hence,

$$\langle A(t)A(t+\tau)\rangle = \sqrt{\frac{\langle I(t)I(t+\tau)\rangle - \langle I\rangle^2}{2}}\,. \tag{4.12}$$

4.2.3 Absorption by an atom

Here we will show how the rate of absorption by an atom depends on the frequency of the field. We can easily make the right guess, but going through the details will help later. The energy level diagram of the two-level atom is shown in Fig. 4.4, defining the notation, using $|e(t)\rangle = \mathrm{e}^{\mathrm{i}\omega_0 t}\,|e\rangle$. The interaction Hamiltonian of the system is

$$V(t) = \hat{d}E(t)\,, \tag{4.13}$$

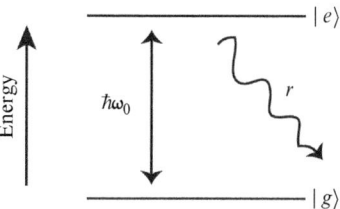

Fig. 4.4 Energy level diagram of a two-level atom, with *ground state* $|g\rangle$ and *excited state* $|e\rangle$. The difference between the levels is given by $\hbar\omega_0$, with the Rabi frequency ω_0.

where $\hat{d} = e\hat{r}$ is the operator for the atomic dipole moment. The transition probability is calculated as

$$P_{ge}(t) = \frac{1}{\hbar^2} \left| \int_0^t e^{-i\omega_0 t'} \langle e|V(t')|g \rangle \, dt' \right|^2$$

$$= \frac{1}{\hbar^2} \int_0^t dt' \, e^{-i\omega_0 t'} \langle e|V(t')|g \rangle \int_0^t dt'' \, e^{i\omega_0 t''} \langle g|V^\dagger(t'')|e \rangle$$

$$= \frac{1}{2\hbar^2} \int_0^t dt' \int_{-t'}^{t-t'} d\tau \, e^{i\omega_0 \tau} \langle g|V^\dagger(t'+\tau)|e \rangle \langle e|V(t')|g \rangle + (t'' \text{ and } t' \text{ interchanged}),$$

$$(4.14)$$

where $t'' = t' + \tau$ and $dt'' = d\tau$ (the interchange of t'' and t' refers to (4.4)). To account for the decay of $|e\rangle$, one replaces $e^{i\omega_0 \tau}$ by $e^{i\omega_0 \tau - \Gamma \tau}$.

For $V(t) = \hat{d}E(t)$, the transition probability is given by

$$P_{ge}(t) = \frac{|\langle e|\hat{d}|g \rangle|^2}{\hbar^2} \mathcal{R}e \left\{ \int_{-t'}^{t-t'} d\tau \, e^{i\omega_0 \tau - \Gamma \tau} \int_0^t dt' \, E^*(t'+\tau)E(t') \right\}. \qquad (4.15)$$

This translates to a transition rate of $R_{ge}(t) = dP_{ge}(t)/dt$:

$$R_{ge}(t) = \frac{|\langle e|\hat{d}|g \rangle|^2}{\hbar^2} \mathcal{R}e \left\{ \int_0^t d\tau \, e^{i\omega_0 \tau - \Gamma \tau} \langle E^*(t+\tau)E(t) \rangle \right\}, \qquad (4.16)$$

where Γ denotes the decay rate of the excited state. With b, we describe the bandwidth of the field correlation function. We can consider two cases:

- **Case 1:** $\Gamma \ll b$, the bandwidth of the optical field $\implies \Gamma \cong 0$
  $\implies R_{ge}$ is the Fourier transformation of the first-order correlation function; in other words R_{ge} is proportional to the spectrum (see Section 4.2.1).
- **Case 2:** $\Gamma \gg b$, and the term $e^{-\Gamma \tau}$ can be replaced by $\delta(\tau)$
  $$\implies R_{ge} = \frac{|\langle e|\hat{d}|g \rangle|^2}{\hbar^2} \langle |E(t)|^2 \rangle \sim \langle I(t) \rangle.$$

4.2.4 Relationship between the different spectral densities

The general relationship is obtained starting from the spectral density of the electric field,

$$S_E(\omega) = \mathcal{R}e\left\{\int_{-\infty}^{\infty} \langle E^*(t)E(t+\tau)\rangle e^{i\omega\tau}\,d\tau\right\}, \tag{4.17}$$

by inserting $E(t) = A(t)e^{i\varphi(t)}e^{-i\omega_0 t}$ with the amplitude $A(t)$, the phase $\varphi(t)$, the frequency $\nu_0 = \omega_0/2\pi$, and the *frequency excursion* $\Delta\nu(t) = \dot{\varphi}(t)$, which leads to

$$S_E(\omega) = \mathcal{R}e\left\{\int_{-\infty}^{\infty} d\tau\,\langle A(t+\tau)e^{i\varphi(t+\tau)-i\omega_0(t+\tau)}A(t)e^{-i\varphi(t)+i\omega_0 t}\rangle e^{i\omega\tau}\right\}$$

$$= \mathcal{R}e\left\{\int_{-\infty}^{\infty} d\tau\,\langle A(t+\tau)A(t)e^{i(\varphi(t+\tau)-\varphi(t))}\rangle e^{i(\omega-\omega_0)\tau}\right\}. \tag{4.18}$$

Assuming the absence of any correlation between amplitude and phase leads to

$$S_E(\omega) = \mathcal{R}e\left\{\int_{-\infty}^{\infty} d\tau\,\langle A(t+\tau)A(t)\rangle\langle e^{i(\varphi(t+\tau)-\varphi(t))}\rangle e^{i(\omega-\omega_0)\tau}\right\}, \tag{4.19}$$

where the averaging has been factorized. The second averaging term can be expanded, taking into account that

$$e^x = 1 + x + \frac{1}{2!}x^2 + \frac{1}{3!}x^3 + \frac{1}{4!}x^4 + \dots \tag{4.20}$$

and with the averages of the odd-order terms being zero,

$$\langle e^{i\Delta\varphi(t)}\rangle = 1 - \frac{1}{2}\langle(\Delta\varphi)^2\rangle + \frac{1}{4!}\langle(\Delta\varphi)^4\rangle - \frac{1}{6!}\langle(\Delta\varphi)^6\rangle + \dots. \tag{4.21}$$

For Gaussian variables (multivariate normal distributions—see above) with

$$\langle\varphi_1\varphi_2\varphi_3\varphi_4\rangle = \langle\varphi_1\varphi_2\rangle\langle\varphi_3\varphi_4\rangle + \langle\varphi_1\varphi_3\rangle\langle\varphi_2\varphi_4\rangle + \langle\varphi_1\varphi_4\rangle\langle\varphi_2\varphi_3\rangle, \tag{4.22}$$

one obtains

$$\langle\varphi^4\rangle = 3\langle\varphi^2\rangle^2 \tag{4.23}$$

or, in general,

$$\langle x^n\rangle = N\int_{-\infty}^{\infty} x^2 e^{-ax^2}\,dx$$

$$= (n-1)\langle x^2\rangle\langle x^{n-2}\rangle \implies \langle x^{2n}\rangle = \frac{(2n-1)!}{2^{n-1}(n-1)!}\langle x^2\rangle^n, \tag{4.24}$$

and thus

$$
\begin{aligned}
\langle e^{i\Delta\varphi(\tau)}\rangle &= 1 - \tfrac{1}{2}\langle(\Delta\varphi)^2\rangle + \tfrac{3}{4!}\langle(\Delta\varphi)^2\rangle^2 - \tfrac{3\cdot5}{6!}\langle(\Delta\varphi)^2\rangle^3 + \ldots \\
&= 1 + \left(-\tfrac{1}{2}\langle(\Delta\varphi)^2\rangle\right) + \tfrac{1}{2}\left(-\tfrac{1}{2}\langle(\Delta\varphi)^2\rangle\right)^2 + \tfrac{1}{3!}\left(-\tfrac{1}{2}\langle(\Delta\varphi)^2\rangle\right)^3 + \ldots \\
&= e^{-\tfrac{1}{2}\langle(\Delta\varphi)^2\rangle} = e^{-\tfrac{1}{2}\langle(\varphi(t+\tau)-\varphi(t))^2\rangle}.
\end{aligned}
\tag{4.25}
$$

Therefore, if higher-order phase moments are Gaussian, the second averaged factor in (4.19) can be replaced by (4.25), $\tfrac{1}{2}\langle(\varphi(t+\tau)-\varphi(t))^2\rangle = \langle(\varphi(t))^2\rangle - \langle\varphi(t+\tau)\varphi(t)\rangle$, with

$$
\langle\varphi(t+\tau)\varphi(t)\rangle = \int d\Omega\, S_\varphi(\Omega)\cos\Omega\tau.
\tag{4.26}
$$

This yields

$$
S_E(\omega) = \mathcal{R}e\left\{\int_{-\infty}^{\infty} d\tau\, \langle A(t+\tau)A(t)\rangle e^{-\int d\Omega\, S_\varphi(\Omega)(1-\cos\Omega\tau)} e^{i(\omega-\omega_0)\tau}\right\}.
\tag{4.27}
$$

Next, we relate $\langle A(t+\tau)A(t)\rangle$ to $S_I(\Omega)$ using

$$
\langle I(t+\tau)I(t)\rangle = \int_0^{\infty} d\Omega\, S_I(\Omega)\cos\Omega\tau,
\tag{4.28}
$$

and, if higher amplitude moments are again Gaussian, we can use (4.12) to obtain

$$
\langle A(t+\tau)A(t)\rangle = \frac{1}{\sqrt{2}}\sqrt{\int_0^{\infty} d\Omega\, S_I(\Omega)\left(\cos\Omega\tau - \langle I\rangle^2\right)}.
\tag{4.29}
$$

Thus, we find

$$
\boxed{
\begin{aligned}
S_E(\omega) = \mathcal{R}e\Bigg\{ &\int_{-\infty}^{\infty} d\tau\, \frac{1}{\sqrt{2}}\left(\sqrt{\int_0^{\infty} d\Omega\, S_I(\Omega)(\cos\Omega\tau - \langle I\rangle^2)}\right) \\
&\times e^{-\int d\Omega\, S_\varphi(\Omega)(1-\cos\Omega\tau)} e^{i(\omega-\omega_0)\tau}\Bigg\}.
\end{aligned}
}
\tag{4.30}
$$

4.2.4.1 *Example: δ-correlated amplitudes*

Let us consider an example with constant phase and *δ-correlated* amplitudes, namely with $\langle A(t+\tau)A(t)\rangle = S_A(0)\delta(\tau)$ and $\varphi(t+\tau) - \varphi(t) = 0$. It follows that

$$
S_E(\omega) = \mathcal{R}e\left\{\int_{-\infty}^{\infty} d\tau\, S_A(0)\delta(\tau)e^{i(\omega-\omega_0)\tau}\right\} = S_A(0),
\tag{4.31}
$$

which describes a *white spectrum.*

4.2.4.2 *Example: white frequency noise*

If the phase diffuses, the frequency $\nu(t)$ is δ-correlated as

$$\langle \nu(t)\nu(t+\tau)\rangle = \delta(\tau)S_\nu(0)\,. \tag{4.32}$$

Thus, $S_\nu(f) = \int_{-\infty}^{\infty} d\tau\,\langle \nu(t)\nu(t+\tau)\rangle \cos(2\pi f\tau) = S_\nu(0)$ is a constant. The phase spectral density is thus

$$S_\varphi(\Omega) = 2\pi \frac{S_\nu(f)}{\Omega^2} \to 2\pi \frac{S_\nu(0)}{\Omega^2}\,. \tag{4.33}$$

This leads to

$$S_E(\omega) = \mathcal{R}e\left\{ \int_{-\infty}^{\infty} d\tau\, A^2 e^{-\int d\Omega\,[2\pi S_\nu(0)/\Omega^2](1-\cos\Omega\tau)} e^{i(\omega-\omega_0)\tau} \right\}. \tag{4.34}$$

We first evaluate the integral in the exponent

$$\int_0^\infty d\Omega\, \frac{S_\nu(0)}{\Omega^2}(1-\cos\Omega\tau) = S_\nu(0)\int_0^\infty d\Omega \frac{1}{\Omega^2}(1-\cos\Omega\tau)$$

$$= S_\nu(0)\left(\int_0^\infty d\Omega\, \frac{1}{\Omega^2} - \int_0^\infty d\Omega\, \frac{1}{\Omega^2}\cos\Omega\tau \right). \tag{4.35}$$

Substituting $u' = 1/\Omega^2$ and $v = \cos\Omega\tau$, with $u = -1/\Omega$ and $v' = -\tau\sin(\Omega\tau)$, by partial integration one finds

$$\int_0^\infty d\Omega\, \frac{S_\nu(f)}{\Omega^2}(1-\cos\Omega\tau) = S_\nu(0)\left(\left[-\frac{1}{\Omega} + \frac{1}{\Omega}\cos\Omega\tau \right]_0^\infty + \int_0^\infty d\Omega\, \frac{\sin\Omega\tau}{\Omega}\tau \right)$$

$$= \frac{\pi}{2}\tau S_\nu(0)\,, \tag{4.36}$$

where the result is obtained by recognizing that $\int_0^\infty dx\,(\sin x)/x = \pi/2$. Thus, the spectral density is given by

$$S_E(\omega) = \mathcal{R}e\left\{ \int_{-\infty}^{\infty} d\tau\, A^2 e^{-\pi^2 S_\nu(0)\tau} e^{i(\omega-\omega_0)\tau} \right\}$$

$$= \mathcal{R}e\left\{ A^2 \frac{1}{-i(\omega-\omega_0) + \pi^2 S_\nu(0)} \right\}$$

$$= \frac{\frac{1}{4}S_\nu(0)}{(\nu-\nu_0)^2 + \left[\frac{1}{2}\pi S_\nu(0)\right]^2}\,. \tag{4.37}$$

The half width at half maximum is given by

$$\Delta\nu_{1/2} = \tfrac{1}{2}\pi S_\nu(0) \tag{4.38}$$

and the full width at half maximum by

$$\Delta \nu = \pi S_\nu(0) \qquad (4.39)$$

(cf. the Schawlow–Townes linewidth of the ideal laser).

4.2.4.3 *Relation between frequency noise spectrum and spectral linewidth*

We now try to give a physical picture of this relation [52, 53]. The value of $S_\nu(f)$ in a narrow frequency interval of width b at noise frequency f gives the variance of the frequency fluctuations in this interval (Fig. 4.5(a)). In order to obtain the total variance, one has to integrate $S_\nu(f)$ from zero to infinity. For $S_\nu(f)$, the value diverges, of course. Therefore, we conclude that from a certain frequency f_c onwards (where 'c' stands for cutoff), $S_\nu(f)$ no longer contributes to the total variance of the frequency fluctuation. The cutoff can be determined by setting the integral equal to the square of the linewidth derived above, (4.39):

$$\int_0^{f_c} S_\nu(f)\, df \stackrel{!}{=} [\pi S_\nu(0)]^2 \,. \qquad (4.40)$$

It follows that

$$S_\nu(0) \cdot f_c = [\pi^2 S_\nu(0)]^2 \,, \qquad (4.41)$$

with

$$f_c = \pi^2 S_\nu(0) \,. \qquad (4.42)$$

In order to understand the meaning of this cutoff, we integrate $S_\varphi(\Omega)$ from $\Omega_c = 2\pi f_c$ to infinity:

$$\int_{2\pi^3 S_\nu(0)}^{\infty} d\Omega\, 2\pi \frac{S_\nu(0)}{\Omega^2} = \frac{1}{\pi^2} = \langle \Delta\varphi^2 \rangle_{\Omega > \Omega_c} \,. \qquad (4.43)$$

This means that the high-frequency tail of the spectral density of the phase fluctuations does not contribute to the linewidth if the integrated variance in this frequency range is significantly less than one radian (Fig. 4.6). This gives us a recipe for the general case: choose f_c such that $\sqrt{\langle \Delta\varphi^2 \rangle_{\Omega > \Omega_c}} \cong 1/\pi$ and then integrate $S_\nu(f)$ up to $f_c = \Omega_c/2\pi$. This determines the contribution of the phase fluctuations (see Fig. 4.5(b)) to the spectral linewidth.

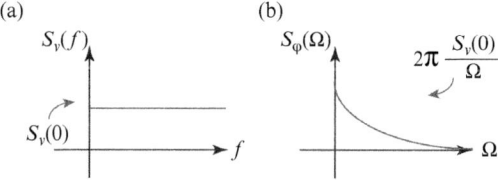

Fig. 4.5 [Colour online] Noise spectrum.

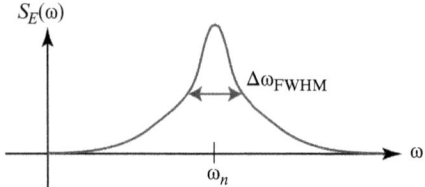

Fig. 4.6 [Colour online] $\langle \Delta\varphi^2 \rangle_{\Omega>\Omega_c}$ changes only the high-frequency wings.

4.2.5 Correlation functions

Next, we consider general electric field correlation functions

$$\left\langle \prod_{j=1}^{n} \mathcal{E}^{(*)}(t_j) \right\rangle, \tag{4.44}$$

with $\langle \ldots \rangle$ denoting a time average, namely $\langle \ldots \rangle = \lim_{T\to\infty} (2T)^{-1} \int_{-T}^{T} \ldots \, dt$. For odd n, the correlation functions of harmonic fields vanish, that is, $\langle \ldots \rangle = 0$. For even n, the non-zero correlation functions have the following general form with an equal number of conjugated and unconjugated fields:

$$G^{(n)}(t_1, t_2, \ldots, t_{2n}) \equiv \langle E^*(t_1)E^*(t_2)\cdots E^*(t_n)E(t_{n+1})\cdots E(t_{2n})\rangle. \tag{4.45}$$

The order of the correlation function is given by n.[1] For $n = 1$, we have $\langle E^*(t_1)E(t_2)\rangle \equiv G^{(1)}(t_2 - t_1)$, where $G^{(1)}$ is second order in the electric field and first order in intensity. For $t_1 = t_2$, it follows that $G^{(1)}(0) = \langle I(t)\rangle$. The correlation function $G^{(1)}(t_2-t_1)$ determines the signal of the output of an interferometer. This has already been discussed in Sections 4.2.1 and 4.2.3. Next, we derive a similar relation for a correlation function with $n = 2$, which is given by $\langle E^*(t_1)E^*(t_2)E(t_3)E(t_4)\rangle$.

1. **Special case.** Choosing $t_1 = t_3$ and $t_2 = t_4$ yields the intensity correlation function:

$$\langle |E(t_1)|^2 |E(t_2)|^2 \rangle = \langle I(t_1)I(t_2)\rangle \equiv G_I^{(2)}. \tag{4.46}$$

   The subscript 'I' refers to the arrangement of a field describing the intensity correlation functions.

2. **Special case.** For two-photon absorption, we find

$$V(t) = \hat{d} \cdot E(t) = \sum_i \frac{\hat{d}E(t)\,|i\rangle\,\langle i|\,\hat{d}E(t)}{W_g + \hbar\omega_L - W_i}$$

$$\sim E^2(t), \tag{4.47}$$

---

[1] Some authors use $2n$. For example, $\langle E(t_2)E^*(t_2)\rangle$ is sometimes called a second-order correlation function since it involves two field amplitudes. Here, we use n in the superscript such that photon antibunching is described by a $G^{(2)}$ function.

where W_g and W_i denote the energies of the atomic ground and intermediate states, respectively. In analogy to Section 4.2.3, i.e. replacing $V(t) \sim E(t)$ by $V(t) \sim E^2(t)$, we have [37]

$$R_{ge} \sim \int_0^t \mathrm{d}\tau \, \mathrm{e}^{\mathrm{i}\omega_0 \tau} \langle E^{*2}(t'+\tau)E^2(t') \rangle$$

$$= \int_0^t \mathrm{d}\tau \, \mathrm{e}^{\mathrm{i}(\omega_0 - 2\omega_L)\tau} \underbrace{\langle E^{*2}(t'+\tau)E^2(t') \rangle}_{G_{\mathrm{TP}}^{(2)}(\tau)} . \tag{4.48}$$

The subscript 'TP' refers to the particular arrangement of the electric fields typical of two-photon absorption. Thus, the second-order correlation function

$$\langle E^{*2}(t_1)E^2(t_3) \rangle \equiv G_{\mathrm{TP}}^{(2)}(t_3 - t_1) \tag{4.49}$$

determines a two-photon absorption process.

In general, the $G^{(n)}(t_1, \ldots, t_{2n})$ are complex-valued functions. The full set of all $G^{(n)}(t_1, \ldots, t_{2n})$ fully characterizes the field (cf. the fact that all higher moments fully characterize a distribution).

4.2.5.1 *Example*

The three fields $E(t)$ shown in Fig. 4.7 have the same spectrum, but they are not identical. The Fourier transforms of $E_1(t)$ and $E_2(t)$ are complex and the Fourier transform of $E_3(t)$ is real:

$$\tilde{E}_1(\omega) = \mathrm{FT}\{\mathcal{E}_1(t)\mathrm{e}^{-\mathrm{i}\omega_L t}\} = \int_{-\infty}^{\infty} \mathrm{d}t \, \mathrm{e}^{\Gamma t}\theta(-t)\mathrm{e}^{\,\mathrm{i}\omega_L t}\mathrm{e}^{\mathrm{i}\omega t}$$

$$= \int_{-\infty}^0 \mathrm{d}t \, \mathrm{e}^{[\Gamma + \mathrm{i}(\omega - \omega_L)]t} = \frac{1}{\Gamma + \mathrm{i}(\omega - \omega_L)}$$

$$= \frac{\Gamma - \mathrm{i}(\omega - \omega_K)}{\Gamma^2 + (\omega - \omega_L)^2} , \tag{4.50}$$

$$\tilde{E}_2(\omega) = \mathrm{FT}\{\mathcal{E}_2(t)\mathrm{e}^{-\mathrm{i}\omega_L t}\} = \int_{-\infty}^{\infty} \mathrm{d}t \, \mathrm{e}^{-\Gamma t}\theta(t)\mathrm{e}^{-\mathrm{i}\omega_L t}\mathrm{e}^{\mathrm{i}\omega t}$$

$$= \int_0^{\infty} \mathrm{d}t \, \mathrm{e}^{[-\Gamma + \mathrm{i}(\omega - \omega_L)]t} = -\frac{1}{-\Gamma + \mathrm{i}(\omega - \omega_L)}$$

$$= \frac{\Gamma + \mathrm{i}(\omega - \omega_K)}{\Gamma^2 + (\omega - \omega_L)^2} , \tag{4.51}$$

and so

$$\tilde{E}_3(\omega) = \tilde{E}_1(\omega) + \tilde{E}_2(\omega)$$

$$= \frac{2\Gamma}{\Gamma^2 + (\omega - \omega_L)^2} . \tag{4.52}$$

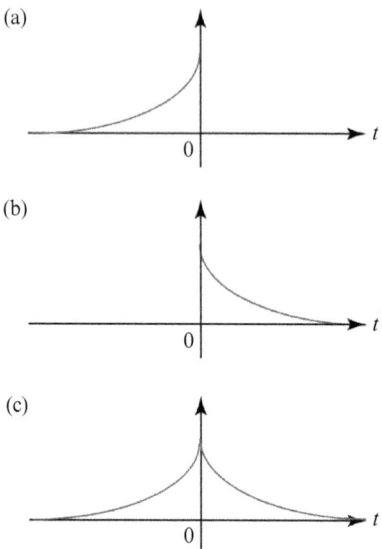

Fig. 4.7 [Colour online] (a) $E_1(t) = e^{\Gamma t}\theta(-t)$. (b) $E_2(t) = e^{-\Gamma t}\theta(t)$. (c) $E_3(t) = E_1(t) + E_2(t)$. Here, $\theta(t) = 1$ for $t \leq 0$ and $\theta(t) = 0$ for $t < 0$.

This shows that these three fields with different temporal shapes all have the same Lorentzian spectrum $I(\omega) = |E(\omega)|^2$. These fields differ in their imaginary parts in the frequency representation. It is obvious that time reversal in the time domain is related to complex conjugation in the frequency domain ($E_1(t)$ is the time-reversed version of $E_2(t)$). Figure 4.8 compares the second-order correlation functions for two different stationary fields having identical first-order correlation functions: a thermal field (Fig. 4.9(a)) and a phase-diffusing field (Fig. 4.9(b)). This underlines the importance of higher-order correlation functions when characterizing light fields [23, 37].

4.3 Quantum optics of a single mode

In this section, we describe the quantum properties of a *single* light mode. We will focus on Gaussian states of a light mode, i.e. states for which the Wigner function has a two-dimensional Gaussian shape. Experimental setups to measure these quantum properties will be explained. Moreover, we will give an overview of quasi-probability distributions and review correlation functions. We will begin with the general description of the phase space for a quantum state in analogy with the harmonic oscillator, since the excitation within each mode is quantized.

4.3.1 Phase space of a single-mode light field

The energy per mode of the light field is quantized, as it is in the case of the harmonic oscillator. The **classical harmonic oscillator** is described as a pure sine wave:

$$x(t) = x(0)\cos\omega t + \frac{p(0)}{m\omega}\sin\omega t. \tag{4.53}$$

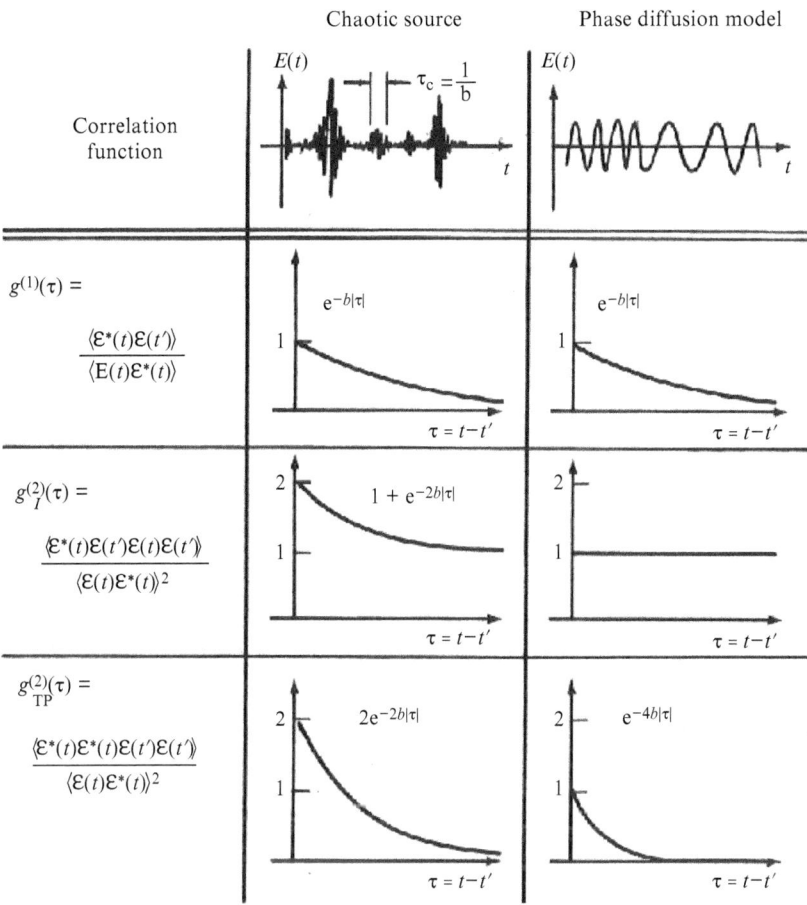

Fig. 4.8 Influence of photon statistics on correlation functions [37].

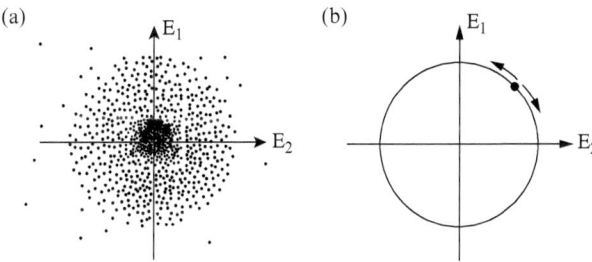

Fig. 4.9 (a) Chaotic source: thermal field. (b) Phase-diffusing laser field.

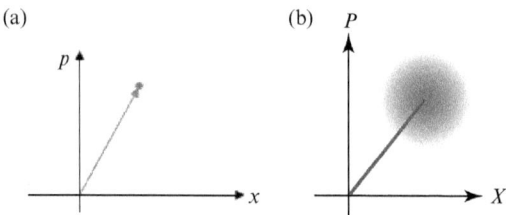

Fig. 4.10 [Colour online] Phase-space diagrams. (a) Classical harmonic oscillator; $x \equiv x(0)$ and $p \equiv p(0)/m\omega$. (b) Quantum harmonic oscillator. As described in the text, the shaded area represents a phase-space distribution function.

Its energy is continuous and its position and momentum have precise values, i.e. values without any uncertainty; see Fig. 4.10(a).

The **quantum harmonic oscillator** instead is described by a diffuse sine wave with discrete energy values but uncertain position and momentum values; see Fig. 4.10(b). The uncertainty (variance of the state) is described by $\langle \Delta X^2 \rangle \neq 0$ and $\langle \Delta P^2 \rangle \neq 0$. The energy levels of the quantum harmonic oscillator are given by:

$$E_n = \left(n + \frac{1}{2} \right) \hbar\omega \,, \tag{4.54}$$

where the lowest energy level has a positive value. The quantized description of a light mode is mathematically equivalent. In the case of a light field, the variables X and P become two conjugate field operators. In a specific case, they may represent the amplitude and the phase quadrature of the light field in a single mode.

A quantum optical field can be expressed in different ways and with different variables: **discrete** variables and **continuous** ones. The wavefunction of a single-mode quantum field can be written for example as a superposition of photon number states (Fock basis), which corresponds to a discrete-variable description. For an infinite-dimensional complex Hilbert space, we write a quantum state as

$$|\Psi\rangle = \sum_{i=0}^{\infty} \alpha_i \,|i\rangle \,, \tag{4.55}$$

where $|i\rangle$ represent photon number states. One specific example is a state description by discrete dichotomic variables:

$$|\Psi\rangle = \alpha_0 \,|0\rangle + \alpha_1 \,|1\rangle = \sum_{i=0}^{1} \alpha_i \,|i\rangle \,, \tag{4.56}$$

where $|\Psi\rangle$ is called a *qubit*. This state consists of no more than one photon. In the continuous case, there are the continuous variables X and P, which span the phase space of the mode.[2] The values of X and P can be estimated, for example,

---

[2] An excellent introduction to the topic is given by Schleich [48].

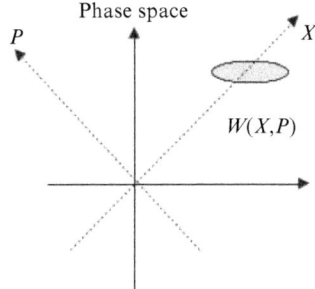

Fig. 4.11 [Colour online] Phase-space diagram including the Wigner function $W(X, P)$.

by the Wigner function $W(X, P)$, a particular phase-space distribution function (Fig. 4.11):

$$W(X, P) = \frac{1}{2\pi\hbar} \int_{-\infty}^{\infty} d\xi \exp\left(-\frac{\mathrm{i}p\xi}{\hbar}\right) \Psi^*\left(X - \frac{1}{2}\xi\right) \Psi\left(X + \frac{1}{2}\xi\right), \qquad (4.57)$$

where $\Psi(x) = \langle x|\Psi\rangle$ holds. It is instructive to look at Wigner's original paper [56].[3]

Different types of continuous quantum variables are the field quadratures $\hat{X}$ and $\hat{P}$ (canonical operators) and the Stokes variables that describe the polarization of the quantum field. It is practical to define the field operators $\hat{a}$ and $\hat{a}^\dagger$, which are related to the canonical ones as follows:

$$\hat{a} = \hat{X} + \mathrm{i}\hat{P}, \qquad (4.58)$$

$$\hat{a}^\dagger = X - \mathrm{i}\hat{P}, \qquad (4.59)$$

or, vice versa,

$$\hat{X} = \frac{1}{2}(\hat{a} + \hat{a}^\dagger), \qquad (4.60)$$

$$\hat{P} = \frac{1}{2\mathrm{i}}(\hat{a} - \hat{a}^\dagger). \qquad (4.61)$$

For the field operators $\hat{a}$ and $\hat{a}^\dagger$, the commutation relation

$$[\hat{a}, \hat{a}^\dagger] = \hat{a}\hat{a}^\dagger - \hat{a}^\dagger\hat{a} = 1 \qquad (4.62)$$

holds, as well as

$$\hat{a}|n\rangle = \sqrt{n}\,|n - 1\rangle, \qquad (4.63)$$

$$\hat{a}^\dagger|n\rangle = \sqrt{n + 1}\,|n + 1\rangle. \qquad (4.64)$$

---

[3] The function later named after Wigner comes out of the blue. The only hint at how this function was found is given by Wigner in his footnote 2. Asked later of what this other purpose was for which he and Szilard used this function, Wigner is reported to have answered that he made this up. His fellow countryman Szilard was looking for a job and Wigner thought such a footnote might help him.

They are also related to the photon number operator $\hat{n} = \hat{a}^\dagger \hat{a}$:

$$\hat{a}^\dagger \hat{a} \left| n \right\rangle = n \left| n \right\rangle , \tag{4.65}$$

where $\left| n \right\rangle$ are the Fock states.

A single quantum mode in general is described by one pair of variables $\hat{X}, \hat{P}$, where their mean varies continuously and their variances obey the uncertainty relation. Furthermore, $\hat{X}$ and $\hat{P}$ are non-commuting operators.[4] For equal and minimum uncertainty of $\hat{X}$ and $\hat{P}$, we have a *coherent state*, which defines the most 'classical' quantum state of light (Fig. 4.12), with amplitude and phase fairly well defined simultaneously, as much as can be allowed by quantum effects.

A coherent state $\left| \alpha \right\rangle$, ideally emitted by a laser, is an eigenstate of the field operator $\hat{a}$, namely

$$\hat{a} \left| \alpha \right\rangle = \alpha \left| \alpha \right\rangle . \tag{4.66}$$

It is possible to expand the coherent state $\left| \alpha \right\rangle$ in its Fock basis $\left| n \right\rangle$:

$$\left| \alpha \right\rangle = \sum_{n=0}^{\infty} c_n \left| n \right\rangle . \tag{4.67}$$

We then have

$$\hat{a} \left| \alpha \right\rangle = \sum_{n=0}^{\infty} c_n \hat{a} \left| n \right\rangle = \sum_{n=1}^{\infty} c_n \sqrt{n} \left| n-1 \right\rangle = \sum_{n=0}^{\infty} c_{n+1} \sqrt{n+1} \left| n \right\rangle \overset{!}{=} \alpha \sum_{n=0}^{\infty} c_n \left| n \right\rangle . \tag{4.68}$$

It follows that

$$c_n = \frac{\alpha}{\sqrt{n}} c_{n-1} = \frac{\alpha}{\sqrt{n}\sqrt{n-1}} c_{n-2} = \ldots = \frac{\alpha^n}{\sqrt{n!}} c_0 . \tag{4.69}$$

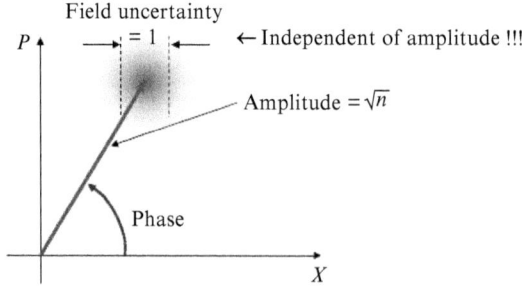

Fig. 4.12 [Colour online] Phase-space diagram for a light field in a coherent state.

---

[4] The corresponding commutator for $\hat{X}$ and $\hat{P}$ is $[\hat{X}, \hat{P}] = \frac{1}{2}\mathrm{i}$.

Normalizing the expression,

$$\sum_{n=0}^{\infty} \left| \frac{\alpha^n}{\sqrt{n!}} c_0 \right|^2 \overset{!}{=} 1 \,, \tag{4.70}$$

gives the coherent state $|\alpha\rangle$, written in its Fock basis $|n\rangle$:

$$|\alpha\rangle = e^{-\frac{1}{2}|\alpha|^2} \sum_{n} \frac{\alpha^n}{\sqrt{n!}} |n\rangle \,. \tag{4.71}$$

The Wigner function of a coherent state is given by

$$W_\alpha(\beta) = \frac{2}{\beta} e^{-2|\alpha-\beta|^2} \,, \tag{4.72}$$

with $\alpha = X_0 + iP_0$ defining the position of the maximum of the Wigner function and $\beta = X + iP$ being the phase-space vector. The Wigner function of a coherent state represented in phase space is visualized in Fig. 4.13.

4.3.2 Wigner function and quasi-probability distributions in general

A quantum state can be fully described by different concepts, for example the wavefunction, the density operator, or the following quasi-probability distributions: the Glauber–Sudarshan distribution $P(\alpha)$, the Wigner distribution $W(\alpha)$, and the Husimi–Kano distribution $Q(\alpha)$. These all contain the entire information about the state, but are different mathematical descriptions of the quantum phase space and, more importantly, represent different ways of measuring a quantum state.

The wavefunction can easily be written for pure states such as coherent states, Fock states, squeezed states, or the vacuum state, but a mixed or thermal state cannot be expressed by its wavefunction. Therefore, the representation of a state with the help of a quasi-probability distribution is often more convenient.

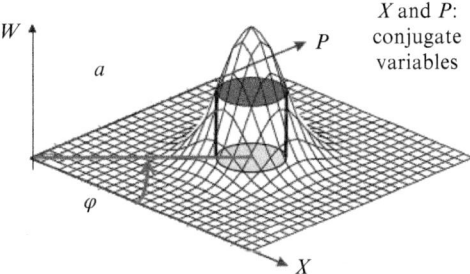

Fig. 4.13 [Colour online] $\hat{X}$ and $\hat{P}$ are quadratures of the field, with $\hat{X} = \frac{1}{2}(\hat{a} + \hat{a}^\dagger)$ and $\hat{P} = (1/2i)(\hat{a} - \hat{a}^\dagger)$. The displaced Gaussian represents a coherent state with eigenvalue $\alpha = X_0 + iP_0$.

In order to describe a measurement process, we can choose the description with the field operators $\hat{a}$ and $\hat{a}^\dagger$. Depending on the measurement, there are different ways of ordering the field operators:

- **normal** ordering $\hat{a}^\dagger \hat{a}$: **direct detection** (photodiode/camera);
- **symmetric** ordering $\frac{1}{2}(\hat{a}^\dagger \hat{a} + \hat{a}\hat{a}^\dagger)$: **homodyne detection**;
- **antinormal** ordering $\hat{a}\hat{a}^\dagger$: **heterodyne detection** (or double-homodyne detection).

Direct detection corresponds to the **Glauber–Sudarshan distribution** $P(\alpha)$. The operators are **normally ordered**, which can be explained with the annihilation operator acting first, diminishing the photon number by 1, when one photon is detected [23]. For a coherent state, the P-distribution is a δ-distribution with $P(\alpha) = \infty$ at the mean value of the state. For a thermal state, the P-distribution is a Gaussian function, and for Fock states and squeezed states, it evolves to a distribution that is singular and contains finite (Fock) and infinite (squeezed) derivatives of the δ-distribution [48]. Thus, it might be complicated to use this description in general.

The **Wigner distribution** $W(\alpha)$, which was already mentioned for a coherent state in (4.57), is formally the P-distribution convoluted with the vacuum state. Experimentally, it can be measured by **homodyne detection**, with the setup shown in Fig. 4.14. To measure the Wigner function, the angle φ, which is the phase shift between the local oscillator and the signal, has to be varied from 0 to π. At each angle ϕ, the homodyne measurement yields a histogram that corresponds to a projection of the Wigner function onto the axis that is defined by φ. The phase-space distribution is then obtained by tomographic reconstruction [43, 54], namely an inverse Radon transform.

For the operators describing the properties of the Wigner function, such as $\hat{X}^2 + \hat{P}^2$, one has to arrange the field operator accordingly, for this case by **symmetric ordering**. The Wigner function has the advantage that it has a Gaussian shape for

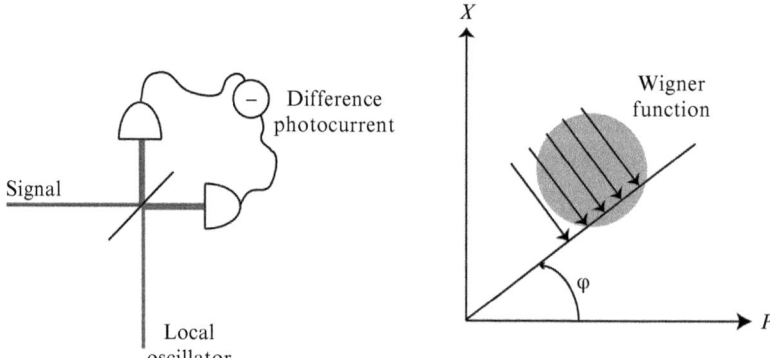

Fig. 4.14 [Colour online] Homodyne detection and the principle of Wigner function reconstruction.

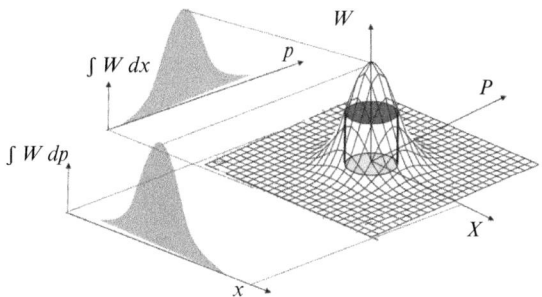

Fig. 4.15 [Colour online] Vacuum state, with $W_0(\alpha) = (2/\pi)e^{-2|\alpha|^2}$.

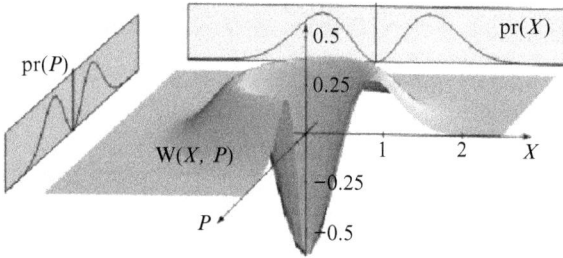

Fig. 4.16 [Colour online] Photon number state in phase space, with $W_1(\alpha) = (2/\pi)(4|\alpha|^2 - 1)e^{-2|\alpha|^2}$.

several commonly used states such as coherent states, squeezed states, thermal states, and many more.

The Wigner function for the vacuum state is depicted in Fig. 4.15 and that for a photon number state in Fig. 4.16. As can be seen in the latter, a phase-space distribution function may have negative values. Therefore, it is called a *quasi-probability distribution*. The *marginal distributions* instead are positive, i.e. any projection onto a line involving a one-dimensional integration such as in the case of homodyne detection.

Yet another convolution, namely that of the Wigner function with the vacuum state, gives us access to the **Husimi–Kano distribution** $Q(\alpha)$. In the experiment, we also have to add a second time the vacuum state, which results in the **heterodyne or double-homodyne measurement** setup, shown in Fig. 4.17. We have direct access to the continuous variables $\hat{X}$ and $\hat{P}$, but the setup is also more complicated. It involves two more beam splitters and a second local oscillator, which is shifted by $\pi/2$ relative to the first one. The problem of simultaneous measurement of two conjugate variables was discussed early by Arthurs and Kelly [2]. The Q-function is based on **antinormal ordering** of the field operators.

Any quasi-probability distribution can be calculated from any other via convolution or deconvolution with the vacuum state. However, deconvolution is often mathematically not a very stable operation, because the experimentally determined Wigner or Husimi–Kano functions are affected by fundamental quantum and technical noise. So it

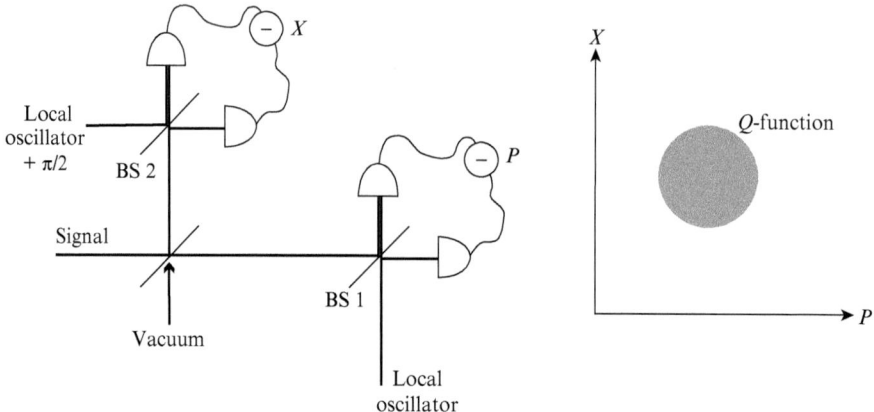

Fig. 4.17 [Colour online] Heterodyne measurement setup.

would be the best to directly measure the P-function—which is, unfortunately, in general not straightforward. There is, however, a way to determine the Wigner function based on the measurement of the photon number distribution $P(\alpha)$ using a photon number resolving detector [5, 33, 35, 41, 42, 45]. The sum $\sum_n (-1)^n P(n) = W(0,0)$ gives the Wigner function at the origin $(0,0)$. By first displacing the distribution in phase space, $W(X,P)$ can be mapped. But even this normal-ordering-based measurement yields $P(\alpha)$ only by way of $W(X,P)$. Table 4.1 provides an overview of the different quasi-probability distributions and the corresponding measurement setups.

Table 4.1 Overview of quasi-probability distributions.

Ordering	Normal	Symmetric	Antinormal
Energy	$\langle n| \hat{a}^\dagger \hat{a} |n\rangle = n$	$\langle n| \frac{1}{2}(\hat{a}^\dagger \hat{a} + \hat{a}\hat{a}^\dagger) |n\rangle = n + \frac{1}{2}$	$\langle n| \hat{a}\hat{a}^\dagger |n\rangle = n+1$
Detection scheme	Direct detection: click detector, photon number resolving	Homodyne: 4-port detection	Double-homodyne: 8-port detection
Determining phase-space distribution	Reconstruction by deconvoluting the Wigner function	Tomographic reconstruction from homodyne data	Phase-space distribution directly measured with heterodyne detection
Corresponding representation	P-distribution	Wigner function	Q-function

4.3.2.1 Superposition of coherent states

A more complex example of a Wigner function with negative eigenvalues is a *cat state*:

$$|\Psi_{\text{even cat}}\rangle = \frac{1}{\sqrt{2(1 + e^{-2|\alpha|^2})}} \left(|\alpha\rangle + |-\alpha\rangle\right), \tag{4.73}$$

which is depicted in Fig. 4.18. It is the superposition of two coherent states. By means of this example, the analogy of quantum phase space with diffraction optics is easily understandable. Figure 4.19 shows the top view of a three-dimensional Wigner function of a cat state. Integration along the X-axis gives two maxima, which can be compared with the optical field through a double slit. When integrating along the P-axis instead, we find an interference pattern, as it is observed on a screen far away from a double slit. In the double-slit experiment, the interference pattern and the field distribution of the double slit are mathematically connected by the Fourier transform. Thus, we can understand, by means of analogy, that the conjugate variables X and P are likewise

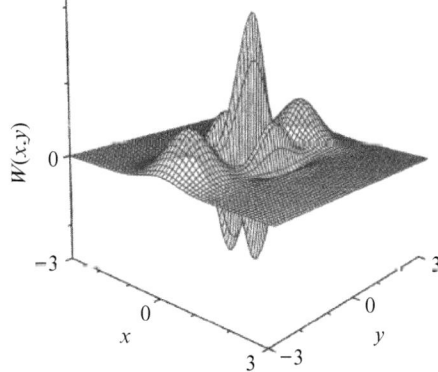

Fig. 4.18 [Colour online] Wigner function of a cat state.

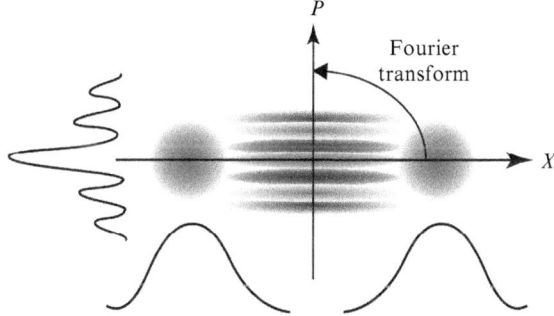

Fig. 4.19 [Colour online] Top view of the Wigner function of a cat state, also showing the marginal distributions. Note the close analogy with diffraction in classical optics.

connected by Fourier transform. If we integrate along another axis with a certain skewed angle in phase space, we obtain all the different interference patterns that can be found when observing the light field at different distances from the double slit.

4.3.2.2 Thermal state

The Wigner function is a helpful tool to represent also more complicated states, such as mixed states, which cannot be described by a wavefunction. As an example, we discuss a thermal state. The Wigner function of the vacuum with added thermal noise is

$$W_{\text{th}}(\alpha) = \frac{1}{\pi(\langle n\rangle + \frac{1}{2})} \exp\left(-\frac{1}{\langle n\rangle + \frac{1}{2}} |\alpha|^2\right). \tag{4.74}$$

The phase-space representation of such a thermal state is shown in Fig. 4.20. It is similar to a coherent state, but the width is larger (higher variance) because, as well as the quantum noise, classical or thermal noise is also added.

4.3.3 Detection

When measuring a light field, we need to take into account the specifications of the detection scheme and the laser source. The spectral density of a light field can be measured by a spectrum analyser:

$$S_I(f) = \int_{-\infty}^{\infty} \underbrace{\langle I(t)I(t+\tau)\rangle}_{\equiv G^{(2)}(\tau)} \cos(2\pi f\tau)\,\mathrm{d}\tau\,, \tag{4.75}$$

First of all, the electronic noise of the spectrum analyser and the detector have to be low enough if one wants to detect signals in the quantum regime.[5]

In the lower-frequency range, thermal noise from the laser cavity is added to the quantum noise, which is detrimental to measurements in the quantum regime

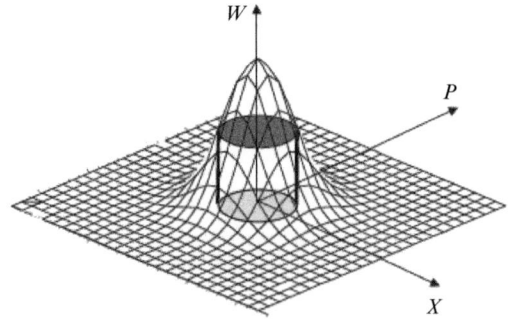

Fig. 4.20 [Colour online] Phase-space representation of a thermal state. The distribution is Gaussian, as for the coherent state, but the variances of X and P are larger.

---

[5] For a given electronic amplifier, the gain is inversely proportional to the bandwidth: the higher the bandwidth, the smaller the gain, and hence the sensitivity of the detector.

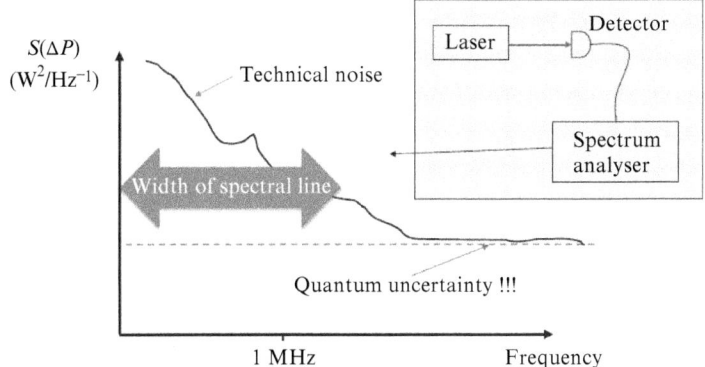

Fig. 4.21 [Colour online] Spectrum of the power fluctuations of laser light.

(Fig. 4.21). For quantum noise, the variance of the photon number fluctuations is proportional to the mean intensity of the signal:

$$\langle \Delta n^2 \rangle = \langle n \rangle, \tag{4.76}$$

corresponding to Poisson statistics of the photon number, as detailed below. Instead, for thermal noise, the variance of the photon number is proportional to the squared mean intensity of the signal

$$\langle n^2 \rangle = 2\langle n \rangle^2 + \langle n \rangle, \tag{4.77}$$

yielding the root mean squared photon number fluctuation.

$$\langle \Delta n^2 \rangle = \langle n^2 \rangle - \langle n \rangle^2 = 2\langle n \rangle^2 + \langle n \rangle - \langle n \rangle^2 = \langle n \rangle^2 + \langle n \rangle. \tag{4.78}$$

For high photon numbers, the quadratic term $\langle n \rangle^2$ is dominant and the variance is proportional to the square of the mean value:

$$\langle \Delta n^2 \rangle \approx \langle n \rangle^2. \tag{4.79}$$

Dependent on the laser source, thermal noise can go up to 1–10 MHz. Thus, quantum optics measurements have to be carried out in a frequency range above this limit.

A quantum von Neumann measurement projects the state onto one of its eigenstates. Note that a coherent state, with its intrinsic quantum uncertainty, is a stationary state. But measuring it in a defined time interval projects the state to a fixed value that fluctuates over time. The quantum uncertainty is thus translated to fluctuations in time via the measurement process (Fig. 4.22).

4.3.3.1 Direct detection

If one converts the signal photons directly into electrons in a photodetector without any admixture of auxiliary light beams, this detection process is called 'direct detection' (Fig. 4.23).

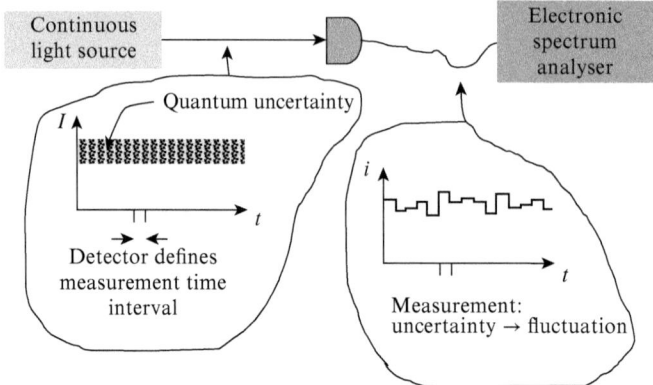

Fig. 4.22   [Colour online] Quantum uncertainty in measurement.

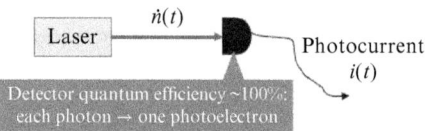

Fig. 4.23   [Colour online] Direct detection via a photodiode.

Coherent states refer to ideal laser radiation when ignoring additional classical noise and the inherent phase diffusion. We assume that the photon number is $\langle n \rangle \gg 1$. For the mean value of the photon number, we get

$$n = \langle n \rangle + \Delta n, \qquad \langle \Delta n \rangle = 0 \,. \tag{4.80}$$

But what is the variance of the photon number, $\langle \Delta n^2 \rangle$? We can write the photon number as (see Fig. 4.24)

$$\langle n \rangle + \Delta n = [\langle X(\theta) \rangle + \Delta X(\theta)]^2$$
$$= \langle X(\theta) \rangle^2 + 2 \langle X(\theta) \rangle \Delta X(\theta) + \Delta X^2(\theta) \,. \tag{4.81}$$

From $\langle X \rangle^2 = \langle n \rangle$, $2\langle X \rangle \Delta X = \Delta n$, and, neglecting ΔX^2, it follows that

$$\langle \Delta n^2 \rangle = 4 \langle n \rangle \times \tfrac{1}{4} = \langle n \rangle \,. \tag{4.82}$$

Thus, the photon noise in a laser beam underlies *Poissonian statistics*, as stated above. Here it becomes clear that this Poissonian statistics is the result of the uncertainty in phase space being independent of the quadrature amplitude.

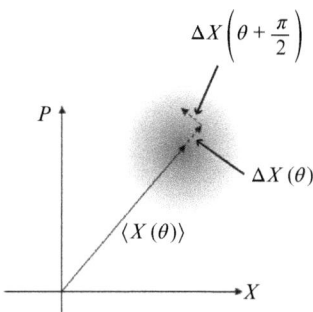

Fig. 4.24 [Colour online] Coherent state in phase space. θ indicates the direction of the displacement in phase space.

The P-distribution, which is the underlying representation of the quantum state for direct detection, cannot be derived directly with help of a photon number resolving setup [50]. It has to be reconstructed from the Wigner function, which itself can be measured by homodyne detection, as explained in Section 4.3.3.2. But the reconstruction causes many problems with real data owing to the presence of singularities. Nevertheless, the P-distribution is used to verify non-classicality by negativity. With data from a homodyne measurement, an estimation of this distribution (e.g. by using so-called pattern functions) may be sufficient [32].

4.3.3.2   Homodyne detection

The convolution of the signal state with the vacuum state will give us access to the Wigner function. The corresponding experimental setup for this convolution is homodyne detection. A sketch can be found in Fig. 4.25. As already mentioned, the integration of the Wigner function along a certain angle in two-dimensional phase space will yield a one-dimensional marginal distribution. This marginal distribution is measured in homodyne detection, the angle being determined by the phase of the coherent auxiliary beam called the local oscillator. The experimentally determined marginal

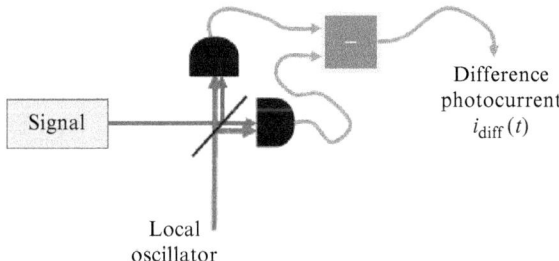

Fig. 4.25 [Colour online] Homodyne detection. The local oscillator describes a mode in an intense coherent state.

distributions taken for various angles are the input data needed for tomographic reconstruction [43].

Let us assume a strong local oscillator (Fig. 4.26):

$$\sqrt{\langle n_{\text{lo}} \rangle} = \langle X_{\text{lo}}(\phi) \rangle \gg X_{\text{signal}}, \Delta X . \tag{4.83}$$

Then we can neglect all terms quadratic in the small signal for the difference photocurrent:

$$
\begin{aligned}
i_{\text{diff}} &\propto [\langle X_{\text{lo}}(\varphi) \rangle + \Delta X_{\text{lo}}(\varphi) + \langle X_{\text{signal}}(\theta) \rangle \cos(\varphi - \theta) + \Delta X_{\text{signal}}(\varphi)]^2 \\
&\quad - [\langle X_{\text{lo}}(\varphi) \rangle + \Delta X_{\text{lo}}(\varphi) - \langle X_{\text{signal}}(\theta) \rangle \cos(\varphi - \theta) - \Delta X_{\text{signal}}(\varphi)]^2 \qquad (4.84) \\
&\approx 4 \langle X_{\text{lo}}(\varphi) \rangle [\langle X_{\text{signal}}(\varphi) \rangle + \Delta X_{\text{signal}}(\varphi)] ,
\end{aligned}
$$

where the minus signs inside the square brackets on the second line result from the phase change imposed by the beam splitter (for details, see Section 4.4). By variation of φ, we can measure projections of the signal on all directions, providing the input data for *quantum state tomography* [43].

Another approach to reconstruct the Wigner function of a quantum state has been presented using data obtained in direct detection with photon number resolving detectors [5, 33, 35, 41, 42, 45]. Using direct detection, the probability of each photon number $P(n)$ is estimated by measurement, and it has been found that the sum over

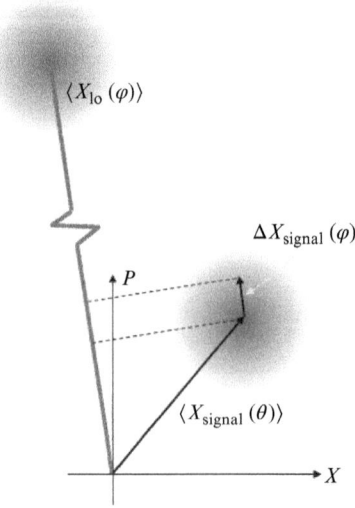

Fig. 4.26 [Colour online] Local oscillator and signal in phase space.

these $P(n)$ with alternating signs gives the value of the Wigner function at position (0,0) in phase space:

$$\sum_n (-1)^n P(n) = W(0,0) \,. \tag{4.85}$$

Shifting the quantum state via an asymmetric beam splitter with a local oscillator prior to detection—in much the same as in homodyne detection—one can also reconstruct the Wigner function in the whole phase space. Note that the measurement requires convolution with a vacuum state (symmetric ordering) and not only direct detection (normal ordering), and hence the procedure yields the Wigner function.

4.3.3.3 *Heterodyne detection*

For many applications, such as quantum key distribution, it is convenient to directly measure the Q-function of the quantum states. This can be performed with a heterodyne measurement or with so-called double-homodyne measurement or eight-port homodyne detection [20]. For this purpose, a second local oscillator (LO) is inserted into the setup (see Fig. 4.17), which is phase-shifted by $\pi/2$ in order to measure both quadratures X and P simultaneously. In real experiments, it is difficult to stabilize the phase between the LO at beam splitter 1 and the LO $+ \pi/2$ phase shift at beam splitter 2. Therefore, another degree of freedom of the optical mode is often taken into account: the polarization (Fig. 4.27). Only one local oscillator is mixed to the signal, as orthogonally polarized beams. With the help of a quarter-wave plate, the relative phase between the orthogonally polarized signal and the local oscillator field is changed by $\pi/2$, entering one of the two polarization beam splitters in order to measure the two conjugate quadratures at the two detector pairs. This setup is used for example as a receiver in quantum key distribution systems [16, 30].

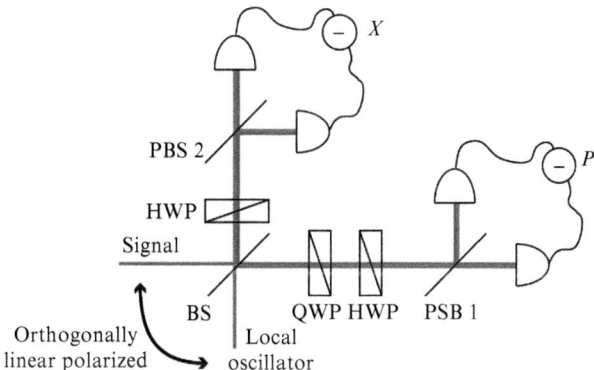

Fig. 4.27 [Colour online] Heterodyne or double-homodyne measurement with the help of the polarization of the light field. BS, beam splitter; PBS, polarization beam splitter; HWP, half-wave plate, used for aligning the polarization of the signal and the LO; QWP, quarter-wave plate, aligned with one of the polarization directions of the signal or the LO.

4.3.4 Squeezing the quantum uncertainty

A single-mode state of a light field can generally be described as an expansion of photon number states, known as the Fock basis:

$$|\Psi\rangle = \sum_{n=0}^{\infty} c_n |n\rangle \,, \tag{4.86}$$

with an n-photon *Fock state* $|n\rangle$. The vacuum state is written as $|0\rangle$ and its uncertainty for both quadratures $\hat{X}$ and $\hat{P}$ is

$$\langle \Delta \hat{X}^2 \rangle = \tfrac{1}{4} = \langle \Delta \hat{P}^2 \rangle \,, \tag{4.87}$$

whereas the mean value is zero: $\langle \hat{X} \rangle = \langle \hat{P} \rangle = 0$.

Let us consider a state that consists of the vacuum state and a small admixture of a one-photon state $|1\rangle$, written as $|\Psi\rangle = (|0\rangle + \varepsilon |1\rangle)/\sqrt{1 + |\varepsilon|^2}$, with $|\varepsilon| \ll 1$. The mean value of this state can be calculated as

$$\langle \hat{X} \rangle = \langle \Psi | \hat{X} | \Psi \rangle = \frac{1}{2(1 + |\varepsilon|^2)} (\varepsilon + \varepsilon^*) \approx \tfrac{1}{2}(\varepsilon + \varepsilon^*) \,, \tag{4.88}$$

with the help of (4.60). Analogously, it follows that

$$\langle \hat{P} \rangle = \frac{1}{2\mathrm{i}(1 + |\varepsilon|^2)} (\varepsilon - \varepsilon^*) \approx \frac{1}{2\mathrm{i}} (\varepsilon - \varepsilon^*) \,. \tag{4.89}$$

Deriving the variance of both quadratures and neglecting terms of $|\varepsilon|^2$, we obtain

$$\langle \Delta \hat{X}^2 \rangle = \langle \Psi | \hat{X}^2 | \Psi \rangle - \langle \Psi | \hat{X} | \Psi \rangle^2 \approx \tfrac{1}{4} \,, \tag{4.90}$$

$$\langle \Delta \hat{P}^2 \rangle = \langle \Psi | \hat{P}^2 | \Psi \rangle - \langle \Psi | \hat{P} | \Psi \rangle^2 \approx \tfrac{1}{4} \,. \tag{4.91}$$

We see that in this case, the variance is independent of the parameter ε. This means that the small admixture of a one-photon state causes a shift of the vacuum state without changing its shape in phase space, i.e. its uncertainty (Fig. 4.28).

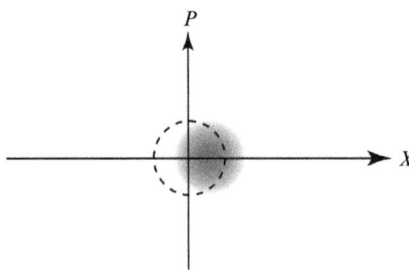

Fig. 4.28 [Colour online] Phase space: small admixture of a one-photon state to the vacuum state.

Proceeding in the same way for a small admixture of a two-photon state $|2\rangle$, written as $|\Psi\rangle = (|0\rangle + \varepsilon |2\rangle)/\sqrt{1 + |\varepsilon|^2}$ with $\varepsilon \ll 1$, gives a different result. The mean values are calculated to be

$$\langle \hat{X} \rangle = \langle \Psi | \hat{X} | \Psi \rangle \approx 0 \,, \tag{4.92}$$

$$\langle \hat{P} \rangle = \langle \Psi | \hat{P} | \Psi \rangle \approx 0 \,, \tag{4.93}$$

but the variance becomes (again neglecting terms of $|\varepsilon|^2$)

$$\langle \Delta \hat{X}^2 \rangle = \langle \Psi | \hat{X}^2 | \Psi \rangle - \langle \Psi | \hat{X} | \Psi \rangle^2 \approx \tfrac{1}{4} \left[1 + \sqrt{2}(\varepsilon + \varepsilon^*) \right] \,, \tag{4.94}$$

$$\langle \Delta \hat{P}^2 \rangle = \langle \Psi | \hat{P}^2 | \Psi \rangle - \langle \Psi | \hat{P} | \Psi \rangle^2 \approx \tfrac{1}{4} \left[1 - \sqrt{2}(\varepsilon + \varepsilon^*) \right] \,. \tag{4.95}$$

Thus, adding a two-photon state does not change the mean amplitude of a state but only the variance of both quadratures in an inverse way: when $\hat{X}$ decreases, $\hat{P}$ will increase, and vice versa. One quadrature is squeezed at the expense of a larger variance of the other quadrature (Fig. 4.29).

We see that the generation of such a state, a *squeezed* state, is based on a process where the simultaneous creation of **two photons** is involved, such as parametric down-conversion or other nonlinear effects. Note also that the area in phase space stays constant, with

$$\langle \Delta \hat{X}^2 \rangle \langle \Delta \hat{P}^2 \rangle - 1 - 2(\varepsilon + \varepsilon^*)^2 \tag{4.96}$$

in the approximation we used throughout this derivation, namely neglecting terms at order ε^2. Without this approximation, the area will still be constant.

The Wigner function of a squeezed state is depicted in Fig. 4.30. In the Hamilton operator, squeezing follows from the '$(\hat{a}^\dagger)^2$' term:

$$\hat{H} = \hbar\omega \left(a^\dagger a + \tfrac{1}{2} \right) + \hbar\gamma (\hat{a}^\dagger)^2 \,. \tag{4.97}$$

The Hamilton operator determines the time evolution operator $U = \mathrm{e}^{\mathrm{i}\hat{H}t/\hbar}$. Under this time evolution operator, the vacuum state evolves as follows:

$$\mathrm{e}^{\mathrm{i}\hat{H}t/\hbar} |0\rangle \approx \left[1 + \mathrm{i}\omega \left(\hat{a}^\dagger \hat{a} + \tfrac{1}{2} \right) t + \mathrm{i}\gamma (\hat{a}^\dagger)^2 t \right] |0\rangle$$

$$= \left(1 + \tfrac{1}{2}\mathrm{i}\omega t \right) |0\rangle + \mathrm{i}\gamma t \sqrt{2} \, |2\rangle \,. \tag{4.98}$$

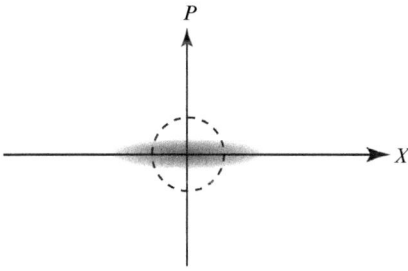

Fig. 4.29 [Colour online] Phase space: squeezed vacuum state.

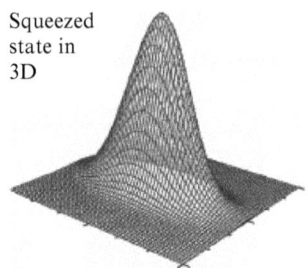

Squeezed
state in
3D

Fig. 4.30 [Colour online] Wigner function of a squeezed state. The height represents the amplitude of the Wigner function and the base plane is the phase space spanned by the two quadratures.

The latter equation suggests that squeezing can always be observed in a nonlinear interaction containing a quadratic term in the field operator, but one has to take account of the losses, which are detrimental for the observation of squeezing.

4.3.4.1   *Types of squeezing*

In general, the squeezing axis can be oriented under a skewed angle with respect to the quadratures axes. If the centre of gravity is displaced away from the origin, this displacement defines the mean excitation. It is often convenient to use rotated quadratures $X(\theta)$, $P(\theta)$ such that one represents the amplitude and the other one the phase quadrature (Fig. 4.31):

$$\langle\Delta X^2(\theta)\rangle\langle\Delta X^2(\theta+\pi/2)\rangle \geq \tfrac{1}{16}\,. \tag{4.99}$$

Squeezing along the direction θ_{sq} means

$$\langle\Delta X^2(\theta_{\mathrm{sq}})\rangle < \tfrac{1}{4} \quad\text{and}\quad \langle\Delta X^2(\theta_{\mathrm{sq}}+\pi/2)\rangle > \tfrac{1}{4}\,, \tag{4.100}$$

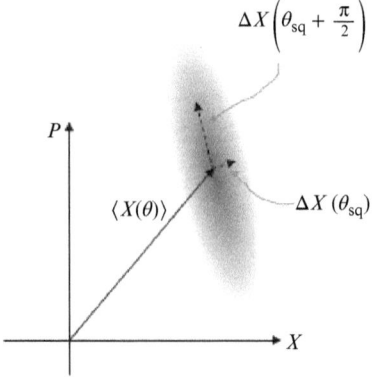

Fig. 4.31 [Colour online] Squeezing along the squeezing angle

which allows a classification of three different squeezing operations:

- **skewed squeezing** if $\theta_{sq} \neq \theta, \theta + \pi/2$, such as the Kerr effect [49] (cf. Fig. 4.31);
- **amplitude squeezing** if $\theta_{sq} = \theta$, such as the Kerr effect in combination with an asymmetric fibre-Sagnac interferometer [14, 31, 47, 49];
- **phase squeezing** if $\theta_{sq} + \pi/2 = \theta$, such as with an optical parametric amplifier, four-wave mixing [4].

4.3.5 Intensity correlations

The correlation functions are generally derived from the wavefunction, which requires knowledge of the wavefunction of a state, which is not always straightforward or even possible, for example for a thermal state. In Sections 4.3.2–4.3.4, we convinced ourselves that the Wigner function is a convenient representation of a quantum state in phase space, especially for the common states, pure or mixed, for which the Wigner function is Gaussian (thermal, coherent, or squeezed). The Wigner function is measured by a homodyne setup, projecting the state onto one axis, where the projection is always positive. Thus, it would be practical to also use the Wigner function for calculating the correlation functions.

Let us first start with a direct detection setup (Fig. 4.32). The detection process requires *normal ordering*. Hence, all dagger operators have to be on the left! The normal ordering is justified in direct detection, where the annihilation operator decreases the photon number by 1, when one photon is detected.

Given an initial state $|i\rangle$ and a final state $|f\rangle$, the probability amplitude for the detection of a photon at time t is

$$\langle f| \, \hat{a}(t) \, |i\rangle \,, \tag{4.101}$$

whereas the probability amplitude for detecting a second photon at time t' conditioned on the first detection at time t is

$$\langle f| \, \hat{a}(t')\hat{a}(t) \, |i\rangle \,. \tag{4.102}$$

Using this probability amplitude, the probability of a correlation, i.e. for jointly detecting one photon at time t and the other photon at time t', can be derived by taking the absolute squares of (4.102) and summing over all final states $|f\rangle$:

$$\sum_f |\,\langle f| \, \hat{a}(t')\hat{a}(t) \, |i\rangle\,|^2 = \sum_f \langle i| \, \hat{a}^\dagger(t)\hat{a}^\dagger(t') \, |f\rangle \, \langle f| \, \hat{a}(t')\hat{a}(t) \, |i\rangle$$

$$= \langle i| \, \hat{a}^\dagger(t)\hat{a}^\dagger(t') \left(\sum_f |f\rangle \, \langle f| \right) \hat{a}(t')\hat{a}(t) \, |i\rangle$$

$$= \langle i| \, \hat{a}^\dagger(t)\hat{a}^\dagger(t')\hat{a}(t')\hat{a}(t) \, |i\rangle \,. \tag{4.103}$$

Fig. 4.32 [Colour online] Direct detection scheme.

The ordering is thus determined by the type of measurement. Using the correspondence between the electric fields and the field operators $\hat{a}$ and $\hat{a}^\dagger$ gives

$$EE^* \to \hat{a}^\dagger \hat{a} \,, \tag{4.104}$$

$$E^* E E^* E \to \hat{a}^\dagger \hat{a}^\dagger \hat{a} \hat{a} \,. \tag{4.105}$$

One recognizes the classical correlation function of Section 4.2, with the difference that now the ordering matters.

Let us take the example of the intensity correlation function at $E = 0$:

$$g_I^{(2)} = \frac{\langle \hat{a}^\dagger \hat{a}^\dagger \hat{a} \hat{a} \rangle}{\langle \hat{a}^\dagger \hat{a} \rangle^2} \,. \tag{4.106}$$

Normal ordering is indicated by colons:

$$\langle : \hat{a}^\dagger \hat{a} \hat{a}^\dagger \hat{a} : \rangle = \langle \hat{a}^\dagger \hat{a}^\dagger \hat{a} \hat{a} \rangle \,. \tag{4.107}$$

If we erroneously assume that the intensity correlation is given by $\langle \hat{n}^2 \rangle / \langle \hat{n} \rangle^2$, we quickly see that a problem arises when we apply this to a coherent state, for which $\langle \Delta \hat{n}^2 \rangle = \langle \hat{n} \rangle$. Using this photon number variance, we get

$$\frac{\langle \hat{n}^2 \rangle}{\langle \hat{n} \rangle^2} = 1 + \frac{1}{\langle \hat{n} \rangle} \,. \tag{4.108}$$

But the $\tau = 0$ value of the intensity correlation function of a coherent state is 1. The error is that we did not normally order $\hat{n}^2$, as is appropriate for direct detection. For the proper correlation function, we have

$$g_I^{(2)} = \frac{\langle : \hat{n}^2 : \rangle}{\langle \hat{n} \rangle^2} = \frac{\langle \hat{a}^\dagger \hat{a}^\dagger \hat{a} \hat{a} \rangle}{\langle \hat{a}^\dagger \hat{a} \rangle^2} \,. \tag{4.109}$$

Since $\hat{a} |\alpha\rangle = \alpha |\alpha\rangle$ and $\langle \alpha | \hat{a}^\dagger = \langle \alpha | \alpha^*$ for a coherent state, we can conclude that the correct value for the correlation function is

$$\frac{\langle \hat{a}^\dagger \hat{a}^\dagger \hat{a} \hat{a} \rangle}{\langle \hat{a}^\dagger \hat{a} \rangle^2} = \frac{\alpha^* \alpha^* \alpha \alpha}{(\alpha^* \alpha)^2} \quad \Longrightarrow \quad g_I^{(2)} = 1 \,. \tag{4.110}$$

For the intensity correlation function defined for direct detection, we always have to use normal ordering, which corresponds to the P-distribution. But if we want to use the Wigner function, as stated at the beginning of this subsection, we need field operators that are ordered *symmetrically*. We will derive in the following the $g^{(2)}$ correlation function based on the Wigner function, for a coherent, a thermal, and a squeezed state (Fig. 4.33). We already know the results from textbooks, [6, 22, 40, 55]. These are the references with which the alternative derivation discussed will have to be compared [39].

We first calculate the photon number operator while using symmetric ordering; that is, we sum the terms with all possible different orderings and divide by the number

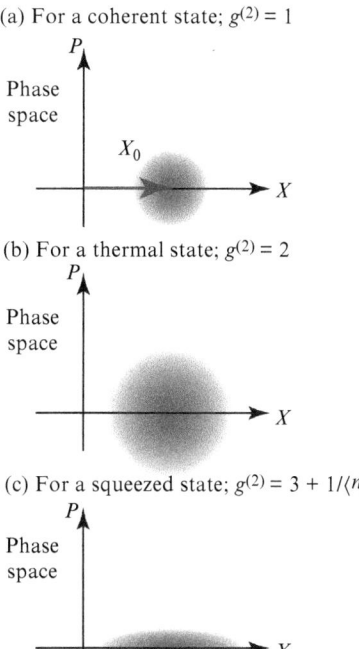

(a) For a coherent state; $g^{(2)} = 1$

(b) For a thermal state; $g^{(2)} = 2$

(c) For a squeezed state; $g^{(2)} = 3 + 1/\langle n \rangle$

Fig. 4.33 [Colour online] $g^{(2)}$ correlation functions for (a) a coherent, (b) a thermal, and (c) a squeezed state.

of terms. As a result, we obtain a relation between the photon number operators in normal and symmetric ordering, with the latter being indicated by 'Sym':

$$\text{Sym}(\hat{n}) = \tfrac{1}{2}(\hat{a}\hat{a}^\dagger + \hat{a}^\dagger\hat{a}) = \hat{n} + \tfrac{1}{2}$$

$$\implies \quad \hat{n} = \text{Sym}(\hat{n}) - \tfrac{1}{2}. \tag{4.111}$$

The operators $\hat{a}$ and $\hat{a}^\dagger$ have been replaced with help of the commutator: $[\hat{a}, \hat{a}^\dagger] = 1 = \hat{a}\hat{a}^\dagger - \hat{a}^\dagger\hat{a}$. In order to obtain $\text{Sym}(\hat{n}^2)$, we need to take into account the symmetric product of four field modes, twice the $\hat{a}$ and $\hat{a}^\dagger$ operators in all possible constellations, which results in

$$\text{Sym}(\hat{n}^2) = \tfrac{1}{6}(\hat{a}^\dagger\hat{a}^\dagger\hat{a}\hat{a} + \hat{a}\hat{a}\hat{a}^\dagger\hat{a}^\dagger + \hat{a}\hat{a}^\dagger\hat{a}\hat{a}^\dagger + \hat{a}^\dagger\hat{a}\hat{a}^\dagger\hat{a} + \hat{a}^\dagger\hat{a}\hat{a}\hat{a}^\dagger + \hat{a}\hat{a}^\dagger\hat{a}^\dagger\hat{a}). \tag{4.112}$$

The factor is $\tfrac{1}{6}$, since we have 6 different terms. This approach can be generalized to

$$\langle \text{Sym}(\hat{n})^k \rangle - \frac{1}{\binom{2k}{k}} \langle \hat{a}^k \hat{a}^{\dagger k} \mid \ldots \rangle, \tag{4.113}$$

but this is not straightforward at all. One can imagine (4.112) as the average of all possible constellations of $\hat{a}$, $\hat{a}$, $\hat{a}^\dagger$, and $\hat{a}^\dagger$. The reader can refer to a theoretical paper for details about the product of operators in symmetric ordering [10]. Using again the known operator constellations such as (4.62) and (4.65), the non-symmetric components can be rewritten for example as

$$\mathrm{Sym}(\hat{n}^2) = \hat{a}^\dagger \hat{a}^\dagger \hat{a}\hat{a} + 2\hat{a}^\dagger \hat{a} + \tfrac{1}{2}$$

$$= \hat{n}^2 + \hat{n} + \tfrac{1}{2}. \tag{4.114}$$

Note that $\mathrm{Sym}(\hat{n}^2)$ is not simply the square of $\mathrm{Sym}(\hat{n})$! This would give $\mathrm{Sym}(\hat{n})^2 = \hat{n}^2 + \hat{n} + \tfrac{1}{4}$. Hence, the second-order correlation function can be written with the photon number operator in symmetric ordering as

$$g^{(2)}(0) = \frac{\langle \hat{a}^\dagger \hat{a}^\dagger \hat{a}\hat{a} \rangle}{\langle \hat{a}^\dagger \hat{a} \rangle^2} = \frac{\langle \hat{n}^2 \rangle - \langle \hat{n} \rangle}{\langle \hat{n} \rangle^2} = \frac{\langle \mathrm{Sym}(\hat{n}^2) \rangle - 2\langle \mathrm{Sym}(\hat{n}) \rangle + \tfrac{1}{2}}{\left[\langle \mathrm{Sym}(\hat{n}) \rangle - \tfrac{1}{2} \right]^2}. \tag{4.115}$$

With the help of this expression, we will derive $g^{(2)}(0)$ for different states. We have already stressed that the Wigner function corresponds to symmetric ordering of the field operators. This fact greatly facilitates the calculation of the expectation values of symmetrized powers of the photon number operator, such that one just has to perform an integration, averaging the Wigner function. In phase space, the photon number n is related to the sum of the squares of two orthogonal quadratures:

$$\mathrm{Sym}(\hat{n}) \quad \rightarrow \quad X^2 + P^2, \tag{4.116}$$

and the expectation value of a power of $\hat{n}$ is

$$\langle \mathrm{Sym}(\hat{n}^k) \rangle = \int (X^2 + P^2)^k W(X, P) \, \mathrm{d}X \, \mathrm{d}P. \tag{4.117}$$

Evaluation of this integral is straightforward and, for large k, (4.117) is very convenient. The calculation is particularly simple if the Wigner function is a Gaussian $W(X, P) = N\mathrm{e}^{-aX^2 - bP^2}$. To give an example,

$$\langle \mathrm{Sym}(\hat{n}) \rangle = N \int (X^2 + P^2)\mathrm{e}^{-aX^2 - bP^2} \, \mathrm{d}X \, \mathrm{d}P = \langle X^2 + P^2 \rangle. \tag{4.118}$$

In analogy to (4.23) and (4.24), we find

$$\langle x^4 \rangle = 3\langle x^2 \rangle^2. \tag{4.119}$$

Therefore, we can rewrite $\mathrm{Sym}(\hat{n}^2)$ as

$$\langle \mathrm{Sym}(\hat{n}^2) \rangle = \langle (X^2 + P^2)^2 \rangle$$

$$= \langle X^4 \rangle + 2\langle X^2 P^2 \rangle + \langle Y^4 \rangle$$

$$= \langle X^4 \rangle + 2\langle X^2 \rangle \langle P^2 \rangle + \langle P^4 \rangle$$

$$= 3\langle X^2 \rangle^2 + 2\langle X^2 \rangle \langle P^2 \rangle + 3\langle P^2 \rangle^2. \tag{4.120}$$

Let us first consider a **squeezed** vacuum state where the variances are as follows (refer to (4.90) and (4.91)):

$$\langle X^2 \rangle = \frac{\varepsilon}{4}\,, \qquad \langle P^2 \rangle = \frac{1}{4\varepsilon}\,, \tag{4.121}$$

with ε being the squeezing parameter. Thus,

$$\langle \mathrm{Sym}(\hat{n}^2) \rangle = 3 \times \frac{\varepsilon^2}{16} + 2 \times \frac{1}{16} + 3 \times \frac{1}{16\varepsilon^2} = \frac{3}{16}\left(\varepsilon^2 + \frac{1}{\varepsilon^2}\right) + \frac{1}{8} \tag{4.122}$$

and since

$$\langle \mathrm{Sym}(\hat{n}) \rangle = \frac{1}{4}\left(\varepsilon + \frac{1}{\varepsilon}\right) \implies \varepsilon^2 + \frac{1}{\varepsilon^2} = 16\langle \mathrm{Sym}(\hat{n}) \rangle^2 - 2\,, \tag{4.123}$$

we have

$$\langle \mathrm{Sym}(\hat{n}^2) \rangle = 3\langle \mathrm{Sym}(\hat{n}) \rangle^2 - \tfrac{1}{4}\,. \tag{4.124}$$

The intensity correlation function given in (4.115) then becomes

$$\begin{aligned}
g^{(2)}(0) &= \frac{3\langle \mathrm{Sym}(\hat{n}) \rangle^2 - 2\langle \mathrm{Sym}(\hat{n}) \rangle + \frac{1}{4}}{\left[\langle \mathrm{Sym}(\hat{n}) \rangle - \frac{1}{2}\right]^2} \\
&= \frac{3\langle \mathrm{Sym}(\hat{n}) \rangle^2 \quad 3\langle \mathrm{Sym}(\hat{n}) \rangle + \frac{3}{4} + \langle \mathrm{Sym}(\hat{n}) \rangle - \frac{1}{2}}{\left[\langle \mathrm{Sym}(\hat{n}) \rangle - \frac{1}{2}\right]^2} \\
&= 3 + \frac{1}{\langle \mathrm{Sym}(\hat{n}) \rangle - \frac{1}{2}}\,. \tag{4.125}
\end{aligned}$$

For experimental evidence for the factor 3, see [8, 28].

For a **thermal** state, we know that $\langle \Delta \hat{X}^2 \rangle = \langle \Delta \hat{P}^2 \rangle$, and with (4.120) and $\mathrm{Sym}(\langle \hat{n} \rangle) = 2\langle \hat{X}^2 \rangle$, we can write

$$\langle \mathrm{Sym}(\hat{n}^2) \rangle = 2\langle \mathrm{Sym}(\hat{n}) \rangle^2\,, \tag{4.126}$$

and the intensity correlation function given in (4.115) becomes

$$g^{(2)}(0) = 2\,. \tag{4.127}$$

For a **coherent** state $X = X_0 + \Delta X$, $P = \Delta P$ with $\langle \Delta \hat{X}^2 \rangle = \frac{1}{4}$ and $\langle \Delta \hat{P}^2 \rangle = \frac{1}{4}$, we find

$$\begin{aligned}
\langle \mathrm{Sym}(\hat{n}) \rangle &= \langle X_0^2 + \Delta X^2 + 2X_0 \Delta X + \Delta P^2 \rangle \\
&= X_0^2 + \tfrac{1}{2}\,, \tag{4.128}
\end{aligned}$$

$$\begin{aligned}
\langle \mathrm{Sym}(\hat{n}^2) \rangle &= X_0^4 + 2X_0^2 + \tfrac{1}{2} \\
&= \langle \mathrm{Sym}(\hat{n}) \rangle^2 + \langle \mathrm{Sym}(\hat{n}) \rangle - \tfrac{1}{4}\,, \tag{4.129}
\end{aligned}$$

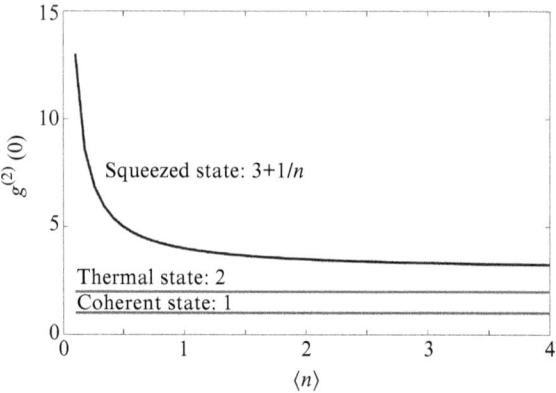

Fig. 4.34 [Colour online] Correlation function $g^{(2)}(0)$ for a coherent state, a squeezed state, and a thermal state.

and the intensity correlation function given in (4.115) becomes

$$g^{(2)}(0) = 1 \,. \tag{4.130}$$

These cases are shown in Fig. 4.34. Obviously, it is not possible to define a consistent value of $g^{(2)}(0)$ for the vacuum state, because different ways of approaching the limit suggests different values. This may seem disturbing at first sight, but it is not, because it is impossible to do direct detection of the vacuum state. The general procedure for calculating quantities such as the intensity correlation function through averaging with the help of the Wigner function is generally applicable and is not limited to Gaussian states.

4.4 Quantum optics of several modes

In this section, the interaction between modes will be considered. Modes can differ in their spatial and their temporal properties. An important component for the interaction is the beam splitter, which will be discussed in detail at the beginning. Moreover, the interference of a carrier with vacuum frequency sidebands as an essential aspect of squeezing the quantum uncertainty will be explained. Finally, we will have a look on the Bogoliubov transformation and how it can describe different processes such as squeezing, amplification, phase conjugation, and attenuation in a unified way.

4.4.1 Continuous variables and beam splitters

Quantum systems can often be described by a stochastic model. For certain quantum states, a stochastic description is completely sufficient, for others it cannot be used owing to their specific nature. A famous example is the Bell inequality [7], a violation of which in experiment is a clear sign that a stochastic model will not work. However, a coherent state with its mean value presents a 'simple' quantum state whose Wigner function is not negative and it can easily be described with a Gaussian state. Hence, a stochastic interpretation is appropriate and practical for the purpose of describing several quantum modes and their correlation. For a field in a coherent state, the uncertainties in amplitude and phase direction are the same and the contour is circular. Figure 4.35 shows a phase diagram where the contour of the Wigner function at half maximum is indicated. Let us now introduce two orthogonal arrows that span the circular region of uncertainty of the field. In order to mimic the full uncertainty area, one has to form a linear superposition of these two arrows with stochastic coefficients of Gaussian statistics. For the coherent state, for example, there will be no correlation between the two stochastic variables. The two-dimensional phase space is hence spanned by two independent variables X and P with all possible values of the Gaussian distribution. When we multiply the basis vectors with a Gaussian distribution, we get all positions in the phase space with a certain probability. One such possible position is indicated in Fig. 4.35 by the dashed line. This approach holds in particular for all states with Gaussian Wigner function.

4.4.1.1 Beam splitter: phase relations

A beam splitter has two input and two output ports [38]. In a single-mode picture, each of these ports is associated with a spatial mode of the quantized electromagnetic field. The corresponding field operators are $\hat{a}_{\mathrm{in1}}$, $\hat{a}_{\mathrm{in2}}$, $\hat{a}_{\mathrm{out1}}$, and $\hat{a}_{\mathrm{out2}}$. They are related by the transmission and reflection coefficients r_1, r_2, t_1, and t_2 (Fig. 4.36):

$$\hat{a}_{\mathrm{out1}} = t_1\hat{a}_{\mathrm{in1}} + r_1\hat{a}_{\mathrm{in2}}\,, \tag{4.131}$$

$$\hat{a}_{\mathrm{out2}} = r_2\hat{a}_{\mathrm{in1}} + t_2\hat{a}_{\mathrm{in2}}\,. \tag{4.132}$$

Note that the coefficients are complex valued to account for both amplitude and phase.

The general relation between the corresponding coefficients for the electric fields was derived by G. G. Stokes (1849) in a remarkable paper that he based upon the

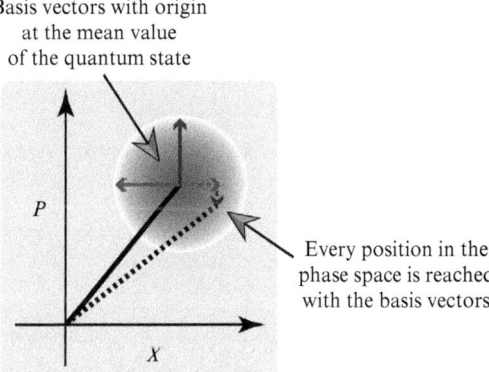

Fig. 4.35 [Colour online] ←, ↑ are the basis vectors. Multiplied by *stochastic Gaussian variables*, they span the *phase-space distribution*.

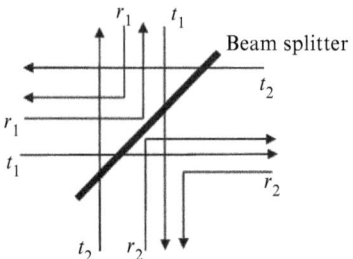

Fig. 4.36 Transmission and reflection at a beam splitter.

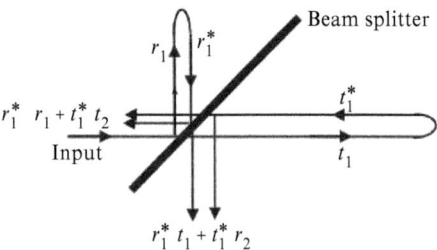

Fig. 4.37 Scenario for time reversal, transmission, and reflection at a beam splitter.

principle of time-reversal symmetry [51]. He considered one incoming ray being split by the beam splitter. Then he argued that if the two rays exiting the beam splitter are time-reversed, they will have to interfere upon re-arrival at the beam splitter such as to generate the time-reversed incoming beam (Fig. 4.37). In full generality, Stokes allowed for the reflection and transmission coefficients r_1, r_2, t_1, and t_2 to be all different in both amplitude and phase. Stokes' argument leads to the following formulae when using the complex notation (which he did not use). In this notation, time reversal corresponds to taking the complex conjugate of the wave amplitude. We offer a short excursion into the correspondence between time reversal and phase conjugation at the end of this subsection. Time reversal means that all the light goes back to input port 1 and no light emerges from input port 2. This leads to the following expressions at both inputs of the beam splitter (on the left and bottom in Fig. 4.37):

$$1 = r_1^* r_1 + t_1^* t_2 \,, \tag{4.133}$$

$$0 = r_1^* t_1 + t_1^* r_2 \,. \tag{4.134}$$

We rewrite $t_k = |t|e^{i\tau_k}$ and $r_k = |r|e^{i\rho_k}$, for $k = 1, 2$. Taking into account energy conservation $|r_k^2| + |t_k^2| = 1$, for $k = 1, 2$, we conclude that $t_1 = t_2 = t$ and $|r_1| = |r_2| = |r|$, and therefore

$$\tau_1 - \tau_2 = 0 + 2m\pi, \qquad m \in \mathbb{N} \tag{4.135}$$

and

$$0 = |r|e^{-i\rho_1} \cdot |t|e^{i\tau_1} + |t|e^{-i\tau_1} \cdot |r|e^{i\rho_2} \tag{4.136}$$

$$\implies \quad e^{-i\rho_1+i\tau_1} = -e^{-i\tau_1+i\rho_2} \tag{4.137}$$

$$\implies \quad \tau_1 - \rho_1 - \rho_2 = \pi + 2m\pi, \qquad m \in \mathbb{N}. \tag{4.138}$$

Without loss of generality, we can claim $\tau_1 = 0$ and $m = -1$, so

$$\rho_1 + \rho_2 = \pi. \tag{4.139}$$

This leads to two cases:

1. symmetric beam splitter, $\rho_1 = \rho_2 = \pi/2$, e.g. a lamella, a beam-splitting cube;
2. asymmetric beam splitter, $\rho_1 = 0, \rho_2 = \pi$, e.g. a glass plate and a coated plate.

In the following, we will use the first version. For further references, see [25].

Short excursion into time reversal Let us assume an electromagnetic wave with real and imaginary parts:

$$E(\mathbf{r}, t) = Ae^{i\phi}e^{i\omega t - i\mathbf{k}\cdot\mathbf{r}} + \text{c.c.} \tag{4.140}$$

If we apply a time-reversal operator $\hat{T}$, this changes the sign of each t:

$$\hat{T}\big(E(\mathbf{r}, t)\big) - \hat{T}\big(Ae^{i\phi}e^{i\omega t - i\mathbf{k}\cdot\mathbf{r}} + A^*e^{-i\phi}e^{-i\omega t + i\mathbf{k}\cdot\mathbf{r}}\big) \tag{4.141}$$

$$= Ae^{i\phi}e^{-i\omega t - i\mathbf{k}\cdot\mathbf{r}} + A^*e^{-i\phi}e^{i\omega t + i\mathbf{k}\cdot\mathbf{r}} \tag{4.142}$$

$$= A^*e^{-i\phi}e^{i\omega t + i\mathbf{k}\cdot\mathbf{r}} + \text{c.c.} \tag{4.143}$$

The conjugate amplitude $A^*e^{i\mathbf{k}\cdot\mathbf{r}}$ is now with the term oscillating as $e^{i\omega t}$ and $Ae^{i\mathbf{k}\mathbf{r}}$ is now with the complex conjugate. Thus, we can easily understand that time reversal corresponds to phase conjugation in the mathematical description.

4.4.1.2 *Beam splitter: correlation of the quantum uncertainty*

The action of a beam splitter on states of light described by a positive-valued Wigner function will be to transfer each arrow from an input port to both output ports with reduced amplitudes, keeping in mind that in the end each arrow will have to be multiplied with its own stochastic variable. In the model, we have to properly take into account the Stokes relations. If the same stochastically varying input arrow contributes to two output ports, one may expect correlations between these two output fields. We will show that this is not the case for every quantum state.

To get used to the arrow description, we consider the interference of two coherent states at a beam splitter. Coherent states have a circular region of uncertainty in phase space, and as a result all four arrows describing the two coherent states have the same size. The situation is shown graphically in Fig. 4.38. Note that the angles in phase space correspond to the classical optical phase of the light beams. The Stokes relations are obeyed by associating a 90° phase shift to each reflection, i.e. the factor 'i',

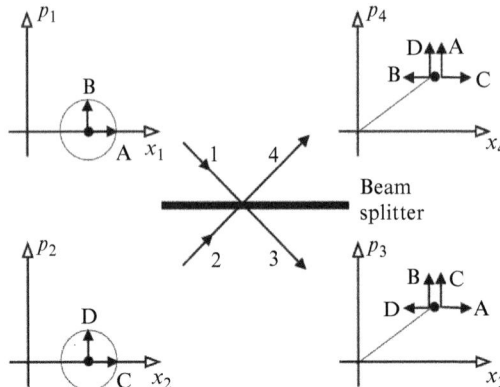

Fig. 4.38 Beam splitter with two coherent input states in the pictorial presentation.

and $0°$ phase shift to each transmission. In practice, this corresponds to a symmetric beam splitter such as a lamella or a beam-splitting cube. For simplicity, the amplitude reduction is not shown here. The four input arrows A, B, C, and D determine the field uncertainties in the two output ports 3 and 4. The amplitude uncertainty in output 3 is determined by the projections of all arrows onto the amplitude direction: $A+B+C-D$. Note that C and D reach output 3 by reflection and acquire a phase shift of $90°$. Each arrow would then still have to be multiplied with its individual stochastic coefficient. This and the additional amplitude reduction due to the projection are again not shown, since the factor will be the same for all arrows in the cases considered here. Likewise, the amplitude uncertainty at output 4 is determined by $A - B + C + D$.

Although the uncertainties in both output ports are governed by the same four arrows, they are **not correlated**. The reason for this lack of correlation can be traced back to the sum of two statistically independent stochastic variables, with their difference again being statistically independent. Output 3 is governed by the sum of two statistically independent variables $(A + C) + (B - D)$, and output 4 is determined by their difference $(A + C) - (B - D)$. To be specific, one and the same particular value at output 3 can be the result of infinitely many sums of $(A + C)$ and $(B - D)$ values. At output 4, the difference of the same values will span a wide range and not result in just one value. Consequently, there is no correlation between the two outputs, although they are determined by the same quantities A, B, C, and D.

If, however, the two input states were amplitude-squeezed states with close to zero amplitude uncertainty A, $C \approx 0$, and correspondingly larger uncertainty in the phase direction, the amplitude uncertainties of the two output ports would be anticorrelated: port 3 '$(B - D)$' and port 4 '$-(B - D)$'. The uncertainties in the phase directions are likewise correlated, as can be seen by going through similar arguments for the orthogonal projection[6] [36].

---

[6] Actually, one squeezed input field is enough to obtain quantum correlations between the two output fields but the correlations will be less strong.

The operators $\hat{X}$ and $\hat{P}$ do not commute, because

$$[\hat{X}, \hat{P}] = \hat{X}\hat{P} - \hat{P}\hat{X} \tag{4.144}$$

$$= \frac{1}{4\mathrm{i}}\left(\hat{a}^2 - \hat{a}\hat{a}^\dagger + \hat{a}^\dagger\hat{a} - (\hat{a}^\dagger)^2 - \hat{a}^2 - \hat{a}\hat{a}^\dagger + \hat{a}^\dagger\hat{a} + (\hat{a}^\dagger)^2\right) \tag{4.145}$$

$$= \frac{1}{4\mathrm{i}}(-2\hat{a}\hat{a}^\dagger + 2\hat{a}^\dagger\hat{a}) = \frac{\mathrm{i}}{2}. \tag{4.146}$$

Nevertheless, we can combine the two output amplitude operators $\hat{X}_3$ and $\hat{X}_4$ as $\hat{X}_3 + \hat{X}_4$ and the phase amplitude operators $\hat{P}_3$ and $\hat{P}_4$ as $\hat{P}_3 - \hat{P}_4$. We calculate the commutator of these combined operators and obtain

$$\left[\hat{X}_3 + \hat{X}_4, \hat{P}_3 - \hat{P}_4\right] = \left[\hat{X}_3, \hat{P}_3\right] - \left[\hat{X}_4, \hat{P}_4\right] - \left[\hat{X}_3, \hat{P}_4\right] + \left[\hat{X}_4, \hat{P}_3\right] = 0, \tag{4.147}$$

with

$$\left[\hat{X}_3, \hat{P}_3\right] = \tfrac{1}{2}\mathrm{i}, \qquad \left[\hat{X}_4, \hat{P}_4\right] = \tfrac{1}{2}\mathrm{i}, \qquad \left[\hat{X}_3, \hat{P}_4\right] = 0, \qquad \left[\hat{X}_4, \hat{P}_3\right] = 0. \tag{4.148}$$

Thus, the operators $\hat{X}_3 + \hat{X}_4$ and $\hat{P}_3 - \hat{P}_4$ commute! Hence, it follows that the uncertainties for these operators can be simultaneously well defined, the variances being zero:

$$\langle(\Delta(\hat{X}_3 + \hat{X}_4))^2\rangle = 0 \quad \text{and} \quad \langle(\Delta(\hat{P}_3 - \hat{P}_4))^2\rangle = 0. \tag{4.149}$$

Note that for the single quadratures, the following Heisenberg uncertainty relation is valid:

$$\langle(\Delta\hat{X}_i)^2\rangle\langle(\Delta\hat{P}_i)^2\rangle \geq \tfrac{1}{16} \qquad \text{for} \quad i = 1, \ldots, 4. \tag{4.150}$$

The fact that the uncertainty for the combination of two operators vanishes seems to be paradoxical, and hence it is also called the **Einstein–Podolsky–Rosen paradox (EPR)**, since it violates a basic quantum mechanical assumption, the Heisenberg uncertainty relation. This 'Gedankenexperiment' was formulated in 1935 by Einstein, Podolsky, and Rosen, with the aim of demonstrating the incompleteness of quantum theory [15]. They pointed out that if a quantum system has two parts, possibly separated by some distance, then special relativity forbids one from measuring the state of one subsystem in such a way that it is possible to determine precisely the outcome of an equivalent measurement on the other subsystem, which they called 'spooky' action at a distance, and they proposed the existence of additional 'local hidden variables'. Only in 1964 did John Bell resolve this conflict, formulating his famous theorem, in which he proved that it is possible to distinguish experimentally between local theories that claim unobservable local properties and non-local theories [7]. These theoretical assumptions were confirmed in the 1970s [3, 19]. While measuring the state of one subsystem may well determine precisely the outcome of an equivalent measurement on the other subsystem, it is indeed impossible to notice this first measurement on one subsystem just by observing the other subsystem. This is enough to prevent communication faster than the speed of light.

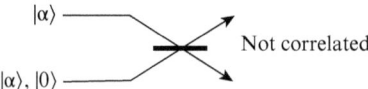

Fig. 4.39 [Colour online] Interference of two coherent states: uncorrelated outputs 3 and 4.

We conclude this subsection with a short overview of different interactions at a beam splitter.

As already discussed, the interference of two coherent states $|\alpha\rangle$ (or a coherent state and the vacuum state $|0\rangle$) gives uncorrelated outputs at port 3 and 4 (Fig. 4.39). This finds application, for example, in cryptography when using post selection.

When squeezing the two input states (or preparing them with other nonlinear processes) and letting them interact linearly at the beam splitter, the outputs of the beam splitter are entangled (Fig. 4.40), and this property is used, for example, in quantum teleportation, secret sharing, and quantum erasing [6].

Another way to obtain two entangled states is by using an interaction that is itself nonlinear (Fig. 4.41), for example a nonlinear crystal, as was done by Ou et al. [46] for non-degenerate parametric amplification.

4.4.2 Squeezing and sideband entanglement

Let us go one step back to single-mode squeezing. For amplitude squeezing, the situation is similar to that in Fig. 4.22. The only difference is that the uncertainty band is reduced, resulting in sub-shot-noise photocurrent fluctuations. But there is an alternative way. Measuring the uncertainty or squeezing of a quantum state in the experiment can be realized with a spectral measurement on a spectrum analyser. As already mentioned, the technical noise of the laser does not allow measurements close to the carrier frequency but only above about 10 MHz. Thus, we measure at a certain distance f from the carrier frequency ν_0, where an uncertainty equivalent to half a photon (zero-point

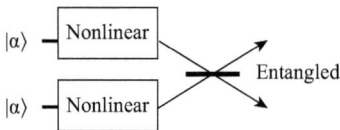

Fig. 4.40 [Colour online] Interference of two squeezed states (or states prepared by some other nonlinear process): entangled outputs 3 and 4.

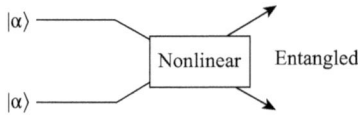

Fig. 4.41 [Colour online] Interaction of two coherent states via a nonlinear process: entangled outputs 3 and 4.

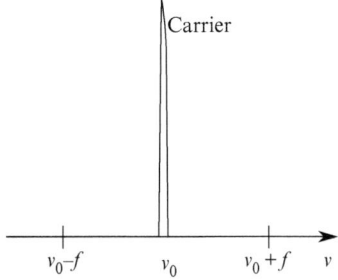

Fig. 4.42 Carrier (laser signal, local oscillator) and sidebands at distance f.

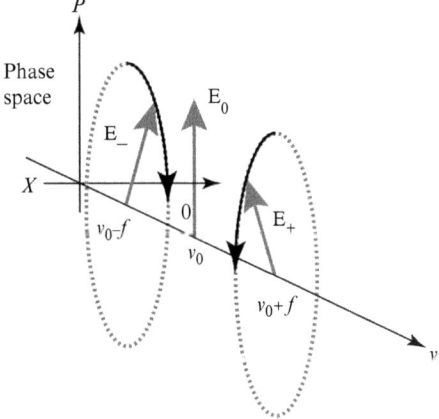

Fig. 4.43 [Colour online] Phase-space diagrams at upper and lower sidebands: measuring at the carrier frequency, the phase vectors of the sidebands rotate either clockwise or counterclockwise (higher or lower frequency).

uncertainty) has to be assumed, in order to explain the shot noise as detailed below. In fact, two sidebands equally spaced above and below the carrier will contribute to the shot noise, and equivalently we can easily imagine that quantum noise squeezing at the sidebands $\nu_0 + f$ and $\nu_0 - f$ arises from the interference of the carrier with the vacuum sidebands (Fig. 4.42). Figure 4.43 depicts the phase space of the carrier and two sidebands, which are shifted in frequency. The fast rotation of the carrier is taken out and thus the phase space vectors of the two sidebands rotate in opposite directions, because one is at higher frequency and the other at lower frequency than the carrier.

The total electric field contains the carrier $\mathcal{E}_0$ and the upper and lower sidebands (with indices '+' and '−'):

$$\mathcal{E}(t)\mathrm{e}^{\mathrm{i}\omega_0 t} = \mathcal{E}_0(t)\mathrm{e}^{\mathrm{i}\omega_0 t} + \mathcal{E}_+(t)\mathrm{e}^{\mathrm{i}(\omega_0 + \Omega)t} + \mathcal{E}_-(t)\mathrm{e}^{\mathrm{i}(\omega_0 - \Omega)t}, \tag{4.151}$$

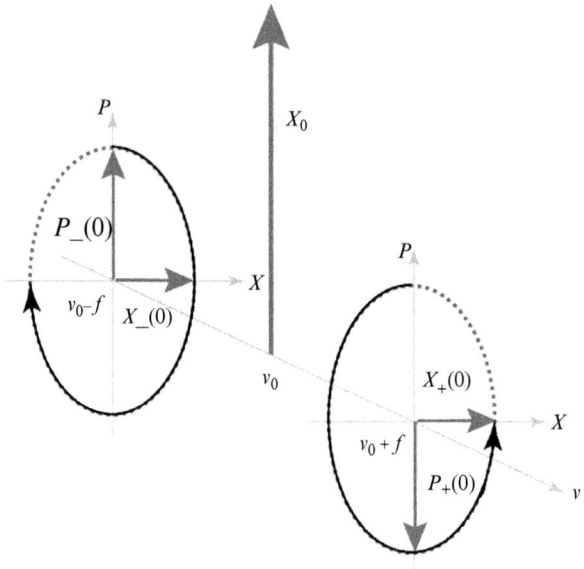

Fig. 4.44 [Colour online] Correlation of phase-space vector in two corresponding sidebands.

where $2\pi f \equiv \Omega$. The field operators are given by

$$a(t) = a_0 + a_+(t)e^{i\Omega t} + a_- e^{-i\Omega t} \qquad (4.152)$$

and the canonical variables are (Fig. 4.44)

$$X_+(t) = \frac{1}{2}\left(ae^{i\Omega t} + a^\dagger e^{-i\Omega t}\right), \qquad (4.153)$$

$$P_+(t) = \frac{1}{2i}\left(ae^{i\Omega t} - a^\dagger e^{-i\Omega t}\right). \qquad (4.154)$$

Analogous equations hold for X_- and P_-.

For $\Omega t = \pi/2$, we find

$$X_+\left(\frac{\pi}{2\Omega}\right) = \frac{1}{2}[ai + a^\dagger(-i)] = -\frac{1}{2i}(a - a^\dagger) = -P_+(0), \qquad (4.155)$$

$$P_+\left(\frac{\pi}{2\Omega}\right) = \frac{1}{2i}[ai - a^\dagger(-i)] = X_+(0), \qquad (4.156)$$

$$X_-\left(\frac{\pi}{2\Omega}\right) = \frac{1}{2}[a(-i) - a^\dagger i] = P_-(0), \qquad (4.157)$$

$$P_-\left(\frac{\pi}{2\Omega}\right) = \frac{1}{2i}[a(-i) - a^\dagger i] = -X_-(0). \qquad (4.158)$$

We find the total intensity with signal X_0 at ω_0 as

$$[X_0 + X_+(0) + X_-(0)]^2 = X_0^2 + 2X_0[X_+(0) + X_-(0)] + \ldots, \qquad (4.159)$$

$$\left[X_0 + X_+\left(\frac{\pi}{2\Omega}\right) + X_-\left(\frac{\pi}{2\Omega}\right)\right]^2 = X_0^2 + 2X_0[-P_+(0) + P_-(0)] + \ldots \qquad (4.160)$$

The noise is reduced if $X_+(0) + X_-(0) \to 0$ and $P_+(0) - P_-(0) \to 0$. In this case, the two sidebands are entangled, in analogy to our earlier discussion of the EPR paradox. The result, possibly surprising at first sight, is that any single-mode squeezed state is actually entangled. The entangled quantities are the two frequency sideband modes. This entanglement becomes apparent if one zooms into a spectral decomposition of the field. The correlation of sidebands in quantum optics was first proposed theoretically [12] and was called 'two-mode squeezing'. Nowadays, the term 'two-mode squeezing' is used for any modes that are correlated with each other—not only vacuum sidebands.

There have been different approaches to measure the entanglement between these vacuum sidebands [24, 27]. The setup shown in Fig. 4.45, an unbalanced interferometer, is one solution to measure the correlation of the sidebands. The signal, carrier plus the two entangled sidebands, is injected at one input of the first beam splitter and interferes with the vacuum state, entering the other input. We can write for the field at output 1 (and analogously for output 2)

$$
\begin{aligned}
\text{``out 1''} = \tfrac{1}{2}\big[& a_0 + a_0 \mathrm{e}^{\mathrm{i}\omega_0 L/c} \\
& + (a_+ + a_{V+}) + (a_+ + a_{V+})\mathrm{e}^{\mathrm{i}(\omega_0+\Omega)L/c} \\
& + (a_- + a_{V-}) + (a_- + a_{V-})\mathrm{e}^{\mathrm{i}(\omega_0-\Omega)L/c}\big],
\end{aligned} \qquad (4.161)
$$

where L is the difference in arm lengths. The goal is to separate both correlated sidebands at the outputs of the interferometer in order to measure the noise of the sidebands. Let us denote the phase differences for the different frequencies as follows:

$$\Delta\phi = \Omega\frac{L}{c}, \qquad \Delta\varphi = \omega_0\frac{L}{c}. \qquad (4.162)$$

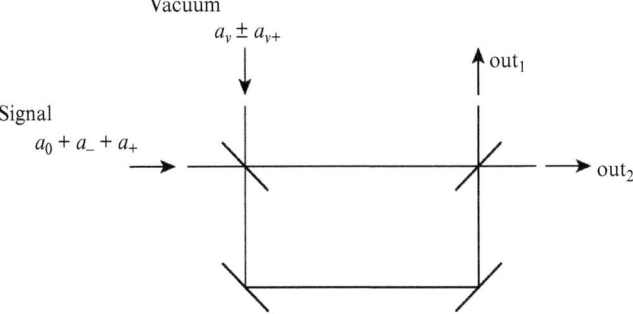

Fig. 4.45 Setup of an unbalanced interferometer.

Since Ω and ω_0 are separated by several orders of magnitude, it is possible to choose the arm length L, up to a certain precision, such as to fulfil the conditions for constructive and destructive interference at the respective outputs:

$$\Omega\frac{L}{c} = \frac{\pi}{2} \implies t = \frac{\pi}{2\Omega}, \tag{4.163}$$

$$\omega_0\frac{L}{c} = \frac{\pi}{2} + 2\pi m, \quad t = \frac{\pi}{2\omega_0} + \frac{2\pi m}{\omega_0}, \quad m \in \mathbb{N}. \tag{4.164}$$

The first condition determines L, and the second determines m.

We can imagine that tuning the arm length L will cause a fast rotation of the phase for ω_0 and a slow rotation of the phase for Ω. Therefore, we obtain, at output 1,

$$\begin{aligned}
\text{``out 1''} &= \tfrac{1}{2}[(X_0 + X_+ + X_- + iP_+ + iP_- + \ldots) \\
&\quad + i(X_0 - P_+ + P_- + iX_+ - iX_- + \ldots)] \\
&= \tfrac{1}{2}[(X_0 + iX_0) + 2X_- + 2iP_- + \ldots].
\end{aligned} \tag{4.165}$$

The interferometer thus splits the sidebands coming from one input port (Fig. 4.46). It can be shown that the noise for each of the separated entangled sidebands is higher than for the input beam with combined sidebands and even higher than it would be for a coherent state, which is an indicator for a correlation between the sidebands. Instead, if the difference signal of both outputs is measured, the noise is suppressed and even lower than the shot-noise limit; hence we observe the squeezing as a result of the interplay between the two sidebands. We should keep in mind that each subsystem of an entangled state possesses its own increased noise, and the squeezed quantum noise is revealed only in their difference signal. When tracing over one subsystem, the other is projected into a mixed state. Conversely, systems in pure states can never be entangled with another system.

4.4.3 Bogoliubov transformation

In the preceding subsections, we have discussed the interaction at a beam splitter, where modes are split into several modes, or vice versa, and modes can interfere with

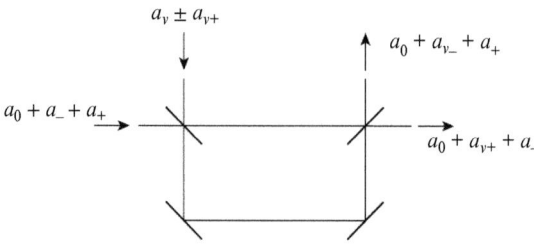

Fig. 4.46 Setup of an unbalanced interferometer with separated sidebands at the outputs.

each other. Generally speaking, the beam splitter introduces a Bogoliubov transformation. Several phenomena can be reduced to such a model, not only attenuation of a state, but also amplification of a state, the aforementioned squeezing, and phase conjugation. In order to find a general formula for these two-mode interactions, we shall examine the different processes.

4.4.3.1 *Attenuation*

Let us first consider the attenuation of a coherent state, which is a pure quantum state, depicted in phase space in Fig. 4.47. The coherent state has its uncertainty in both quadratures and while the amplitude of the state is reduced, the uncertainty does not decrease because of the Heisenberg uncertainty relation being state independent. The noise figure (NF) is defined as the relation of the signal-to-noise ratio (SNR) at the output to the SNR at the input,

$$\text{NF} = \frac{\text{SNR}_\text{out}}{\text{SNR}_\text{in}} = \frac{\text{signal}_\text{out}/\text{noise}_\text{out}}{\text{signal}_\text{in}/\text{noise}_\text{in}}, \tag{4.166}$$

and it increases for the attenuation of a pure, coherent state.

The field operator $\hat{a}$ corresponds to the electric field E, and for finding the field operator in the case of attenuation, we can guess a new field operator $\hat{c} = \sqrt{\eta}\hat{a}$ in order to obtain a lower field amplitude. However, calculating the commutator gives

$$[\hat{c}, \hat{c}^\dagger] = [\sqrt{\eta}\hat{a}, \sqrt{\eta}\hat{a}^\dagger] = \eta[\hat{a}, \hat{a}^\dagger] = \eta \neq 1. \tag{4.167}$$

Thus, $\sqrt{\eta}\hat{a}$ is not a valid field operator and one needs to add an ancillary operator $\hat{L}$:

$$\hat{c} = \sqrt{\eta}\hat{a} + \hat{L}. \tag{4.168}$$

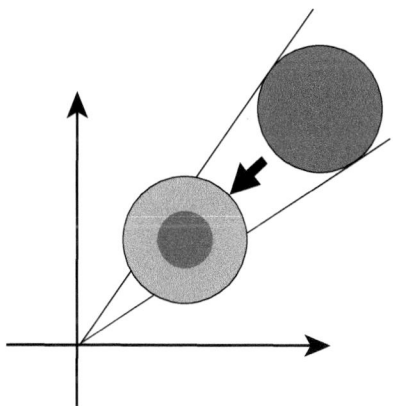

Fig. 4.47 [Colour online] Attenuation: pure state $\rightarrow$ pure state.

We derive again the commutator

$$[\hat{c}, \hat{c}^\dagger] = [\sqrt{\eta}\,\hat{a} + \hat{L}, \sqrt{\eta}\,\hat{a}^\dagger + \hat{L}^\dagger]$$
$$= (\sqrt{\eta}\,\hat{a} + \hat{L})(\sqrt{\eta}\,\hat{a}^\dagger + \hat{L}^\dagger) - (\sqrt{\eta}\,\hat{a}^\dagger + \hat{L}^\dagger)(\sqrt{\eta}\,\hat{a} + \hat{L})$$
$$= \eta\,[\hat{a}, \hat{a}^\dagger] + [\hat{L}, \hat{L}^\dagger] = 1\,. \tag{4.169}$$

Under the condition

$$[\hat{L}, \hat{L}^\dagger] = 1 - \eta\,, \tag{4.170}$$

$\hat{c}$ does indeed fulfil the commutation relation required for a field operator. It then follows, with $\eta < 1$, that the operator $\hat{L}$ is an annihilation operator and can be written as

$$\hat{L} = \begin{cases} \sqrt{1-\eta}\,\hat{a} & \text{(case 1)}, \\ \sqrt{1-\eta}\,\hat{b} & \text{(case 2)}, \end{cases} \tag{4.171}$$

with $\hat{a}$ the same mode operator and $\hat{b}$ another mode operator. Case 1 is not possible, because it cannot satisfy the commutation relation $[\hat{c}, \hat{c}^\dagger] = 1$ in general, except for $\eta = 0, 1$. Hence, we can conclude that the field operator $\hat{c}$ in the case of **attenuation** has to be written in the following form:

$$\textbf{Attenuation:} \quad \hat{c} = \sqrt{\eta}\,\hat{a} + \sqrt{1-\eta}\,\hat{b}\,. \tag{4.172}$$

We note that attenuation always accompanies the interaction with a second mode; thus, the noise addition by the second mode is the reason for the decrease in the signal-to-noise ratio [17]. We also note that we recover the operator relation at the beam splitter.

4.4.3.2 Amplification and squeezing: phase-insensitive and phase-sensitive amplifiers

In the case of amplification, we need to write the field operator $\hat{c}$ as before, but with an amplification factor G:

$$\hat{c} = \sqrt{G}\,\hat{a} + \hat{L}\,. \tag{4.173}$$

An ancillary mode operator is again needed, for the same reasons as in the process of attenuation. The commutator can be derived analogously to (4.169), with

$$[\hat{L}, \hat{L}^\dagger] = 1 - G \tag{4.174}$$

and $G > 1$, which we rewrite as

$$[(\hat{L}^\dagger), (\hat{L}^\dagger)^\dagger] = G - 1\,. \tag{4.175}$$

Thus, the operator $\hat{L}^\dagger$ has to be an annihilation operator and $\hat{L}$ a creation operator, and we can again deduce two cases [11]:

$$\hat{c} = \sqrt{G}\hat{a} + \hat{L}, \qquad \hat{L} = \begin{cases} \sqrt{G-1}\,\hat{b}^\dagger & \text{(case 1)}, \\ \sqrt{G-1}\,a^\dagger & \text{(case 2)}. \end{cases} \tag{4.176}$$

The operators $a^\dagger$ and $b^\dagger$ are creation operators; $\hat{a}$ is for the same optical mode as the signal and $\hat{b}$ is for another optical mode. Both cases are in principle possible, since the commutation relation can be satisfied. Let us start with case 1, where an interaction with a second mode $\hat{b}$ takes place.

Case 1   For the field operator $\hat{a}$, we use the subscript a and recall the relation to the canonical phase-space quadrature operators:

$$\hat{X}_a = \frac{1}{2}(\hat{a} + \hat{a}^\dagger), \qquad \hat{P}_a = \frac{1}{2\mathrm{i}}(\hat{a} - \hat{a}^\dagger). \tag{4.177}$$

For the field operator [17]

$$\hat{c} = \sqrt{G}\,\hat{a} + \sqrt{G-1}\,\hat{b}^\dagger, \tag{4.178}$$

we can write the continuous variable $\hat{X}_c$ and its square as

$$\hat{X}_c = \tfrac{1}{2}\left[\sqrt{G}\,(\hat{a} + \hat{a}^\dagger) + \sqrt{G-1}\,(\hat{b} + \hat{b}^\dagger)\right] = \sqrt{G}\hat{X}_a + \sqrt{G-1}\hat{X}_b, \tag{4.179}$$

$$\hat{X}_c^2 = G\hat{X}_a^2 + 2\sqrt{G(G-1)}\,\hat{X}_a\hat{X}_b + (G-1)\hat{X}_b^2. \tag{4.180}$$

It follows that the mean values are

$$\langle \hat{X}_c^2 \rangle = G\langle \hat{X}_a^2 \rangle + (G-1)\langle \hat{X}_b^2 \rangle + 2\sqrt{G(G-1)}\,\langle \hat{X}_a \rangle \langle \hat{X}_b \rangle, \tag{4.181}$$

$$\langle \Delta\hat{X}_c^2 \rangle = G(\langle \hat{X}_a^2 \rangle - \langle \hat{X}_a \rangle^2) + (G-1)(\langle \hat{X}_b^2 \rangle - \langle \hat{X}_b \rangle^2)$$
$$+ 2\sqrt{G(G-1)}\,\langle \hat{X}_a \rangle \langle \hat{X}_b \rangle - 2\sqrt{G(G-1)}\,\langle \hat{X}_a \rangle \langle \hat{X}_b \rangle, \tag{4.182}$$

$$\langle \Delta\hat{X}_c^2 \rangle = G\langle \Delta\hat{X}_a^2 \rangle + (G-1)\langle \Delta\hat{X}_b^2 \rangle, \tag{4.183}$$

with $\langle \hat{X}_b \rangle = 0$. Analogously, we find for the other quadrature $\Delta\hat{P}_c$,

$$\langle \Delta\hat{P}_c^2 \rangle = G\langle \Delta\hat{P}_a^2 \rangle + (G-1)\langle \Delta\hat{P}_b^2 \rangle. \tag{4.184}$$

Obviously, there is excess noise in both quadratures (Fig. 4.48). The SNR before amplification is the signal over the noise:

$$\mathrm{SNR}_\mathrm{in} = \frac{\mathrm{signal}_\mathrm{in}}{\mathrm{noise}_\mathrm{in}} = \frac{\langle \hat{X}_a \rangle^2}{\langle \Delta\hat{X}_a^2 \rangle}. \tag{4.185}$$

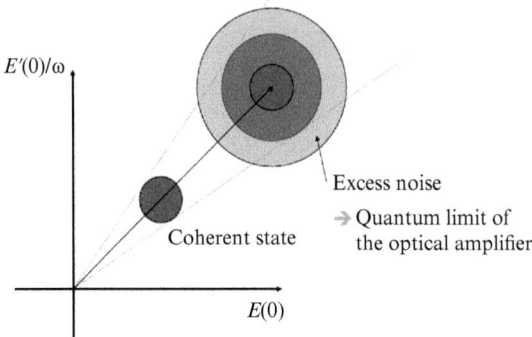

Fig. 4.48   [Colour online] Amplification with a phase-insensitive amplifier.

The noise figure for this process, as defined before, can then be found as

$$\text{NF} = \frac{G\langle \hat{X}_a \rangle^2}{G\langle \Delta \hat{X}_a^2 \rangle + (G-1)\langle \Delta \hat{X}_b^2 \rangle} \frac{\langle \Delta \hat{X}_a^2 \rangle}{\langle \hat{X}_a \rangle^2} \tag{4.186}$$

$$= \frac{G}{2G-1} \quad \text{if } \Delta \hat{X}_b^2 = \Delta \hat{X}_a^2 \tag{4.187}$$

(and analogously for $\hat{P}_a$). The condition applies if the signal is a coherent state and the ancillary mode is in the vacuum state. For high G, the noise figure tends to

$$\text{NF} \geq \tfrac{1}{2}. \tag{4.188}$$

This is known as the 3 dB quantum limit of an amplifier and means that noise is always added to both quadratures, as long as the amplification is phase-insensitive. Thus, this case is called the **phase-insensitive amplifier**. It can be described with

$$\textbf{Amplification:} \quad \hat{c} = \sqrt{G}\,\hat{a} + \sqrt{G-1}\,\hat{b}^\dagger. \tag{4.189}$$

For low gain $G > 1$, the quantum-limited noise figure is closer to 1. It turns out that for continuous quantum variables, the optimal cloning operation, i.e. two clones of one original, can be implemented by a quantum-optimized $G = 2$ amplifier, followed by a beam splitter. After being proposed theoretically [9, 18], optimal cloning was implemented experimentally [1]. Quantum-limited amplification was also demonstrated using a quantum electro-optic feedforward amplifier [29] based on the development of more general quantum electro-optic feedforward techniques [34].

Case 2   In the second case, we discuss the interaction with the complex conjugate of the same mode $\hat{a}^\dagger$:

$$\hat{c} = \sqrt{G}\,\hat{a} + \sqrt{G-1}\,\hat{a}^\dagger. \tag{4.190}$$

In order to obtain the noise figure NF, we again derive $\hat{X}_c$, $\langle\hat{X}_c^2\rangle$, and $\langle\Delta\hat{X}_c^2\rangle$:

$$\hat{X}_c = \tfrac{1}{2}\left[\sqrt{G}\,(\hat{a}+\hat{a}^\dagger) + \sqrt{G-1}\,(\hat{a}+\hat{a}^\dagger)\right]$$
$$= \left(\sqrt{G}+\sqrt{G-1}\right)\hat{X}_a\,, \tag{4.191}$$

$$\langle\hat{X}_c^2\rangle = \left[G+(G-1)+2\sqrt{G(G-1)}\right]\langle\hat{X}_a^2\rangle\,, \tag{4.192}$$

$$\langle\Delta\hat{X}_c^2\rangle = \left[G+(G-1)+2\sqrt{G(G-1)}\right]\langle\Delta\hat{X}_a^2\rangle$$
$$= \left(\sqrt{G}+\sqrt{G-1}\right)^2\langle\Delta\hat{X}_a^2\rangle\,. \tag{4.193}$$

Analogous expressions hold for the other quadrature $\hat{P}_c$:

$$\hat{P}_c = \left(\sqrt{G}-\sqrt{G-1}\right)\hat{P}_a\,, \tag{4.194}$$

$$\langle\hat{P}_c^2\rangle = \left(\sqrt{G}-\sqrt{G-1}\right)^2\langle\hat{P}_a^2\rangle\,, \tag{4.195}$$

$$\langle\Delta\hat{P}_c^2\rangle = \left(\sqrt{G}-\sqrt{G-1}\right)^2\langle\Delta\hat{P}_a^2\rangle\,. \tag{4.196}$$

We are now able to calculate the noise figure for both quadratures, and we obtain, for the quadrature $\hat{X}_c$,

$$\mathrm{NF} = \frac{\left(\sqrt{G}+\sqrt{G-1}\right)^2\langle\hat{X}_a\rangle^2}{\left(\sqrt{G}+\sqrt{G-1}\right)^2\langle\Delta\hat{X}_a^2\rangle}\frac{\langle\Delta\hat{X}_a^2\rangle}{\langle\hat{X}_a\rangle^2} - 1 \tag{4.197}$$

and, for the quadrature $\hat{P}_c$,

$$\mathrm{NF} = \frac{\left(\sqrt{G}-\sqrt{G-1}\right)^2\langle\hat{P}_a\rangle^2}{\left(\sqrt{G}-\sqrt{G-1}\right)^2\langle\Delta\hat{P}_a^2\rangle}\frac{\langle\Delta\hat{P}_a^2\rangle}{\langle\hat{P}_a\rangle^2} = 1\,. \tag{4.198}$$

In this case, no additional noise is added to the state during amplification, and the SNR remains the same throughout the process. However, this does not mean that the noise is the same in both quadratures (that would be unphysical). In order to understand this, we look at the mean values of each quadrature: for the quadrature $\hat{X}_c$,

$$\langle\hat{X}_c\rangle = \left(\sqrt{G}+\sqrt{G-1}\right)\langle\hat{X}_a\rangle\,, \tag{4.199}$$

and for the quadrature $\hat{P}_c$,

$$\langle\hat{P}_c\rangle = \left(\sqrt{G}-\sqrt{G-1}\right)\langle\hat{P}_a\rangle\,. \tag{4.200}$$

For $\hat{X}_c$, the amplitude is increasing, and the state is amplified. But since the SNR remains the same, the noise is also increasing. For $\hat{P}_c$, the opposite is the case. For high G, the amplitude tends to a minimum. Also here, the SNR does not change,

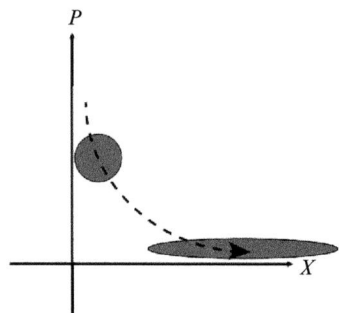

Fig. 4.49 [Colour online] Amplification with a phase-sensitive amplifier.

which means that the noise in the direction of the quadrature $\hat{P}_c$ is reduced, and is even smaller than the shot-noise limit (Fig. 4.49)! This is known as a **phase-sensitive amplifier** and corresponds to **squeezing**, as discussed already in Section 4.3.4. The general description is then

$$\textbf{Squeezing:} \quad \hat{c} = \sqrt{G}\,\hat{a} + \sqrt{G-1}\,\hat{a}^{\dagger}. \tag{4.201}$$

4.4.3.3 Phase conjugation

In order to obtain an operator for phase conjugation, we can assume the following form, where the use of $\hat{a}^{\dagger}$ is required:

$$\hat{c} = \sqrt{G}\,\hat{a}^{\dagger} + \hat{L}. \tag{4.202}$$

We again use the commutation relation

$$\left[\hat{c},\hat{c}^{\dagger}\right] = G\left[\hat{a}^{\dagger},\hat{a}\right] + \left[\hat{L},\hat{L}^{\dagger}\right] = 1. \tag{4.203}$$

With $[\hat{a}^{\dagger},\hat{a}] = -1$, this gives

$$\left[\hat{L},\hat{L}^{\dagger}\right] = G+1. \tag{4.204}$$

The ancillary operator $\hat{L}$ is thus an annihilation operator, and so we can write

$$\hat{L} = \sqrt{G+1}\,\hat{b}. \tag{4.205}$$

As before, we also have to take into account the possibility of using the same mode, such as $\hat{L} = \sqrt{G+1}\,\hat{a}$. But calculating the commutator $[\hat{c},\hat{c}^{\dagger}]$ for this case will give a value of -1; hence, we need a second independent mode and we can write for the process of **phase conjugation** the following transformation [13]:

$$\textbf{Phase conjugation:} \quad \hat{c} = \sqrt{G}\,\hat{a}^{\dagger} + \sqrt{G+1}\,\hat{b}. \tag{4.206}$$

Again, we can calculate the variance of the quadrature operators at the output and obtain the noise figure as

$$\mathrm{NF} = \frac{G}{2G+1} . \tag{4.207}$$

4.4.3.4 *General form of the Bogoliubov transformation*

In the preceding subsections, we expressed the different processes, such as attenuation, amplification, squeezing, and phase conjugation, in terms of the respective mode operators. We can combine them into one transformation, which is called the **Bogoliubov transformation** for two modes:

$$\hat{c} = \alpha_1 \hat{a} + \alpha_2 \hat{b} + \alpha_3 \hat{a}^\dagger + \alpha_4 \hat{b}^\dagger . \tag{4.208}$$

The respective processes are shown in Fig. 4.50.

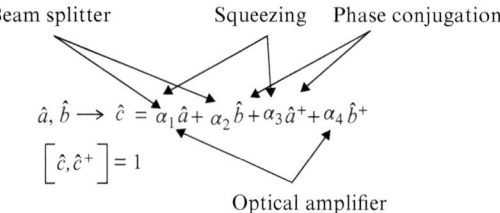

Fig. 4.50 [Colour online] Bogoliubov transformation for two modes.

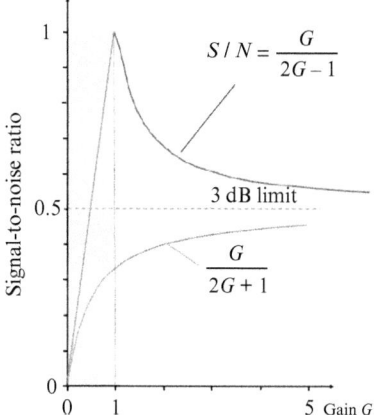

Fig. 4.51 [Colour online] The upper [blue] curve shows the behaviour of the SNR with respect to the gain for the ideal amplifier. The lower [red] curve shows the performance of a phase-conjugating mirror that is equal in performance to an amplifier with destructive measurement and re-creation.

Plotting the noise figures as a function of gain for the different processes gives some insight (Fig. 4.51). When measuring the Q-function of a signal state (Fig. 4.17), one can amplify electronically the photocurrent measured simultaneously for the two quadratures $\hat{X}$ and $\hat{P}$ and modulate a laser with these amplified quadratures. This is the technique used in electronic repeater stations in optical telecommunication. It is interesting to note that the noise figure for this electronic amplification of optical signals is identical to the noise figure of the phase conjugation. It is furthermore worth noting that the limiting performance of the optimum optical quantum amplifier for large gain is essentially the same as that of the electronic amplifier.

Acknowledgements

The authors would like to thank Gunnar Björk, Luis Sanchez-Soto, and Lev Plimak for helpful discussions and proofreading of parts of this chapter.

References

[1] U. L. Andersen, V. Josse, and G. Leuchs, Unconditional quantum cloning of coherent states with linear optics. *Phys. Rev. Lett.* **94**, 240503 (2005).

[2] E. Arthurs and J. L. Kelly. On the simultaneous measurement of a pair of conjugate observables. *Bell Syst. Tech. J.* **44**, 725 (1965).

[3] A. Aspect, P. Grangier, and G. Roger. Experimental tests of realistic local theories via Bell's theorem. *Phys. Rev. Lett.* **47**, 460 (1981).

[4] H.-A. Bachor and T. C. Ralph. *A Guide to Experiments in Quantum Optics*, 2nd edn. Wiley-VCH, Berlin (2004).

[5] K. Banaszek and K. Wodkiewicz. Direct probing of quantum phase space by photon counting. *Phys. Rev. Lett.* **76**, 4344 (1996).

[6] S. M. Barnett and P. M. Radmore. *Methods in Theoretical Quantum Optics*. Oxford University Press, Oxford (1997).

[7] J. Bell. On the Einstein–Podolsky–Rosen paradox. *Physics* **1**, 195 (1964).

[8] F. Boitier, A. Godard, E. Rosencher, and C. Fabre. Measuring photon bunching at ultrashort timescale by two-photon absorption in semiconductors. *Nat. Phys.* **5**, 267 (2009).

[9] S. L. Braunstein, N. J. Cerf, S. Iblisdir, P. van Loock, and S. Massar. Optimal cloning of coherent states with a linear amplifier and beam splitters. *Phys. Rev. Lett.* **86**, 4938 (2001).

[10] W. B. Case. Wigner functions and Weyl transforms for pedestrians. *Am. J. Phys.* **76**, 937 (2008).

[11] C. M. Caves. Quantum limits on noise in linear amplifiers. *Phys. Rev. D* **26**, 1817 (1982).

[12] C. M. Caves and B. L. Schumaker. New formalism for two-photon quantum optics. I. Quadrature phases and squeezed states. *Phys. Rev. A* **31**, 3068 (1985).

[13] N. J. Cerf and S. Iblisdir. Phase conjugation of continuous quantum variables. *Phys. Rev. A* **64**, 032307 (2001).

[14] J. F. Corney, P. D. Drummond, J. Heersink, V. Josse, G. Leuchs, and U. L. Andersen. Many-body quantum dynamics of polarization squeezing in optical fibers. *Phys. Rev. Lett.* **97**, 023606 (2006).

[15] A. Einstein, B. Podolsky, and N. Rosen. Can quantum-mechanical description of physical reality be considered complete? *Phys. Rev. Lett.* **47**, 777 (1935).

[16] D. Elser, T. Bartley, B. Heim, Ch. Wittmann, D. Sych, and G. Leuchs. Feasibility of free space quantum key distribution with coherent polarization states. *New J. Phys.* **11**, 045014 (2009).

[17] C. Fabre. Quantum fluctuations in light beams. In: S. Reynaud, E. Giacobino, and J. Zinn-Justin (eds.). *Quantum Fluctuations. Les Houches Summer School, Session LXIII*, p. 181. Elsevier, Amsterdam (1997).

[18] J. Fiurasek. Optical implementation of continuous-variable quantum cloning machines. *Phys. Rev. Lett.* **86**, 4942 (2001).

[19] S. Freedman and J. Clauser. Experimental test of local hidden-variable theories. *Phys. Rev. Lett.* **28**, 938 (1972).

[20] M. Freyberger, K. Vogel, and W. P. Schleich. From photon counts to quantum phase. *Phys. Lett. A* **176**, 41 (1993).

[21] A. L. Gaeta and R. Boyd. Quantum noise in phase conjugation. *Phys. Rev. Lett.* **60**, 2618 (1988).

[22] C. Gerry and P. Knight. *Introductory Quantum Optics.* Cambridge University Press, Cambridge (2005).

[23] R. J. Glauber. Optical coherence and photon statistics. In: C. DeWitt, A. Blandin, and C. Cohen-Tannoudji (eds.). *Quantum Optics and Electronics. Les Houches Summer School, Session XIV*, p. 63. Gordon and Breach, New York (1965).

[24] B. Hage, A. Samblowski, and R. Schnabel. Towards Einstein–Podolsky–Rosen quantum channel multiplexing. *Phys. Rev. A* **81**, 062301 (2010); selected as a research highlight in *Nat. Photonics* **4**, 500 (2010).

[25] M.W. Hamilton. Phase shifts in multilayer dielectric beam splitters. *Am. J. Phys.* **68**, 186 (2000).

[26] H. A. Haus and J. A. Mullen. Quantum noise in linear amplifiers. *Phys. Rev.* **128**, 2407 (1962).

[27] E. H. Huntington, G. N. Milford, C. Robilliard, T. C. Ralph, O. Glöckl, U. L. Andersen, S. Lorenz, and G. Leuchs. Demonstration of the spatial separation of the entangled quantum sidebands of an optical field. *Phys. Rev. A* **71**, 041802 (2005).

[28] T. Sh. Iskhakov, A. Perez, K. Yu. Spasibko, M. V. Chekhova, and G. Leuchs. Superbunched bright squeezed vacuum state. *Opt. Lett.* **37**, 1919 (2012).

[29] V. Josse, M. Sabuncu, N. Cerf, G. Leuchs, and U.L. Andersen. Universal optical amplification without nonlinearity. *Phys. Rev. Lett.* **96**, 163602 (2006).

[30] I. Khan, C. Wittmann, N. Jain, N. Killoran, N. Lütkenhaus, Ch. Marquardt, and G. Leuchs. Optimal working points for continuous-variable quantum channels. *Phys. Rev. A* **88**, 010302(R) (2013).

[31] M. Kitagawa and Y. Yamamoto. Number-phase minimum-uncertainty state with reduced number uncertainty in a Kerr nonlinear interferometer. *Phys. Rev. A* **34**, 3974 (1986).

[32] T. Kiesel, W. Vogel, B. Hage, and R. Schnabel. Direct sampling of negative quasiprobabilities of a squeezed state. *Phys. Rev. Lett.* **107**, 113604 (2011).

[33] K. Laiho, K. N. Cassemiro, D. Gross, and C. Silberhorn. Probing the negative Wigner function of a pulsed single photon point by point. *Phys. Rev. Lett.* **105**, 253603 (2010).

[34] P. K. Lam, T. C. Ralph, E. H. Huntington, and H.-A. Bachor. Noiseless signal amplification using positive electro-optic feedforward. *Phys. Rev. Lett.* **79**, 1471 (1997).

[35] D. Leibfried, D. M. Meekhof, B. E. King, C. Monroe, W. M. Itano, and D. J. Wineland. Experimental determination of the motional quantum state of a trapped atom. *Phys. Rev. Lett.* **77**, 4281 (1996).

[36] U. Leonhard. *Measuring the Quantum State of Light.* Cambridge University Press, Cambridge (1997).

[37] G. Leuchs. Photon statistics, antibunching and squeezed states. In: G. Moore and M. O. Scully (eds.). *Frontiers in Non-Equilibrium Quantum Statistical Physics*, p. 329. Plenum, New York (1986).

[38] G. Leuchs. Precision in length. In: H. Figger, D. Meschede, and C. Zimmermann (eds.). *Laser Physics at the Limits*, p. 209. Springer-Verlag, Berlin (2002).

[39] G. Leuchs, W. P. Schleich, and R. J. Glauber. Dimension of quantum space measured by photon correlations. *Physica Scr.* **90**, 074066 (2015).

[40] R. Loudon. *The Quantum Theory Of Light*, 3rd edn, Oxford University Press, Oxford (2000).

[41] L.G. Lutterbach and L. Davidovich. Method for direct measurement of the Wigner function in cavity QED and ion traps. *Phys. Rev. Lett.* **78**, 2547 (1997).

[42] L. G. Lutterbach and L. Davidovich. Non-classical states of the electromagnetic field in cavity QED. *Opt. Express* **3**, 147 (1998).

[43] A. I. Lvovsky and M. G. Raymer. Continuous-variable optical quantum-state tomography. *Rev. Mod. Phys.* **81**, 299 (2009).

[44] L. Mandel and E. Wolf. *Optical Coherence and Quantum Optics.* Cambridge University Press, Cambridge (1995).

[45] G. Nogues, A. Rauschenbeutel, S. Osnaghi, P. Bertet, M. Brune, J. M. Raimond, S. Haroche, L. G. Lutterbach, and L. Davidovich. Measurement of a negative value for the Wigner function of radiation. *Phys. Rev. A* **62**, 054101 (2000).

[46] Z. Y. Ou, S. F. Pereira, H. J. Kimble, and K. C. Peng. Realization of the Einstein–Podolsky–Rosen paradox for continuous variables. *Phys. Rev. Lett.* **68**, 3663 (1992).

[47] S. Schmitt, J. Ficker, M. Wolff, F. König, A. Sizmann, and G. Leuchs. Photon-number squeezed solitons from an asymmetric fiber-optic Sagnac interferometer. *Phys. Rev. Lett.* **81**, 2446 (1998).

[48] W. P. Schleich. *Quantum Optics in Phase Space.* Wiley-VCH, Berlin (2001).

[49] A. Sizmann and G. Leuchs. The optical Kerr effect and quantum optics in fibers. *Prog. Opt.*, **39**, 373 (1999).

[50] S. Stenholm. The Wigner function. I. The physical interpretation. *Eur. J. Phys.* **1**, 244 (1980).

[51] G. G. Stokes. On the perfect blackness of the central spot in Newton's rings, and on the verification of Fresnel's formulae for the intensities of reflected and refracted rays. *Camb. Dublin Math. J.* **4**, 1 (1849).

[52] H. R. Telle. Absolute measurement of optical frequencies. In: M. Ohtsu (ed.). *Frequency Control of Semiconductor Lasers*, p. 137. Wiley, Hoboken, NJ (1996).

[53] H. R. Telle. Stabilization and modulation schemes of laser diodes for applied spectroscopy. *Spectrochim. Acta Rev.* **15**, 301 (1993).

[54] S. Wallentowitz and W. Vogel. Unbalanced homodyning for quantum state measurements. *Phys. Rev. A* **53**, 4528 (1996).

[55] D. F. Walls and G. J. Milburn. *Quantum Optics*, 2nd edn. Springer-Verlag, Berlin (2008).

[56] E. Wigner. On the quantum correction for thermodynamic equilibrium. *Phys. Rev.* **40**, 749 (1932).

5

Quantum optics with nitrogen-vacancy centres in diamond

Yiwen CHU and Mikhail D. LUKIN

Department of Physics
Harvard University
Cambridge, Massachusetts, USA

Yiwen Chu and Mikhail D. Lukin: *Quantum optics with nitrogen-vacancy centres in diamond*. In: *Quantum Optics and Nanophotonics*. Edited by: Claude Fabre, Vahid Sandoghdar, Nicolas Treps, and Leticia F. Cugliandolo. Oxford University Press (2017), pp. 229–270. © Oxford University Press 2017. DOI: 10.1093/oso/9780198768609.003.0005

Chapter Contents

Colour figures. For those figures in this chapter that use colour, please see
the version of these lecture notes at arXiv:1504.05990 [quant-ph]. These figures
are indicated by '[Colour online]' at the start of the caption.

5.1 Background

The study of classical mechanics, thermodynamics, and electromagnetism has led to great technological advances over the course of history. We now live in an exciting era where quantum mechanics begins to join these other fields of physics in enabling new technologies with unprecedented capabilities. Theoretical and experimental efforts have moved towards harnessing the power of various different physical systems for applications in metrology and quantum information and to further our understanding of quantum mechanics itself.

Some of the most successful applications of quantum mechanical systems to date have been in the field of metrology, or the accurate sensing and measurement of forces, fields, frequencies, etc. [11, 20, 46, 63]. At the same time, the field of quantum information and communications has become the focus of much effort in recent years. There have been many theoretical proposals for using quantum systems for efficient computation, simulation of complex systems, secure dissemination of information, etc. [47, 52, 78]. However, the practical implementation of these proposals in real physical systems remains an open challenge. A variety of different avenues are being actively explored, ranging from photons, neutral atoms, ions, impurities, and structures in solid state systems, to more exotic materials such as electrons on the surface of liquid helium [5, 21, 58, 67, 84]. A common theme for the implementation of quantum technologies involves addressing the seemingly contradictory need for controllability and isolation from external effects. Undesirable effects of the environment must be minimized, while at the same time techniques and tools must be developed that enable us to interact with the system in a controllable and well-defined manner. These lectures address several aspects of this theme with regard to a particularly promising candidate for developing applications in both metrology and quantum information, namely the nitrogen-vacancy (NV) centre in diamond. In particular, we explore how the NV centre can be investigated, controlled, and efficiently coupled with optical photons, and tackle the challenges of controlling the optical properties of these emitters inside a complex solid-state environment. Some recent reviews on NV centres can be found in [24, 26, 44].

As with other solid state systems such as quantum dots, rare earth dopants in inorganic crystals, and superconducting qubits, NV centres present the attractive feature of eliminating the need for trapping and cooling, as is usually required for atomic systems. In addition to greatly reducing the complexity of experiments, solid state systems offer the possibility of straightforward integration with fabricated structures and development of scalable quantum devices. The issue of scalability, however, is not simply a matter of being able to create a large number of physical devices in a small volume. In order to truly move beyond proof-of-principle experiments, the properties of many qubits and devices must all be well controlled in a reproducible, reliable, and efficient manner. In other words, while it is important to demonstrate that a certain quantum operation, coherence time, metrological sensitivity, etc. is possible in one particular device with favourable characteristics, it must also be possible to reproduce these characteristics for a large number of systems.

Diamond is a particularly promising host material for optically active quantum emitters owing to its wide bandgap (5.5 eV) and large transparency window.

It supports a huge variety of impurities and defects that give rise to optical bands at energies throughout the infrared and visible spectrum [91]. The NV centre is unique in that it combines well-defined optical transitions with long-lived spin sublevels in the ground state, much like alkali atoms with transitions involving hyperfine sublevels. The present lectures focus on quantum optical experiments with NV centres. In recent years, a comprehensive theoretical model has been developed to describe the spin and optical properties of the NV centre, and has been reviewed in [24, 25, 56]. We begin by summarizing this theoretical framework, which allows us to understand the underlying physics of the subsequent experimental results. We then describe the experimental techniques that are used to investigate and address the optical properties of NV centres. As part of this description, we will encounter some general characteristics of the system that will become important issues to be addressed later, such as the Debye–Waller factor and the sensitivity of the excited state to the local environment. Based on our theoretical understanding and spectroscopic results, we show that the NV centre offers a rich set of possibilities for experiments demonstrating coherent optical control of the system, several of which are summarized in this chapter.

5.2   Level structure and polarization properties of the NV centre

As is the case in any system, the electronic structure of the NV centre can be analysed by considering its symmetry, in addition to the various interactions that are present. Our description starts with electronic states that obey the C_{3v} symmetry of the NV centre. We then consider the effect of various interactions one by one, starting from the the Coulomb interaction, which sets the energy scale of the optical transitions. Spin–orbit and spin–spin interactions act to lift the degeneracy of the different states in the ground- and excited-state manifolds. Finally, we consider the effect of external perturbations such as electric and magnetic fields that change the nature and energies of the unperturbed states. More detailed analyses of many of the concepts presented in this section can be found in [56].

5.2.1   Electronic states of the NV centre

The negatively charged NV centre has six electrons, five of which are from the nitrogen and the three carbons surrounding the vacancy. They occupy the orbital states a'_1, a_1, e_x, and e_y, which are combinations of the four dangling bonds surrounding the vacancy that satisfy the symmetry imposed by the nuclear potential and can be written as

$$a'_1 = \alpha \frac{\sigma_1 + \sigma_2 + \sigma_3}{3} + \beta \sigma_n \,, \tag{5.1}$$

$$a_1 = \beta \frac{\sigma_1 + \sigma_2 + \sigma_3}{3} + \alpha \sigma_n \,, \tag{5.2}$$

$$e_x = \frac{2\sigma_1 - \sigma_2 - \sigma_3}{\sqrt{6}} \,, \tag{5.3}$$

$$e_y = \frac{\sigma_2 - \sigma_3}{\sqrt{2}} \,, \tag{5.4}$$

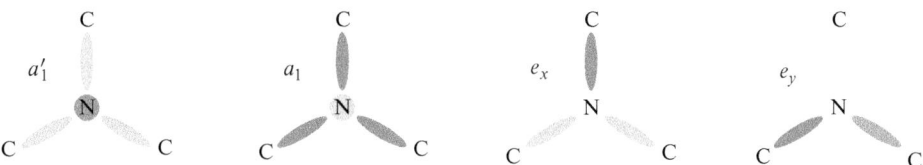

Fig. 5.1 [Colour online] Single-electron orbitals of the NV centre. The nitrogen and three carbons surrounding the vacancy are shown, and the perspective is along the NV axis. The grayscale [colour scale online] roughly represents the occupation of each orbital.

where σ_1, σ_2, σ_3, and σ_n are the dangling bonds from the three carbons and the nitrogen, respectively. These orbitals transform as the irreducible representations of the C_{3v} group and have been extensively used by many authors [36, 39, 51, 53]. They are schematically illustrated in Fig. 5.1 to give an idea of their symmetries. These pictorial representations allow one to deduce, for example, that $\langle e_y | \hat{\mathbf{y}} \cdot \mathbf{r} | e_x \rangle = \langle e_x | \hat{\mathbf{x}} \cdot \mathbf{r} | e_x \rangle = -\langle e_y | \hat{\mathbf{x}} \cdot \mathbf{r} | e_y \rangle \neq 0$. This means that the $|e_x\rangle$ and $|e_y\rangle$ states have permanent electric dipole moments, which will become important later for understanding the electric field sensitivity of the NV centre's excited states and spectral diffusion of the optical transitions. In addition, one also sees that $\langle a | \hat{\mathbf{x}} \cdot \mathbf{r} | e_x \rangle = \langle a | \hat{\mathbf{y}} \cdot \mathbf{r} | e_y \rangle \neq 0$, which gives rise to the transition dipole moments between the ground and excited states. These properties will be discussed in more detail in the following sections.

In the ground state of the NV centre, the a_1' and a_1 states, which are lowest in energy, are filled by four electrons, as shown in Fig. 5.2. The remaining two electrons occupy the degenerate orbitals c_x and c_y. The orbitals e_x and e_y can be viewed as p-type orbitals and $e_+ = -e_x - ie_y$, $e_- = e_x - ie_y$ are analogous to p states with definite orbital angular momentum. An antisymmetric combination of the orbital states minimizes the Coulomb energy and results in a spin-triplet ground-state manifold

$$|^3A_2\rangle = |E_0\rangle \otimes \begin{cases} |{+}1\rangle \\ |0\rangle \\ |{-}1\rangle \end{cases}, \tag{5.5}$$

where $|{\pm}1\rangle, |0\rangle$ correspond to the $m_s = \pm 1, 0$ states, respectively. The notation $|E_0\rangle = |e_x e_y - e_y e_x\rangle$ indicates that the orbital state has 0 orbital angular momentum projection along the NV axis. A_2 denotes the orbital symmetry of the state, which is determined by the symmetries of the e_x and e_y orbitals. From here on, we will often denote these ground states simply as $|0\rangle$ and $|{\pm}1\rangle$ when it is clear that we are referring to the full electronic state and not just the spin projections.

The relevant excited state for the optical transitions of the NV centre is a pair of triplets that arises from the promotion of one of the electrons occupying the a_1 orbital to the e_x or e_y orbital [39]. Note that this state can be modelled by one hole in the orbital e and another hole in the orbital a_1, i.e. a triplet in the ae electronic configuration (similarly, the ground state can be modelled by two holes in the e^2

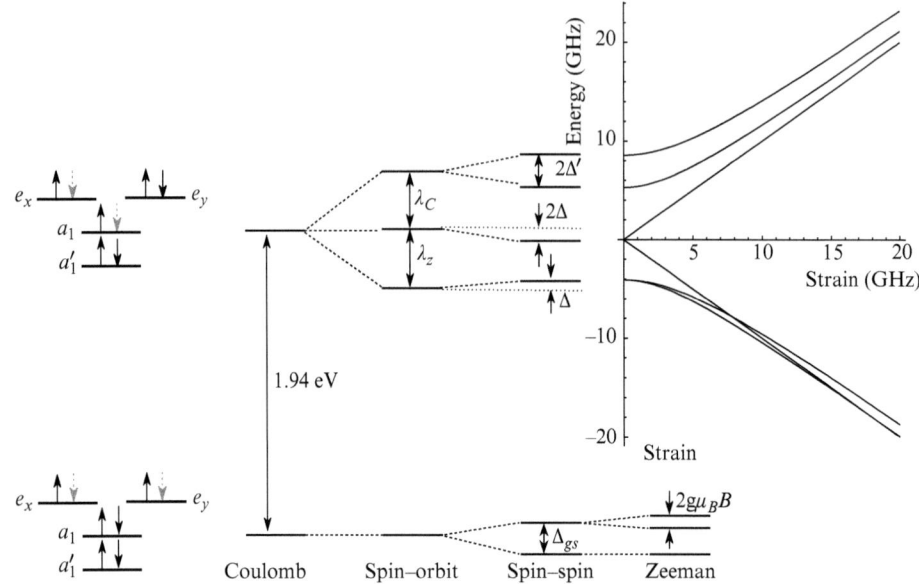

Fig. 5.2 Ground and excited states associated with the NV centre's optical transitions. The four symmetrized states are filled by six electrons, which can also be viewed as two holes (indicated with dashed arrows). The Coulomb interaction defines the optical transition energy of the NV centre. The effect of the spin–orbit and spin–spin interactions are indicated for both the ground and excited states. In the ground state, the application of a magnetic field further splits the $|m_s = \pm 1\rangle$ states due to the Zeeman interaction. In the excited state, crystal strain modifies the energies of sublevels. See [56] for details.

electronic configuration). A total of six states can be formed in this configuration, and their symmetries are determined by a group theoretical analysis [56]:

$$|A_1\rangle = |E_-\rangle \otimes |+1\rangle - |E_+\rangle \otimes |-1\rangle \,,$$
$$|A_2\rangle = |E_-\rangle \otimes |+1\rangle + |E_+\rangle \otimes |-1\rangle \,,$$
$$|E_x\rangle = |X\rangle \otimes |0\rangle \,,$$
$$|E_y\rangle = |Y\rangle \otimes |0\rangle \,, \tag{5.6}$$
$$|E_1\rangle = |E_-\rangle \otimes |-1\rangle - |E_+\rangle \otimes |+1\rangle \,,$$
$$|E_2\rangle = |E_-\rangle \otimes |-1\rangle + |E_+\rangle \otimes |+1\rangle \,,$$

where we have named the first four states as A_1, A_2, E_x, and E_y according to their symmetries and named the last two states as E_1 and E_2 since they also transform according to the irreducible representation E. Here, $|E_\pm\rangle = |ae_\pm - e_\pm a\rangle$ and $|X(Y)\rangle = |ae_{x(y)} - e_{x(y)}a\rangle$.

As spin–orbit and spin–spin interactions are invariant under any operation of the C_{3v} group, the states given in (5.6) are eigenstates of the full Hamiltonian including these interactions and in the absence of any perturbation such as magnetic field and/or crystal strain. The Hamiltonians for the spin–orbit and spin–spin interactions can be written in the basis of the six states listed in (5.6) as

$$H_{so} = \lambda_z(|A_1\rangle\langle A_1| + |A_2\rangle\langle A_2| - |E_1\rangle\langle E_1| - |E_2\rangle\langle E_2|),\qquad(5.7)$$

$$H_{ss} = \Delta(|A_1\rangle\langle A_1| + |A_2\rangle\langle A_2| + |E_1\rangle\langle E_1| + |E_2\rangle\langle E_2|)$$
$$- 2\Delta(|E_x\rangle\langle E_x| + |E_y\rangle\langle E_y|) + \Delta'(|A_2\rangle\langle A_2| - |A_1\rangle\langle A_1|)$$
$$+ \Delta''(|E_1\rangle\langle E_y| + |E_y\rangle\langle E_1| - i|E_2\rangle\langle E_x| + i|E_x\rangle\langle E_2|),\qquad(5.8)$$

where λ_z is the axial spin–orbit interaction and $3\Delta \approx 1.42\,\text{GHz}$ and $\Delta' \approx 1.55\,\text{GHz}$ characterize the spin–spin induced zero-field splittings [51, 56, 71]. It can be seen that spin–orbit interaction splits states with different total angular momentum, i.e. the pairs (A_1, A_2), (E_x, E_y), and (E_1, E_2) are split from each other by about 5.5 GHz [9, 51, 54]. The Δ'' term leads to non-spin-preserving cross transitions, which results in optical pumping of the spin states and plays a role in, for example, limiting the efficiency of spin readout using resonant excitation. The spin–spin interaction also plays a crucial role in the stability of the states $|A_1\rangle$ and $|A_2\rangle$. This interaction shifts up the non-zero spin states (A_1, A_2, E_1, and E_2) by $\sim \frac{1}{3} \times 1.42$ GHz [9, 34] and shift the states $|E_x\rangle$ and $|E_y\rangle$ down by $\sim \frac{2}{3} \times 1.42\,\text{GHz}$. Thus, the gap between (A_1, A_2) and (E_x, E_y) is increased, but the gap between the states (E_x, E_y) and (E_1, E_2) is reduced. In addition, the spin–spin interaction splits the states $|A_2\rangle$ and $|A_1\rangle$ by $\sim 3.3\,\text{GHz}$ [9]. Thus, for relatively low strain, we obtain a very robust $|A_2\rangle$ state with stable symmetry properties that are protected by an energy gap arising from the spin–orbit and spin–spin interactions.

We note here that the zero-field splitting in the ground state also arises from the spin–spin interaction, and gives an energy difference of $\Delta_{gs} \sim 2.88\,\text{GHz}$ between the $|m_s = 0\rangle$ and $|m_s = \pm 1\rangle$ states.

5.2.2 Properties of optical transitions

Once the wavefunctions are known, it is possible to calculate the selection rules for optical transitions between the triplet excited state and the triplet ground state. The dipole moment between the ground and excited states is produced by the hole left in the a orbital under optical excitation. As mentioned earlier, the matrix elements $\langle a|\,\hat{\mathbf{x}}\cdot\mathbf{r}\,|e_x\rangle$ and $\langle a|\,\hat{\mathbf{y}}\cdot\mathbf{r}\,|e_y\rangle$ are non-zero, where $\hat{\mathbf{x}}$ and $\hat{\mathbf{y}}$ represent the polarization of the involved photon. We can then calculate the selection rules for transitions between every pair of ground and excited states, as shown in Table 5.1. As expected, these selection rules conserve the total angular momentum of the photon–NV centre system.

The allowed optical transitions and their polarization properties indicate several possible Λ schemes in the NV centre. However, although these properties are relatively robust in the $|A_1\rangle$ and $|A_2\rangle$ states, the $|A_1\rangle$ state is coupled non-radiatively to a metastable singlet state, which then decays to the ground state $|0\rangle$. This results in

Table 5.1 Selection rules for optical transitions between the triplet excited state (ae) and the triplet ground state (e^2). Linear polarizations are represented by $\hat{\mathbf{x}}$ and $\hat{\mathbf{y}}$, while circular polarizations are represented by $\hat{\sigma}_\pm = \hat{\mathbf{x}} \pm i\hat{\mathbf{y}}$. As an example, a photon with σ_+ polarization is emitted when the electron decays from state A_2 to state $^3A_{2-}$.

Polarization	A_1	A_2	E_1	E_2	E_x	E_y
$^3A_{2-}$	$\hat{\sigma}_+$	$\hat{\sigma}_+$	$\hat{\sigma}_-$	$\hat{\sigma}_-$		
$^3A_{20}$					$\hat{\mathbf{y}}$	$\hat{\mathbf{x}}$
$^3A_{2+}$	$\hat{\sigma}_-$	$\hat{\sigma}_-$	$\hat{\sigma}_+$	$\hat{\sigma}_+$		

leakage out of the Λ system consisting of the $|A_1\rangle$ state and the $|\pm 1\rangle$ states. Thus, we find that the $|A_2\rangle$ state provides the an almost ideal closed Λ scheme for experiments such as spin–photon entanglement, as described in Section 5.6 and [88]. On the other hand, the open Λ system involving $|A_1\rangle$ can be useful for efficient detection of dark states during coherent population trapping, as shown in Section 5.7 and [87].

5.2.3   The effect of strain

We now discuss the effect of local strain on the properties of the optical transitions in order to understand variations between different NV centres and deviations from the unperturbed system described above. This perturbation splits the degeneracy between the e_x and e_y orbitals and results in their mixing. In the limit of high strain (larger than the spin–orbit splitting), the excited-state manifold splits into two triplets, each with a particular well-defined spatial wavefunction. Orbital and spin degrees of freedom separate in this regime and spin-preserving transitions are excited by linearly polarized light. The strain Hamiltonian is given by [56, 71]

$$H_{\text{strain}} = \delta_1(|e_x\rangle\langle e_x| - |e_y\rangle\langle e_y|) + \delta_2(|e_x\rangle\langle e_y| + |e_y\rangle\langle e_x|), \qquad (5.9)$$

where δ_1 and δ_2 are different parameters describing the crystal strain. Figure 5.3 shows an example of how δ_1 strain splits and mixes the excited states, and as a result changes the polarization properties of the optical transitions. A similar effect occurs for δ_2 strain, where $|A_2\rangle$ is mixed with $|E_1\rangle$.

The full Hamiltonian for the excited-state manifold of the NV centre is then

$$H = H_{ss} + H_{so} + H_{\text{strain}}, \qquad (5.10)$$

which can be written in the set of basis states $\{|A_1\rangle, |A_2\rangle, |E_x\rangle, |E_y\rangle, |E_1\rangle, |E_2\rangle\}$ as

$$H = \begin{pmatrix} \Delta - \Delta' + \lambda_z & 0 & 0 & 0 & \delta_1 & -i\delta_2 \\ 0 & \Delta + \Delta' + \lambda_z & 0 & 0 & i\delta_2 & -\delta_1 \\ 0 & 0 & -2\Delta + \delta_1 & \delta_2 & 0 & i\Delta'' \\ 0 & 0 & \delta_2 & -2\Delta - \delta_1 & \Delta'' & 0 \\ \delta_1 & -i\delta_2 & 0 & \Delta'' & \Delta - \lambda_z & 0 \\ i\delta_2 & -\delta_1 & -i\Delta'' & 0 & 0 & \Delta - \lambda_z \end{pmatrix}. \qquad (5.11)$$

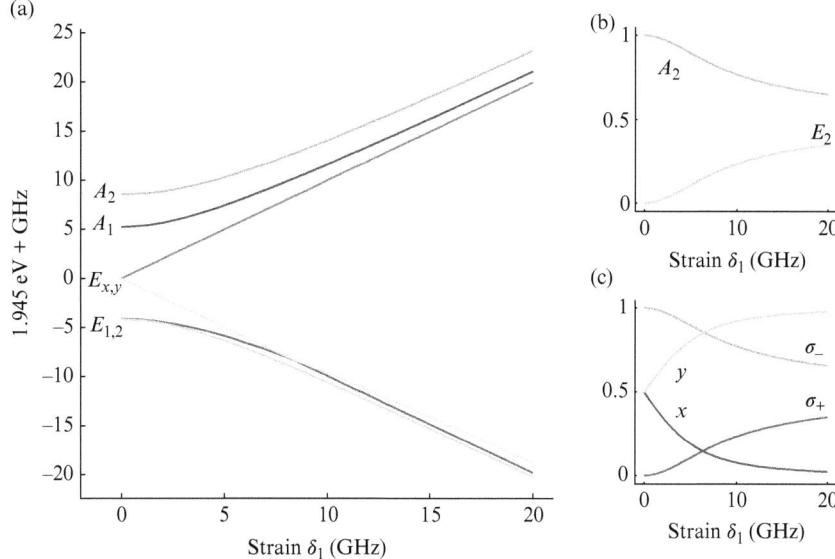

Fig. 5.3 [Colour online] Effect of strain on the properties of the NV centre. (a) Energies of the excited states as a function of strain, expressed in units of the linear strain induced splitting between the $|E_x\rangle$ and $|E_y\rangle$ states. (b) Strain induced mixing of the $|A_2\rangle$ and $|E_2\rangle$ states, showing the fraction of A_2 and E_2 character of the highest-energy excited state. (c) Polarization character of the optical transition between the $|A_2\rangle$ state and the ground state $|+1\rangle$ as a function of strain. As strain increases, the polarization changes from circularly to linearly polarized. The transition between $|A_2\rangle$ and $|-1\rangle$ shows the same behaviour, but with σ_- and σ_+ switched. Therefore, at high strain, decay from the $|A_2\rangle$ results in a separable state of the photon polarization and spin rather than an entangled state.

It is worth mentioning that while the excited-state configuration is greatly affected by strain, the ground-state configuration is unaffected to first order owing to its anti-symmetric combination of e_x and e_y orbitals. It is also protected by the large optical gap between the ground and excited states to second-order perturbation in strain.

5.3 Experimental techniques

Having developed a theoretical model for the optical transitions of the NV centre, we now move on to experimental explorations of these properties. While many descriptions of setups and procedures for room-temperature experiments with NV centres are available [19, 40], we will focus here on some techniques relevant for low-temperature studies of the optical properties.

5.3.1 Diamond materials

We begin this section with a description of different types of single-crystal diamond samples and their properties with regard to NV centre properties. The first type of commonly available diamond material is classified as type Ib. In these diamonds, substitutional nitrogen atoms (also called P1 centres) are the dominant defects, and cause the samples to have a yellow colour [91]. Type Ib diamonds can occur naturally and are typically grown by the high-pressure/high-temperature (HPHT) method by companies such as Element Six and Sumitomo. Element Six specifies their type Ib diamond as having substitutional nitrogen concentrations of less than 200 ppm. In our experience, the concentration of NV centres in type Ib samples varies widely among different samples, but NV centres in bulk samples are usually unresolvable under confocal microscopy. There have been no reports of coherent optical transitions from type Ib diamonds, except in one nanocrystal sample where a narrow, but spectrally unstable line was observed under photoluminescence excitation (PLE) spectroscopy [77]. However, as far as we know, this result has never been reproduced. In our experiments, type Ib samples are typically used for developing and fabricating nanophotonic structures and studies of NV centres in these structures at room temperature.

The remaining diamond samples that are suitable for the experiments described here can be classified as type IIa, which is typically defined as having nitrogen impurity concentrations of less than 1 ppm. Type IIa diamonds are rare in nature, but one such sample was found containing a resolvable concentration of single NV centres that show optical linewidths that are close to the lifetime-limited linewidth of 13 MHz [85]. Many initial experiments on the optical properties of NV centres were performed on several different pieces of this sample, including those described in Sections 5.6 and 5.7 of this chapter, as well as, for example, in [85] and [10]. The unique properties of this sample, and its somewhat exotic origins in the Ural mountains, has earned it the nickname 'The Magic Russian Diamond'. The piece used in our experiments is shown in Fig. 5.4(a). Clearly, such specialized samples cannot be used for wide-scale experiments or for the development of diamond-based devices. An alternative is to work with synthetic type IIa diamonds grown using microwave-assisted chemical vapour deposition (CVD) at companies such as Element Six Inc. Standard type IIa grade samples typically also show a unresolvable concentration of NV centres. However, 'electronic grade' diamonds developed by Element Six, which are specified to have substitutional nitrogen concentrations of less than 5 ppb, have NV centre concentrations that range from barely resolvable single NVs to less than 1 NV per tens of μm^3. Native NV centres found in these samples often have linewidths that are broadened by spectral diffusion to several hundred megahertz [1, 12, 81]. However, they have been used in a variety of experiments involving optical transitions of NV centres [12, 13, 79], and are also excellent starting materials for introducing optically coherent NV centres through ion implantation, as described in Section 5.8.

5.3.2 Experimental setup for optical characterization of NV centres

Figure 5.4(b) shows a schematic diagram of a multipurpose setup used for optical investigation of NV centres and diamond-based photonic devices. Depending on the type of experiment, only parts of this comprehensive setup are needed. A typical basic confocal microscope for room-temperature studies of NV centres is shown in black. An objective with high numerical aperture (NA) is used to focus laser light

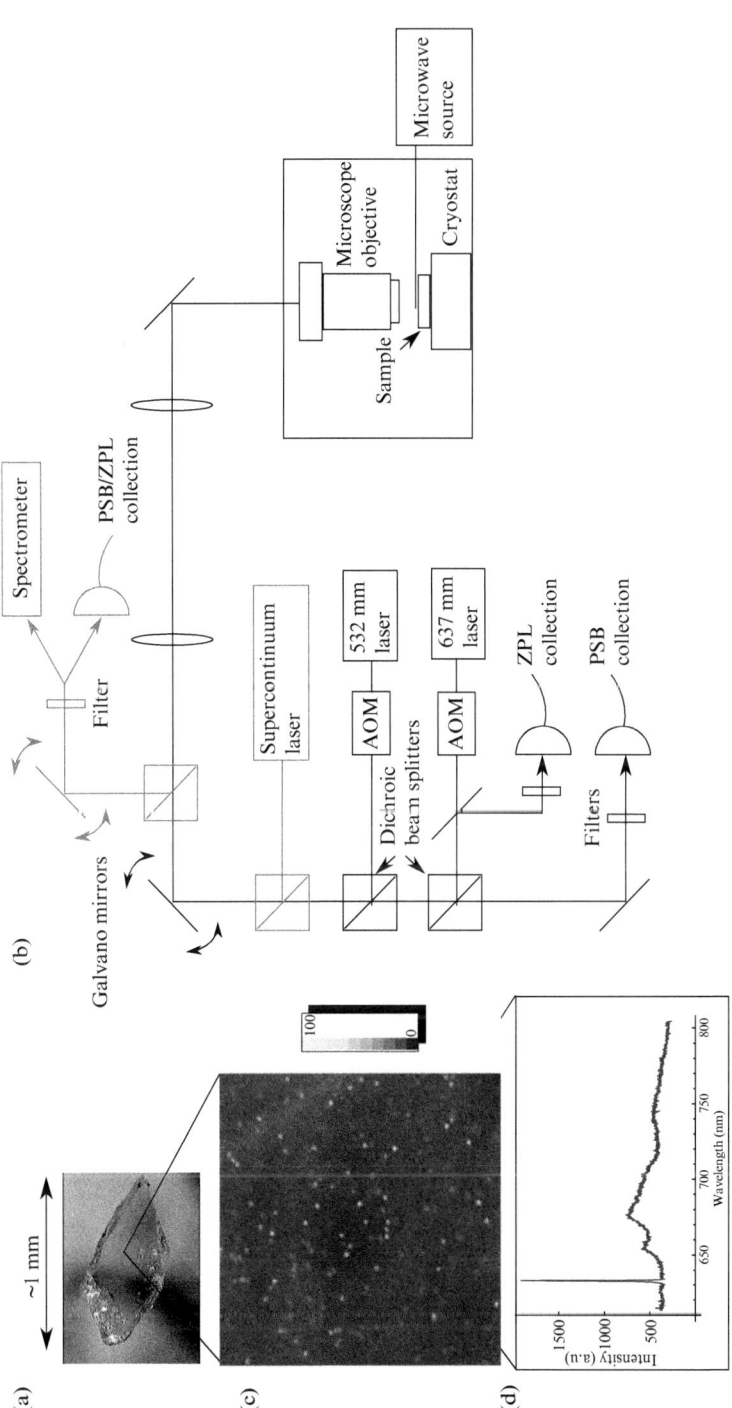

Fig. 5.4 [Colour online] Experimental setup and characterization of NV centres. (a) The type IIa natural diamond sample used for the experiments in Sections 5.6 and 5.7. (b) Confocal microscope setup. The off-resonant 532 nm laser is needed for a basic room-temperature apparatus for investigating the ground-state spin physics of the NV centre. A cryostat and resonant excitation laser are additionally needed for low-temperature studies of coherent optical properties. A supercontinuum laser is added for characterization of photonic devices such as cavities and waveguides, and an additional, independent, collection channel is used for collecting light from a spatial location differing from that of the excitation. In addition, the spectrometer can also be used for photoluminescence studies for the NV centre. (c) Confocal image of single NV centres obtained using 532 nm excitation and PSB collection. (d) Low-temperature photoluminescence spectrum of an NV centre. The sharp peak is the zero-phonon line (ZPL) at 637 nm.

onto the sample. In order to scan over the sample and locate individual centres in a confocal image such as that shown in Fig. 5.4(c), either the diamond is mounted on a high-resolution positioning stage or the optical path is scanned using a galvano mirror imaged onto the back of the objective with a pair of lenses comprising a $4f$ imaging system. The vibronic sidebands of the NV centre's optically excited states can be addressed using a laser of higher energy than the 637 nm zero-phonon line (ZPL). Typically, this is done using a 532 nm frequency-doubled Nd:YAG laser, which is the most common method for controlling the charge state of the NV centre. Continuous-wave (CW) 532 nm excitation can ionize both the negatively charged NV centre and surrounding charge donors such as substitutional nitrogens, resulting in a equilibrium charge state of the NV centre that is $\sim$70% NV$^-$ and $\sim$30% NV^0 [3]. Off-resonant excitation is also used in-room temperature experiments to polarize the spin of the NV centre into the $|m_s = 0\rangle$ ground state. This process happens through non-spin-preserving transitions from the optically excited state into spin-singlet states of the NV centre, which then preferentially decay to the $|m_s = 0\rangle$ state [54]. Finally, microwave and radiofrequency (RF) fields for driving spin transitions of the NV centres can be applied with a variety of methods, including copper wires, coplanar waveguides, and resonator structures.

For low-temperature experiments, the diamond sample is placed inside a helium flow cryostat (Janis ST-500) and cooled to $\sim$4 K. Alternative low-temperature systems such as closed-cycle pulse tube coolers and bath cryostats can also used. Diamond samples are typically mounted onto the cold finger, or some intermediate substrate such as a silicon wafer, using indium or silver paste. For resonant excitation of the NV centres, a external cavity diode laser around 637 nm is introduced into the optical path. The New Focus TLB 6304 laser is a popular choice because of its large mode-hop-free scanning range of 60–80 GHz, wavelength tunability over several nanometres, and ease of operation. Both the 532 and 637 nm lasers need to be pulsed for most experiments, which can be accomplished using acousto–optical modulators (AOMs) or electro-optical modulators (EOMs).

Fluorescence from the NV centre is collected through the same optical path, coupled into single-mode fibres, and detected using avalanche photodiodes (APDs) or a spectrometer. Filters are placed in the collection path to remove reflected excitation light and separate different parts of the emission spectrum. As shown in Fig. 5.4(d), the NV centre spectrum consists of a broad separate phonon sideband (PSB) and a sharp ZPL at low temperatures. The percentage of emission into the ZPL is typically 3–5%, which is a major limitation on the efficiency of applications based on coherent photons from the NV centre. In Section 5.9, we will address this issue by enhancing the emission into the ZPL using cavity QED with nanophotonic structures. In order to characterize these structures, we use an additional supercontinuum laser as a broadband light source to perform transmission measurements of cavities and waveguides. The light is coupled into one port of the optical device, and the transmitted light must be collected from an output port at a spatially separate location. For this purpose, a second collection channel can be added with its own set of galvano mirrors. In addition, the signal from the two channels can be combined on a time-correlated single-photon-counting (TCSPC) device for measurements of the second-order autocorrelation function $g^2(\tau)$

of the emission from individual NV centres to show that they are single-photon emitters [50]. This can also be accomplished by splitting the light collected in a single channel using a fibre beam splitter.

5.4   PLE spectroscopy of NV centres

The optical transitions of the NV centre can be identified using PLE spectroscopy by scanning the resonant excitation laser across the ZPL and collecting photons in the PSB. Since the $|0\rangle \to |E_x\rangle$ and $|0\rangle \to |E_y\rangle$ are the best cycling transitions, they are usually the ones observed in PLE spectroscopy in the absence of microwave driving or laser modulation. However, repeated resonant excitation eventually ionizes the NV$^-$ and optically pumps the NV centre into a different ground-state spin sublevel owing to the small non-spin-conserving cross transitions out of the $|m_s = 0\rangle$ excited states into the $|\pm 1\rangle$ ground states [86]. Therefore, some form of charge and spin repumping must eventually be performed. As shown in Fig. 5.5(b), the first method we use involves scanning the 637 nm laser across the NV transitions and turning on the laser while simultaneously collecting PSB counts for ~ 2 ms at each frequency. At the end of the scan, the 637 nm laser is turned off and a 532 nm repumping pulse is applied. Any duration of repumping longer than a few microseconds is sufficient to reinitialize the NV centre. Depending on the charge environment of the NV centre and its strain-dependent level structure, however, it may be difficult to obtain good signal from this simple method. For example, in the presence of a large number of other charge traps in the environment, the NV centre may be ionized before a sufficient number of photons have been scattered on resonance. Another potential issue is that the $|0\rangle \to |E'_x\rangle$ and $|0\rangle \to |E'_y\rangle$ transitions have some branching ratio into the $m_s = \pm 1$ states, causing a decrease in fluorescence on the timescale of several microseconds. However, fluorescence can potentially still be detected owing to off-resonant excitation of transitions from the $|\pm 1\rangle$ spin states. For example, at high strain, the $|\pm 1\rangle \to |A_1\rangle$ transition becomes almost degenerate with the $|\pm 1\rangle \to |E_x\rangle$ transition, allowing a steady-state fluorescence to be established. However, this may not be the case for all NV centres. More robust methods of obtaining the PLE spectrum without repumping during the scan involve either CW microwave mixing of the $|0\rangle$ and $|\pm 1\rangle$, or modulation of the resonant excitation laser by the ground-state zero-field splitting, which also allows the $|m_s = \pm 1\rangle$ transitions to be observed [33, 72, 86].

Figure 5.5(b) shows the results of PLE scans taken using this first method. In each scan, a clear resonance can be seen, whose frequency then jumps during each successive scan. This is an indication of spectral diffusion, where changes in the local electric field environment of the NV centre cause the energies of its excited states to fluctuate over time. As mentioned earlier, from the electronic orbitals $|e_x\rangle$ and $|e_y\rangle$ involved in the excited states, one can see that the $|E_x\rangle$ and $|E_y\rangle$ states have a permanent electric dipole moment. This dipole moment has been measured to be ~ 0.8 debye, which gives rise to energy shifts of $\sim 4\,\mathrm{GHz\,MV^{-1}\,m^{-1}}$ [56]. It is now well established that 532 nm repumping causes complex charge dynamics in the diamond environment by photoionizing impurities that act as charge donors [8]. Therefore, while

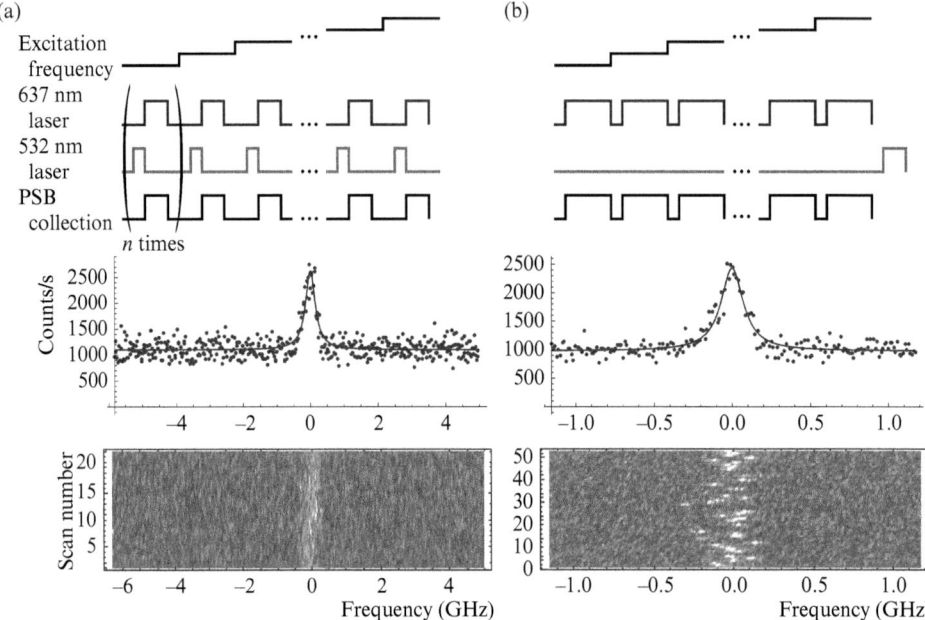

Fig. 5.5 [Colour online] Methods for PLE spectroscopy of NV centres. (a) Pulse sequence for repumping during each PLE scan (top), along with an example showing the result of successive scans (bottom) and the averaged spectrum (middle). (b) Pulse sequence for repumping at the end of each PLE scan (top), along with corresponding results for the same transition as in (a), but zoomed in over a smaller frequency range.

very narrow single-scan linewidths can be obtained using this method [33, 77, 85], the long-term, extrinsically broadened linewidth obtained by averaging many such scans can be much broader [1, 12, 81]. This issue is addressed in more detail in Section 5.8.

A second method of obtaining the PLE spectrum involves a pulse sequence consisting of a 1 μs 532 nm repumping pulse, followed by a 10 μs resonant excitation pulse, during which PSB counts are collected. As the 637 nm laser is scanned across the NV transitions, this sequence is repeated many times for each frequency, as illustrated in Fig. 5.5(a). This method has the advantage that it is robust against ionization and spin polarization, and should in principle give a good signal-to-noise ratio, regardless of the charge environment and strain-dependent level structure of the NV centre. It also shows the overall lineshape of the transition, including effects of spectral diffusion, in a single scan. Again, microwave fields are used to obtain a spectrum of all the possible transitions, including those that involve the $|m_s = \pm 1\rangle$ states.

5.5   Optical properties of NV centres

Figure 5.6(a) shows a PLE spectrum taken with CW microwave excitation that mixes the ground states during the 637 nm laser pulse. All resonances expected from the

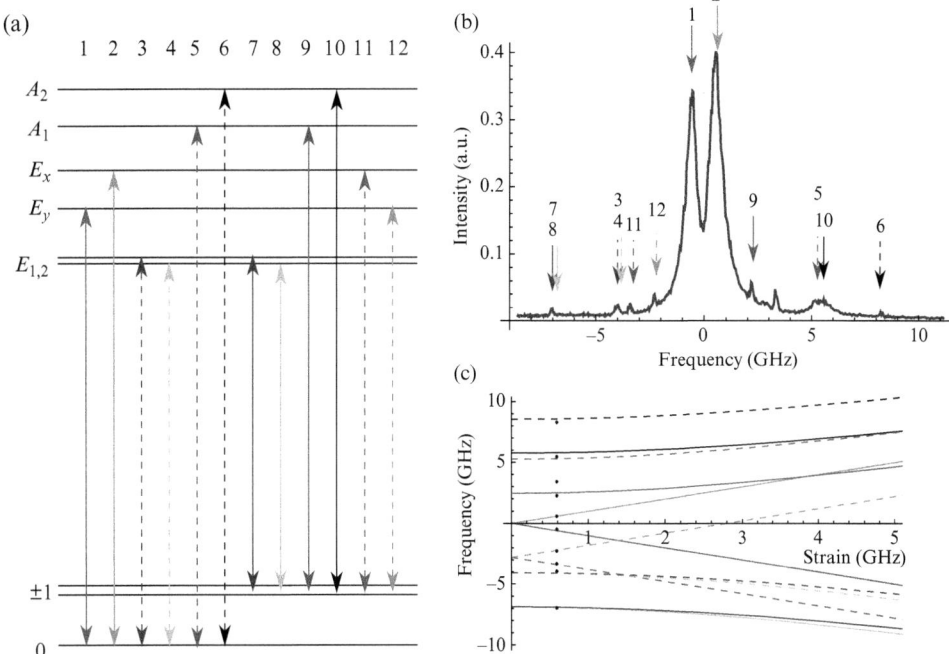

Fig. 5.6 [Colour online] Optical transitions of the NV centre. (a) All possible transitions between the ground and excited states, with direct transitions indicated by solid lines, and spin-non-conserving cross transitions by dashed lines. (b) PLE spectrum taken with CW microwave excitation. (c) Frequencies of all possible transitions shown in (a) as a function of strain. The frequencies of the peaks in (b) are matched to a particular strain value. The extra unidentified peak may be due to a two-photon transition from the $|0\rangle$ state to the $|E_x\rangle$ state through absorption of an optical photon and emission of a microwave photon.

direct and cross transitions are observed, as shown in Fig. 5.6(b), along with some possible microwave–optical two-photon transitions.

The strongest transitions in Fig. 5.6(a) are the $|0\rangle \rightarrow |E_x\rangle$ and $|0\rangle \rightarrow |E_y\rangle$ cycling transitions. Figure 5.7(a) shows an example of optical Rabi oscillations on the $|0\rangle \rightarrow |E_y\rangle$ transition, measured by turning on a 40 ns resonant laser pulse a rise time of $\sim$1 ns using an EOM (Guided Color Technologies). We then accumulate a histogram of photon arrival times on a TCSPC device with a timing resolution of 195 ps. The photon count rate is proportional to the population in the $|E_y\rangle$ state, which, in the absence of decay or decoherence, is given by $P_{E_y}(t) = \cos^2(|\Omega|t)$, where $\Omega = \boldsymbol{\mu} \cdot \mathbf{E}/\hbar$ is the Rabi frequency. In this case, however, the oscillations are damped by the finite lifetime of the excited states, which can be measured by fitting to the exponentially decaying tail after the pulse turns off. In addition to population decay, additional decoherence, such as frequency fluctuations of the transition frequency relative to the laser frequency caused by, for example, spectral diffusion, will also contribute to the

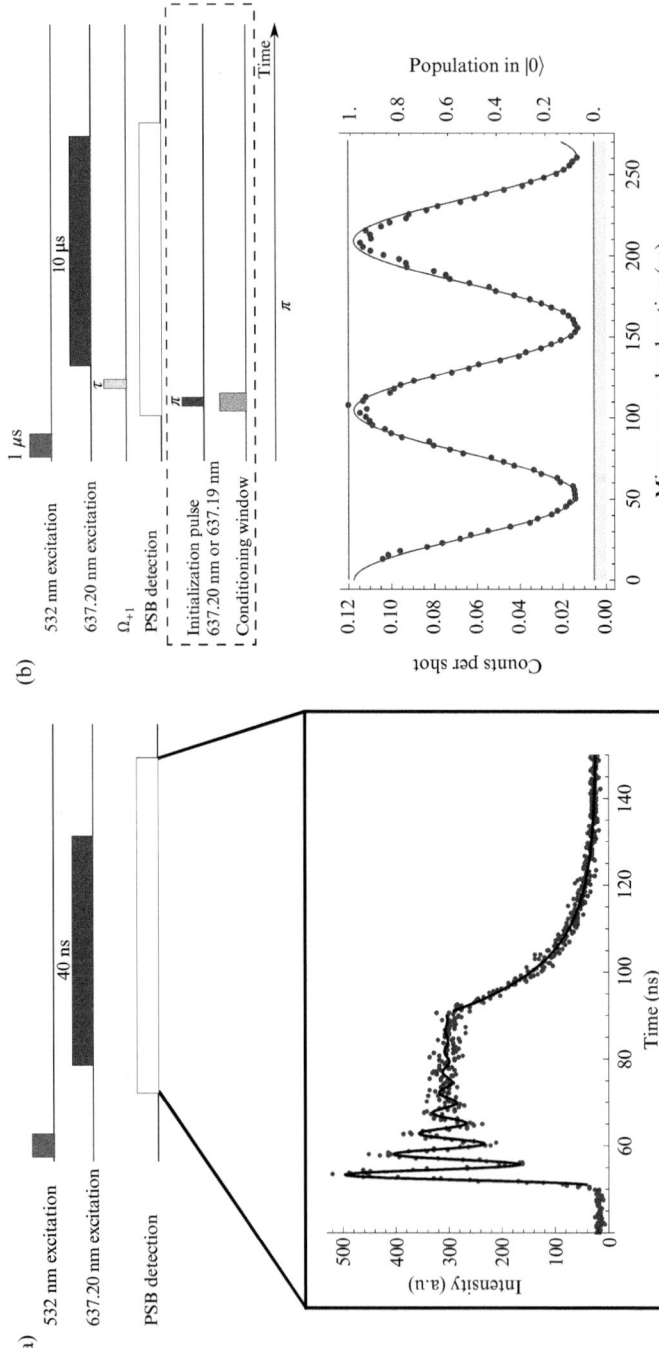

Fig. 5.7 [Colour online] (a) Optical Rabi oscillations detected using PSB fluorescence and resonant excitation, along with the corresponding pulse sequence. The 637.2 nm laser is tuned to the $|0\rangle \rightarrow |E_y\rangle$ transition and is on from 50 to 90 ns. (b) Rabi oscillations between $|0\rangle$ and $|+1\rangle$ detected using resonant spin readout and the corresponding pulse sequence. Horizontal lines are calibration levels for the $|0\rangle$ and $|+1\rangle$ states. To initialize the spin states for this calibration, we use a measurement-based technique where the $|0\rangle$ is prepared conditioned on the detection of a PSB photon after excitation to the $|E_y\rangle$ state. To prepare the $|+1\rangle$ state, we then apply a microwave π pulse before readout.

damping of Rabi oscillations. We note here, however, that this additional dephasing alone does not always give a good estimate of the spectral diffusion of the NV centre, in contrast to what is sometimes claimed in the literature. For example, if the transition frequency jumps well outside of the power-broadened linewidth of the transition, then the NV centre will effectively not be excited, and there will simply be no contribution to the Rabi oscillation data.

By measuring Rabi oscillations, we can determine the length of the laser pulse needed to, for example, perform an optical π pulse to transfer population to the $|A_2\rangle$ state for entanglement generation, as described in Section 5.6. Note that, while optical Rabi oscillations can also be observed between the $|m_s = \pm 1\rangle$ ground and excited states, the dynamics are more complicated because these are Λ-type transitions. In the absence of a magnetic field, instead of settling to a steady-state value corresponding to 50% population in the excited state, the photon count rate continually decreases after a few optical cycles as population is pumped into a 'dark' ground state that is not excited by the by the laser. However, the observed Rabi oscillations are usually sufficient for determining the optical π pulse length.

The cycling transitions of the NV centre can also be used to efficiently read out the spin state of the NV centre. As shown in Fig. 5.7, when a resonant laser pulse is turned on for 10 µs, the number of photons scattered is proportional to the population in the $|m_s = 0\rangle$ states, and microwave-driven Rabi oscillations between the $|0\rangle$ and $|+1\rangle$ ground-state sublevels can be observed. A detailed description of the resonant spin-readout procedure used in Section 5.6 can be found in [88]. The efficiency and fidelity of this spin readout method are limited by the photon collection efficiency, the temperature, and the branching ratio out of the $|m_s = 0\rangle$ manifold. As can be predicted by the Hamiltonian given in (5.11), this branching ratio depends on the local strain. In our first experiments, the spin readout contrast was limited to $c_M = 0.11 \pm 0.0022$ counts/shot when the spin was prepared in the $|0\rangle$ state and a background-limited $c_B = 0.0057 \pm 0.0010$ counts/shot when the spin was prepared in the $|\pm 1\rangle$ states. Therefore, to determine the spin of the NV centre, the state had to be prepared and read out many times to obtain an average number of counts per shot C, which then allows us to calculate the population in the $|0\rangle$ state as $P = (C - c_B)/(c_M - c_B)$. By increasing the collection efficiency with a solid immersion lens and working with low-strain NV centres, single-shot resonant readout of the spin state has subsequently been demonstrated [70].

The rest of this chapter gives several examples of applications based on the optical properties of NV centres that we have just described.

5.6 Quantum entanglement between an optical photon and a solid state spin qubit

5.6.1 Introduction

A quantum network [47] consists of several nodes, each containing a long-lived quantum memory and a small quantum processor, that are connected via entanglement. Its potential applications include long-distance quantum communication and distributed quantum computation [27]. Several recent experiments have demonstrated on-chip entanglement of solid-state qubits separated by nanometre [61] to millimetre [2, 23] lengthscales. However, realization of long-distance entanglement based on solid state systems coupled to single optical photons [22] is an outstanding challenge. The NV centre is a promising candidate for implementing a quantum node. The ground state of the negatively charged NV centre is an electronic spin triplet with a 2.88 GHz zero-field splitting between the $|m_s = 0\rangle$ and $|m_s = \pm 1\rangle$ states (from here on denoted by $|0\rangle$ and $|\pm 1\rangle$, respectively). With long coherence times [7], fast microwave manipulation, and optical preparation and detection [35], the NV electronic spin presents a promising qubit candidate. Moreover, it can be coupled to nearby nuclear spins that provide exceptional quantum memories and allow for the robust implementation of few-qubit quantum registers [28, 61]. In this section, we demonstrate the preparation of quantum entangled states between a single photon and the electronic spin of a NV centre:

$$|\Psi\rangle = \frac{1}{\sqrt{2}} (|\sigma_-\rangle |+1\rangle + |\sigma_+\rangle |-1\rangle), \qquad (5.12)$$

where $|\sigma_+\rangle$ and $|\sigma_-\rangle$ are orthogonal circularly polarized single-photon states.

The scheme we use here is inspired by previous work done with trapped ions [15]. The key idea of our experiment is illustrated in Fig. 5.8. The NV centre is prepared in a specific excited state ($|A_2\rangle$ in Fig. 5.8) that decays with equal probability into two different long-lived spin states ($|\pm 1\rangle$) by the emission of orthogonally polarized optical photons at 637 nm. The entangled state given by (5.12) is created because photon polarization is uniquely correlated with the final spin state. This entanglement is verified by spin state measurement using a cycling optical transition following the detection of a 637 nm photon of chosen polarization.

5.6.2 Experimental demonstration of spin–photon entanglement

We now turn to the experimental demonstration of spin–photon entanglement. The experimental setup is outlined in [88]. To create the entangled state, we use coherent

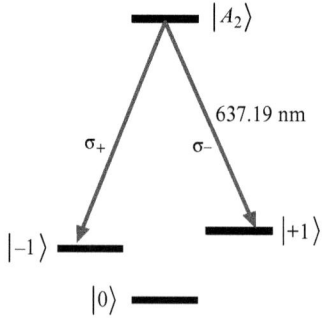

Fig. 5.8 [Colour online] Scheme for spin–photon entanglement. Following selective excitation to the $|A_2\rangle$ state, the Λ system decays to two different spin states through the emission of orthogonally polarized photons, resulting in spin–photon entanglement.

emission within the narrowband ZPL, which includes only 4% of the NV centre's total emission. The remaining optical radiation occurs in the frequency-shifted PSB, which is accompanied by phonon emission that causes the spin–photon entanglement to deteriorate [45]. Isolating the weak ZPL emission presents a significant experimental challenge owing to strong reflections of the resonant excitation pulse reaching the detector. By exciting the NV centre with a circularly polarized 2 ns π pulse that is shorter than the emission timescale, we can use detection timing to separate reflection from fluorescence photons. A combination of confocal rejection, modulators, and finite transmittivity of our optics suppresses the reflections sufficiently to clearly detect the NV centre's ZPL emission in a 20 ns region.

For photon state determination, ZPL photons in either the $|\sigma_\pm\rangle$ or the $|H\rangle = \frac{1}{\sqrt{2}}(|\sigma_+\rangle + |\sigma_-\rangle)$, $|V\rangle = \frac{1}{\sqrt{2}}(|\sigma_+\rangle - |\sigma_-\rangle)$ basis are selected by a polarization analysis stage and detected after an optical path of $\sim$2 m. Spin readout then occurs after a 0.5 μs spin memory interval following photon detection by transferring population from either the $|\pm 1\rangle$ states or from their appropriately chosen superposition into the $|0\rangle$ state using microwave pulses $\Omega_{\pm 1}$. The pulses selectively address the $|0\rangle \leftrightarrow |\pm 1\rangle$ transitions with resonant frequencies $\omega_\pm$ that differ by $\delta\omega = w_+ - w_- = 122$ MHz owing to an applied magnetic field. For superpositions of $|\pm 1\rangle$ states, an echo sequence is applied before the state transfer to extend the spin coherence time. The transfer is followed by resonant excitation of the $|0\rangle \leftrightarrow |E_y\rangle$ transition and collection of the PSB fluorescence. We carefully calibrate the transferred population measured in the $|0\rangle$ state using the procedure detailed in [88]. Figure 5.9(a) shows the populations in the $|\pm 1\rangle$ states, measured conditionally on the detection of a single circularly polarized ZPL photon. Excellent correlations between the photon polarization and NV spin states are observed.

To complete the verification of entanglement, we now show that correlations persist when ZPL photons are detected in a rotated polarization basis. Upon detection of a linearly polarized $|H\rangle$ or $|V\rangle$ photon at time t_d, the entangled state in (5.12) is projected to $|\pm\rangle = \frac{1}{\sqrt{2}}(|+1\rangle \pm |-1\rangle)$, respectively. These states subsequently evolve in time t according to

$$|\pm\rangle_t = \frac{1}{\sqrt{2}}\left(e^{-i\omega_+(t-t_d)}|+1\rangle \pm e^{-i\omega_-(t-t_d)}|-1\rangle\right). \tag{5.13}$$

In order to read out the relative phase of superposition states between $|+1\rangle$ and $|-1\rangle$, we use two resonant microwave fields with frequencies ω_+ and ω_- to coherently transfer the state $|M\rangle = \frac{1}{\sqrt{2}}\left(e^{-i\omega_+ t}|+1\rangle + e^{-i[\omega_- t - (\phi_+ - \phi_-)]}|-1\rangle\right)$ to $|0\rangle$, where the initial relative phase $\phi_+ - \phi_-$ is set to the same value for each round of the experiment. Thus, the conditional probability of measuring the state $|M\rangle$ is

$$p_{M|H,V}(t_d) = \frac{1 \pm \cos\alpha(t_d)}{2}, \tag{5.14}$$

where $\alpha(t_d) = (\omega_+ - \omega_-)t_d + (\phi_+ - \phi_-)$. Equation (5.14) indicates that the two conditional probabilities should oscillate with a π phase difference as a function of the photon detection time t_d. This can be understood as follows. In the presence of

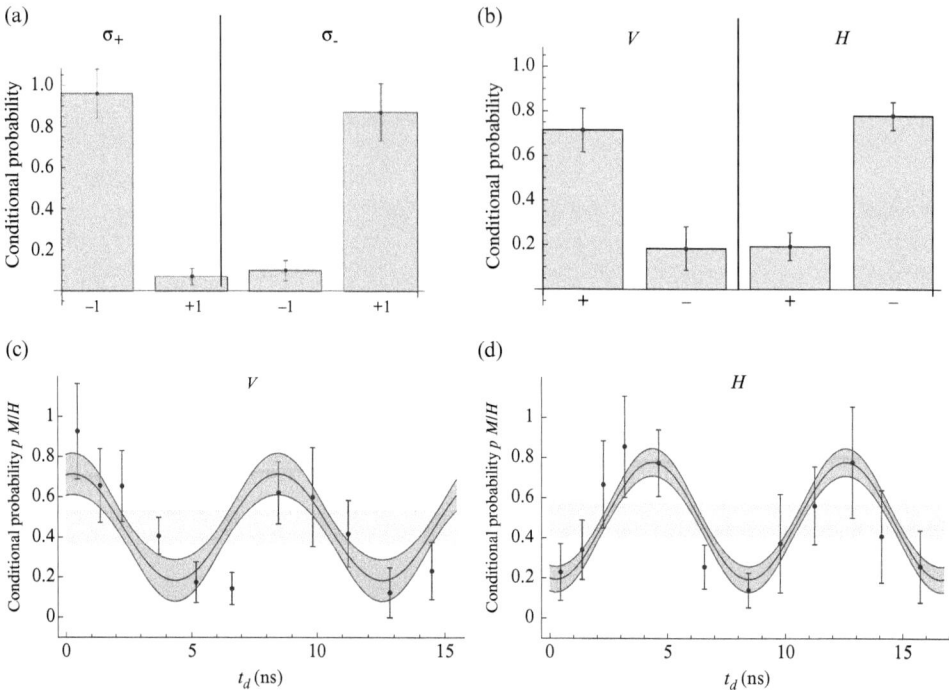

Fig. 5.9 [Colour online] Measurement of spin–photon correlations in two bases. (a) Conditional probability of measuring $|\pm1\rangle$ after the detection of a σ_+ or σ_- photon. (b) Conditional probability of measuring $|\pm\rangle$ after the detection of an H or V photon, extracted from a fit to data shown in (c) and (d). (c, d) Measured conditional probability of finding the electronic spin in the state $|M\rangle$ after detection of a V (c) or H (d) photon at time t_d. The shaded region around the fit (solid line) to the time-binned data is the 68% confidence interval. Errors on data points are one standard deviation. Combined with the data shown in (a), oscillations with amplitude outside of the horizontal shaded regions result in fidelities greater than 0.5. The visibilities of the measured oscillations are 0.59 ± 0.18 (c) and 0.60 ± 0.11 (d). (Reprinted from Togan, E., Chu, Y., Trifonov, A. S., et al. (2010). Quantum entanglement between an optical photon and a solid-state spin qubit. Nature, 466(7307), 730–734. doi:10.1038/nature09256. Copyright 2010 Nature Publishing Group.)

Zeeman splitting ($\delta\omega \neq 0$), the NV centre's spin state is entangled with both the polarization and frequency of the emitted photon. The photon's frequency provides which-path information about its decay. In the spirit of quantum eraser techniques, the detection of $|H\rangle$ or $|V\rangle$ at t_d with high time resolution ($\sim300\,\text{ps} \ll 1/\delta\omega$) erases the frequency information [73, 90]. When the initial relative phase between the microwave fields $\Omega_{\pm1}$ is kept constant, the acquired phase difference $(\omega_+ - \omega_-)t_d$ gives rise to oscillations in the conditional probability and produces an effect equivalent to varying the relative phase in the measured superposition, allowing us to verify the coherence of the spin–photon entangled state.

The detection times of ZPL photons are recorded during the experiment without any time gating, which allows us to study spin–photon correlations without reducing the count rate. The resulting data are analysed in two different ways. First, we time-bin the data and use it to evaluate the conditional probabilities of measuring the spin state $|M\rangle$ as a function of the $|H\rangle$ or $|V\rangle$ photon detection time (Fig. 5.9(c, d)). Off-diagonal elements of the spin–photon density matrix are evaluated from a simultaneous fit to the binned data and the time bins are chosen to minimize fit uncertainty as described in [88]. The resulting conditional probabilities (Fig. 5.9(b)) are used to evaluate a lower bound on the entanglement fidelity of $F \geq 0.69 \pm 0.068$, above the classical limit of 0.5, indicating the preparation of an entangled state.

We further reinforce our analysis using the method of maximum likelihood estimation. As described in [88], this method is applied to raw, unbinned ZPL photon detection and spin measurement data and yields a probability distribution of a lower bound on the fidelity. Consistent with the time-binned approach, we find that our data are described by a near-Gaussian probability distribution associated with a fidelity of $F \geq 0.70 \pm 0.070$. Significantly, the cumulative probability distribution directly shows that the measured lower bound on the fidelity is above the classical limit with a probability of 99.7%.

Several experimental imperfections reduce the observed entanglement fidelity. First, the measured strain and magnetic field slightly mixes the $|A_2\rangle$ state with the other excited states. We estimate that $|A_2\rangle$ state imperfection and photon depolarization in the setup together reduce the fidelity by 12%, the latter being the dominant effect. Imperfections in readout and echo microwave pulses decrease the fidelity by 3%. Other error sources include finite signal to noise ratio in the ZPL channel (fidelity decrease 11%), as well as timing jitter (another 4%). The resulting expected fidelity (73%) is consistent with our experimental observations. Finally, the entanglement generation succeeds with probability $p \sim 10^{-6}$, which is limited by low collection and detection efficiency as well as the small probability of ZPL emission.

The demonstration of spin–photon entanglement is the first step in implementing a quantum network architecture for long-distance quantum communications using NV centres [47]. Such an entangled state connects the solid state, spin-based local quantum register with optical photons, which can then be disseminated over free space or through optical fibres. Following the above experiment, tremendous progress has been made in achieving the next steps in building a diamond-based quantum network. Hong–Ou–Mandel interference was demonstrated with indistinguishable photons from remote NV centres hosted in separate diamond samples [12, 79]. Combined with spin–photon entanglement, this recently allowed for the remote entanglement of two NV centres separated by a distance of 3 m through entanglement swapping [13].

5.7 Laser cooling and real-time measurement of nuclear spin environment of a solid state qubit

In this section, we describe the use of the NV centre's optical transitions and coherent population trapping (CPT) to sense and manipulate its surrounding spin bath, which consists of the spin of the ^{14}N nucleus associated with the NV centre itself and the ^{13}C nuclei in the diamond lattice. Over the past two decades, CPT has been employed for laser cooling of neutral atoms and ions [4], the creation of ultracold molecules [62], optical magnetometry [17, 74], and atomic clocks [89], as well as for slowing and stopping light pulses [31]. The electronic spin of the NV centre is a promising system for extending these techniques to the solid state. The NV centre has a long-lived spin triplet as its electronic ground state [54], whose $m_s = \pm 1, 0$ sublevels are denoted by $|\pm 1\rangle$ and $|0\rangle$. In pure samples, the electron spin dynamics are governed by interactions with the spin-1 ^{14}N nucleus of the NV centre and spin-$\frac{1}{2}$ ^{13}C nuclei present in 1.1% natural abundance in the diamond lattice (Fig. 5.10(a)). Control over nuclear spins [28, 60] is of interest both for fundamental studies and for applications such as nanoscale magnetic sensing [6, 57] and realization of quantum networks [19, 88]. Here we achieve such control via two complementary methods: effective cooling of nuclear spins through nuclear-state-selective CPT [43] and conditional preparation based on fast measurements of the nuclear environment and subsequent post-selection [38].

While most previous work has involved the use of microwave and RF fields for manipulating both the electronic and nuclear spin states, we utilize all-optical control of the electronic spin [16, 69, 72]. Specifically, we make use of Λ-type level configurations involving the NV centre's $|A_1\rangle$ and $|A_2\rangle$ optically excited electronic states and the $|\pm 1\rangle$ ground states (Fig. 5.10(a)) [56, 88]. At low temperatures (<10 K) and in the limit of zero strain, $|A_1\rangle$ and $|A_2\rangle$ are entangled states of spin and orbital momentum coupled to the $|+1\rangle$ ($|-1\rangle$) state with σ_- ($\sigma+$) circularly polarized light. Correspondingly, excitation with linearly polarized light drives the NV centre into a so-called dark superposition state when the two-photon detuning is zero [75]. In the present case, the two-photon detuning is determined by the Zeeman splitting between the $|\pm 1\rangle$ states due to the combined effect of the Overhauser field originating from the nuclear spin environment and any externally applied magnetic field [43, 83]. When the external field exactly compensates the Overhauser field, the electronic spin of the NV centre is pumped into the dark state after a few optical cycles and remains in the dark state, resulting in vanishing fluorescence. This is the essence of the dark resonances and CPT.

5.7.1 Coherent population trapping with NV centres

In our experiments, the $|A_1\rangle$ and $|A_2\rangle$ states are separated by approximately 3 GHz and are addressed individually with a single linearly polarized laser at near zero magnetic field. Since there is a finite branching ratio from the $m_s = \pm 1$ manifold of the electron spin into the $|0\rangle$ state, we use a recycling laser that drives the transition between $|0\rangle$ and the $|E_y\rangle$ excited state, which decays with a small but non-vanishing probability ($\sim 10^{-2}$) back to the $|\pm 1\rangle$ states. Figure 5.10(b) presents experimental observations of the CPT spectrum as a function of an external magnetic field at three different powers of a laser tuned to the $|\pm 1\rangle \rightarrow |A_2\rangle$ transition. While a broad resonance is observed at high power levels, as the power is reduced, we clearly resolve three features in the spectrum separated by 4.4 MHz, which is twice the hyperfine splitting between three ^{14}N nuclear spin states. This separation corresponds to the magnetic field required

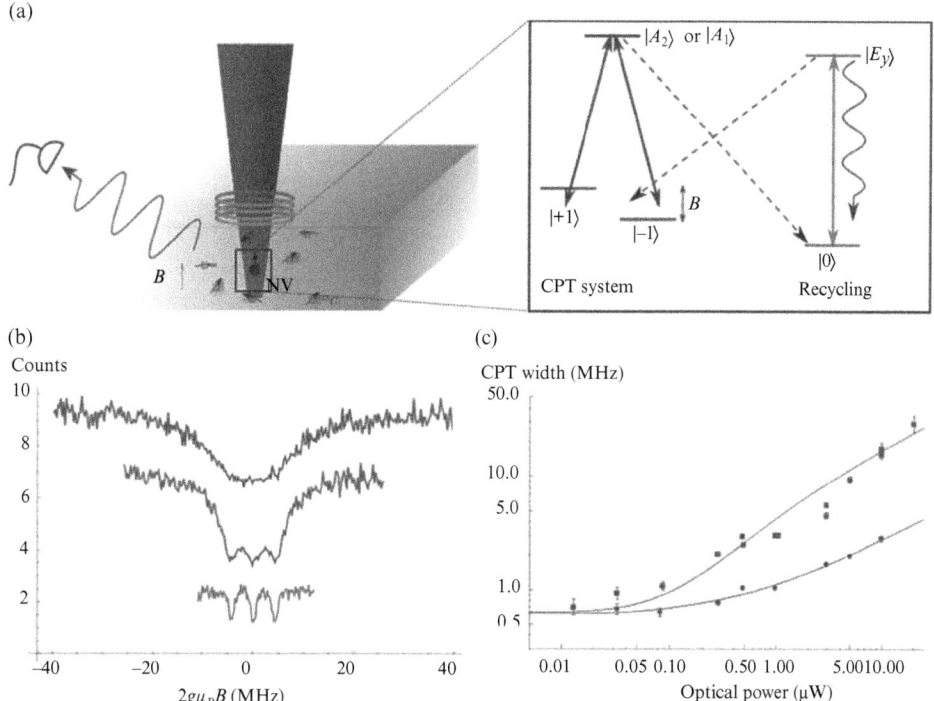

Fig. 5.10 [Colour online] Coherent population trapping in NV centres. (a) The Λ-type transitions between the ground states $|\pm1\rangle$ and excited states $|A_{1,2}\rangle$ of a single NV centre are addressed with a CPT laser, while a recycling laser drives the $|0\rangle$ to $|E_y\rangle$ transition. An external magnetic field is applied using a solenoid. (b) Photon counts from NVs in a $300\,\mu$s window are plotted versus the applied field for $10\,\mu$W (top), $3\,\mu$W (middle), and $0.1\,\mu$W (bottom) of laser power addressing the $|A_2\rangle$ state. Top and middle datasets are shifted vertically by 5 and 2 counts for clarity. (b) Width of individual ^{14}N CPT lines versus CPT laser power when the $|A_1\rangle$ (circles) or $|A_2\rangle$ (squares) state is used. Error bars in all figures show ±1 standard deviation. Solid curves represent theoretical models. (Reprinted from Togan, E., Chu, Y., Imamoglu, A., & Lukin, M. D. (2011). Laser cooling and real-time measurement of the nuclear spin environment of a solid-state qubit. Nature, 478(7370), 497–501. doi:10.1038/nature10528. Copyright 2011 Nature Publishing Group.)

to bring the electronic $m_s = \pm1$ hyperfine states with equal nuclear spin projection ($m_I = \pm1, 0$) into two-photon resonance.

The dependence of the CPT resonance width upon the laser power, shown in Fig. 5.10(c), reveals an important role played by repumping on the near-cycling $|0\rangle \leftrightarrow |E_y\rangle$ transition. In contrast to a conventional, closed three-level system, this recycling transition can be used to enhance the utility of our CPT system by both decreasing the width of the CPT resonance and increasing the signal-to-noise ratio. The $|A_1\rangle$ state decays into the $m_s = 0$ ground state through the singlet with a substantial probability of $\sim$40%. However, the population is returned to the $m_s = \pm 1$ state from $|E_y\rangle$ only after $\sim$100 optical excitation cycles. As a result, away from the two-photon resonance, the NV quickly decays to the $|0\rangle$ state after being excited, where it then scatters many photons through the $|0\rangle \leftrightarrow |E_y\rangle$ cycling transition before returning to the Λ system. If the NV centre is not in a dark state, this process effectively increases the number of photons we collect by $2/\eta = \gamma_{s1}/\gamma_{ce}$, where γ_{ce} is the cross transition rate from $|E_y\rangle$ into $|\pm 1\rangle$ and γ_{s1} is the rate from $|A_1\rangle$ to the singlet. The cycling effect also reduces the width of the CPT line, since the $|0\rangle \leftrightarrow |E_y\rangle$ transition is quickly saturated away from two-photon resonance, provided that the CPT laser excitation rate exceeds the leakage rate out of the recycling transition. Significantly, both of these effects lead to improved sensitivity of dark resonances to small changes in two-photon detuning.

To demonstrate this effect, the widths of dark resonances observed via excitation of $|A_1\rangle$ and $|A_2\rangle$ are compared in Fig. 5.10(c). Through an independent measurement of the branching ratios [87], we determined that $|A_{1(2)}\rangle$ corresponds to an open (nearly closed) Λ system with $\eta_{A_1} \sim 3.1 \times 10^{-2}$ ($\eta_{A_2} \sim 2.6$). These experimental results are compared with a theoretical model described in [87], which predicts that the resonance linewidth δ_0 is given by

$$\delta_0 = \sqrt{R_A^2 \Big/ \left[1 + \frac{1}{\eta}\left(\frac{R_A}{R_E} + \frac{2R_A}{\gamma}\right)\right]}$$

$$\sim \sqrt{\frac{R_A R_E \eta \gamma}{R_E + \gamma}} \qquad \text{for small } \eta, \tag{5.15}$$

where $R_{A(E)}$ corresponds to the optical excitation rate by a laser tuned to the $A(E)$ state and γ is the decay rate of excited states. The width at low powers is determined by the random magnetic field associated with surrounding ^{13}C nuclear states. When this line-broadening mechanism is taken into account, the experimental results are in excellent agreement with these predictions, plotted as solid lines in Fig. 5.10(c), showing that a high degree of optical control over the electronic spins of NV centres can be achieved.

Subsequent experiments [87] showed that CPT can be used not only for probing the nuclear spin environment of the NV centre, but also for controlling and preparing the state of the spin bath. For the ^{14}N spins, it is possible to optically cool the nuclear spin states using dark resonances. The physics is reminiscent of laser cooling of atomic motion via velocity-selective CPT [4, 43]. A redistribution of the ^{14}N spin state population upon optical excitation takes place because the hyperfine coupling in the excited electronic state of the NV centre is enhanced by a factor of $\sim$20 compared with the ground state [34]. If the external field is set such that, for example, the $m_I = 0$ hyperfine states are in two photon resonance, only the states with nuclear configuration $m_I = \pm 1$ will be promoted to the excited states, where flip-flops with the electron spin will change the nuclear spin state to $m_I = 0$. When the NV centre spontaneously decays into the dark superposition of electronic spin states, optical excitation will cease, resulting in effective polarization (cooling) of nuclear spin into the $m_I = 0$ state.

We have also extended this technique to control the many-body environment of the NV centre, consisting of ^{13}C nuclei distributed throughout the diamond lattice. The large number of nuclear spin configurations associated with an unpolarized environment results in a random Overhauser field (B_{ov}) with unresolved hyperfine lines. It produces a finite CPT linewidth in measurements that average over all configurations of the ^{13}C spin bath (Fig. 5.10(b, c)). However, one can make use of fast measurements and the long correlation time (T_1^{nuc}) associated with evolution of the nuclear bath to observe its instantaneous state and and manipulate its dynamics. For example, such a fast measurement is illustrated in Fig. 5.11(a), where the externally applied field

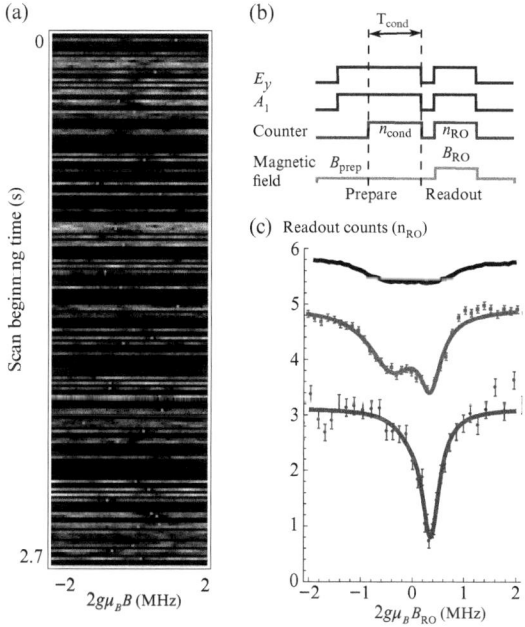

Fig. 5.11 [Colour online] Fast measurement and preparation of a ^{13}C spin bath. (a) Counts from 200 successive fast magnetic field ramps are shown on horizontal lines. The centres of Lorentzian fits to individual runs are indicated by [green] dots. (b) Pulse sequence for preparation and subsequent measurement of the ^{13}C configuration. B_{prep} is set within the central ^{14}N line while B_{RO} is varied to cover all associated ^{13}C states. Counts during a conditioning window of length T_{cond} (n_{cond}) at the end of preparation and the readout window (n_{RO}) are recorded for each run. The data presented represent an average of many such experimental runs. (c) n_{RO} versus B_{RO} is shown in the middle plot with a double Lorentzian fit. The same dataset analysed by keeping only events with $n_{cond} = 0$ is shown in the bottom plot. The unprepared ^{13}C distribution is shown in the top plot for comparison (shifted by 4.3 counts for clarity). (Reprinted from Togan, E., Chu, Y., Imamoglu, A., & Lukin, M. D. (2011). Laser cooling and real-time measurement of the nuclear spin environment of a solid-state qubit. *Nature*, 478(7370), 497–501. doi:10.1038/nature10528. Copyright 2011 Nature Publishing Group.)

is ramped across a single ^{14}N $m_I = 0$ line while the CPT lasers are on, and counts collected in 80 μs time bins are plotted for successive individual runs, many of which distinctly show a narrow dark region. Lorentzian fits to selected experimental scans reveal 'instantaneous' CPT resonances with linewidths that are over a factor three less than those of the averaged measurement. The motion of the dark line centres (dots [green online] in Fig. 5.11(a)) indicates that the instantaneous field evolves in time. Fast measurements can also be used to conditionally prepare the ^{13}C environment of the NV centre in a desired state with post-selection. Here, the nuclear spins are first prepared though optical excitation at a fixed value of the external magnetic field B_{prep}, as shown in Fig. 5.11(b). By conditionally selecting zero-photon-detection events during this preparation step, we can select the states of the ^{13}C environment with vanishing two-photon detuning $\delta = 2g\mu_B(B_{\mathrm{prep}} + B_{\mathrm{ov}}) = 0$, where μ_B is the Bohr magneton. The resulting modified distribution of the Overhauser field is then probed in a readout step by rapidly the changing magnetic field value to B_{RO}, which is scanned across the resonance. The middle curve in Fig. 5.11(c) shows (unconditioned) readout counts recorded following the preparation step, while the bottom curve shows the result of measurement-based preparation. The measured width of such a conditionally prepared distribution is significantly smaller than the width corresponding to individual ^{14}N resonances obtained without preparation.

5.8   Coherent optical transitions in implanted NV centres

Since one-of-a-kind natural diamond samples such as that investigated in the experiments described in the preceding sections cannot be used as reproducible starting materials for scalable systems, recent experiments have made use of NV centres incorporated into high quality synthetic diamond during the growth process [1, 12, 81]. In addition to imperfections in their optical coherence, NV centres introduced during the diamond growth process are limited in their usefulness owing to their low concentration and random locations. For example, in order to incorporate NV centres into nanophotonic or nanomechanical devices, it is necessary to develop a method to create a controllable concentration of emitters in a well-defined device layer while maintaining their excellent optical properties.

This section describes how NV centres that exhibit coherent optical transitions with nearly lifetime-limited linewidths, even under the effects of spectral diffusion, can be created in a surface layer. We accomplish this by first introducing NV centres at a well-defined and controllable depth using ion implantation [65]. We then demonstrate how to suppress the fundamental cause of spectral diffusion by creating a diamond environment that is nearly free of defects that contribute to charge fluctuations. This is achieved through a combination of annealing and surface treatments. Finally, we further improve the optical properties of the NV centres by combining our approach with a newly developed technique for preventing photoionization of the remaining charge traps [80].

Figure 5.12(a) summarizes the experimental procedure to create shallow implanted NV centres with narrow optical lines. As our starting material, we use single-crystal diamond from Element Six grown using microwave-assisted chemical vapour deposition

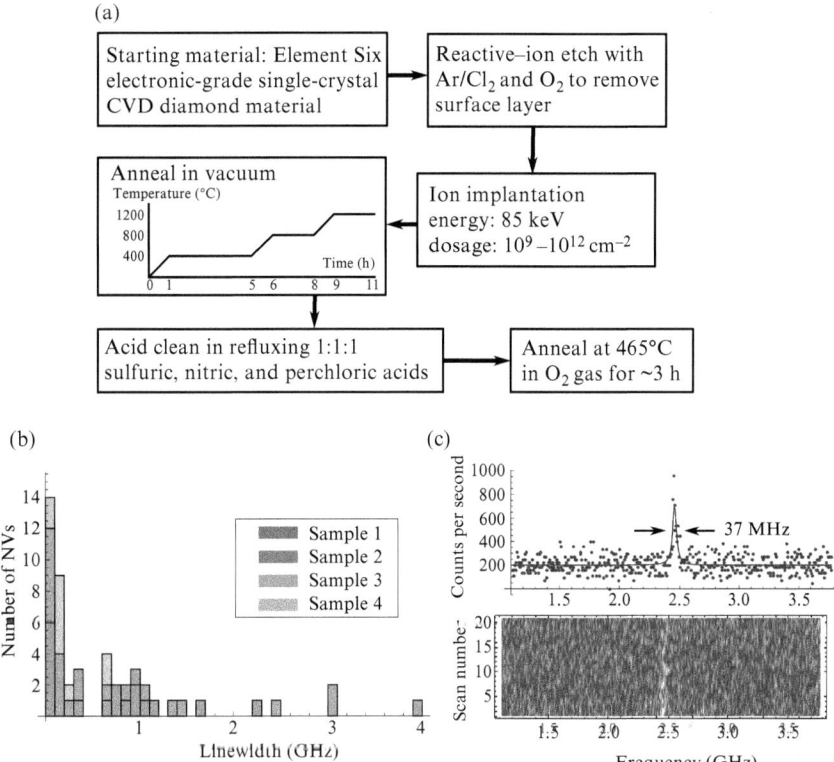

Fig. 5.12 [Colour online] Procedure for creating spectrally stable NV centres through ion implantation, annealing, and surface treatment. The Ar/Cl_2 (O_2) etch was done using an ICP power of 400 W (700 W), an RF power of 250 W (100 W), a DC bias of 423 V (170 V), a chamber pressure of 8 mTorr (10 mTorr), and a gas flow rate of 25/40 sccm (30 sccm). The sample temperature was set to 17°C. (Reprinted with permission from Chu, Y., Leon, N. D., Shields, B., et al. (2014). Coherent optical transitions in implanted nitrogen vacancy centers. Nano Lett., 14(4), 1982–1986. doi:10.1021/nl404836p. Copyright 2014 American Chemical Society.)

(CVD). The single crystal samples are homoepitaxially grown on specially prepared ⟨100⟩-oriented synthetic diamond substrates, with care being taken to ensure that the surface quality of the substrates is as high as possible to reduce sources of dislocations. The CVD synthesis is performed using conditions as described in [42, 55]. Electron paramagnetic resonance (EPR) measurement shows that the resulting diamond material has a nitrogen concentration of less than 5 ppb. The surface layer of these samples is generally damaged and highly strained because of mechanical polishing. Therefore, we first remove the top several micrometres of the sample using a 30-minute Ar/Cl_2 etch followed by a 20-minute O_2 etch in an UNAXIS Shuttleline inductively coupled plasma-reactive ion etcher. We then implant $^{15}N^+$ ions at an energy of 85 keV at a variety of doses from 10^9 to 10^{12} cm^{-2}. According to simulations done using Stopping and Range of Ions in Matter (SRIM), this should result in a mean nitrogen stopping depth of ~100 nm with a straggle of ~20 nm [92]. Subsequently, we anneal the implanted samples under high vacuum ($P < 10^{-6}$ Torr). Our standard annealing recipe consists of a four-hour 400°C step, a two-hour 800°C step, and a two-hour 1200°C step, with a one-hour ramp up to each temperature. The motivation behind these temperature steps will be explained later. It is important that a good vacuum is maintained throughout the annealing process, since diamond etches and graphitizes under residual oxidizing gases at such high temperatures [76]. Following the anneal, we remove graphitic carbon and other surface contaminants by cleaning the diamonds for one hour in a 1:1:1 refluxing mixture of sulfuric, nitric, and perchloric acids. Finally, we perform a 465°C anneal in an O_2 atmosphere in a rapid thermal processor (Modular Process Technology, RTP-600xp) in three 48-minute steps following the recipe used in [32]. This oxygen anneal is believed to remove sp^2 hybridized carbon and result in a more nearly perfect oxygen termination of the surface than acid cleaning alone. We have observed that it enhances charge stability in the NV^- state and helps to reduce blinking and photobleaching.

Let us focus on samples implanted with a $^{15}N^+$ ion dose of 10^{10} cm^{-2}. Figure 5.12(b) shows a histogram of linewidths for 50 NV centres in four such samples. While there is a range of linewidths, a majority of NV centres exhibit linewidths less than 500 MHz, which is comparable to naturally occurring NV centres deep inside electronic-grade CVD diamond [12]. This is an indication that we have eliminated most impurities with fluctuating charge states in the vicinity of the NV centres, which may include both impurities on the nearby surface and defects inside the diamond introduced by the implantation process. In Fig. 5.12(c), we present an example spectrum of the narrowest linewidth we have observed, along with the results of 21 successive PLE scans to demonstrate the long-term stability of the transition line. Our measured linewidth of 37 ± 4.8 MHz under conventional 532 nm repumping is, to the best of our knowledge, a record for the extrinsically broadened linewidth of NV centres.

5.9 Diamond-based nanophotonic devices for quantum optics

5.9.1 Cavity QED with NV centres: the idea

In the preceding sections, we have demonstrated the potential of using the NV centre's optical properties for a variety of different applications. However, the NV centre is not a perfect optical emitter for several reasons. First, the Debye–Waller factor ξ, defined as the ratio of emission into the ZPL to the total emission, is only 3–5%. This is dictated by the relatively large shift in nuclear coordinates between the NV centre's ground and excited states, which, by the Franck–Condon principle, results in decay from the excited state into phonon-coupled higher vibrational levels of the ground state [37]. In addition to such intrinsic properties of the NV centre, there are also external effects such as spectral diffusion that result in non-ideal optical properties. While we can eliminate these effects to a large degree, as demonstrated in Section 5.8, some residual imperfections in spectral stability usually remain. Additional techniques to stabilize the transition frequency, such as feedback techniques with applied electric fields, reduce the repetition rate of the experiments. Finally, it is generally difficult to extract photons from bulk diamond material with high efficiency. Owing to the high index of refraction, total internal reflection confines most of the light inside the diamond at an air interface. In general, as shown in [66], the collection efficiency of a lens or objective with numerical aperture NA in a medium with index n_1 for light from a dipole source in index n_2 is given by

$$C = \tfrac{1}{8}\left[4 - 3\cos\theta_{\max} - \cos^3\theta_{\max} + 3\left(\cos^3\theta_{\max} - \cos\theta_{\max}\right)\cos^2\phi_{\mathrm{em}}\right], \qquad (5.16)$$

where ϕ_{em} is the angle of the dipole axis with the optical axis, and $\theta_{\max} = \sin^{-1}(\mathrm{NA} \cdot n_1/n_2)$ is the effective NA of the collection optics. For reference, the collection efficiency is plotted against several parameters in Fig. 5.13. As can be seen, even for optimal dipole orientation and a 0.95 NA objective in air, only around 6% of the emitted photons are collected from an NV centre in bulk diamond.

The combination of these factors results in a very low rate of indistinguishable, coherent photons from single NV centres. One promising path towards alleviating all of these imperfections is to incorporate NV centres with photonic structures. For example, optical cavities can be used to enhance the emission into the ZPL and overcome spectral diffusion, while coupling to waveguide structures can increase the efficiency of photon collection and detection.

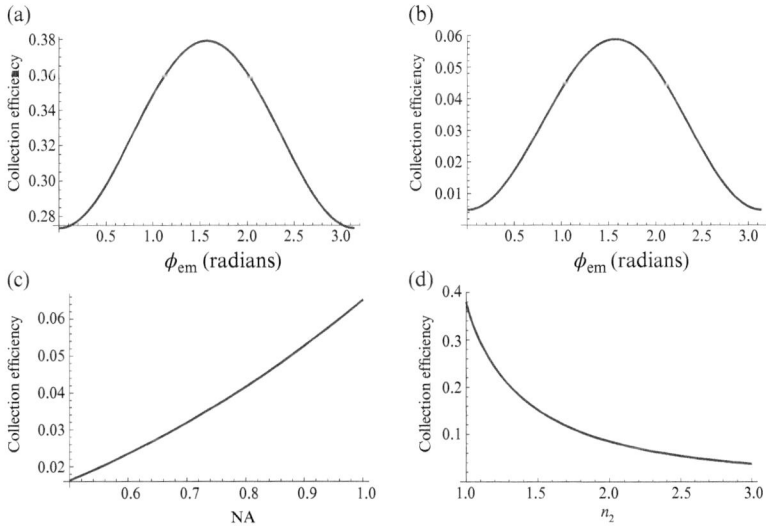

Fig. 5.13 Collection efficiency of light from a dipole emitter. (a) Collection efficiency versus dipole orientation for NA = 0.95 and $n_2 = 1$. (b) Collection efficiency versus dipole orientation for NA = 0.95 and $n_2 = 2.4$. (c) Collection efficiency versus NA for $\phi_{\mathrm{em}} = \pi/2$ and $n_2 = 2.4$. (d) Collection efficiency versus n_2 for $\phi_{\mathrm{em}} = \pi/2$ and NA = 0.95.

The idea of using cavity quantum electrodynamics (QED) to improve the emission properties of a NV centre is as follows. When an NV centre is placed inside the an optical cavity, its emission into the cavity mode is enhanced. By tuning the cavity mode to the NV centre's ZPL, the ZPL emission is therefore enhanced relative to the PSB emission. When this enhancement becomes larger than the Debye–Waller factor, the ZPL becomes the dominant emission channel of the NV centre. This enhancement of the emission also means that the lifetime of the NV centre is decreased and the lifetime-limited linewidth of the ZPL is increased. When the lifetime-limited linewidth becomes much larger than the spectral diffusion, the NV centre can be viewed as a perfect emitter of indistinguishable photons. In other words, the energy uncertainty of the excited state caused by fast decay into the cavity mode is large enough that the energy uncertainty introduced by spectral jumps is negligible.

We can calculate the lifetime of an NV centre coupled to the cavity (τ) compared with the lifetime of one that is decoupled from the cavity (τ_0). They are related through a quantity called the Purcell factor, given by

$$P = \frac{3}{4\pi^2} \left(\frac{\lambda_0}{n}\right)^3 \frac{Q}{V} \frac{\left|\mathbf{E}_{\mathrm{NV}} \cdot \boldsymbol{\mu}_{\mathrm{ZPL}}\right|^2}{\left|\mathbf{E}_{\max}\right|^2 \left|\boldsymbol{\mu}_{\mathrm{NV}}\right|^2} = \frac{\tau_0}{\tau} - 1, \tag{5.17}$$

where λ_0/n is the wavelength of the resonant mode inside the cavity and Q is the quality factor of the cavity. $\mathbf{E}_{\mathrm{NV}}$ and $\mathbf{E}_{\max}$ are the electric fields at the NV and the maximum field inside the cavity, respectively. The mode volume of the cavity is defined as

$$V = \frac{\int \left|\mathbf{E}(\mathbf{x})\right|^2 dV}{\left|\mathbf{E}_{\max}\right|^2}. \tag{5.18}$$

Finally, $\boldsymbol{\mu}_{\mathrm{ZPL}}$ is the effective dipole moment of the ZPL transition, which is the only partly coupled to the cavity, while $\boldsymbol{\mu}_{\mathrm{NV}}$ is the total dipole moment of the NV centre. The Debye–Waller factor is then given by $\xi = |\boldsymbol{\mu}_{\mathrm{ZPL}}|^2/|\boldsymbol{\mu}_{\mathrm{NV}}|^2$. The Purcell factor gives the enhancement of emission into the cavity mode over the emission when the NV is decoupled from the cavity. Here, we define 'decoupled' as an NV that is still spatially located in the cavity, but detuned from the cavity mode in frequency, as is the case in our experiments. This means that n in (5.17) is the effective index including all possible modes of the structure. Although this quantity can be difficult to estimate, it is somewhere between 1 and 2.4, the index of bulk diamond.

While the Purcell factor expresses the relevant enhancement of emission in terms of experimentally relevant quantities, it is equivalent to the cooperativity, which gives more physical intuition as simply the ratio between the coupling strength of the emitter to the cavity and the decay rates of the system. The Purcell factor or cooperativity can also be written as

$$P = \frac{2g^2}{\kappa\gamma}, \tag{5.19}$$

where $\kappa = \omega/Q$ is the decay rate from the cavity, with ω being the cavity frequency. γ is the decay rate of the NV centre into modes other than the resonant cavity mode, and is given by

$$\gamma = \frac{\omega^3 |\boldsymbol{\mu}_{\mathrm{NV}}|^2}{3\pi\epsilon\hbar c^3} = \frac{8\pi^2 |\boldsymbol{\mu}_{\mathrm{NV}}|^2}{3\epsilon\hbar} \left(\frac{n}{\lambda_0}\right)^3. \tag{5.20}$$

The single-photon Rabi frequency g is given by $g = \mathbf{E}_{\mathrm{NV}} \cdot \boldsymbol{\mu}_{\mathrm{ZPL}}/\hbar$. We emphasize here that since the cavity mode couples only to the ZPL, we use the effective dipole moment of that transition only. Inserting these expressions into (5.19) and using the fact that $\epsilon \int |\mathbf{E}(\mathbf{x})|^2 \, dV = \hbar\omega$, we recover the expression for the Purcell factor given in (5.17).

We note here that there are several alternative definitions of the Purcell factor that might also relevant here. One can also consider the lifetime change relative to emission into bulk diamond or free space, in which case the index n in (5.17) should be 2.4 or 1, respectively. Another quantity is the enhancement of the ZPL emission into the cavity relative to the ZPL emission when the NV is decoupled from the cavity. This is given by $P' = P/\xi$, and it is a useful quantity when comparing the Purcell enhancement with other cavity QED systems where the emitter has a single decay channel. In other words, it gives a better comparison for the quality of the emitter–cavity coupling without factoring in the internal imperfections of the emitter itself, and is the quantity usually quoted in the literature. It is also the relevant quantity for estimating the enhancement of the ZPL photon rate from the cavity compared with, say, bulk diamond.

As explained above, in order to radiatively broaden the NV centre through Purcell enhancement to overcome spectral diffusion, we require that $P \gg \Gamma/\gamma$, or $P' \gg (1/\xi)(\Gamma/\gamma)$, where Γ is the extrinsically broadened linewidth. From (5.17), we see that to obtain a large Purcell factor, we need large Q and small mode volume. This can be achieved using photonic crystal cavities, where the mode volume can be designed to be $\sim (\lambda/n)^3$. Then, for a NV centre with $\Gamma \sim 100\,\mathrm{MHz}$, a cavity with $Q \sim 3000$ is required. We now discuss some directions for the physical implementation of such cavities in diamond.

5.9.2 Diamond-based nanophotonic devices

Over the past several decades, a large range of techniques have been developed for the fabrication of nanoscale devices in semiconductor materials ranging from silicon-based materials to III–V compounds such as gallium arsenide [14, 29]. While many of these techniques can be transferred to the fabrication of diamond, one major difference between diamond and most semiconductor materials presents a significant challenge to the fabrication of nanoscale devices for confining light. Because of the large lattice mismatch of diamond with most other crystalline materials, it is not possible to grow high-quality thin films of single-crystal diamond on top of a different substrate. This means that we cannot take advantage of the index mismatch between the device and substrate to confine light through total internal reflection. Here, we will briefly describe two methods for creating diamond-based nanophotonic devices, either by creating a thin film that is placed on a substrate or by directly fabricating suspended structures on the surface of a bulk diamond sample.

The first method for fabricating diamond-based photonic crystal cavities begins with a mechanically polished diamond membrane that is thinned down using a deep reactive-ion etch (RIE) in inductively coupled plasma (ICP) of oxygen to the device thickness [30, 41]. Devices are then defined on these thin films by electron-beam lithography (EBL) using a hydrogen silsesquioxane (HSQ) resist, which forms a mask that is transferred into the diamond film during another RIE etching step.

The second technique for scalable fabrication of nanophotonic structures involves undercut structures suspended above the surface of bulk diamond. This is achieved using a technique described in [18], where the diamond is placed inside a Faraday cage during RIE so that the ions are redirected and impinge on the surface at an angle. As a result, instead of etching the diamond in the direction perpendicular to the surface, it undercuts the material beneath the mask from two directions, resulting in suspended structures with a triangular cross-section that are suspended from or supported by anchors. As illustrated in Fig. 5.14 and described in detail in [18], we can make a variety of devices, including ring resonators, cantilevers, and one-dimensional photonic crystal cavities using this method. In addition, one-dimensional waveguide structures can be incorporated into hybrid cavities where a Bragg structure is defined by another material around an NV inside the waveguide, such as poly(methyl methacrylate) (PMMA).

5.9.3 Example: Purcell enhancement from a coupled NV–cavity system

As an example, we present here a measurement demonstrating the Purcell enhancement of an NV centre's ZPL emission by a nanoscale hybrid diamond–PMMA cavity. These measurements were made on an NV centre with a 3 GHz linewidth inside a hybrid cavity with $Q \sim 600$.

In order to tune the cavity mode frequency, we use a combination of two techniques. The first method uses an O_2 plasma etcher (Technics, Model 220) to remove a layer of material from the structure [41]. This tunes the cavity resonance to lower wavelengths because more of the cavity field resides in air than before and therefore has a higher energy. We have been able to achieve a wavelength shift of $\sim$15 nm without significant degradation in the Q factor using this method. Complementary to this blue-tuning process, we can also red-tune the cavity resonance by depositing neon gas on the structure when it has been cooled inside the cryostat [59]. We note here that gas deposition is a common method for tuning photonic crystal cavities, and a tuning range of $\sim$23 nm was observed in [41]. However, it is somewhat less effective for our hybrid structures since the high aspect ratio of the PMMA slabs prevent the diffusion of gas onto the diamond waveguide, where the field is concentrated. Despite this, we have occasionally been able to obtain tuning ranges of $\sim$6 nm.

To measure the change in lifetime when the NV centre becomes coupled with the cavity, we first tune the cavity resonance using plasma etching until it is to the blue of the NV centre's ZPL. We then gradually red-tune the cavity onto resonance with

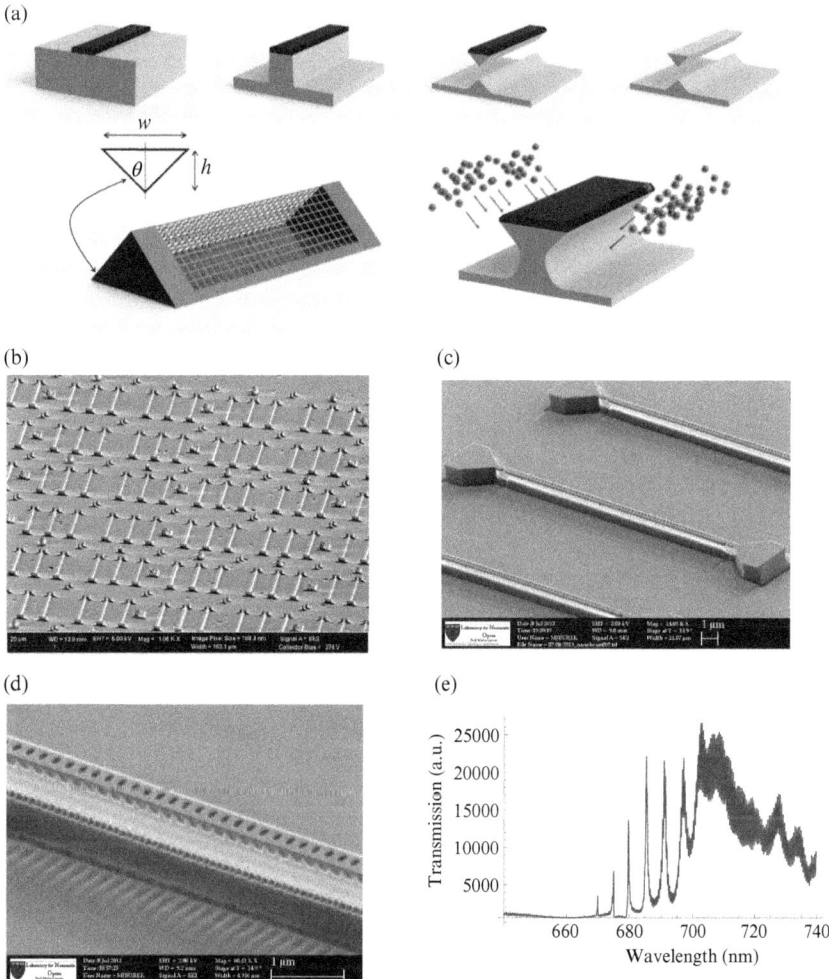

Fig. 5.14 [Colour online] Suspended diamond-based nanophotonic structures. (a) Procedure for angled RIE etching. A mask is defined using EBL on the surface of the diamond. A top-down etching step is performed, followed by an angle-etching step where the diamond is placed in the centre of a machined aluminium Faraday cage shown on the bottom left. The mask is removed, leaving a suspended structure attached at both ends to anchors (not shown) (b) Scanning electron microscope (SEM) image of a large array of ~1000 angle-etched waveguides. (c, d) SEM images of photonic crystal cavities formed in one-dimensional waveguides. (e) Transmission spectrum of one such device. (Based on Burek, M. J., Leon, N. P., Shields, B. J., et al. (2012). Free-standing mechanical and photonic nanostructures in single-crystal diamond. Nano Lett., 12(12), 6084–6089. doi:10.1021/nl302541e.)

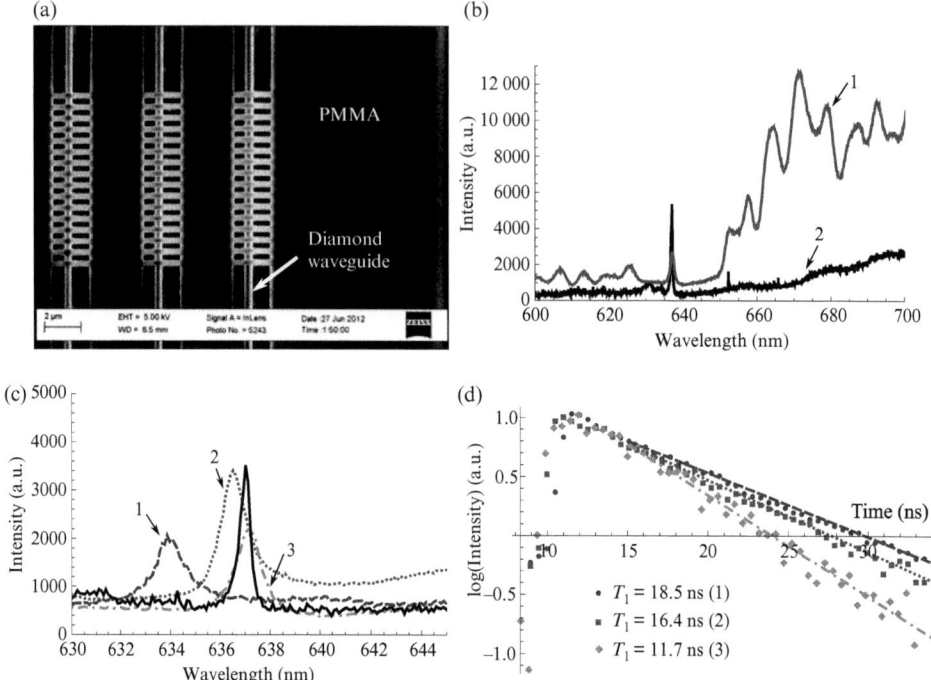

Fig. 5.15 [Colour online] Purcell enhancement of the NV centre ZPL emission using hybrid photonic crystal cavities. (a) SEM image of hybrid diamond–PMMA cavities. Visible are the top surface of the suspended triangular beams and the top of the the PMMA layer and slabs, along with the exposed diamond surface underneath. (b) Transmission spectrum of hybrid cavity (1) and emission spectrum of the NV centre (2). The spectra have been scaled relative to each other to fit on the same scale. (c) Zoomed-in spectra showing the ZPL of the NV centre (solid line) and the cavity resonance tuned to three different frequencies (dashed, dotted, and dash-dotted lines). (d) Lifetime measurements corresponding to the detunings in (c), taken by exciting the NV centre with a pulsed laser and collecting the PSB emission.

the ZPL while measuring the lifetime using a filtered supercontinuum laser (NKT Photonics, SuperK EXTREME) as a pulsed laser source around 532 nm. As shown in Fig. 5.15, the lifetime decreases from 18.5 ns when the cavity is completely off-resonant to 11.6 ns when the cavity is tuned to the ZPL. This gives a Purcell factor $P = 0.59$, or a ZPL enhancement $P' = 20$ for $\xi = 0.03$. We emphasize here that this enhancement factor implies that the ZPL emission is now $\sim 40\%$ of the total emission from the NV centre.

5.10 Outlook

Following the experiments presented in Section 5.6, there has been tremendous progress in using optical photons to connect remote NV centres. Future work will explore the possibility of combining the few-qubit local spin register with the resource of remote entanglement through, for example, teleportation of quantum information between remote, long-lived nuclear spin memories. Distribution of quantum information by teleportation has been demonstrated in several different physical systems [48, 64, 82]. However, remote teleportation over long distances is an outstanding challenge with solid state systems, which, in the long run, offer the possibility of implementing more scalable local quantum nodes. In particular, one major factor in determining the ultimate bandwidth and fidelity of a remote quantum channel is the efficiency with which single photons can be made to interact with the stationary qubits and then extracted into, for example, a optical fibre. Current photon collection efficiencies and entanglement rates may be sufficient for proof-of-principle experiments, but certainly must be greatly improved for the implementation of practical quantum networks. The work described in Sections 5.8 and 5.9 begins to address this issue. The spectral stability of NV centres is an important factor determining the repetition rate of experiments requiring indistinguishable optical photons, such as remote entanglement and teleportation. The availability of NV centres with lifetime-limited linewidths would eliminate the need for active stabilization and post-selection of the the transition frequency in future experiments. Cavity QED has been used to enhance the interaction of photons with atomic systems and improve their collection efficiency, leading to demonstrations of quantum teleportation with high fidelity and success rates [64]. Diamond-based nanophotonics offers the possibility of achieving efficient photonic quantum devices in a solid state system.

Finally, future quantum technologies will most likely be implemented with a combination of several types of physical systems, each with its own strengths and weaknesses. There are already many theoretical proposals and experimental demonstrations of interfacing NV centres with other quantum devices [49, 68]. Most of these demonstrations make use of the NV centre's unique spin properties, effectively turning its interaction with the magnetic environment from a detrimental factor into a resource. Analogous to this, a potential direction for future exploration would be to take advantage of the NV centre's optical transitions to interface it with other systems. Combined with the ability to communicate information with remote systems through optical photons, the NV centre becomes a quantum system that not only has uniquely attractive properties as an isolated system, but also can be made to interact with other systems in a controlled manner. Therefore, the ideas and techniques described in this chapter will likely enable researchers to use the optical properties of the NV centre as a powerful resource for building complex and useful quantum technologies.

Acknowledgements

The work described in this chapter was carried out by several generations of Harvard Quantum Optics Group members, including Lillian Childress, M. V. Gurudev Dutt, Ruffin Evans, Liang Jiang, Nathalie de Leon, Jeronimo Maze, Brendan Shields, Emre Togan, Alexei Trifonov, and Alexander Zibrov. We additionally acknowledge many discussions and collaborations with Michael Burek, Birgit Hausmann, Atac

Imamoglu, Fedor Jelezko, Marko Loncar, Philip Hemmer, Patrick Maletinsky, Matthew Markham, Hongkun Park, Anders Sorensen, Daniel Twitchen, Alastair Stacey, Ron Walsworth, Jeorg Wrachtrup, and Amir Yacoby. Support for the work was provided by the NSF, CUA, DARPA, MURI, the Packard Foundation, an ERC Advanced Investigator Grant, and NDSEG, NSF GRFP, Element Six, and Harvard Quantum Optics Center Fellowships.

References

[1] Acosta, V. M., Santori, C., Faraon, A., Huang, Z., Fu, K.-M. C., Stacey, A., Simpson, D. A., Ganesan, K., Tomljenovic-Hanic, S., Greentree, A. D., Prawer, S., and Beausoleil, R. G. (2012). Dynamic stabilization of the optical resonances of single nitrogen-vacancy centers in diamond. *Phys. Rev. Lett.*, **108**, 206401.

[2] Ansmann, M., Wang, H., Bialczak, R. C., Hofheinz, M., Lucero, E., Neeley, M., O'Connell, A. D., Sank, D., Weides, M., Wenner, J., Cleland, A. N., and Martinis, J. M. (2009). Violation of Bell's inequality in Josephson phase qubits. *Nature*, **461**, 504–506.

[3] Aslam, N., Waldherr, G., Neumann, P., Jelezko, F., and Wrachtrup, J. (2013). Photo-induced ionization dynamics of the nitrogen vacancy defect in diamond investigated by single-shot charge state detection. *New J. Phys.*, **15**, 013064.

[4] Aspect, A., Arimondo, E., Kaiser, R., Vansteenkiste, N., and Cohen-Tannoudji, C. (1988). Laser cooling below the one-photon recoil energy by velocity-selective coherent population trapping. *Phys. Rev. Lett.*, **61**, 826–829.

[5] Awschalom, D. D., Bassett, L. C., Dzurak, A. S., Hu, E. L., and Petta, J. R. (2013). Quantum spintronics: engineering and manipulating atom-like spins in semiconductors. *Science*, **339**, 1174–1179.

[6] Balasubramanian, G., Chan, I. Y., Kolesov, R., Al-Hmoud, M., Tisler, J., Shin, C., Kim, C., Wjcik, A., Hemmer, P. R., Krueger, A., Hanke, T., Leitenstorfer, A., Bratschitsch, R., Jelezko, F., and Wrachtrup, J. (2008). Nanoscale imaging magnetometry with diamond spins under ambient conditions. *Nature*, **455**, 648–651.

[7] Balasubramanian, G., Neumann, P., Twitchen, D., Markham, M., Kolesov, R., and *et al.* (2009). Ultralong spin coherence time in isotopically engineered diamond. *Nat. Mater.*, **8**, 383.

[8] Bassett, L. C., Heremans, F. J., Yale, C. G., Buckley, B. B., and Awschalom, D. D. (2011, Dec). Electrical tuning of single nitrogen-vacancy center optical transitions enhanced by photoinduced fields. *Phys. Rev. Lett.*, **107**, 266403.

[9] Batalov, A., Jacques, V., Kaiser, F., Siyushev, P., Neumann, P., Rogers, L. J., McMurtrie, R. L., Manson, N. B., Jelezko, F., and Wrachtrup, J. (2009). Low temperature studies of the excited-state structure of negatively charged nitrogen-vacancy color centers in diamond. *Phys. Rev. Lett.*, **102**, 195506.

[10] Batalov, A., Zierl, C., Gaebel, T., Neumann, P., Chan, I.-Y., Balasubramanian, G., Hemmer, P. R., Jelezko, F., and Wrachtrup, J. (2008). Temporal coherence of photons emitted by single nitrogen-vacancy defect centers in diamond using optical Rabi-oscillations. *Phys. Rev. Lett.*, **100**, 077401.

[11] Benko, C., Ruehl, A., Martin, M. J., Eikema, K. S. E., Fermann, M. E., Hartl, I., and Ye, J. (2012). Full phase stabilization of a Yb:fiber femtosecond frequency comb via high-bandwidth transducers. *Opt. Lett.*, **37**, 2196–2198.

[12] Bernien, H., Childress, L., Robledo, L., Markham, M., Twitchen, D., and Hanson, R. (2012). Two-photon quantum interference from separate nitrogen vacancy centers in diamond. *Phys. Rev. Lett.*, **108**, 043604.

[13] Bernien, H., Hensen, B., Pfaff, W., Koolstra, G., Blok, M. S., Robledo, L., Taminiau, T. H., Markham, M., Twitchen, D. J., Childress, L., and Hanson, R. (2013). Heralded entanglement between solid-state qubits separated by three metres. *Nature*, **497**, 86–90.

[14] Blanco, A., Chomski, E., Grabtchak, S., Ibisate, M., John, S., Leonard, S. W., Lopez, C., Meseguer, F., Miguez, H., and Mondia, J. P. (2000). Large-scale synthesis of a silicon photonic crystal with a complete three-dimensional bandgap near 1.5 micrometres. *Nature* **405**, 437–440.

[15] Blinov, B. B., Moehring, D. L., Duan, L. M., and Monroe, C. (2004). Observation of entanglement between a single trapped atom and a single photon. *Nature*, **428**, 153.

[16] Buckley, B. B., Fuchs, G. D., Bassett, L. C., and Awschalom, D. D. (2010). Spin–light coherence for single-spin measurement and control in diamond. *Science*, **330**, 1212–1215.

[17] Budker, D. and Romalis, M. (2007). Optical magnetometry. *Nat. Phys.*, **3**, 227–234.

[18] Burek, M. J., de Leon, N. P., Shields, B. J., Hausmann, B. J. M., Chu, Y., Quan, Q., Zibrov, A. S., Park, H., Lukin, M. D., and Loncar, M. (2012). Free-standing mechanical and photonic nanostructures in single-crystal diamond. *Nano Lett.*, **12**, 6084–6089.

[19] Childress, L., Gurudev Dutt, M. V., Taylor, J. M., Zibrov, A. S., Jelezko, F., Wrachtrup, J., Hemmer, P. R., and Lukin, M. D. (2006). Coherent dynamics of coupled electron and nuclear spin qubits in diamond. *Science*, **314**, 281–285.

[20] Chou, C. W., Hume, D. B., Koelemeij, J. C. J., Wineland, D. J., and Rosenband, T. (2010). Frequency comparison of two high-accuracy Al$^+$ optical clocks. *Phys. Rev. Lett.*, **104**, 070802.

[21] Clarke, J. and Wilhelm, F. K. (2008). Superconducting quantum bits. *Nature*, **453**, 1031–1042.

[22] de Riedmatten, H., Afzelius, M., Staudt, M. U., Simon, C., and Gisin, N. (2008). A solid-state light–matter interface at the single-photon level. *Nature*, **456**, 773.

[23] DiCarlo, L., Chow, J. M., Gambetta, J. M., Bishop, L. S., Johnson, B. R., Schuster, D. I., Majer, J. , Blais, A., Frunzio, L., Girvin, S. M., and Schoelkopf, R. J. (2009). Demonstration of two-qubit algorithms with a superconducting quantum processor. *Nature*, **46**, 240–244.

[24] Doherty, M. W., Dolde, F., Fedder, H., Jelezko, F., Wrachtrup, J., Manson, N. B., and Hollenberg, L. C. L. (2012). Theory of the ground-state spin of the NV center in diamond. *Phys. Rev. B*, **85**, 205203.

[25] Doherty, M. W., Manson, N. B., Delaney, P., and Hollenberg, L. C. (2011). The negatively charged nitrogen-vacancy centre in diamond: the electronic solution. *New J. Phys.*, **13**, 025019.

[26] Doherty, M. W., Manson, N. B., Delaney, P., Jelezko, F., Wrachtrup, J., and Hollenberg, L. C. (2013). The nitrogen-vacancy colour centre in diamond. *Phys. Rept.*, **528**, 1 – 45.

[27] Duan, L.-M. and Monroe, C. (2008). Robust quantum information processing with atoms, photons, and atomic ensembles. *Adv. At., Mol. Opt. Phys.*, **55**, 419–464.

[28] Dutt, M. V. G., Childress, L., Jiang, L., Togan, E., Maze, J., Jelezko, F., Zibrov, A. S., Hemmer, P. R., and Lukin, M. D.(2007). Quantum register based on individual electronic and nuclear spin qubits in diamond. *Science*, **316**, 1312–1316.

[29] Englund, D., Faraon, A., Fushman, I., Stoltz, N., Petroff, P., and Vučković, J. (2007). Controlling cavity reflectivity with a single quantum dot. *Nature* **450**, 857–861.

[30] Faraon, A., Barclay, P., Santori, C., Fu, K., and Beausoleil, R. (2011). Resonant enhancement of the zero-phonon emission from a colour centre in a diamond cavity. *Natu. Photon.*, **5**, 301–305.

[31] Fleischhauer, M., Imamoglu, A., and Marangos, J. P. (2005). Electromagnetically induced transparency: optics in coherent media. *Rev. Mod. Phys.*, **77**, 633.

[32] Fu, K.-M. C., Santori, C., Barclay, P. E., and Beausoleil, R. G. (2010). Conversion of neutral nitrogen-vacancy centers to negatively charged nitrogen-vacancy centers through selective oxidation. *Appl. Phys. Lett.*, **96**, 121907.

[33] Fu, K.-M. C., Santori, C., Barclay, P. E., Rogers, L. J., Manson, N. B., and Beausoleil, R. G. (2009). Observation of the dynamic Jahn–Teller effect in the excited states of nitrogen-vacancy centers in diamond. *Phys. Rev. Lett.*, **103**, 256404.

[34] Fuchs, G. D., Dobrovitski, V. V., Hanson, R., Batra, A., Weis, C. D., Schenkel, T., and Awschalom, D. D. (2008). Excited-state spectroscopy using single spin manipulation in diamond. *Phys. Rev. Lett.*, **101**, 117601.

[35] Fuchs, G. D., Dobrovitski, V. V., Toyli, D. M., Heremans, F. J., and Awschalom, D. D. (2009). Gigahertz dynamics of a strongly driven single quantum spin. *Science*, **326**, 1520.

[36] Gali, A., Fyta, M., and Kaxiras, E. (2008). Ab initio supercell calculations on nitrogen-vacancy center in diamond: electronic structure and hyperfine tensors. *Phys. Rev. B*, **77**, 155206.

[37] Gali, A., Simon, T., and Lowther, J. E. (2011). An ab initio study of local vibration modes of the nitrogen-vacancy center in diamond. *New J. Phys.*, **13**, 025016.

[38] Giedke, G., Taylor, J. M., D'Alessandro, D., Lukin, M. D., and Imamoglu, A. (2006). Quantum measurement of a mesoscopic spin ensemble. *Phys. Rev. A*, **74**, 032316.

[39] Goss, J., Jones, R., Breuer, S., Briddon, P., and Oberg, S. (1996). The twelve-line 1.682 eV luminescence center in diamond and the vacancy–silicon complex. *Phys. Rev. Lett.*, **77**, 3041.

[40] Gruber, A., Dräbenstedt, A., Tietz, C., Fleury, L., Wrachtrup, J., and von Borczyskowski, C. (1997). Scanning confocal optical microscopy and magnetic resonance on single defect centers. *Science*, **276**, 2012–2014.

[41] Hausmann, B. J. M., Shields, B. J., Quan, Q., Chu, Y., de Leon, N. P., Evans, R., Burek, M. J., Zibrov, A. S., Markham, M., Twitchen, D. J., Park, H., Lukin, M. D., and Lončar, M. (2013). Coupling of NV centers to photonic crystal nanobeams in diamond. *Nano Lett.*, **13**, 5791–5796.

[42] Isberg, J., Hammersberg, J., Johansson, E., Wikström, T., Twitchen, D. J., Whitehead, A. J., Coe, S. E., and Scarsbrook, G. A. (2002). High carrier mobility in single-crystal plasma-deposited diamond. *Science*, **297**, 1670–1672.

[43] Issler, M., Kessler, E. M., Giedke, G., Yelin, S., Cirac, I., Lukin, M. D., and Imamoglu, A. (2010). Nuclear spin cooling using Overhauser-field selective coherent population trapping. *Phys. Rev. Lett.*, **105**, 267202.

[44] Jelezko, F. and Wrachtrup, J. (2006). Quantum information processing in diamond. *J. Phys.: Condens. Matter*, **18**, S807–S824.

[45] Kaiser, F., Jacques, V., Batalov, A., Siyushev, P., Jelezko, F., and Wrachtrup, J. (2009). Polarization properties of single photons emitted by nitrogen-vacancy defect in diamond at low temperature. arXiv:0906.3426 [quant-ph].

[46] Kessler, T., Hagemann, C., Grebing, C., Legero, T., Sterr, U., Riehle, F., Martin, M. J., Chen, L., and Ye, J. (2012). A sub-40-mHz-linewidth laser based on a silicon single-crystal optical cavity. *Nat. Photon.* **6**, 687–692.

[47] Kimble, H. J. (2008). The quantum Internet. *Nature*, **453**, 1023.

[48] Krauter, H., Salart, D., Muschik, C. A., Petersen, J. M., Shen, H., Fernholz, T., and Polzik, E. S. (2013). Deterministic quantum teleportation between distant atomic objects. *Nat. Phys.* **9**, 400–404.

[49] Kubo, Y., Grezes, C., Dewes, A., Umeda, T., Isoya, J., Sumiya, H., Morishita, N., Abe, H., Onoda, S., Ohshima, T., Jacques, V., Dréau, A., Roch, J.-F., Diniz, I., Auffeves, A., Vion, D., Esteve, D., and Bertet, P. (2011). Hybrid quantum circuit with a superconducting qubit coupled to a spin ensemble. *Phys. Rev. Lett.*, **107**, 220501.

[50] Kurtsiefer, C., Mayer, S., Zarda, P., and Weinfurter, H. (2000). Stable solid-state source of single photons. *Phys. Rev. Lett.*, **85**, 290–293.

[51] Lenef, A. and Rand, S. (1996). Electronic structure of the NV center in diamond: theory. *Phys. Rev. B*, **53**, 13441.

[52] Lloyd, S. (1996). Universal quantum simulators. *Science*, **273**, 1073–1078.

[53] Loubser, J. and Wyk, J. V. (1978). Electron spin resonance in the study of diamond. *Rep. Prog. Phys.*, **91**, 1201–1248.

[54] Manson, N. B., Harrison, J. P., and Sellars, M. J. (2006). Nitrogen-vacancy center in diamond: model of the electronic structure and associated dynamics. *Phys. Rev. B*, **74**, 104303.

[55] Markham, M., Dodson, J., Scarsbrook, G., Twitchen, D., Balasubramanian, G., Jelezko, F., and Wrachtrup, J. (2011). CVD diamond for spintronics. *Diamond Relat. Mater.*, **20**, 134–139.

[56] Maze, J., Gali, A., Togan, E., Chu, Y., Trifonov, A., Kaxiras, E., and Lukin, M. (2011). Properties of nitrogen-vacancy centers in diamond: the group theoretic approach. *New J. Phys.*, **13**, 025025.

[57] Maze, J., Stanwix, P. L., Hodges, J. S., Hong, S., Taylor, J. M., Cappellaro, P., Jiang, L., Dutt, M. V. G., Togan, E., Zibrov, A. S., Yacoby, A., Walsworth, R. L., and Lukin, M. D. (2008). Nanoscale magnetic sensing with an individual electronic spin in diamond. *Nature*, **455**, 644–647.

[58] Monroe, C. and Kim, J. (2013). Scaling the ion trap quantum processor. *Science*, **339**, 1164–1169.

[59] Mosor, S., Hendrickson, J., Richards, B. C., Sweet, J., Khitrova, G., Gibbs, H. M., Yoshie, T., Scherer, A., Shchekin, O. B., and Deppe, D. G. (2005). Scanning a photonic crystal slab nanocavity by condensation of xenon. *Appl. Phys. Lett.*, **87**, 141105.

[60] Neumann, P., Beck, J., Steiner, M., Rempp, F., Fedder, H., Hemmer, P. R., Wrachtrup, J., and Jelezko, F. (2010). Single-shot readout of a single nuclear spin. *Science*, **329**, 542.

[61] Neumann, P., Mizuochi, N., Rempp, F., Hemmer, P., Watanabe, H., Yamasaki, S., Jacques, V., Gaebel, T., Jelezko, F., and Wrachtrup, J. (2008). Multipartite entanglement among single spins in diamond. *Science*, **320**, 1326–1329.

[62] Ni, K.-K., Ospelkaus, S., de Miranda, M. H. G., Pe'er, A., Neyenhuis, B., Zirbel, J. J., Kotochigova, S., Julienne, P. S., Jin, D. S., and Ye, J. (2008). A high phase-space-density gas of polar molecules. *Science*, **322**, 231–235.

[63] Nicholson, T. L., Martin, M. J., Williams, J. R., Bloom, B. J., Bishof, M., Swallows, M. D., Campbell, S. L., and Ye, J. (2012). Comparison of two independent Sr optical clocks with 1×10^{-17} stability at 10^3 s. *Phys. Rev. Lett.*, **109**, 230801.

[64] Nölleke, C., Neuzner, A., Reiserer, A., Hahn, C., Rempe, G., and Ritter, S. (2013, Apr). Efficient teleportation between remote single-atom quantum memories. *Phys. Rev. Lett.*, **110**, 140403.

[65] Pezzagna, S., Naydenov, B., Jelezko, F., Wrachtrup, J., and Meijer, J. (2010). Creation efficiency of nitrogen-vacancy centres in diamond. *New J. Phys.*, **12**, 065017.

[66] Plakhotnik, T., Moerner, W., Palm, V., and Wild, U. P. (1995). Single molecule spectroscopy: maximum emission rate and saturation intensity. *Opt. Commun.*, **114**, 83–88.

[67] Platzman, P. M. and Dykman, M. I. (1999). Quantum computing with electrons floating on liquid helium. *Science*, **284**, 1967–1969.

[68] Rabl, P., Kolkowitz, S. J., Koppens, F. H. L., Harris, J. G. E., Zoller, P., and Lukin, M. D. (2010). A quantum spin transducer based on nanoelectromechanical resonator arrays. *Nat. Phys.* **6**, 602–608.

[69] Robledo, L., Bernien, H., van Weperen, I., and Hanson, R. (2010). Control and coherence of the optical transition of single nitrogen vacancy centers in diamond. *Phys. Rev. Lett.*, **105**, 177403.

[70] Robledo, L., Childress, L., Bernien, H., Hensen, B., Alkemade, P. F. A., and Hanson, R. (2011). High-fidelity projective read-out of a solid-state spin quantum register. *Nature*, **477**, 574–578.

[71] Rogers, L., McMurtrie, R., Sellars, M., and Manson, N. B. (2009). Time-averaging within the excited state of the nitrogen-vacancy centre in diamond. *New J. Phys.*, **11**, 063007.

[72] Santori, C., Tamarat, P., Neumann, P., Wrachtrup, J., Fattal, D., Beausoleil, R. G., Rabeau, J., Olivero, P., Greentree, A. D., Prawer, S., Jelezko, F., and Hemmer, P. (2006). Coherent population trapping of single spins in diamond under optical excitation. *Phys. Rev. Lett.*, **97**, 247401.

[73] Scully, M. O. and Druhl, K. (1982). Quantum eraser: a proposed photon correlation experiment concerning observation and 'delayed choice' in quantum mechanics. *Phys. Rev. A.*, **25**, 2208–2213.

[74] Scully, M. O. and Fleischhauer, M. (1992, Aug). High-sensitivity magnetometer based on index-enhanced media. *Phys. Rev. Lett.*, **69**, 1360–1363.

[75] Scully, M. O. and Zubairy, M. S. (1997). *Quantum Optics.* Cambridge University Press, Cambridge.

[76] Seal, M. (1963). The effect of surface orientation on the graphitization of diamond. *Phys. Stat. Sol. (b)*, **3**, 658–664.

[77] Shen, Y., Sweeney, T. M., and Wang, H. (2008, Jan). Zero-phonon linewidth of single nitrogen vacancy centers in diamond nanocrystals. *Phys. Rev. B*, **77**, 033201.

[78] Shor, P. (1994). Algorithms for quantum computation: discrete logarithms and factoring. In: *Proceedings of the 35th Annual Symposium on Foundations of Computer Science, (SFCS 1994)*, pp. 124–134. IEEE Computer Society, Washington, DC.

[79] Sipahigil, A., Goldman, M. L., Togan, E., Chu, Y., Markham, M., Twitchen, D. J., Zibrov, A. S., Kubanek, A., and Lukin, M. D. (2012). Quantum interference of single photons from remote nitrogen-vacancy centers in diamond. *Phys. Rev. Lett.*, **108**, 143601.

[80] Siyushev, P., Pinto, H., Vörös, M., Gali, A., Jelezko, F., and Wrachtrup, J. (2013r). Optically controlled switching of the charge state of a single nitrogen-vacancy center in diamond at cryogenic temperatures. *Phys. Rev. Lett.*, **110**, 167402.

[81] Stacey, A., Simpson, D. A., Karle, T. J., Gibson, B. C., Acosta, V. M., Huang, Z., Fu, K. M. C., Santori, C., Beausoleil, R. G., McGuinness, L. P., Ganesan, K., Tomljenovic-Hanic, S., Greentree, A. D., and Prawer, S. (2012). Near-surface spectrally stable nitrogen vacancy centres engineered in single crystal diamond. *Adv. Mater.*, **24**, 3333–3338.

[82] Steffen, L., Salathe, Y., Oppliger, M., Kurpiers, P., Baur, M., Lang, C., Eichler, C., Puebla-Hellmann, G., Fedorov, A., and Wallraff, A. (2013). Deterministic quantum teleportation with feed-forward in a solid state system. *Nature* **500**, 319–322.

[83] Stepanenko, D., Burkard, G., Giedke, G., and Imamoglu, A. (2006). Enhancement of electron spin coherence by optical preparation of nuclear spins. *Phys. Rev. Lett.*, **96**, 136401.

[84] Stern, A. and Lindner, N. H. (2013). Topological quantum computation—from basic concepts to first experiments. *Science*, **339**, 1179–1184.

[85] Tamarat, P., Gaebel, T., Rabeau, J. R., Khan, M., Greentree, A. D., Wilson, H., Hollenberg, L. C. L., Prawer, S., Hemmer, P., Jelezko, F., and Wrachtrup, J. (2006, Aug). Stark shift control of single optical centers in diamond. *Phys. Rev. Lett.*, **97**, 083002.

[86] Tamarat, P., Manson, N. B., Harrison, J. P., McMurtrie, R. K., Nizovtsev, A., Santori, C., Beausoleil, R. G., Neumann, P., Gaebel, T., Jelezko, F., Hemmer, P., and Wrachtrup, J. (2008). Spin-flip and spin-conserving optical transitions of the nitrogen-vacancy centre in diamond. *New J. Phys*, **10**, 045004.

[87] Togan, E., Chu, Y., Imamoglu, A., and Lukin, M. D. (2011). Laser cooling and real-time measurement of the nuclear spin environment of a solid-state qubit. *Nature*, **478**, 497–501.

[88] Togan, E., Chu, Y., Trifonov, A. S., Jiang, L., Maze, J., Childress, L., Dutt, M. V. G., Sørensen, A. S., Hemmer, P. R., Zibrov, A. S., and Lukin, M. D. (2010). Quantum entanglement between an optical photon and a solid-state spin qubit. *Nature*, **466**, 730–734.

[89] Vanier, J. (2005). Atomic clocks based on coherent population trapping: a review. *Appl. Phys. B: Lasers Opt.*, **81**, 421–442.

[90] Wilk, T., Webster, S. C., Kuhn, A., and Rempe, G. (2007). Single-atom single-photon quantum interface. *Science*, **317**, 488–490.

[91] Zaitsev, A. M. (2010). *Optical Properties of Diamond: A Data Handbook.* Springer-Verlag, Berlin.

[92] Ziegler, J. F. (2013). Interactions of Ions with Matter. http://www.srim.org.

6
Single-molecule spectroscopy

Michel ORRIT

Institute of Physics
Leiden University, The Netherlands

Michel Orrit: *Single-molecule spectroscopy*. In: *Quantum Optics and Nanophotonics*. Edited by:
Claude Fabre, Vahid Sandoghdar, Nicolas Treps, and Leticia F. Cugliandolo. Oxford University
Press (2017), pp. 271–314. © Oxford University Press 2017.
DOI: 10.1093/oso/9780198768609.003.0006

Chapter Contents

We give an overview of the main optical methods used to detect and study single molecules and other small objects (nano-objects). Much of the work so far has exploited the excellent sensitivity and selectivity of fluorescence, but several new techniques, mostly based on nonlinear optics, have recently reached the single-molecule or single-nanoparticle regime. We briefly discuss some results by referring to published reviews. Single-molecule techniques have now been incorporated into the arsenal of the physico-chemist and the cell biologist. However, the recent development of super-resolution techniques and of new labels suggests that further progress can be expected from measurements on single nano-objects in the next decade or so.

6.1 Introduction

At the end of the nineteenth century, evidence had accumulated for the existence of atoms and molecules. Final hard proofs appeared at the beginning of the twentieth century, with the quantization of charge, Brownian motion, and, most importantly, the diffraction of X-rays by crystals. Already at that time, Jean Perrin [1] suggested that it should be possible to observe single fluorescent molecules directly by eye in an optical microscope. He attempted to do this in experiments in which, to decrease background, he used black soap films, which scatter only minute amounts of light, mostly through Rayleigh scattering, with low concentrations of fluorescent molecules. Unfortunately, because of the limited responsivity of the human eye, the low brightness of the light sources used, and the poor quality of the optical filters and the fluorescent dyes available at that time, Perrin could not detect single molecules by eye. Instead, he performed a very interesting study of fluorescent black films, where he found other evidence for molecular structures—the quantized thickness of soap multilayers. Thanks to spectacular progress in all the above-mentioned aspects, including optical microscopes, light sources and detectors, filters, and dyes, Perrin's suggested observations can nowadays be made in a straightforward manner, and even single molecules can be seen in an optical microscope.

Even long after Perrin's first heroic attempt, it was thought that molecules were much too weak as emitters to be detected optically. In the 1950s, progress in electron microscopy allowed observations of single rows of atoms in crystals, but the resolution was too low, and the irradiation damage too high, to observe individual molecules under the electron microscope. More recently, however, higher selectivity and better collection of secondary electrons have made possible electron-microscope observations of individual isolated heavy atoms in a sample made of light atoms. The detection of single atoms and single molecules on surfaces first became a reality in the early 1980s with the invention of the scanning tunnelling microscope (STM) [2], and later of its variant for insulating substrates, the atomic force microscope (AFM) [3]. These two techniques have opened the way to microscopic investigations of matter at the single-atom and single-molecule levels, and are widely used today in many laboratories.

Meanwhile, thanks to lasers and general progress in optical techniques, it had become possible in the 1970s to observe single atoms in the gas phase, in dilute atomic beams. The atoms were detected via the short bursts of fluorescence light they emitted when crossing the laser focus. Then, single trapped ions were also detected by their fluorescence. However, observing a single fluorescent atom in the gas phase is a very different problem from observing a single molecule in a condensed phase, mainly for the following two reasons:

1. Condensed matter produces a strong optical background via Rayleigh and Raman scattering. Moreover, other fluorescent impurities in the sample may easily dominate the fluorescence from a single molecule, and therefore the purity of the sample is crucial. This problem obviously does not arise in vacuum.

2. A single atom usually can fluoresce millions of photons per second, without any degradation. An atom in vacuum is a stable system, even in its excited state. It can thus perform indefinitely many excitation–emission cycles, which facilitates detection considerably. This convenient property of atoms is unfortunately not shared by most fluorescing systems in condensed matter, the vast majority of which are subject to irreversible transformations putting an end to their luminescence. Those processes are called photobleaching and will be discussed briefly later in this chapter. In the late 1980s and early 1990s, progress in sources, optics, and detectors opened the way to detecting single molecules in condensed matter. These experiments have been extended to other single luminescent objects, including semiconductor nanocrystals or quantum dots, metal particles, colour centres, etc. They are nowadays performed in condensed matter environments as diverse and complex as facets of catalyst crystals or live biological cells. Part of the driving force behind this research domain is the broad current interest in nanoscience, i.e. the structure and dynamics of matter at nanometre scales. By coupling lasers to optical microscopes, the powerful spectroscopic techniques developed in the 1970s and 1980s have been extended not only down to submicrometre scales, but even down to nanometre scales if a single object can be selected. Moreover, as we will see in this chapter, the selection of single objects opens new motivations, methods, and results. This relatively young field of investigation is called single-molecule optics or nano-optics.

In this chapter, we briefly review the principles of optical microscopy (Section 6.2) and of the fluorescence method, since these two techniques are the workhorses of single-molecule studies. Let us stress again how small a molecule really is. If a dewdrop were magnified to the size of the Earth, a single water molecule would be about the size of a person. It may appear hopeless to look for optical signals from such a tiny source in the background of all other molecules in the drop. This is possible, however, owing to the special properties of the fluorescence signal, and to two tricks:

1. The size of the illuminated sample must be reduced as much as possible. For a given light intensity, i.e. for a given signal from the molecule, the background will be proportional to the number of illuminated sample molecules, i.e. proportional to the volume of the illuminated sample. Reducing background will require reducing the illuminated spot, usually by focusing the laser beam very tightly into the sample. This focusing step thus performs a spatial selection.

2. The intrinsic background of the detection method used must in principle be low enough for the single molecule's signal to dominate the background from all sample molecules in the illuminated volume. An illuminated volume of $1\,\mu m^3$ contains about 10^9 sample molecules. The detection method must therefore have a selectivity ratio, i.e. a ratio of signals from the molecule of interest to that of a sample molecule, higher than 10^9. Only very few optical methods have such a high selectivity—but fluorescence does. The high selectivity of fluorescence arises from optical resonance. A fluorescent molecule must first absorb a laser photon, which a sample molecule cannot do. Therefore, the laser frequency performs a spectral selection on the molecules contained in the sample. Only molecules resonantly excited by the laser can emit fluorescence. It is interesting to compare fluorescence with infrared (IR) absorption, for example, which might open up studies of single chemical bonds. Unfortunately, the selectivity of IR absorption is 1000 at best, because of the width and shape of vibrational resonances. Therefore, such a method could only be applied to a single bond in combination with spatial selection of fewer than 1000 molecules. Selection of such a small excited sample cannot be achieved by diffraction-limited optics alone.

We then proceed to discuss particular implementations of the fluorescence technique and give examples of the results that can be obtained with them. Fluorescence of immobilized molecules (Section 6.3) is easiest to discuss, as the signal arises from a single system with a fixed emission dipole. Such recent developments as super-resolution microscopy are based on the fluorescence of immobilized single molecules. When molecules can move and tumble, fluorescence signals are subject to even more dynamical processes. Their evaluation requires statistical methods, in particular autocorrelation of fluorescence time traces. The associated technique, known as fluorescence correlation spectroscopy, is discussed in Section 6.4. In the special case of cryogenic experiments at low temperatures, much information can be obtained from fluorescence excitation spectroscopy, which will be discussed in Section 6.5. These conditions give rise to narrow resonance lines, which make it possible to perform highly accurate quantum optics experiments. Finally, single molecules and single nano-objects can be detected by other far-field optical techniques such as scattering or photothermal detection. Several experiments have been demonstrated in the last years. These new methods will be discussed in Section 6.6 with their applications to single metal nanoparticles and other small objects.

6.2 Microscopy methods

6.2.1 Principles

The basis of single-molecule optical detection is an optical microscope [4], which in principle is nothing other than a simple lens system for magnifying small objects. The first lens, called the objective, has a short focal length (a few millimetres), and creates an image of the object in the intermediate image plane. This image in turn can be looked at with another lens, the eyepiece. Typical magnification ratios are in the range 40–100. The resolution of the image provided by the microscope is limited by diffraction. The Abbe–Rayleigh criterion states that, for a wavelength λ, the smallest distance d resolvable between two point sources as deduced from diffraction theory is

$$d = 0.61 \frac{\lambda}{\text{NA}}, \tag{6.1}$$

where $\text{NA} = n \sin \alpha$, called the numerical aperture of the objective lens, is proportional to the index of refraction n in the object space, and α is half the maximal angle under which the objective lens collects light from the object. In single-molecule experiments, NA should be as large as possible for two reasons:

1. The spatial resolution improves for larger NA, both laterally and axially (see below).
2. The collection efficiency, i.e. the brightness of the image, increases very rapidly with NA, quadratically for small apertures. In addition to angle considerations, the index of refraction of the object medium is a very important factor in the NA. Indeed, because of total reflection at interfaces, a high index enables collection over a wider range of angles. This is the reason why most microscopes use immersion to enhance resolution and collection efficiency.

The main difficulty in manufacturing microscope objectives is to achieve good correction of all aberrations (spherical and chromatic) also for off-axis rays with incidence angles that can be larger than 60°. This is achieved by assembling several lenses (sometimes more than 10) made out of different glasses, and which have to be anti-reflection-coated for good luminosity. Good microscope objectives are therefore quite advanced and expensive pieces of equipment. Immersion oil objectives reach an NA of 1.4, corresponding to collection angles close to 90°.

An ideal microscope lens will image a point source as an Airy pattern if a circular iris or diaphragm limits the aperture. This point-spread function (PSF) therefore has the classical form of a diffraction spot from a round hole. If the light collected is instead a Gaussian beam, then, provided it is fully included in the acceptance pupil of the objective, the PSF will be a Gaussian spot. For many experiments on three-dimensional (3D) samples, it is important to consider the depth of focus, i.e. the length of the PSF in the axial direction. This size L (sometimes called the Rayleigh length) is approximately

$$L = 2n \frac{\lambda}{\text{NA}^2}. \tag{6.2}$$

The 3D appearance of the PSF is thus an elongated (prolate, or cigar-shaped) ellipsoid. Decreasing this length is a third reason to make NA as high as possible, since the volume of the PSF scales as NA^{-4}.

NA is inversely proportional to the index of refraction. Therefore, it is advantageous to collect light through glass, or through high-index oil, when possible. Special water-immersion objectives are used for biological samples. To fully benefit from a high index, the index must of course be matched between sample and objective glass. This is achieved thanks to the immersion oil. A further advantage of immersion is the higher efficiency of fluorescence collection. An air gap leads to light losses by total internal reflection at the interface from high- to low-index media. For emitters placed at an air–glass interface, the situation is even worse, since a large fraction of the emitted light (up to 90%) is radiated into the high-index material. Collecting emission on the air side therefore entails big losses of intensity, in addition to losses in resolution.

6.2.2 Correction of aberrations

The ideal microscope objective would image a planar object onto a plane field, without distortion, and without change in image with wavelength (achromatism). To achieve this, the following aberrations must be corrected:

1. Chromatic aberrations from the dispersion of glasses (their refractive index is larger for blue than for red light). Even with well-corrected objectives, the focus often moves by some micrometres when the wavelength varies over the visible spectrum.
2. Geometrical aberrations: spherical (change of focal point with distance from axis), coma (due to changes of the image point with ray direction), field curvature (the image focus is not obtained on a plane but on a curved surface), field distortion (pincushion or barrel images), etc.

Simple spherical lenses made of ordinary dispersive glass suffer from all these aberrations and cannot fulfil the requirements of the ideal lens for large NA. For example, a plano-convex lens has spherical aberrations, which are minimized by placing the convex surface on the side of the parallel beam. Aspheric singlets as used in CD readers correct, in principle, perfectly for their focus, but are designed to work at one wavelength only and have strong chromatic aberrations. To approximate the requirements of an ideal lens, one uses combinations of spherical lenses possessing various radii of curvature, thicknesses, and materials, and varies their positions. To design objectives and other multilens systems, special codes are used to calculate imaging with rays far from paraxial for arbitrary systems of lenses. A good objective may contain more than 10 lenses, which have to be positioned with specifications as narrow as micrometres for some of them. The various air–glass interfaces have to be anti-reflection-coated to reduce reflection losses, and small gaps between the lenses are often bridged with a high-index medium, usually UV-polymerizable glue, after the respective positions of the lenses has been adjusted by hand. Therefore, objective lenses are expensive and sensitive optical components.

Nowadays, most objectives are infinity-corrected, i.e. they are not calculated to form their image directly in the intermediate image plane, but rather at infinity. Another lens, called a tube lens, then images the plane waves into the intermediate image. Some manufacturers use the tube lens to correct some aberrations. In that case, the tube lens and objective must be used in combination for optimal correction. The advantages of infinity-corrected objectives are essential in confocal microscopy, polarization studies, and spectroscopy, because plane waves are easier to filter and manipulate than spherical waves (for example, they are not distorted by flat windows).

6.2.3 Polarization structure at the focus

The electric field of a laser wave is a vector quantity, which complicates the polarization structure in the case of high NA. We will briefly discuss the polarization of the field at the focus of a linearly polarized laser wave [4]. For low NA, the polarization of the spot is the same as that of the incident beam. At the focal point itself, by symmetry, the polarization of the incident beam is conserved. At high NA, however, and for parts of the PSF away from the centre, interference of the incoming rays leads to deviations from this polarization. For large angle of incidence and a linearly polarized incident beam, simple consideration of the interference of different rays shows that the axial (i.e. polarized along the axis) component of the field has a node at the centre and presents two (weak) lobes in the focal plane. The third transverse component (perpendicular to the axis and to the incoming polarization) is even weaker and presents four lobes. By using annular illumination, and/or introducing phase masks into the incoming beam, the polarization of exciting laser light at the focus can be manipulated, and can even present a PSF with a single lobe for the longitudinal polarization. This is of great interest when determining the full 3D orientation of single absorbers, since their transverse polarization can be easily probed with linearly polarized light and normal illumination.

A similar, but distinct, problem is to find the polarization structure of a wave radiated by a linear dipole at the focus, and collimated into a plane wave by an objective with large NA. In the case of a dipole lying in the focal plane, it can be shown [5] that the polarization is linear throughout the field, parallel to the dipole along the horizontal and vertical directions (N, S, E, W), and significantly tilted in the NW, NE, SW, SE positions. In the case of a dipole perpendicular to the focal plane, the polarization is radial.

6.2.4 Various microscopy methods

There are several ways to record images with a microscope. We briefly mention those that are most important for single-molecule studies.

1. **Confocal microscopy.** In this method, only one point of the sample is imaged onto a photodetector. If the sample is scanned in three dimensions, a 3D image is obtained by recording the optical signal as a function of sample position. To reduce background, i.e. to ensure that the signal arises only from the focus, a diaphragm or pinhole is inserted in one of the image planes. To scan the area to be imaged, one can move either the sample itself with piezoelectric transducers (sample scanning) or the focus by means of tilting mirrors (beam scanning). In the latter case, a telecentric system images the spot of the illumination beam on the scanning mirror onto the input lens of the objective. In this way, the two points remain conjugated when the mirror is tilted during scanning. Fluorescence is collected in the backwards-scattered geometry, and is folded back exactly onto the incoming path by the mirror–telecentric system assembly.

2. **Wide-field imaging.** In this method, a large part of the field is illuminated by an unfocused beam (epi-illumination), and the image is formed on a multichannel

detector such as a CCD camera or an image intensifier. For a 3D sample, this design leads to background arising from emission by planes below and above the imaged plane of the sample.

3. **Total internal reflection illumination.** To reduce the background, the excitation light can be sent at a large incidence angle on the surface, achieving total internal reflection (TIR). Fluorescence and other emissions can be collected either across the interface or on the same side as the illumination. In that case, collection has to be performed in a solid angle where no illumination light is present, which requires a large NA.

The confocal configuration shown schematically in Fig. 6.1 is particularly useful for detecting single molecules because of its simplicity and of its low background. This design has the advantage that only the focus is excited with high efficiency, thus limiting photobleaching in other sample planes. Moreover, fluorescence arising from other points does not reach the detector. The spatial selection of single molecules or single nano-objects is therefore performed in two steps, with equivalent selectivity:

1. Excitation selection, by focusing the laser beam on a small spot.
2. Detection selection, by detecting light arising from the same area only.

Several other optical designs have been developed in the last two decades, with various elements for excitation and collection: single-mode optical fibres, parabolic mirrors, aspheric lenses, gradient index lenses, etc. An overview can be found in [6].

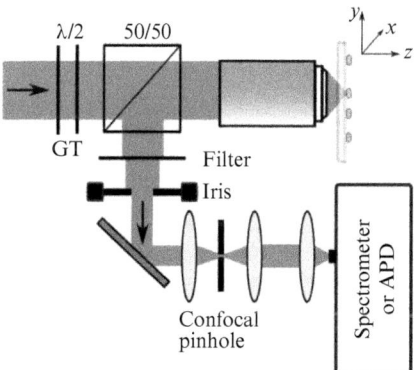

Fig. 6.1 Scheme of a confocal setup for single nano-object spectroscopy. The confocal spot can be scanned with mirrors and a telecentric lens system (not shown) or simply by scanning the sample with a translation stage. (From Naumov, A. V., Gorshelev, A. A., Vainer, Y. G., Kador, L., & Köhler, J. (2009). Far-field nanodiagnostics of solids with visible light by spectrally selective imaging. Angewandte Chemie International Edition, 48(51), 9747–9750. doi:10.1002/anie.200905101, by permission of John Wiley & Sons.)

6.3   Detection of fluorescence

6.3.1   Introduction. Signal-to-noise ratio

If the single molecules are immobilized in or on a solid or in a highly viscous sample, it becomes possible to study the same molecule for long periods of time. This situation is very different from experiments in liquid solutions, where statistics over a large number of molecules usually always give a signal, even in difficult conditions with strong background and noise. For molecules immobilized in solids or on surfaces, it is of crucial importance to be able to 'recognize' the molecule, i.e. to distinguish its signal from background and noise.

The following discussion of the signal-to-noise ratio applies not just to microscopic images, but also to other signals from immobilized single molecules, for example their spectra. Let us consider the background B and the signal S of the molecule as functions of a scanning parameter (typically one position coordinate of the sample). If t is the acquisition time per channel or pixel, then the number of acquired background counts is Bt. Assuming shot noise to be dominant (which is usually the case for weak signals), the noise from the background is $\sqrt{Bt}$, which has to be compared with the signal St. The most favourable case to detect the molecule is when its signal occupies about one pixel or one channel width (a broad structure is more difficult to distinguish from noise than a sharp one). The molecule will be detectable if and only if the signal-to-noise ratio is significantly larger than unity, let us say equal to 3:

$$St > 3\sqrt{Bt}, \qquad \text{or} \quad t > 10\frac{B}{S^2}. \tag{6.3}$$

For a typical background of 100 counts/s arising from the dark counts of the avalanche detector, and for an acquisition time of 1 s, it follows that the signal has to be larger than 30 counts/s to be clearly distinguished from noise. This condition becomes only a lower bound if the molecule's image is spread over more than one pixel.

6.3.2   Sample preparation

In fluorescence experiments, the main source of background is fluorescence from residual impurities or from optical elements. Raman scattering is often negligible. To observe single molecules, it is therefore crucial to reduce contamination by fluorescent impurities, and to work with very clean optical parts. The excited volume must be as small as possible, and the substrate should be non-fluorescent (fused silica, or very pure glass). Hereafter, we cite a few usual methods to prepare samples.

6.3.2.1   Spin-coating

Quick spinning of a flat substrate with a thin layer of solution leads to a uniform film, draining more and more slowly as time goes by. At the same time, evaporation decreases the thickness too, and increases the concentration and the viscosity. In a first phase, viscous draining is dominant in reducing the thickness, then evaporation takes over. The combination of viscous draining and evaporation leads to uniform films of tuneable thickness, for example to prepare photoresists in the semiconductor industry.

Typical spin-coating parameters (1000 rpm and 1% weight of polymer in a solvent like water) lead to thicknesses between 10 nm and a few micrometres. Spin-coating is a very interesting approach to reducing the illuminated volume, and therefore the background, by reducing the thickness.

6.3.2.2 Langmuir–Blodgett films

Even thinner, monomolecular films are obtained by the Langmuir–Blodgett technique, i.e. deposition onto a substrate from a monomolecular layer at the air–water interface. The fluorescent dye molecules are usually introduced at low concentration in the spreading solution of the amphiphilic molecules, but may also be adsorbed from an aqueous solution.

6.3.2.3 Bilayers, membranes, and black lipid films

These thin layers are particular cases of a pair of monomolecular layers, which can be either deposited on a substrate, or suspended to a hole. Cell and vesicle membranes are particularly important matrices for single biomolecules.

6.3.2.4 Other preparation methods

Molecules can be deposited directly on a substrate (in which case they are more sensitive to oxygen and water), included in nanocrystals, etc., or simply selected or imaged within a 3D sample, such as a cell. In this last case, a thin sample is preferred to limit background. Much information can sometimes be obtained from images with diffraction-limited single molecule spots. Counting them can provide concentrations, the stoichiometry of complexes [7], etc. Another important application is co-localization [8], which makes it possible to test interactions between biomolecules labeled with two different fluorophores. However, much progress has been made recently in super-resolution imaging, as will be discussed later in this chapter.

6.3.3 Orientation

The orientation of the in-plane component of the molecular transition moment can be obtained easily by polarization measurements. Two different methods are used predominantly.

6.3.3.1 Polarization analysis in detection

The fluorescence beam is sent to a polarizing beam splitter and the two beams are sent to different detectors. An example of polarization-dependent images is shown in Fig. 6.2. The intensity ratio gives two possible solutions for the in-plane orientation. Discriminating between these requires a third measurement. In combination with polarized excitation, polarized analysis gives the fluorescence anisotropy, related to the angular mobility and rotational diffusion of the fluorophore during fluorescence. This method is very useful to monitor rotational relaxation of slowly moving objects, such as dyes in highly viscous glass formers [9, 10]. These experiments show that different

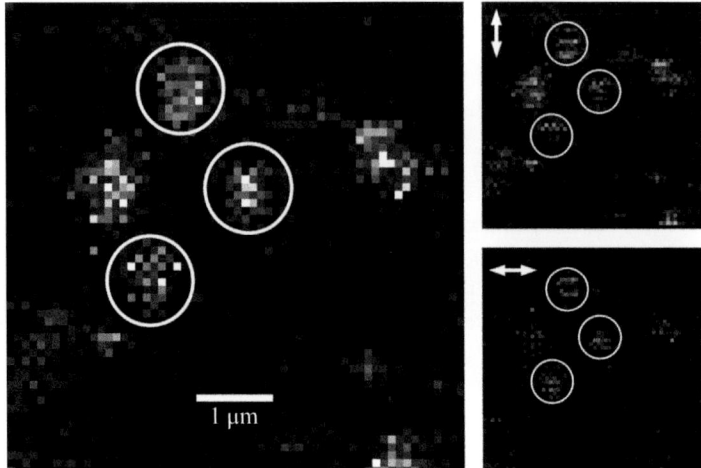

Fig. 6.2 The left panel shows an example of rotational diffusion of single dye molecules in glycerol at low temperature (200 K), monitored via intensity fluctuations in two orthogonal fluorescence polarizations. The right panels show the vertical and horizontal polarizations separately: vertical at the top, horizontal at the bottom. The diffusion times of molecules are comparable to the scan time, about 1 s per line.

single molecules diffuse as Brownian particles on short timescales, but that different molecules diffuse at different rates, and that this diffusion can change in time. This inhomogeneity gives rise to the complex (non-exponential) behaviour of the glass just above the glass transition. Single-molecule experiments directly confirm the spatial heterogeneity of glass dynamics.

6.3.3.2 *Polarization modulation in excitation*

The polarization of the exciting beam can be modulated in time, for example with a rotating polarizer. The phase and amplitude of the sinusoidal variation of the detected intensity gives information about the orientation and diffusion of the fluorophore. For a fixed molecule, the modulation is sinusoidal with maximum contrast. For a rapidly diffusing molecule, however, the intensity modulation depths decreases, indicating diffusion or hampered diffusion [11]. As discussed in Section 6.2.3, the full 3D orientation of the dipole moment can be obtained with a widely opened excitation beam or by measuring the distribution of fluorescence photons as a function of the angles of emission and comparison with the emission pattern of a dipole. This latter method requires a matrix of detectors for fast determination.

6.3.4 Blinking

When the fluorescence intensity of a nano-object is recorded as a function of time (in a so-called fluorescence intensity trace), random variations of the average emission intensity are often observed [12]. The variations can sometimes be progressive, but

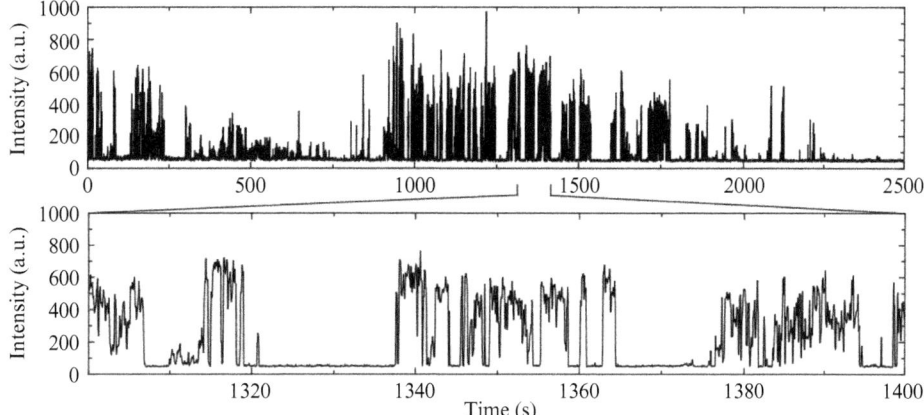

Fig. 6.3 Example of blinking luminescence traces of a semiconductor nanocrystal (a CdSe core with a ZnS shell), showing bright and dark periods with a wide distribution of durations. (Reprinted from Cichos, F., von Borczyskowski, C., & Orrit, M. (2007). Power-law intermittency of single emitters. Current Opinion in Colloid & Interface Science, 12(6), 272–284. doi:10.1016/j.cocis.2007.07.012, with permission from Elsevier.)

often exhibit sudden jumps between bright and dark states. We discuss here only the latter behaviour, which is commonly known as blinking or intermittency. Blinking is a characteristic feature of the emission of single nano-objects, and has been observed for organic molecules, semiconductor nanocrystals, and colour centres. Figure 6.3 presents typical blinking fluctuations in the luminescence signal of a single semiconductor nanocrystal. In large populations, blinking is almost always hidden because the fluctuations of individual objects are not synchronized. In some cases, an external parameter can synchronize the blinking of all objects (an obvious example is switching the exciting laser on). In those cases, blinking will appear as a transient variation of the average signal. It is one of the most powerful features of single-molecule methods that no synchronization is needed to observe such fluctuations directly, as the example of blinking shows.

Several possible mechanisms can lead to blinking. Most of them involve a 'dark' state of the emitter, i.e. a state that does not fluoresce, either because it does not absorb efficiently or because its fluorescence yield is too low. Hereinafter, we examine a few possible sources of blinking.

6.3.4.1 Triplet state

A molecule in its triplet state can in theory absorb and emit photons, but usually with only very low fluorescence yield. Moreover, the laser wavelength is normally not suited to excite triplet–triplet transitions. Therefore, the triplet state is dark. The molecule's fluorescence trace presents dark periods lasting for the triplet lifetime on average. Depending on the molecule, this time can be as short as microseconds and as long as tens of milliseconds for the usual dyes. Atmospheric oxygen has a strong influence on triplet lifetime. Since O_2 has a triplet ground state, it can exchange electrons with a molecular triplet, yielding two singlet states, one the ground state of the dye, the other singlet oxygen, a very reactive and phosphorescent species.

6.3.4.2   *Electron transfer*

The excited molecule may accept an electron from its environment or give one electron to it [12]. The radical-ion left often has very low fluorescence yield. The charge-transfer state lives as long as the electron is away, giving rise to dark states with lifetimes longer than seconds if the electron goes far away, as the example of rhodamine 6G in polyvinyl alcohol shows [13]. This process appears to be very general, as was recently demonstrated by the groups of Sauer and Tinnefeld. Charge-transfer rates can be controlled by adding redox species to the solution. This scheme (known as ROXS) [14], profoundly modifies the lifetimes of the triplet state and of the charged dark states, which can have the added benefit that bleaching (see Section 6.3.5) is also significantly reduced.

6.3.4.3   *Other reversible photochemical reactions*

Excited molecules may change conformation. If the change is reversible, and if the two forms have different optical properties, blinking will follow. An example of intramolecular reactions of this kind is *cis–trans* isomerization, which is common in cyanine dyes. The excited molecule may also react with its environment, for example by abstracting a proton from a nearby molecule or from an acid matrix. Several dyes present protonated 'leuco' forms, in which an additional proton shortens the conjugation path and shifts the absorption spectrum to the blue. The main feature of these photochemical reactions involving large rearrangements of atoms is that they often require large activation energies and are therefore frozen at low temperatures.

6.3.4.4   *Other sources of blinking*

These may exist in principle. For example, the orientation of the molecule could switch between parallel and perpendicular to the excitation polarization. The absorption spectrum of the molecule could also shift between different spectral positions. This spectral diffusion is a prominent effect in low-temperature experiments (see Section 6.5.4), and has also been observed at room temperature in polymers [15]. However, such 'physical' processes do not seem to lead to significant blinking, entailing large intensity fluctuations.

6.3.5   Bleaching

Many photochemical reactions are irreversible. In that case, the molecule stays in the dark state the first time it goes there. Fluorescence is lost for good, and a new single molecule must be found. At room temperature, all fluorescent organic molecules photobleach sooner or later. Much effort has been spent to screen, adapt, and design dyes able to resist photobleaching as far as possible. This has been crucial not only for laser dyes, but also for fluorescent labelling in biology. In spite of several decades of effort, little is known about photobleaching and how to control it. Atmospheric oxygen and small reactive molecules such as water obviously open efficient channels for photobleaching. Production of highly reactive singlet oxygen from the triplet state of the dye is a well-characterized degradation channel. That is the reason why red and infrared dyes with their triplet state below the energy of singlet oxygen are highly photoresistant, even in air. Because of the important role of oxygen, many single-molecule fluorescence studies are done in the presence of reducing agents (oxygen scavengers, mercaptoethanol, Trolox, etc.). The ROXS scheme (see Section 6.3.4.2), by shortening the lifetimes of metastable and reactive states, also improves the resistance of dyes to bleaching. However, even in an inert and dry atmosphere, or under high vacuum, photobleaching still occurs, albeit at a lower rate. Because photobleaching involves barrier crossing in chemical reactions, it is likely that low temperatures considerably decrease the process. Temperature studies of rhodamine in PVA have shown that bleaching is indeed reduced by several orders of magnitude at low temperatures for some, but not all, molecules [16]. Recent articles report dyes of the perylene or terrylene diimide family, which present excellent resistance to bleaching, and can still be made water-soluble by convenient substitutions [17].

6.3.6   Super-resolution

The size of the PSF, for example the image spot of a single molecule, is limited by diffraction. However, as was realized from the very beginning of single-molecule observations, the centre of the image of a single molecule can be found with accuracy much higher than the width of the PSF. If N photons are detected, we can exploit our knowledge that all of them arise from the same single molecule, a point-like region of space. Therefore, the position of their centre of gravity is determined down to roughly $\Delta x/\sqrt{N}$, where Δx is the spatial width of the PSF. Indeed, the probability distribution of the average of N random variables, each of them distributed along the same normal Gaussian probability law, is itself a Gaussian with width reduced by $\sqrt{N}$.

In low-temperature experiments (see Section 6.5), single molecules can be discriminated by their resonance frequency. The selection of many different molecules at many different laser frequencies gives rise to a super-resolved image [18], where each molecule provides one point with $\Delta x/\sqrt{N}$ accuracy. Because a large number of photons can be accumulated for each single molecule, the accuracy can reach as small as a few nanometres. This has been done recently on a sample of millions of molecules in a crystal at low temperature [19], providing a super-resolved image of the crystal. In 2006, three important papers appeared almost simultaneously, demonstrating super-resolution based on photochemical switching of single molecules [20–22]. These three papers proposed the same idea of using photochemical reactions to switch molecules on and off, and localizing the on-molecules with super-resolution accuracy. The proposed photochemical reactions differed, however, ranging from transitions to dark states in autofluorescent proteins to blinking or bleaching reactions in dyes. In all cases, the central idea of super-resolution imaging with photoswitches is to randomly select a small subpopulation of all fluorescent molecules present in the sample by switching them to a bright state, either with an auxiliary laser or by spontaneous blinking in the right conditions [23]. The concentration of these activated switches should be low enough that each one of them gives rise to a spot isolated from the others. Because these spots are well separated from each other, one can find their centres very accurately. For the brightest molecules, providing many photons, the spot centre can be located to an accuracy of a few tens of nanometres, sometimes even better. Another related super-resolution method uses the transient binding of fluorescent molecules to the surface or object to be imaged [24]. Only the immobile molecules give rise to a measurable spot, whose centre determines with super-resolution the source's location.

6.4 Fluorescence excitation spectroscopy

6.4.1 Molecular fluorescence

Most of the optical microscopy and spectroscopy of single objects is based on fluorescence. Therefore, it is important to recall the principal features and properties of fluorescent molecules [25–27]. All fluorescent organic molecules are conjugated, i.e. they have delocalized electronic π-wavefunctions. π-wavefunctions are built from carbon-atom p-orbitals, and have a node in the plane of the bond. The delocalized p-orbital can be seen as a 1D box potential along which π-electrons can move freely. This very simple picture is known as the 1D electron gas model.

When the electron cloud of a molecule is put into an excited state (after absorption of one photon), the π-electron density in this new state is changed, which means that the geometry of the molecule, in particular the distances between carbon atoms, is also (slightly) changed. This leads to coupling of the electronic excitation with vibrations. Thus, photon absorption leads to an electronic transition accompanied by a series of vibronic bands where 0, 1, 2, ... vibrational quanta are created along with the electronic excitation. In the conjugated photon emission process (called fluorescence), one photon is emitted from the relaxed excited state (0 vibration) to vibrational levels of the ground state. Vibronic coupling will give rise to sidebands at higher energy than the 0–0 transition in the case of absorption, to sidebands at lower energy in the case of fluorescence. For purely harmonic vibrations, this leads to an approximate symmetry (mirror image) between normalized absorption and fluorescence spectra as a function of energy (or frequency).

It is important to consider how strongly a molecular transition is coupled to radiation. The quantity describing the strength of this coupling is called the oscillator strength. It is proportional to the square of the transition dipole moment, a vector homogeneous to an electric dipole and fixed relative to the molecular axes. Being conjugated processes, absorption and fluorescence strengths are related (provided the geometry of the molecule does not completely change between absorption and emission). The inverse of the radiative lifetime of the molecule is proportional to the integral of its absorption coefficient over the spectrum (the Strickler–Berg relation). However, channels other than the radiative one can be accessible from the excited state of the molecule, thus decreasing the fluorescence intensity. The yield of emission events compared with excitation events is called the fluorescence quantum yield. Molecules with high radiative rates will tend to be better emitters. This is the case for many laser dyes, which are also good fluorescent probes. The non-radiative channel is governed mainly by radiationless transitions to lower-lying electronic states, in particular the ground state and to a lesser extent to the metastable triplet state (see below). Non-radiative transitions are much reduced for planar, rigid molecules. As a rule, therefore, good fluorophores have to be planar and rigid. Important classes of fluorescent dyes are polycyclic aromatic hydrocarbons, porphyrins, cyanines, and rhodamines.

All strongly fluorescent molecules, called fluorophores, possess conjugated π-bonds with or without heteroatoms participating in the conjugation. They are planar, rigid, and chemically stable enough to sustain many excitation–emission cycles in reactive environments such as air or water. The latter is particularly important for biological labelling.

The ground state of a large majority of fluorescent molecules is a singlet state, i.e. electrons are paired into a zero-spin state [26]. The ground state is called S_0. The first excited singlet state S_1 is the one reached after absorption of a photon. The reason for this is that a photon cannot flip spins, at least in molecules consisting of light atoms, because the magnetic field of a light wave is too weak. But there is another excited state at lower energy than the first excited singlet. It is the triplet state T_1, with total spin 1. This state lies at a lower energy than the singlet S_1 because of a reduced Coulomb repulsion between electrons (exchange interaction). The triplet state has three spin sublevels, which are usually indistinguishable at room temperature. Transitions between singlet and triplet states are called intersystem crossing (ISC). ISC is spin-forbidden, and is therefore caused only by weak interactions such as spin–orbit coupling. ISC transition rates are therefore rather low, and the lifetime of the triplet state is rather long (often microseconds to seconds). However, the triplet states play a central role in the photodynamics of the molecule, because they limit the number of resonant absorption and fluorescence events per unit time that a molecule can perform [28, 29]. We come back to the consequences of these transitions later.

It must be realized that aromatic molecules, being conjugated, are rather reactive: the π-electrons are very polarizable (being far from the carbon nuclei), and therefore readily engage in new chemical bonds. The situation is even worse when the molecule is excited, because the electron cloud has then been given extra energy, which can help cross photochemical barriers and which gives access to more new states for the excited electron. Fluorescent molecules, which have long excited-state lifetimes, are therefore particularly sensitive to photochemistry. Among the possible reactions taking place in excited molecules, we can cite charge (electron or hole) transfer, proton transfer (intra- or intermolecular), *cis–trans* isomerization of double bonds, twisted intramolecular charge transfer, photo-oxidations, photo-additions, etc. The immense majority of these reactions destroy fluorescence. The new product does not absorb light in the same spectral region, or, if it does absorb, no longer emits, because of new non-radiative channels. All fluorescent molecules are therefore subject to photobleaching [28], i.e. the 'death' of fluorescence after a variable average number of excitation–emission cycles. This number can vary between less than 1 and several hundreds of millions, depending on molecules and conditions (atmosphere, temperature, solvent, environment, etc.).

Fluorescence is a very low-background technique. The reason is that fluorescence arises only after real excitation of the molecule in its excited state. If a molecule is not excited resonantly, it may still end up in an excited vibrational level of the ground state: this is a Raman-scattering process. As compared with fluorescence, the probability of this process is reduced by the extremely short dwell time in the 'virtual' excited state. Raman-scattering cross-sections are very small, more than 10 orders of magnitude smaller than the typical fluorescence cross-section of a good fluorescent dye. This explains why spectral selection by fluorescence is so efficient.

6.4.2   Zero-phonon line and phonon wing

Fluorescence excitation is usually equivalent to absorption. It is mostly useful in low-temperature studies of single molecules, where certain lines become extremely narrow and intense. In this case, high resolution is more easily reached by excitation with a narrow single-frequency laser than by other spectroscopic devices such as spectrographs. Before discussing low-temperature spectroscopy, we need some basic knowledge of the optical spectroscopy of impurities in solids. We start with the simple case of a single molecule in a solid at zero temperature. If there were no coupling to vibrations, the absorption spectrum would consist of a single line, with a width given by the excited-state lifetime. However, the electronic system of a molecule is in principle coupled to all possible vibrational modes in the molecule and in the solid matrix around it. In the Born–Oppenheimer picture, this coupling arises from changes in vibrational potential between the ground and excited electronic states: because the nuclear wavefunctions in the excited state differ from those in the ground state, exciting the electron modifies the nuclear movement, i.e. it can bring about vibrations in the molecule. Fortunately, one does not have to consider all possible vibrational modes to understand the absorption and fluorescence spectra. Usually, only a few of these modes are coupled strongly to the optical transition, which simplifies the analysis considerably. We first consider coupling of the electron to a single harmonic vibration mode. Often, a totally symmetric C–C stretching mode (breathing mode) is responsible for most of the coupling. We further assume that only the equilibrium position of the nuclei along the associated vibration coordinate is modified. This case is called linear vibronic coupling. We first consider only one deformation mode R of a molecule, with a harmonic potential $E_g(R)$ in the ground state $|g\rangle$. The minimum of this potential is displaced by ΔR in the excited state $|e\rangle$ with potential $E_g(R)$, without a change in the curvature. In the crude Born–Oppenheimer approximation, we can write the wavevectors as products:

$$|g, m\rangle \equiv |g\rangle |m\rangle \ ,$$
$$|e, n\rangle \equiv |e\rangle |\tilde{n}\rangle \ ,$$

(6.4)

where $\tilde{n}$ means n quanta in the excited well potential and m means m quanta in the ground-state potential. The overlap amplitude

$$f_m^n = \langle \tilde{n} | m \rangle$$

(6.5)

is usually called the Franck–Condon amplitude and the square modulus of this overlap

$$F_m^n = |\langle \tilde{n} | m \rangle|^2 \tag{6.6}$$

the Franck–Condon factor. The Franck–Condon factors represent the intensities of the m–n transitions from the ground to excited states. The first of these are the 0–0 and 0–1 transitions. These factors can be expressed analytically in terms of the overlap of harmonic oscillator wavefunctions. The special case where one of the wavefunctions is the ground vibrational state is most important. It can be shown easily that the Franck–Condon factors have the form

$$F_0^n = \frac{\xi^{2n}}{n!} e^{-\xi^2}, \tag{6.7}$$

where

$$\xi = \Delta R \sqrt{\frac{M\Omega}{2\hbar}} \tag{6.8}$$

is the displacement of the harmonic oscillator upon excitation, expressed in dimensionless units of the spatial spread of the ground state. M is the mass and Ω the angular frequency of the oscillator.

The absorption spectra of a molecule at low enough temperature are recorded starting from the ground vibrational state of the ground electronic state, and going to all vibrational states of the excited electronic state (these states are called vibronic ['vibr'-'onic'] states). The intensities of these absorptions are proportional to the Franck–Condon factors. A similar argument applies to the emission or fluorescence spectrum, starting from the ground vibrational state of the excited electronic state, and going to all vibrational states of the ground electronic state. For linear vibronic coupling, the intensity distributions of the absorption and emission spectra are symmetric with respect to the zero-phonon line (see below). Absorption and fluorescence spectra are mirror images of each other with respect to the 0–0 transition.

We now consider the coupling of a molecule to a large number of modes, for example to a branch of phonons in a host crystal. Each mode is now weakly coupled (weak ξ), and contributes a weak 0–1 absorption sideband shifted to the blue of the main absorption line (which corresponds to the 0–0 transition). Neglecting the 0–2 transitions (although they are weak, there are a large number of them, of order N^2, with N the number of modes), we find a broad band on the blue side of the 0–0 line. The 0–0 line, which is common to all modes, is called the zero-phonon line (ZPL), and the blueshifted broad band is the phonon wing (PW). The intensity of the ZPL is called the Debye–Waller factor, and is analogous to the intensity of spots in X-ray diffraction, or to the intensity of the recoil-free structure in Mössbauer spectroscopy. This intensity decreases very rapidly with temperature. One often uses the following approximate formula for the oscillator strength of the ZPL:

$$I_{\text{ZPL}} \approx \exp\left[-\xi^2 \coth\left(\frac{\hbar\Omega}{2k_B T} \right) \right]. \tag{6.9}$$

The intensity of the ZPL therefore decreases exponentially with temperature. In most organic materials, the intensity of the ZPL becomes negligible at temperatures higher than 30 K. In diamond, which is a much harder material with a high Debye temperature (about 1000 K), the ZPL of certain impurities can still be observed at room temperature.

In the preceding model, the ZPL is a narrow line, with a linewidth given by the excited-state lifetime. However, several processes contribute to its broadening:

- higher-order coupling to phonons, notably via the change in curvature of the potential—the associated process is called quadratic coupling;
- anharmonicity of the vibrations;
- slower processes, called spectral diffusion, which arise from drift and jumps of the resonance frequency due to dynamics in the molecule's environment.

If spectral diffusion is neglected, the homogeneous linewidth of a molecule is composed of two contributions, one from the lifetime T_1 of the excited state, the other from 'pure dephasing'. This process, often called 'decoherence' in quantum optics, arises from interactions with phonons and other dynamical modes activated at finite temperatures. One therefore writes the full width at half maximum of the optical line, measured in units of angular frequency ω, as

$$\gamma_{\text{hom}} = \frac{1}{T_1} + \frac{2}{T_2^*} \tag{6.10}$$

where T_2^* represents the 'pure dephasing' time, i.e. the lifetime of the coherence between the two electronic levels, determined by bath fluctuations at relatively low frequencies (phonons, etc.). At very low temperatures, the pure dephasing rate tends to zero, since no degrees of freedom are activated any more. In many molecular crystals, this limit is reached already at a few Kelvin, which means that the optical width is limited by the lifetime of the excited state only. For an allowed transition the lifetime is of the order of a few nanoseconds, corresponding to a width of 30 MHz, or one-thousandth of a cm^{-1}. When coupling to both phonons and intramolecular vibrations is considered, the absorption spectrum of a single molecule consists of a single very sharp line at the lowest frequency, the ZPL of the 0–0 transition. At higher energies, we find the phonon wing PW of the 0–0 transition, the ZPL of 0–1, 0–2, ... transitions of various intramolecular vibrations, followed by their PWs. Note that the vibronic ZPLs are much broader than the pure electronic ZPL because of the width of the vibrational levels, determined by the lifetime of the molecular vibration, rarely longer than a few picoseconds. Each intramolecular vibration mode gives rise to its own vibronic progression, ZPLs, and PWs, but only the pure electronic ZPL is common to all modes.

6.4.3   Inhomogeneous broadening

In usual samples, billions of molecules are observed at the same time. Because of defects and disorder, the electronic transition of each individual molecule is shifted with respect to the average value. This phenomenon is called inhomogeneous broadening

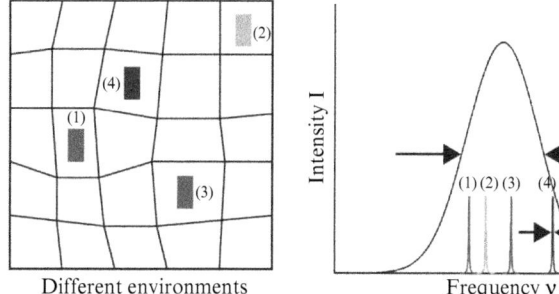

Fig. 6.4 Inhomogeneous broadening of optical lines in disordered solids. The different nano-environments around different molecules give rise to shifts of the optical transition over the distribution profile with width Γ_{inh}. The homogeneous line of each individual molecule is much narrower (width Γ_{inh}). These narrow lines can be used for hole-burning spectroscopy and for photon-echo experiments. (Figure created by N. Verhart, Leiden University.)

(for a schematic rendering, see Fig. 6.4). It is nearly independent of temperature, because it is determined by structure, and structure is nearly frozen at low temperatures (at least as long as the matrix is solid). The size of this inhomogeneous broadening depends strongly on the quality of the sample and on the way in which the impurity molecules are embedded in the matrix. For optical transitions of molecules, the inhomogeneous width varies from about $300\,\text{cm}^{-1}$ ($10\,\text{THz}$) in a polymer or in a frozen solution, to less than $3 \times 10^{-2}\text{cm}^{-1}$ ($1\,\text{GHz}$) in unstressed sublimation-grown crystals, which are the best molecular crystals one can grow. A typical value of inhomogeneous broadening for substitutional impurities in a low-quality molecular crystal is about $10\,\text{cm}^{-1}$. Therefore, in all cases, the inhomogeneous width is several orders of magnitude broader than the lifetime-limited linewidth of a single molecule at low temperature. The broad absorption profiles observed in ensemble experiments result from the superposition of the many randomly shifted narrow lines of individual molecules.

Comparatively little is known about the inhomogeneous distributions and the associated lineshapes. They are determined by the local structure around the impurity. In many cases, the local structure is relatively well defined, and can be considered as perturbed by many independent defects, such as vacancies, grain boundaries, or dislocations in crystals. Assuming the contributions of such defects to be approximately equal, and applying the central limit theorem, we then expect a Gaussian inhomogeneous lineshape. However, other models give different shapes. For example, a uniform distribution of defects interacting with an inverse-cube dependence on distance (dipole–dipole interaction) gives rise to a Lorentzian lineshape. In general, however, the inhomogeneous distribution is neither Gaussian nor Lorentzian, and often is asymmetrical.

In crystals, a guest molecule often may occupy several imbedding positions, called insertion sites. These give rise to multiplets of ZPLs in absorption spectra or in fluorescence spectra recorded with broad excitation, or excited at high energies (several thousands of wavenumbers) above the 0–0 transition. Each one of these sites corresponds to molecules with slightly different atom configurations, giving rise to slightly different vibrational spectra, lifetimes, etc. A particularly important case is that of Shpol'skii matrices, which are crystals of linear n-alkanes. Their crystal structure is layered with the long axis of the molecules nearly perpendicular to the layer. Narrow Shpol'skii lines are often obtained when the guest molecule matches the holes left in the crystal when one or a few host molecules are removed. Shpol'skii systems are of great current interest and are useful in molecular spectroscopy.

6.4.4 Hole-burning

Spectral hole-burning [30] is a nonlinear optical technique applied to large ensembles of molecules, in which a spectral feature (the hole) results from modification by light of the optical properties of a material. Let us consider an ideal hole-burning experiment at zero temperature. An ensemble of molecules absorbing with very narrow homogeneous lines (ZPLs) but a broad inhomogeneous distribution of optical resonance frequencies is irradiated with a monochromatic (or very narrow) laser. In a first approximation, if the laser intensity is not too high, only the resonant molecules will absorb light. The other ones do not see the laser. Now, an excited molecule may undergo a number of possible photophysical and photochemical processes. For example, the molecule can be temporarily stored in the triplet state, with a different (usually lower or zero) absorption. During the lifetime of the triplet, the sample will absorb less at the frequency of the illuminating laser. If an absorption spectrum of the sample is measured, it will show a hole—a transient spectral hole—at the frequency of the laser. But a molecule can also enter a much longer-lived dark state. Some of these product states may have very long (for all practical purposes infinite) lifetimes, in particular if the molecule undergoes a (photo)chemical reaction. If the spectral hole is permanent, one calls the process persistent spectral hole burning (PSHB).

Molecules in solid matrices at low temperatures can undergo a number of different processes leading to photochemical and photophysical hole-burning. Photochemistry may involve electron transfer, proton transfer, or large rearrangements of atoms. Photochemical hole-burning generally leads to very large shifts of the photoproducts, which means that products of the reaction do not absorb in the same spectral regions as the reactants (in other words, no antiholes can be found in the spectrum). In photophysical processes, the conformation of neighbouring atoms or groups of atoms is modified by illumination, leading to a shift of the absorption line, much larger than the homogeneous width, but usually smaller than the inhomogeneous bandwidth. The resulting antihole is much broader than the hole, and can only be measured after very deep and broad holes have been burned. Photophysical hole-burning is also called 'light-induced spectral diffusion' in the context of single-molecule spectroscopy.

The experimental method used to burn persistent holes is very simple, since the holes are long-lived. A single tuneable narrowband laser suffices. It is first used to irradiate the sample somewhere in the inhomogeneous bandwidth. Then, the laser is scanned to record an absorption spectrum, revealing the hole. Contrast enhancement methods can be used to detect narrow holes, for example by holography, polarization, or lock-in detection methods. The hole-burning method is very useful in molecular spectroscopy (determination of lifetimes and vibrational spectroscopy), in the spectroscopy of molecules or ions with spin-multiplet states (via the analysis of satellite holes), for the study of dynamical degrees of freedom in solids at low temperatures (notably glasses), and for investigating the effects of external perturbations such as electric and magnetic fields, pressure, and stress. Being very narrow, a hole enhances the sensitivity of these experiments by several orders of magnitude as compared with bulk experiments. In the past couple of decades, hole-burning materials have also been proposed as optical memories. A hole-burning sample can be seen as a photographic plate sensitive to several millions of different colours. However, practical difficulties, such as the low temperatures required and the problem of burning while reading, have limited their applications so far. Current research aims at using such media as analysers and processors of optical information, which could treat many different wavelengths in a massively parallel way, or as nonlinear media for photonics applications [31].

6.4.5 Single-molecule spectroscopy

We now consider a host–guest system in which molecules are very stable, i.e. in which hole-burning is inefficient or unlikely. If we focus our laser on a very small volume of sample and move it spatially, we will start to see statistical fluctuations of the number of molecules in resonance with the laser. If we scan the frequency of our laser at a fixed spot in the sample, we will see characteristic fluctuations of the optical absorption or of the total fluorescence of the sample. These fluctuations were first detected by Moerner in 1987 and are called statistical fine structure [32]. Analysis of this structure provides the homogeneous width of the molecules. In the same way that fluorescence correlation spectroscopy reveals the average diffusion time in an ensemble of molecules, the autocorrelation of the statistical fine structure provides a peak whose width is related to the homogeneous linewidth. As illustrated in Fig. 6.5, if we further reduce the focal volume and/or the concentration, the relative amplitude of the statistical fine structure increases, since it scales with the inverse square root of the average number of molecules in the focus. In the regime where the average number of molecules is less than unity, the spectrum ideally consists of a set of resolved sharp lines on a low background. Each single peak corresponds to the absorption line of a single molecule.

The first optical signal of a single molecule was detected in 1989 by Moerner and Kador with a complex method, involving a double frequency modulation (the laser frequency and molecular resonance frequency were modulated at different frequencies) of the absorption of a thin sample of pentacene in a p-terphenyl crystal. Although this method is very sensitive, its application to a single molecule is difficult because of photon noise and of optical saturation. The absorption signal is measured as a weak variation of the intensity of the transmitted beam. In order to reduce the relative photon noise on that beam, the intensity has to be large, and consequently the molecular signal saturates (in other words, the absorption cross section of the molecule decreases). Orrit and Bernard showed in 1990 that a fluorescence excitation method provides a much better signal-to-noise ratio. Since then, the fluorescence excitation

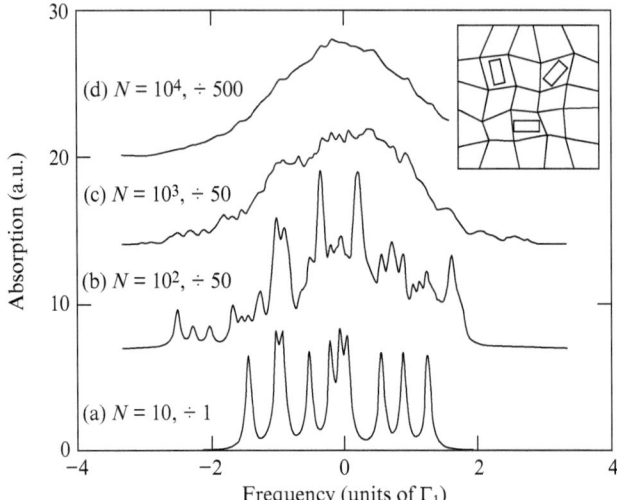

Fig. 6.5 For a large number of molecules (as in (d), for 10^4 molecules), the absorption spectrum appears smooth and reproduces the inhomogeneous band profile. At lower and lower numbers of molecules, spectral fluctuations start to appear, until single-molecule zero-phonon lines are completely resolved (as in (a), for 10 molecules). The natural width of the ZPLs has been greatly exaggerated here for the purpose of illustration. (Adapted from Orrit M. & Moerner, W. E. High-resolution single-molecule spectroscopy in condensed matter. In *Physics and Chemistry at Low Temperatures*, ed. L. Khriachtchev. Pan Stanford, Singapore, 2011, pp. 381–417.)

method has been improved and generalized. It was first used for low-temperature spectroscopy, then for microscopy at room temperature.

6.5 Experiments based on fluorescence excitation spectroscopy

We will not review the whole wealth of new results brought by single-molecule spectroscopy at low temperatures during the past two decades, since such reviews have been published already [33–36]. We give only a brief summary of the different types of experiments that have been done with fluorescence excitation of single molecules.

6.5.1 Two-level system in a laser field

For the discussion of the ZPL of the pure electronic 0–0 transition, we can forget about vibrations and phonons. We consider our molecule as a two-level system, coupled on the one hand to the exciting laser field and on the other to baths causing relaxation:

- the empty photon modes responsible for spontaneous emission;
- the (usually thermally populated) bath of phonons and low-frequency degrees of freedom responsible for coherence loss, i.e. for the decoherence time (often called pure dephasing time in the earlier literature).

This problem is perfectly analogous to that of a spin-$\frac{1}{2}$ in a static magnetic field (the optical excitation frequency replaces the Zeeman splitting), submitted to an oscillating electromagnetic wave (the laser field replaces the radio wave or microwave used to achieve the magnetic resonance for a spin-$\frac{1}{2}$ in nuclear magnetic resonance or electron spin resonance). This analogy can be demonstrated rigorously from a density-matrix description of the two-level system (a theorem due to Feynman, Vernon, and Hellwarth).

Just as in magnetic resonance, the state of the two-level system can be represented as the Bloch vector, a vector restricted to the Bloch sphere. The population difference between the ground and excited states, and the real and imaginary parts of the coherence (i.e. the off-diagonal element of the density matrix) can be seen as the three components of a three-dimensional vector in a fictitious space, the Bloch vector. If there is no relaxation, the Bloch vector remains on the sphere, because it is subject to only rotations. There are two possible rotation vectors in the rotating frame (in magnetic resonance, the reference frame rotating along the static magnetic field at the angular frequency of the oscillating magnetic field):

- a rotation around the vertical axis, at a rate given by the detuning between the laser frequency and the resonance frequency of the molecule,;
- a rotation around a horizontal axis at a rate given by the Rabi frequency, proportional to the laser electric field and to the transition dipole moment of the molecule.

The ground state corresponds to the lowest point of the sphere, and the excited state to the highest point. The points on the vertical axis represent statistical mixtures of ground and excited states, and the points on the equator circle represent coherent states with equal admixtures of ground and excited states.

Coupling a two-level molecule to a bath leads to two qualitatively different processes:

1. **Non-adiabatic processes,** in which the bath exchanges energy with the system. The bath must act via fluctuations at the transition frequency, i.e. at very high frequencies. In the case of optical transitions, the only process is population relaxation from the excited to the ground state. This relaxation can occur via spontaneous emission or via non-radiative relaxation, in which the electronic energy is transformed into heat (usually into high-frequency vibrational modes of the molecule). The population relaxation rate is usually nearly independent of temperature, of the matrix, and of the insertion sites.

2. **Adiabatic processes,** in which the bath slightly changes the transition energy between the two levels, but does not induce transitions between the levels. In that case, the bath acts via its low-frequency fluctuations. This frequency is so low that it is in general a good approximation to use a classical model for the bath (i.e. a model in which bath fluctuations are treated classically). The slight fluctuations of the transition frequency under bath fluctuations lead to a loss of memory of the phase, i.e. to a damping of the Bloch oscillation, and therefore to a broadening of the optical line. This is the origin of the decoherence (or pure

dephasing) contribution to the homogeneous width. The major cause of adiabatic relaxation for molecules in solids is coupling to the phonons, mainly the acoustic phonons, which are still populated at low temperatures, and the low-frequency optical phonons, which are strongly coupled to the optical transition.

The Bloch relaxation times in the optical case are quite similar to those defined for the magnetization.

The relaxation time T_1 is the excited-state lifetime (analogous to the magnetic longitudinal relaxation time). It describes how fast the population relaxes to equilibrium—in this case to the ground state. In general, in contrast to magnetic resonance, there is no relaxation for the ground state in optical two-level systems (the temperature is much lower than the optical transition energy).

The relaxation time T_2 is the coherence lifetime (analogous to the magnetic transverse relaxation time). It is the inverse of the homogeneous width. As we have seen earlier, this time includes a contribution from the excited-level lifetime and a contribution from the inverse of the rate of decoherence (or pure dephasing), due to adiabatic processes.

If we now introduce relaxation into the dynamics of the Bloch vector, we find, in addition to the coherent evolution caused by the laser field, two new processes. The Bloch vector relaxes at the decoherence rate towards the vertical axis of the sphere. In addition, it also relaxes towards the ground state, i.e. the lowest point of the sphere, at the longitudinal relaxation rate. When relaxation is present, the oscillating or rotating movements of the Bloch vector are damped. The Bloch vector relaxes towards the steady-state solution, which is constant in the rotating frame. The differential equations describing the movement of the Bloch vector under the coherent field and with relaxation are called the Bloch equations.

We first discuss optical saturation by looking for steady-state solutions of the optical Bloch equations. This steady state is reached in the frame rotating at the frequency of the laser oscillation. This means that the laser induces a forced oscillation of the system (the system accompanies the laser oscillation with a phase shift and an amplitude that vary according to the laser frequency). The steady state is reached when the variations of the Bloch vector due to relaxation are compensated by the variations due to precession in the laser field. For example, in the case of a resonant laser field, the steady-state solution deviates from the ground state, which means that some coherence is established by the laser. The vertical component of the Bloch vector gives the excited-state population, which for weak powers is proportional to the fluorescence intensity, but which saturates for large intensities. This optical saturation manifests itself by the limitation of the fluorescence or absorption, and by the broadening of the optical line. At high power, it is possible to saturate the system, even with a non-resonant laser. This phenomenon is also called power broadening.

Optical saturation was observed long ago on the very sharp optical lines of atoms. For molecules in the solid state, however, such experimental observations were difficult because of the broad inhomogeneous profile. One could in principle observe saturation via spectral hole-burning, but many other sources of saturation (e.g. the saturation due to the photochemical hole-burning process) mask the optical saturation of the two-level system. Single-molecule observations, in contrarst, do not suffer from chemical saturation effects, and directly display optical saturation.

A second important nonlinear effect observed with single molecules is optical nutation. The dynamics of an optical two-level system in a rapidly varying laser field can often be discussed qualitatively by means of the Bloch vector picture. We examine an important example of a transient solution of the Bloch equations, optical nutation. A constant field amplitude is suddenly applied at time zero. The system goes from the ground state at time zero (before the field was applied) to the steady state under constant illumination at long times. In the rotating frame, the system starts from a downward-pointing Bloch vector (ground state) and begins to precess on a cone around the axis defined by the effective magnetic field resulting from the laser detuning (vertical component) and the laser field amplitude (Rabi frequency, transverse component). If no damping is present, the precession continues at the same rate as long as the laser field remains constant. If damping is present, the transient motion is damped with a characteristic time depending on the Rabi frequency and on the relaxation times, and the Bloch vector tends towards the steady-state vector. A very direct way to observe this transient motion is to select a single optical two-level system, such as a single molecule. The measured quantity is fluorescence, which is proportional to the population of the excited state. Because each observation of a photon means that the molecule has just reached the ground state, the observation projects the molecule into the ground state (in the quantum mechanical sense). Therefore, the system starts from the ground state after each photon has been detected, which is exactly the same problem as suddenly applying a constant laser intensity to a system previously in the ground state. Thus, the probability per unit time of observing the next photon (after each photon observation) gives us the population of the excited state with the ground state as initial state. The histogram of time delays between consecutive photons is a picture of the time evolution of the excited-state population during the optical nutation. As we discussed earlier, this histogram cannot be measured directly with a single detector because of the dead-time of the detector. It is measured in coincidence measurements between two detectors (in a Hanbury-Brown and Twiss setup), each of them detecting photons in a split beam. The dip of the distribution for short times is a signature of the quantum nature of the light emitted by a single molecule. It would be ruled out in a classical description of light (because the correlation of a classical function of time must always be maximal for zero time).

A nutation transient at low temperature often shows oscillations, because the laser is resonant with the transition of the two-level system. Therefore, the laser induces not only absorption, but also stimulated emission (a schematic way to visualize Rabi oscillations is to consider the interplay between these two effects). In room-temperature experiments, in contrast, the molecule is excited via a vibronic band, so that the laser is not resonant with the emission. Therefore, no Rabi oscillations can appear. Moreover, the damping time of these oscillations at room temperature would be of the order of some tens of femtoseconds, much too short to detect in antibunching plots.

A final quantum optical effect is the light shift, or ac-Stark effect. As is well known, the transition frequency of an atomic or molecular system can be shifted by a static electric field (the static Stark effect). A light wave can also cause level shift. This shift (called the light shift) is usually very weak, but it can be dramatically enhanced by resonance. For the alternating electric field of a laser wave, the shift is quadratic, i.e. proportional to the intensity. It has been measured by recording a single-molecule line with a weak probe beam, while a strong pump beam is used to illuminate the molecule. The results are in full agreement with the optical Bloch equations.

6.5.2 External field effects

Just as with spectral hole-burning, the narrow lines of single molecules provide a vastly enhanced sensitivity of optical signals to external perturbations such as electric or magnetic fields, mechanical pressure and stresses, etc. The unique advantage of single molecules for such probing of external actions is that a molecule reports local conditions in its surroundings, free of ensemble averaging. This can be useful in many cases in nanoscience, for example probing trapped charges in organic conductors [37, 38].

6.5.3 Optically detected magnetic resonance and other quantum optical experiments

Electron paramagnetic resonance (EPR) is a powerful technique to explore the electronic structure of materials. However, this technique requires billions of spins to provide measurable magnetic resonance signals. Optically detected magnetic resonance (ODMR) transfers the sensitivity of optical measurements to the magnetic domain and enhances the sensitivity of EPR by orders of magnitude. The optical detection of a single molecule has made it possible to detect a single electron spin with high signal-to-noise ratio [39, 40]. This experiment has recently been extended to single colour centres in diamond, which have paramagnetic ground and excited states. Using the electron spin as a probe for neighbouring nuclear spins, several groups have proposed a broad range of quantum optical manipulations on nuclear spins, ultimately detected through changes in optical signals [41–43].

A single molecule is also a well-defined electronic system, in which the quantum state can be manipulated. Single quantum systems can be applied to deliver single photons on command [44] for a variety of experiments, from near-field imaging [45] to photon number interference [46].

6.5.4 Spectral diffusion

Spectral diffusion of a single-molecule line results from changes in the molecule's optical resonance frequency. These changes are themselves caused by slow motion or transitions in the molecule's neighbourhood, with ranges varying from nanoseconds (comparable to the fluorescence lifetime) to minutes or even longer. Some examples of spectrally diffusing molecules with widely different dynamics in the same samples are presented in Fig. 6.6 [47]. In such a wide time widow, many processes can contribute to spectral diffusion. Single molecules enable a study of these different processes on a truly local scale, without ensemble averaging. Among the most studied degrees of freedom have been the tunnelling systems of glasses and polymers at cryogenic

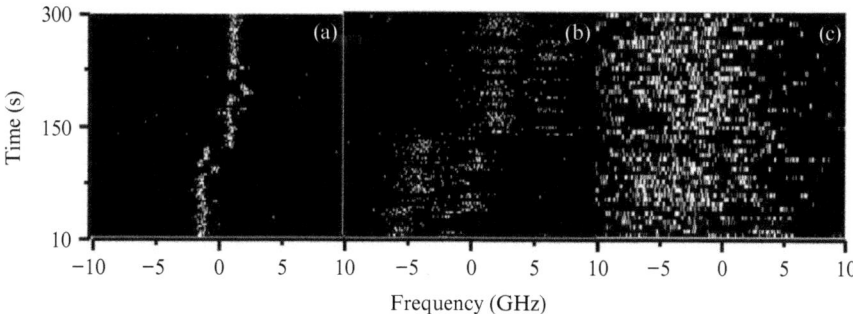

Fig. 6.6 Example of spectral diffusion traces measured for single terrylene molecules in a 2,3-dimethylanthracene crystal. The dynamics is attributed to methyl group conformational changes. (From Tian, Y., Navarro, P., Kozankiewicz, B., & Orrit, M. (2012). Spectral diffusion of single dibenzoterrylene molecules in 2,3-dimethylanthracene. ChemPhysChem, 13(15), 3510–3515. doi:10.1002/cphc.201200463, by permission of John Wiley & Sons.)

temperatures. Single molecules signals show random telegraph blinking assignable to switching of individual two-level systems [48], as well as more complex behaviour [49, 50]. Another well-studied example is the flipping of the central phenyl group of *o*-terphenyl in a crystal at low temperature, studied through the spectral diffusion of probe molecules such as pentacene [51] or terrylene [52]. These studies have proved that host molecules at crystal defects such as domain walls are particularly prone to conformational changes. Flipping phenyl groups leads to nearly independent two-level systems and to a specific pattern of spectral diffusion for the probe molecules [53, 54]. Many other slow degrees of freedom in low-temperature matrices can lead to spectral diffusion, such as subtle reorientations of methyl groups [55] or correlated molecular motions characteristic of incommensurate phases [56].

6.5.5 Interacting single molecules

In regular single-molecule experiments, excited molecules are very far from each other, so as to achieve spatial and spectral selection. Sometimes, however, molecules can be close enough to interact significantly and give rise to new coupled states. One such example is given by the chromophores of individual antenna complexes studied by Van Oijen et al. [57]. These authors found in the same individual complex weakly coupled states from the B800 ring as well as strongly coupled states demonstrating delocalized exciton states in the B850 ring. These systems have recently been studied by ultrafast single-molecule spectroscopy and have been found to exhibit quantum coherence even at room temperature [58]. Applying variable electric fields to single molecules gives access to a wider range of frequencies, and may help reaching the resonance condition for two neighbouring molecules. This effect was used by Hettich et al. [59] to study the nonlinear optical response of two terrylene molecules in a crystal and demonstrate their excitonic coupling. Yet another possibility to enhance the probability of finding coupled chromophores is to link them covalently. This technique has allowed Basché's group to study energy transfer between single donor and acceptor moieties [60].

This short overview shows the wide scope of single-molecule experiments at low temperature. Much remains to be done, in particular concerning the interaction of single molecules with plasmonic structures. For these experiments, the main advantage of low-temperature experiments is that a large number of distinguishable single molecules can be coupled to the same plasmonic structure.

6.6 Other detection methods

In this section, we review the main techniques used to detect individual nano-objects, in particular metal nanoparticles, through their effect on a probing laser beam. The object or particle can be observed in dark field, i.e., with a high contrast in a direction where no intensity is scattered in the absence of the particle. It can also be detected in bright field, by comparing scattered intensities with and without the particle in the beam. This latter scheme amounts to an interferometric detection of the particle. Finally, pump–probe methods exploit a change of the scattered probe wave which is conditional to pump excitation. Therefore, these methods have the power to selectively detect absorbing (e.g., metal) nanoparticles against any background of non-absorbing (dielectric, non-metallic) scatterers. These different optical techniques have become widespread in the past few years and given rise to a large body of work. Here, we only mention the main advances, stressing the underlying detection principles.

6.6.1 General considerations on signal, background, and noise

To detect a small object, usually less than 100 nm in size, in the optical far field, one should optimize its interaction with the light beam. This is achieved by placing the object at the focus of a spherical wave or of a Gaussian beam, or very close to this focus. Indeed, the sensitivity optimum is not always exactly at the wave focus. To a good approximation, a small object in an optical field behaves as an oscillating dipole, radiating a scattered wave. Superposition of this scattered wave with the incoming wave leads to measurable intensity changes in different spatial regions. Here, we briefly discuss the three main ways to measure these changes and to detect and study individual particles.

6.6.1.1 Dark field

The relative change of intensity is strongest in directions where the incident wave, in the absence of a particle, would not produce any intensity. The wave scattered by the particle will thus contribute a non-zero intensity appearing on a dark background. This is called dark-field scattering. For ideal conditions, it can, in principle, provide an arbitrarily large signal-to-background ratio. This property is very similar to that of fluorescence, because no fluorescence background exists when no fluorescent object is present. In practice, however, these ideal low-background conditions are very difficult to achieve for Rayleigh scattering, because small defects in the substrate, impurities, or small structures also scatter light in all directions. This problem is particularly serious in biological cells, because of scattering by organelles, vesicles, and other small cell structures.

6.6.1.2 Bright field

The wave scattered by a small object also has an amplitude in the direction of the transmitted wave, and thereby interferes with it. In the forward direction, the interference always reduces the intensity of the incident wave, leading to extinction of the incident wave. The relation between extinction and the scattering amplitude is given by the optical theorem in scattering theory [61]. The optical theorem expresses the conservation of energy: the energy removed from the incident wave is either locally absorbed or scattered away. There, it appears as new scattered waves and as a simultaneous weakening of the transmitted wave. Local absorption can give rise to heat release or to re-emitted radiation at different (usually lower) frequencies, i.e. as fluorescence or photoluminescence. A similar interference effect occurs in other propagation directions of the incident wave, particularly when this wave is reflected by an interface or scattered by a bigger object. The following reasoning thus applies to those cases as well.

Let us discuss the signal-to-noise ratio in such an extinction experiment. The detected intensity change ΔI results from the sum of a reference (incoming) field r and a small scattered field s:

$$\Delta I = |r + se^{i\phi}| \approx 2\,\mathrm{Re}(r^* se^{i\phi})\,, \tag{6.11}$$

where ϕ is the phase difference between the incident field and the scattered field at a large distance, taking all phases into account (including propagation, Gouy shifts, and the phase shift due to the polarizability of the particle). For example, a pure absorber placed at the focus of a plane wave has a dipole in quadrature with the incident field. The spherical wave it scatters introduces an additional Gouy phase of $\pi/2$. The combination of these shifts gives rise to destructive interference between transmitted and scattered waves, attenuating the incident wave. In dark-field scattering, there is no reference field and no background, and the signal-to-noise ratio can in principle be arbitrarily large. Note, however, that the intensity scattered by a small object is very weak and is easily drowned by background from experimental imperfections. In a bright-field experiment, the scattered intensity $|s|^2$ is negligible for a small object. The signal arises mostly from the interference term $2rs\cos\phi$ (for real reference and scattered fields), which has to be detected against the background $|r|^2$ of the intensity transmitted or reflected by the substrate. This bright-field difference signal therefore scales as the amplitude of the scattered field, i.e. as the volume of the scatterer. For measurements limited by photon noise, the noise on the background scales as the square root of the background, and therefore as the amplitude of the reference field, just as the signal does. Therefore, the signal-to-noise ratio is independent of the intensity of reference field. This useful property allows one to adapt the experimental configuration to available sources and detectors. Choosing a weak reference field (e.g. a weak reflection from the interface between two nearly index-matched media [62]) is a convenient means to suppress laser noise (which scales as $|r|^2$) without losing signal-to-noise ratio. A similar reduction can be achieved by means of an interferometer [63], but at the cost of a more complex setup and more difficult adjustments.

6.6.1.3 Absorption detection

The dissipation of light energy by the single object, due to absorption processes, may give rise to a number of secondary effects, including local heat production, luminescence, and other emissions. These effects can also be used to detect true absorption, as opposed to extinction, and provide a means to distinguish metal nanoparticles from dielectric scatterers. As we have seen already, fluorescence or photoluminescence is a highly selective and sensitive way to detect absorbers, but heat-induced, photothermal, and pump–probe effects are also useful to detect metal particles with a high selectivity. The signal-to-noise ratio in such experiments will depend on the sources of background, which can be very low in luminescence or photothermal measurements. In the following subsections, we successively discuss these three large classes of methods with experiments from the past few years on small metal nanoparticles.

6.6.2 Dark-field scattering and total internal reflection

Dark-field microscopy was first used by Zsigmondy around 1900 to study the scattering of suspensions of gold nanoparticles with his 'ultramicroscope'. His observations helped clarify the structure of colloidal suspensions and convince the last skeptics of the reality of atoms and molecules [64]. In 1998, Feldmann's group published the first experiments in the near field [65], while Yguerabide et al. [66] studied individual nanoparticles in the far-field in the context of labelling in biological applications. Studying the dark-field scattering of silver nanoparticles, Schultz pointed out the high interest of nanoparticles of different sizes and shapes for multicolour labelling [67].

Light scattered by a small object can be detected against a dark background in a variety of ways, often also used in fluorescence microscopy. Microscope companies offer dark-field condensers with associated objectives to specifically observe scattering objects. Alternatively, high-NA objectives can accommodate annular illumination with detection in the centre of the pupil, or vice versa. Total internal reflection offers a convenient way to illuminate particles when they are situated close to an interface on the low-index side. Scattered light can be observed on that same side with very low background. The main advantage of dark-field scattering is its ease of operation, because it requires only a standard microscope with a regular objective. Moreover, it also allows for wide-field imaging, and makes it possible to follow hundreds of particles in parallel at the same time. Scattering spectra are observed directly, without any need for background subtraction or post-processing. Here, we mention some recent results obtained by dark-field scattering of metal nanoparticles. Soennichsen et al. [68] used dark-field scattering images to obtain the spectra of nanorods and nanospheres of various shapes and sizes. These spectra are free from the large broadening resulting from inhomogeneity in ensemble experiments. Soennichsen et al. demonstrated a spectacular reduction in plasmon width for rods and attributed the main channel of plasmon damping to relaxation towards electron–hole pairs for small particles. They recognized the importance of radiative damping for large particles. The radiative channel broadens the plasmon resonance for large sizes, whereas surface-induced broadening becomes noticeable for small sizes, 10 nm and lower, so that a minimum in broadening occurs at about 30 nm.

Dark-field scattering in combination with scanning electron microscopy (SEM) reveals the plasmon bands of more complex assemblies of nanoparticles such as the pairs of nanorods studied by Funston et al. [69]. Various relative positions of the rods give rise to hybridized or coupled plasmon bands, in agreement with calculations. The polarization of scattered light can be obtained by a rotating polarizer. With this technique, a large number of silver and gold nanoparticles can be monitored in parallel during their growth. Dark-field scattering is a very versatile method allowing many correlated measurements in parallel, thanks to its ease of operation. Its limitation, however, is that the scattered intensity scales as the squared volume of the nanoparticles. Therefore, for small particles, dark-field scattering is easily overcome by signals from dielectric impurities, substrate roughness, or even thermodynamic fluctuations of the medium. Its application to small objects requires very clean substrates, and often reaches its limits around 30 nm for gold nanospheres under standard conditions of optical quality and cleanliness.

6.6.3 Absorption, extinction, and interference-based methods

The reduction in transmitted intensity due to a small object is called extinction. In contrast to the dark-field scattering signal, the extinction signal scales with the amplitude of the field scattered by the object, rather than with its intensity. However, this signal has to be detected against a strong background of transmitted light. Similarly, a change in reflection intensity scales with the field amplitude and therefore the volume of the particle, but has to be detected against the background of light reflected by the interface.

The first discussion of this interference effect for a single molecule in reflection was given by Plakhotnik and Palm [70] for a low-temperature spectroscopy experiment. The spectral shape of the interference line can change from absorptive to dispersive, depending on the phase difference between the two fields. This difference was varied by choosing molecules at different distances from the interface. The authors proposed to extend this technique to non-fluorescent objects. Lindfors et al. [62] studied the plasmon resonance of a single gold nanoparticle immobilized on a surface and surrounded by immersion oil with a similar reflection technique. For large particles, the scattered intensity dominated the weak reflection from the nearly index-matched glass–oil interface, but for small particles the reflection was stronger. Destructive interference of the scattered field led to reduced reflection, enabling detection of particles down to 10 nm in diameter against the weak reflected background. The reflection of a water–glass interface was used as the reference signal by Jacobsen et al. [71] to demonstrate fast and wide-field detection of gold nanoparticles immobilized on a glass surface but in contact with an aqueous phase, which is interesting for the detection of biomolecules. Metal nanoparticles were distinguished from dielectric scatterers such as particles or substrate defects by their plasmon resonance probed at two different wavelengths. This scattering method in transmission or reflection has recently been improved so as to reach the threshold of a single molecule's absorption [72]. The main limitation of the scattering method, however, is not its sensitivity. Different principles of subtraction have been used to extract the small extinction signal from the strong transmitted or reflected background. A more serious obstacle to this method of detection of small particles is the lack of specificity of their scattering signals against those from other scatterers. These can be dielectric particles in suspension; impurities, corrugations, or refractive-index inhomogeneities of the substrate; or organelles in biological cells. To distinguish these from metal nanoparticles, one needs specific signals arising from the absorption itself instead of the extinction. As explained in the next subsection, pump–probe methods do provide such specificity.

6.6.4 Pump–probe and photothermal detections

Pump–probe detection techniques achieve selectivity by measuring only that part of the scattered probe signal which is conditional to the presence of the pump. Whereas this usually entails a significant loss in signal amplitude, it has the considerable benefit that only resonantly absorbing objects, such as metal nanoparticles, semiconductor nanocrystals, or dye molecules, have sufficient nonlinear susceptibilities to produce a sizeable signal. Cross talk between pump and probe beams induced by the nonlinear $\chi^{(3)}$ susceptibility of transparent dielectrics is negligible at the intensities used in microscopy of absorbing objects. In pump–probe techniques, one measures the change in intensity of the transmitted or reflected probe beam caused by a pump beam. The wave measured is a superposition of the scattered probe field with the transmitted or reflected probe beam, as explained in Section 6.6.1.2. Now, however, the subtracted intensity is not that of the probe beam without the particle, but rather the probe intensity without pump excitation. This subtraction is usually done by modulating the pump power at high frequency (often around 1 MHz) and detecting the change in probe intensity with a lock-in amplifier. Any nonlinearity of the $\chi^{(3)}$ type will give rise to a change in the scattered probe field due to the presence of the pump, by a signal proportional to the probe amplitude and to the pump intensity (or squared amplitude). The physical origin of the nonlinearity can be the heat dissipation due to pump absorption (photothermal effect), an electronic response (electronic pump–probe, ground-state depletion), or a vibrational response (vibrational pump–probe, stimulated Raman scattering).

6.6.4.1 Pump–probe

For excitations at the plasmon frequency of a metal particle, the pump mainly excites conduction electrons. The energy dumped into these electronic degrees of freedom launches transient responses of, in turn, the collective plasmon oscillations, then incoherent motions of conduction electrons, and then phonons and vibrations of the nanoparticle as a whole. Finally, the energy is dissipated in the environment and diffuses away from the particle. The activation of these degrees of freedom takes place on different timescales, determined by their coupling strengths. Plasmonic excitations are damped in less than 10 fs, the electron bath equilibrates with the lattice within a few picoseconds, the nanoparticle vibrations are damped in some hundreds of picoseconds, and heat dissipation into the environment proceeds between nanoseconds and microseconds, depending on the lengthscale considered. Depending on which optical technique is used and on the timescale, these different degrees of freedom can contribute to a pump–probe signal.

We first discuss pulsed experiments with 200 fs pulses. This pulse duration is too long to resolve the nonlinearity of plasmon oscillations, but it clearly shows a prompt optical response of hot electrons, which decays within a few picoseconds. This response arises from the transient distribution of hot electrons and holes, which causes changes in the real and imaginary parts of the optical permittivity. It is strong enough to allow for detection of gold nanospheres down to 10 nm in diameter. With a laser repetition rate of 80 MHz, however, the signal-to-noise ratio is much lower than that of photothermal detection. Indeed, because of the low duty cycle, many fewer photons are absorbed and used as probes in the pulsed experiments. The advantage of these pulsed experiments, however, is that they give access to the time-resolved response up to 12 ns, and thereby probe vibration frequencies and damping rates. Single-nanoparticle studies have the unique advantage of revealing full distributions of shape and size as well as rare cases of composed particles such as dumbbells, and of providing access to intrinsic damping mechanisms, which are often hidden by inhomogeneity in ensemble measurements. Optical probing of vibrations of single metal nanoparticles has been reviewed in [73]. In metal nanospheres, because the isotropic heating process does not break the spherical symmetry, only radial vibration modes are efficiently excited. The fundamental normal mode and higher-order modes (overtones) can be observed [74]. A non-radial extensional mode appears in the pump–probe trace when the spherical symmetry is broken, either by the shape in slightly elongated particles or by the contact with another spherical particle in pairs of touching nanospheres (dumbbells) [75]. In the dumbbell, optical excitation also leads to low-frequency oscillations of the centres of the two component particles around their equilibrium positions, resembling the stretching mode of a diatomic molecule. Transient vibrational traces of gold nanorods [76] mainly show the breathing mode and the extension mode at a much lower frequency, as the example of Fig. 6.7 shows [77]. The ratio of their frequencies depends on the rod's shape and gives information about the elastic moduli of the single-crystal rods. Silver nanocubes have been studied by Staleva et al. [78]. Again, the fundamental breathing mode is the dominant one in the traces, with sometimes non-radial modes at lower frequency, possibly arising from a deviation from cubic symmetry. The origin of the damping of mechanical vibrations of gold nanoparticles has been explored by experiments in optical traps [79], and confirmed the importance of intrinsic damping [80] pointed out in ensemble experiments on high-quality samples [81].

6.6.4.2 *Photothermal contrast*

In photothermal microscopy, the pump beam is again absorbed by the particle to be detected, whereas the probe beam monitors the change in index of refraction in

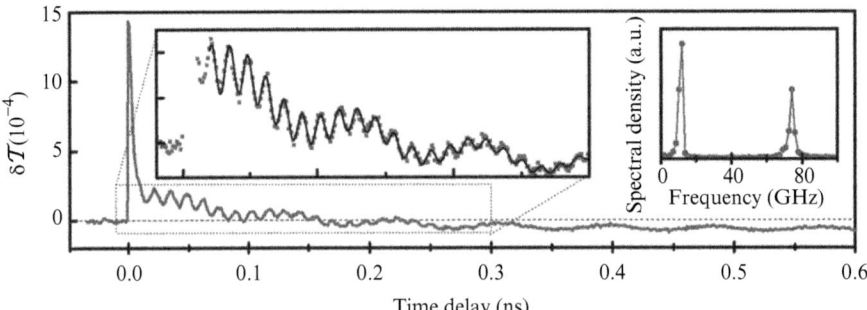

Fig. 6.7 A pump–probe time trace of a single gold nanorod (25 nm × 60 nm), showing the rapid electronic response in the first few picoseconds, followed by coherent oscillations of the diameter (high frequency) and length of the rod (low frequency). (Reprinted with permission from Ruijgrok, P. V., Zijlstra, P., Tchebotareva, A. L., & Orrit, M. (2012). Damping of acoustic vibrations of single gold nanoparticles optically trapped in water. Nano Lett., 12(2), 1063–1069. doi:10.1021/nl204311q. Copyright 2012 American Chemical Society.).

the environment of the absorber. As the probe wavelength is usually chosen to be outside the absorption range of the particle, it is advantageous to increase the probe intensity as much as possible, to reduce photon noise. The heating beam is modulated at a high frequency in the megahertz range, and the ensuing variations of scattered intensity of the continuous-wave probe beam are detected with a lock-in amplifier. The photothermal signal thus makes use of a weak effect, the change in refractive index of the surrounding medium with temperature, but as it accumulates the contributions of many photons, both for pumping and for probing, it can still achieve excellent signal-to-noise ratios. The method has been implemented in microscopy in different ways, first on silver nanoparticles in free solution by Matawari et al. [82], then in a polarization-sensitive configuration by Boyer et al. [83] in a confocal microscope. Berciaud et al. improved on this method and pushed the sensitivity to gold nanoparticles 1.4 nm in diameter [84]. The photothermal method has been adapted to wide-field microscopy by Absil et al. [85] by exploiting image subtraction at video rate, and has been pushed to single-molecule sensitivity by Gaiduk et al. [86] (Fig. 6.8).

Photothermal detection of small gold nanoparticles requires the absorption of many photons and, by its very principle, an associated heating. The ensuing temperature rise at the particle may be a limitation for probing applications in biology or soft matter studies. Moreover, even temperatures as low as 100°C can cause reshaping of gold nanorods and of complex nanoparticles. Therefore, a compromise must be found between particle size, integration time, and maximum allowed temperature rise. In glycerol, an absorbed power of 3 nW can be detected with a signal-to-noise ratio of 8 in an integration time of 10 ms, corresponding to a temperature increase of less than 0.1 K for a 20 nm gold particle [87]. Such high sensitivities allow the photothermal detection of single non-fluorescent molecules, as shown in Fig. 6.8. In water, this signal is reduced by a factor of about 5.

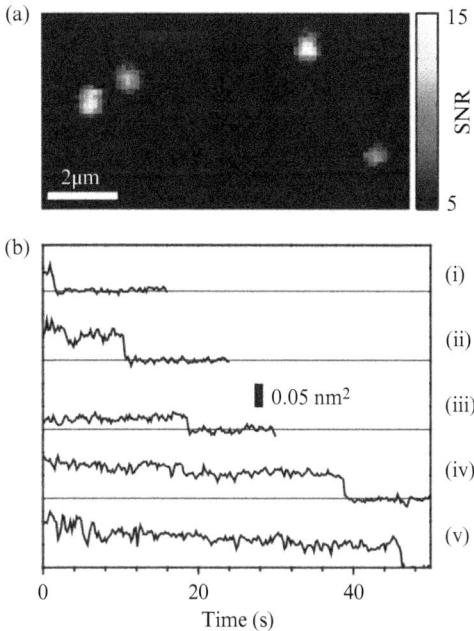

Fig. 6.8 (a) Photothermal image showing signals originating from single absorbing molecules. (b) Time traces recorded on some of the spots, showing one-step photobleaching after several seconds of illumination at high intensity. (From Gaiduk, A., Yorulmaz, M., Ruijgrok, P. V., & Orrit, M. (2010). Room-temperature detection of a single molecule's absorption by photothermal contrast. Science, 330(6002), 353–356. doi:10.1126/science.1195475. Reprinted with permission from AAAS.)

6.6.4.3   *Electronic response and ground-state depletion*

This detection technique closely resembles photothermal contrast, but relies on an electronic nonlinearity instead of a thermal one. A pump beam is modulated at high frequency (megahertz or higher) and the modulation of a confocal probe beam is detected with a lock-in amplifier. However, the probe frequency is now chosen in an absorption band of the object to be detected, which may be the same as the one addressed by the pump. The pump and probe frequencies should be far enough apart to be efficiently separated before detection of the probe. For a molecule, the depletion of the ground-state population by the pump obviously leads to a decrease in probe absorption. For a metal particle, the hot electrons and holes created by the pump change the optical response to the probe in a manner similar to the time-resolved experiments described above. Chong et al. [88] used this method to image immobilized gold nanoparticles (diameter 20 nm). Moreover, the sensitivity of the technique was enough to detect single Atto647 dye molecules. This technique will be a useful alternative to photothermal microscopy in cases where heating is undesirable or in media where temperature-induced refractive index changes are small.

6.7 Conclusions

In this chapter, we have described the main optical methods used for detection and studies of single small objects, from molecules to metal nanoparticles. Fluorescence is still the workhorse method of single-molecule detection, and has recently been augmented by fascinating applications in super-resolution imaging. Fluorescence-related methods have been immediately extended to a broad range of emitting objects, including semiconductor nanocrystals [89] and metal nanoparticles [90]. Although the photoluminescence yield of gold nanoparticles is very low, of the order of 10^{-6}, their coupling to light is dramatically improved by their plasmonic properties, generating very large absorption cross sections. Therefore, the fluorescence (or photoluminescence) signals of plasmonic particles can be comparable to or even stronger than those of fluorescent molecules. However, gold particles are much more stable and durable than molecules. Their main disadvantage is their rather large size.

In addition to fluorescence, during the last few years a number of new optical signals have been proposed, some of which we have reviewed here. The most direct of these are based on scattering in dark or bright field in linear optics. They are very general and experimentally simple, but often lack sensitivity and selectivity for the objects of interest. By relying on nonlinear spectroscopy, more refined methods can enhance selectivity through one or several resonant steps. One example is photothermal detection, which provides contrast for absorbing objects only, or the closely related methods of ground-state depletion and stimulated Raman scattering. Pump–probe spectroscopy gives time-resolved information, which can be exploited to obtain phase-sensitive information [91].

Among the new developments that have hardly been mentioned in this chapter is the coupling of optically active nano-objects to plasmonic nanostructures [92]. Large field enhancement in the vicinity of metallic nanostructures can lead to spectacular enhancement of the coupling to light of molecules and other nano-objects, in absorption as well as in emission. These lead to surface-enhanced Raman scattering of single molecules [93], tip-enhanced Raman scattering [94, 95], enhanced fluorescence by bowtie antennas [96], nanorods [97], and antennas in boxes [98]. Enhanced electric fields in metallic nanostructures will certainly raise many fundamental questions and find a broadening range of applications in the coming years.

Acknowledgements

We acknowledge financial support by ERC (Advanced Grant SiMoSoMa). Work presented in this chapter was done within the research program of the Stichting voor Fundamenteel Onderzoek der Materie, which is financially supported by the Netherlands Organization for Scientific Research (NWO).

Parts of this chapter have been published in Chapter 2 of *Photonics*, Vol. 4: *Biomedical Photonics, Spectroscopy, and Microscopy*, ed. D. L. Andrews. Wiley, Hoboken, NJ, 2015.

References

[1] Perrin, J. La fluorescence. *Ann. Physique* **9**, 133 (1918).

[2] Binnig, G., Rohrer, H., and Gerber, C. Surface studies by scanning tunneling microscopy. *Phys. Rev. Lett.* **49**, 57–61 (1982).

[3] Binnig, G. and Rohrer, H. Scanning tunneling microscopy—from birth to adolescence. *Rev. Mod. Phys.* **59**, 615–625 (1987).

[4] Novotny, L. and Hecht, B. *Principles of Nano-Optics.* Cambridge University Press, Cambridge, 2006.

[5] Fourkas, J. T. Rapid determination of the three-dimensional orientation of single molecules. *Opt. Lett.* **26**, 211–213 (2001).

[6] Fromm, D. P. and Moerner, W. E. Methods of single-molecule fluorescence spectroscopy and microscopy. *Rev. Sci. Instrum.* **74**, 3597–3619 (2003).

[7] Schmidt, T, Schütz, G. J., and Gruber, H. J. Local stoichiometries determined by counting individual molecules. *Anal. Chem.* **68**, 4397–4401 (1996).

[8] Schütz, G. J., Trabesinger, W., and Schmidt, T. Direct observation of ligand colocalization on individual receptor molecules. *Biophys. J.* **74**, 2223–2226 (1998).

[9] Deschenes, L. A. and Vanden Bout, D. A. Heterogeneous dynamics and domains in supercooled *o*-terphenyl: a single molecule study. *J. Phys. Chem. B* **106**, 11438–11445 (2002).

[10] Zondervan, R., Kulzer, F., Berkhout, G. C. G., and Orrit, M. Local viscosity of supercooled glycerol near T_g probed by rotational diffusion of ensembles and single dye molecules. *Proc. Natl Acad. Sci. USA* **194**, 12628–12633 (2007).

[11] Ha, T., Enderle, T., Chemla, D. S., Selvin, P. R., and Weiss, S. Single molecule dynamics studied by polarization modulation. *Phys. Rev. Lett.* **77**, 3979–3982 (1996).

[12] Cichos, F., von Borczyskowski, C., and Orrit, M. Power-law intermittency of single emitters. *Curr. Opin. Coll. Interf. Sci.* **12**, 272–284 (2007).

[13] Zondervan, R., Kulzer, F., Orlinskii, S. B., and Orrit, M. Photoblinking of rhodamine 6G in poly(vinyl alcohol): radical dark state formed through the triplet. *J. Phys. Chem. A* **107**, 6770–6776 (2003).

[14] Vogelsang, J., Kasper, R., Steinhauer, C., Person, B., Heileman, M., Sauer, M., Tinnefeld, P. A reducing and oxidizing system minimizes photobleaching and blinking of fluorescent dyes. *Angew. Chem. Intl. Ed.* **47**, 5465–5469 (2008).

[15] Lu, H.P. and Xie, X. S. Single-molecule spectral fluctuations at room temperature. *Nature* **385**, 143–146 (1997).

[16] Zondervan, R., Kulzer, F., Kol'chenko, M. A., and Orrit, M. Photobleaching of rhodamine 6G in poly(vinyl alcohol) at the ensemble and single-molecule levels. *J. Phys. Chem. A* **108**, 1657–1665 (2004).

[17] Davies, M., Jung, C., Wallis, P., Schnitzler, T., Li, C., Müllen, K., and Bräuchle, C. Photophysics of new photostable rylene derivatives: applications in single-molecule studies and membrane labelling. *ChemPhysChem* **12**, 1588–1595 (2011).

[18] van Oijen, A. M., Köhler, J., Schmidt, J., Muller, M., and Brakenhoff, G. J. 3-dimensional super-resolution by spectrally selective imaging. *Chem. Phys. Lett.* **292**, 183–187.

[19] Naumov, A. V., Gorshelev, A. A., Vainer, Y. G., Kador, L., and Köhler, J. Far-field nanodiagnostics of solids with visible light by spectrally selective imaging. *Angew. Chem. Int. Ed.* **48**, 9747–9750 (2009).

[20] Betzig, E., Patterson, G. H., Sougrat, R., Lindwasser, O. W., Olenych, S., Bonifacino, J. S., Davidson, M. W., Lippincott-Schwartz, J., and Hess, H. F. Imaging intracellular fluorescent proteins at nanometer resolution. *Science* **313**, 1642–1645 (2006)

[21] Rust, M. J., Bates, M., and Zhuang, X.W. Sub-diffraction-limit imaging by stochastic optical reconstruction microscopy (STORM). *Nat. Methods* **3**, 793–795 (2006).

[22] Hess, S.T., Girirajan, T. P. K., and Mason, M. D. Ultra-high resolution imaging by fluorescence photoactivation localization microscopy. *Biophys. J.* **91**, 4258–4272 (2006).

[23] Heilemann, M., van de Linde, S., Mukherjee, A., d anSauer, M. Superresolution imaging with small organic fluorophores. *Angew. Chem. Int. Ed.* **48**, 6903–6908 (2009).

[24] Sharonov, A. and Hochstrasser, R.M. Wide-field subdiffraction imaging by accumulated binding of diffusing probes. *Proc. Natl Acad. Sci. USA* **103**, 18911–18916 (2006).

[25] Atkins, P. W. and Friedman, R. *Molecular Quantum Mechanics*, 5th edn. Oxford University Press, Oxford, 2010.

[26] Steinfeld, J. I. *Molecules and Radiation: An Introduction to Modern Molecular Spectroscopy*, 2nd edn. MIT Press, Cambridge, MA, 1985 (reprinted by Dover Publications, New York, 2005).

[27] Lakowicz, J. R. *Principles of Fluorescence Spectroscopy*, 3rd edn. Springer-Verlag, New York, 2006.

[28] Eggeling, C., Widengren, J., Rigler, R., and Seidel, C. A. M. Photobleaching of fluorescent dyes under conditions used for single-molecule detection: evidence of two-step photolysis. *Anal. Chem.* **70**, 2651–2659 (1998).

[29] Bernard, J., Fleury, L., Talon, H., and Orrit, M. Photon bunching in the fluorescence from single molecules: a probe for intersystem crossing. *J. Chem. Phys.* **98**, 850–859 (1993).

[30] Moerner, W. E., ed. *Persistent Spectral Hole-Burning: Science and Applications*. Springer-Verlag, Berlin, 1988.

[31] Heinze, G., Hubrich, C., and Halfmann, T. Stopped light and image storage by electromagnetically induced transparency up to the regime of one minute. *Phys. Rev. Lett.* **111**, 033601 (2013).

[32] Moerner, W. E. and Carter, T. P. Statistical fine structure of inhomogeneously broadened absorption lines. *Phys. Rev. Lett.* **50**, 2705–2708 (1987).

[33] Moerner, W. E. and Orrit, M. Illuminating single molecules in condensed matter. *Science* **283**, 1670–1676 (1999).

[34] Tamarat, P., Maali, A., Lounis, B., and Orrit, M. Ten years of single molecule spectroscopy. *J. Phys. Chem. A* **104**, 1–16 (2000).

[35] Plakhotnik, T., Donley, E. A., and Wild, U. P. Single-molecule spectroscopy. *Annu. Rev. Phys. Chem.* **48**, 181–212 (1997).

[36] Orrit, M. and Moerner, W. E. High-resolution single-molecule spectroscopy in condensed matter. In *Physics and Chemistry at Low Temperatures*, ed. L. Khriachtchev. Pan Stanford, Singapore, 2011, pp. 381–417.

[37] Kol'chenko, M. A., Nicolet, A. A. L., Galouzis, M. D., Hofmann, C., Kozankiewicz, B., and Orrit, M. Single molecules detect ultra-slow oscillators in a molecular crystal excited by ac-voltages. *New J. Phys.* **11**, 023037 (2009).

[38] Nicolet, A. A. L., Kol'chenko, M. A., Hofmann, C., Kozankiewicz, B., and Orrit, M. Nanoscale probing of charge transport in an organic field-effect transistor at cryogenic temperatures. *Phys. Chem. Chem. Phys.* **15**, 4415–4421 (2013).

[39] Köhler, J., Disselhorst, J. A. J. M., Donckers, M. C. J. M., Groenen, E. J. J., Schmidt, J., and Moerner, W. E. Magnetic resonance of a single molecular spin. *Nature* **363**, 242–244 (1993).

[40] Wrachtrup, J., von Borczyskowski, C., Bernard, J., Orrit, M., and Brown, R. Optical detection of magnetic resonance in a single molecule. *Nature* **363**, 244–245 (1993).

[41] Childress, L., Dutt, M. V. G., Taylor, J. M., Zibrov, A. S., Jelezko, F., Wrachtrup, J., Hemmer, P. R., and Lukin, M. D. Coherent dynamics of coupled electron and nuclear spin qubits in diamond. *Science* **314**, 281–285 (2006).

[42] Hanson, R., Dobrovitski, V. V., Feiguin, A. E., Gywat, O., Awschalom, D. D. Coherent dynamics of a single spin interacting with an adjustable spin bath. *Science* **320**, 352–355 (2008).

[43] van der Sar, T., Wang, Z. H, Blok, M. S., Bernien, H., Taminiau, T. H., Toyli, D. M., Lidar, D. A., Awschalom, D. D., Hanson, R., and Dobrovitski, V. V. Decoherence-protected quantum gates for a hybrid solid-state spin register. *Nature* **484**, 82–86 (2012).

[44] Brunel, C., Lounis, B., Tamarat, P., and Orrit, M. Triggered source of single photons based on controlled single molecule fluorescence. *Phys. Rev. Lett.* **83**, 2722–2725 (1999).

[45] Michaelis, J., Hettich, C., Mlynek, J., and Sandoghdar, V. Optical microscopy using a single-molecule light source. *Nature* **405**, 325–328 (2000).

[46] Lettow, R., Rezus, Y. L. A., Renn, A., Zumofen, G., Ikonen, E., Götzinger, S., Sandoghdar, V. Quantum interference of tunably indistinguishable photons from remote organic molecules. *Phys. Rev. Lett.* **104**, 123605 (2010).

[47] Tian, Y., Navarro, P., Kozankiewicz, B., and Orrit, M. Spectral diffusion of single dibenzoterrylene molecules in 2,3-dimethylanthracene. *ChemPhysChem* **13**, 3510–3515 (2012).

[48] Zumbusch, A., Fleury, L., Brown, R., Bernard, J., and Orrit, M. Probing individual two-level systems in a polymer by correlation of single molecular fluorescence. *Phys. Rev. Lett.* **70**, 3584–3587 (1993).

[49] Boiron, A. M., Tamarat, P., Lounis, B., Brown, R., and Orrit, M. Are the spectral trails of single molecules consistent with the standard two-level system model of glasses at low temperatures? *Chem. Phys.* **247**, 119–132 (1999).

[50] Naumov, A. V., Vainer, Y. G., and Kador, L. Does the standard model of low-temperature glass dynamics describe a real glass? *Phys. Rev. Lett.* **98**, 145501 (2007).

[51] Ambrose, W. P., Basché, T., Moerner, W. E. Detection and spectroscopy of single pentacene molecules in a *para*-terphenyl crystal by means of fluorescence excitation. *J. Chem. Phys.* **95**, 7150–7163 (1991).

[52] Kulzer, F., Kummer, S., Matzke, R., Bräuchle, C., Basché, T. Single-molecule optical switching of terrylene in *p*-terphenyl. *Nature* **387**, 688–691 (1997).

[53] Reilly, P.D., Skinner, J.L. Spectral diffusion of individual pentacene molecules in *p*-terphenyl crystal. Stochastic theoretical model and analysis of experimental data *J. Chem. Phys.* **102**, 1540–1552 (1995).

[54] Geva, E., Reilly, P. D., and Skinner, J.L. Spectral dynamics of individual molecules in glasses and crystals. *Acc. Chem. Res.* **29**, 579–584 (1996).

[55] Tian, Y. X., Navarro, P., Kozankiewicz, B., and Orrit, M. Spectral diffusion of single dibenzoterrylene molecules in 2,3-dimethylanthracene. *ChemPhysChem* **13**, 3510–3515 (2012).

[56] Paers, M., Palm, V., and Kikas, J. Single-molecule probing of incommensurate biphenyl. *Low Temp. Phys.* **36**, 448–450 (2010).

[57] van Oijen, A. M., Ketelaars, M., Köhler, J., Aartsma, T. J., and Schmidt, J. Unraveling the electronic structure of individual photosynthetic pigment–protein complexes. *Science* **285**, 400–402 (1999).

[58] Hildner, R., Brinks, D., Nieder, J. B., Cogdell, R.J., and van Hulst, N.F. Quantum coherent energy transfer over varying pathways in single light-harvesting complexes. *Science* **340**, 1448–1451 (2013).

[59] Hettich, C., Schmitt, C., Zitzmann, J., Kuhn, S., Gerhardt, I., and Sandoghdar, V. Nanometer resolution and coherent optical dipole coupling of two individual molecules. *Science* **298**, 385–389 (2002).

[60] Fückel, B., Hinze, G., Nolde, F., Müllen, K., and Basché, T. Control of the electronic energy transfer pathway between two single fluorophores by dual pulse excitation. *Phys. Rev. Lett.* **103**, 103003 (2009).

[61] van de Hulst, H. *Light Scattering by Small Particles*. Wiley, New York, 1957 (reprinted by Dover Publications, New York, 1981).

[62] Lindfors, K., Kalkbrenner, T., Stoller, P., and Sandoghdar, V. Detection and spectroscopy of gold nanoparticles using supercontinuum white light confocal microscopy. *Phys. Rev. Lett.* **93**, 037401 (2004).

[63] Ignatovich, F. V. and Novotny, L. Real-time and background-free detection of nanoscale particles. *Phys. Rev. Lett.* **96**, 013901 (2006)

[64] Zsigmondy, R. *Colloids and the Ultramicroscope: A Manual of Colloid Chemistry and Ultramicroscopy.* Wiley, New York, 1909.

[65] Klar, T., Perner, M., Grosse, S., von Plessen, G., Spirkl, W., and Feldmann, J. Surface-plasmon resonances in single metallic nanoparticles. *Phys. Rev. Lett.* **80**, 4249–4252 (1998).

[66] Yguerabide, J. and Yguerabide, E. E. Light scattering submicroscopic particles as highly fluorescent analogs and their use as tracer labels in clinical and biological applications. *Anal. Biochem.* **262**, 137–156 (1998).

[67] Schultz, S., Smith, D. R., Mock, J. J., and Schultz, D.A. Single-target molecule detection with nonbleaching multicolor optical immunolabels. *Proc. Natl Acad. Sci. USA* **97**, 996–1001 (2000).

[68] Sönnichsen, C., Franzl, T., Wilk, T., von Plessen, G., Feldmann, J., Wilson, O., and Mulvaney, P. Drastic reduction of plasmon damping in gold nanorods. *Phys. Rev. Lett.* **88**, 077402 (2002).

[69] Funston, A. M., Novo, C., Davis, T. J., and Mulvaney, P. Plasmon coupling of gold nanorods at short distances and in different geometries. *Nano Lett.* **9**, 1651–1658 (2009).

[70] Plakhotnik, T. and Palm, V. Interferometric signatures of single molecules. *Phys. Rev. Lett.* **87**, 183602 (2001).

[71] Jacobsen, V., Stoller, P., Brunner, C., Vogel, V., and Sandoghdar, V. Interferometric optical detection and tracking of very small gold nanoparticles at a water–glass interface. *Opt. Express* **14**, 405–414 (2006).

[72] Kukura, P., Celebrano, M., Renn, A., and Sandoghdar, V. Single-molecule sensitivity in optical absorption at room temperature. *J. Phys. Chem. Lett.* **1**, 3323–3327 (2010).

[73] Tchebotareva, A. L., Ruijgrok, P. V., Zijlstra, P., and Orrit, M. Probing the acoustic vibrations of single metal nanoparticles by ultrashort laser pulses. *Laser Photonics Rev.* **4**, 581–597 (2010).

[74] van Dijk, M. A., Lippitz, M., and Orrit, M. Detection of acoustic oscillations of single gold nanoparticles by time-resolved interferometry. *Phys. Rev. Lett.* **95**, 267406 (2005).

[75] Tchebotareva, A. L., van Dijk, M. A., Ruijgrok, P. V., Fokkema, V., Hesselberth, M. H. S., Lippitz, M., and Orrit, M. Acoustic and optical modes of single dumbbells of gold nanoparticles. *ChemPhysChem* **9**, 111–114 (2009).

[76] Zijlstra, P., Tchebotareva, A. L., Chon, J. M. W., Gu, M., and Orrit, M. Acoustic oscillations and elastic moduli of single gold nanorods. *Nano Lett.* **8**, 3493–3497 (2008).

[77] Ruijgrok, P. V., Zijlstra, P., Tchebotareva, A., and Orrit, M. Damping of acoustic vibrations of single gold nanoparticles. *Nano Lett.* **12**, 1063–1069 (2012).

[78] Staleva, H. and Hartland, G. Transient absorption studies of single silver nanocubes. *J. Phys. Chem. C* **112**, 7535–7539 (2008).

[79] Ruijgrok, P. V., Verhart, N. R., Zijlstra, P., Tchebotareva, A. L., and Orrit, M. Brownian fluctuations and heating of an optically aligned gold nanorod. *Phys. Rev. Lett.* **107**, 037401 (2011).

[80] Ruijgrok, P. V., Zijlstra, P., Tchebotareva, A. L., and Orrit, M. Damping of acoustic vibrations of single gold nanoparticles optically trapped in water. *Nano Lett.* **12**, 1063–1069 (2012).

[81] Pelton, M., Sader, J. E., Burgin, J., Liu, M., Guyot-Sionnest, P., and Gosztola, D. Damping of acoustic vibrations in gold nanoparticles. *Nat. Nanotechnol.* **4**, 492–495 (2009).

[82] Matawari, K., Kitamori, T., and Sawada, T. Individual detection of single nanometer-sized particles in liquid by photothermal microscopy. *Anal. Chem.* **70**, 5037–5041 (1998).

[83] Boyer, D., Tamarat, P., Maali, A., Lounis, B., and Orrit, M. Photothermal imaging of nanometer-sized metal particles among scatterers. *Science* **297**, 1160–1163 (2002).

[84] Berciaud, S., Cognet, L., Blab, G. A., and Lounis, B. Photothermal heterodyne imaging of individual nonfluorescent nanoclusters and nanocrystals. *Phys. Rev. Lett.* **93**, 257402 (2004).

[85] Absil, E., Teissier, G., Gross, M., Atlan, M., Warnasooriya, N., Suck, S., Coppey-Moisan, M., and Fournier, D. Photothermal heterodyne holography of gold nanoparticles *Opt. Express* **18**, 780–786 (2010).

[86] Gaiduk, A., Yorulmaz, M., Ruijgrok, P. V., and Orrit, M. Room-temperature detection of a single molecule's absorption by photothermal contrast. *Science* **330**, 353-356 (2010).

[87] Gaiduk, P., Ruijgrok, P. V., Yorulmaz, M., and Orrit, M. Detection limits in photothermal microscopy. *Chem. Sci.* **1**, 343-350 (2010).

[88] Chong, S., Min, W., and Xie, X. S. Ground-state depletion microscopy: detection sensitivity of single-molecule optical absorption at room temperature. *J. Phys. Chem. Lett.* **1**, 3316–3322 (2010).

[89] Schwartz, O. and Oron, D. A present understanding of colloidal quantum dot blinking. *Isr. J. Chem.* **52**, 992–1001 (2012).

[90] Zijlstra, P. and Orrit, M. Single metal nanoparticles: optical detection, spectroscopy and applications. *Rep. Prog. Phys.* **74**, 106401 (2011).

[91] Brinks, D., Stefani, F. D., Kulzer, F., Hildner, R., Taminiau, T. H., Avlasevich, Y., Müllen, K., and van Hulst, N. F. Visualizing and controlling vibrational wave packets of single molecules. *Nature* **465**, 905–908 (2010).

[92] Ringe, E., Sharma, B., Henry, A. I., Marks, L. D., and Van Duyne, R. P. Single nanoparticle plasmonics. *Phys. Chem. Chem. Phys.* **15**, 4110–4129 (2013).

[93] Etchegoin, P. G. and Le Ru, E. C. A perspective on single molecule SERS: current status and future challenges. *Phys. Chem. Chem. Phys.* **10**, 6079–6089 (2008).

[94] Lantman, E. M. V., Deckert-Gaudig, T., Mank, A. J. G., Deckert, V., and Weck-huysen, B.M. Catalytic processes monitored at the nanoscale with tip-enhanced Raman spectroscopy. *Nat. Nanotechnol.* **7**, 583–586 (2012).

[95] Zhang, R., Zhang, Y., Dong, Z. C., Jiang, S., Zhang, C., Chen, L. G., Zhang, L., Liao, Y., Aizpurua, J., Luo, Y., Yang, J. L., and Hou, G.J. Chemical mapping of a single molecule by plasmon-enhanced Raman scattering. *Nature* **498**, 82–86 (2013).

[96] Kinkhabwala, A., Yu, Z. F., Fan, S. H., Avlasevich, Y., Mullen, K., and Moerner, W. E. Large single-molecule fluorescence enhancements produced by a bowtie nanoantenna. *Nat. Photonics* **3**, 654–657 (2009).

[97] Yuan, H. F., Khatua, S., Zijlstra, P., Yorulmaz, M., and Orrit, M. Thousand-fold enhancement of single-molecule fluorescence near a single gold nanorod. *Angew. Chem. Int. Ed.* **52**, 1217–1221 (2013).

[98] Punj, D., Mivelle, M., Moparthi, S. B., van Zanten, T. S., Rigneault, H., van Hulst, N. F., Garcia-Parajo, M. F., and Wenger, J. A plasmonic 'antenna-in-box' platform for enhanced single-molecule analysis at micromolar concentrations. *Nat. Nanotechnol.* **8**, 512–516 (2013).

7

Quantum measurements: a modern view for quantum optics experimentalists

Aephraim M. STEINBERG

Centre for Quantum Information and Quantum Control
and Department of Physics
University of Toronto, Canada

Aephraim M. Steinberg: *Quantum measurements: a modern view for quantum optics experimentalists*. In: *Quantum Optics and Nanophotonics*. Edited by: Claude Fabre, Vahid Sandoghdar, Nicolas Treps, and Leticia F. Cugliandolo. Oxford University Press (2017), pp. 315–363. © Oxford University Press 2017. DOI: 10.1093/oso/9780198768609.003.0007

Chapter Contents

Colour figures. For those figures in this chapter that use colour, please see
the version of these lecture notes at arXiv:1406.5535 [quant-ph]. These figures are
indicated by '[Colour online]' at the start of the caption.

In these lectures, I have tried to give a brief survey of some important approaches and modern tendencies in quantum measurement. I wish it to be clear from the outset that I shy explicitly away from the 'quantum measurement problem', and that the present treatment aims to elucidate the theory and practice of various ways in which measurements can, in the light of quantum mechanics, be carried out, as well as various formalisms for describing them. While the treatment is by necessity largely theoretical, the emphasis is meant to be on an experimental 'perspective' on measurement—that is, to place the priority on the possibility of gaining information through some process, and then attempting to model that process mathematically and consider its ramifications, rather than stressing a particular mathematical definition as the *sine qua non* of measurement. The textbook definition of measurement as being a particular set of mathematical operations carried out on particular sorts of operators has been so well drilled into us that many have the unfortunate tendency of saying 'that experiment can't be described by projections onto the eigenstates of a Hermitian operator, so it is not really a measurement'—when of course any practitioner of an experimental science such as physics should instead say 'that experiment allowed us to measure something, and if the standard theory of measurement does not describe it, the standard theory of measurement is incomplete'. Idealizations are important, but when the real world breaks the approximations made in the theory, it is the theory that must be fixed, and not the real world.

7.1 Information from measurement: update rules, density matrices, and POVMs

The defining characteristic of measurement—whether in classical or in quantum physics—is that it increases our information about a system (or a process). From the perspective that views the (e.g. quantum) state as no more than an expression of our information, it is therefore tautological that measurements disturb systems' states. While the disturbances occasioned by quantum measurement are richer and perhaps more surprising than the trivial classical observation that a 50% chance of rain 'collapses' into either 100% or 0% once the fact of the matter is observed, much of their behaviour can be derived by first thinking about the how to modify one's description of a state upon successful gain of information via a measurement. This is a well-trod topic in classical statistics, but I treat it (briefly) here for several reasons. First, while such 'update rules' in some sense are the fundamental basis of all experimental sciences, they are widely misunderstood by physicists, presumably because the approximate rules we learn in our laboratory courses suffice in so many cases. Second, a clear perspective on the axioms of quantum measurement can be derived from these considerations, and leads naturally to some of the modern generalizations of the quantum measurement formalism that I will discuss here. Finally, a great deal of work in recent years has focused on quantum state and process estimation ('tomography') and on quantum metrology, and some controversies in those areas in fact stem from older questions in classical statistics.

7.1.1 Classical information update and Bayes's rule

Let us consider a classic example of parameter estimation, whose relationship to quantum measurement should become obvious. Suppose we are given a coin with probabilities p_H of coming up heads and $p_T = 1 - p_H$ of coming up tails. Lacking prior knowledge of p_H, how would we estimate it? If we flip the coin N times, say that we find H heads and $T = N - H$ tails. Our intuition that p_H is best estimated to be H/N is well founded. This is, in the technical sense of the term (which I shall discuss presently) the 'most likely' solution. But what uncertainty should we report on this estimate? We know that a sequence of fair coin tosses with probability of heads equal to p_H will lead to binomial statistics, that is, $p_H N \pm \sqrt{N(p_H)(1 - p_H)}$, for an *observed* ratio of $p_H \pm \sqrt{p_H(1 - p_H)/N}$. Usual (which is to say careless, but oftentimes acceptable) practice is to report this latter quantity as the uncertainty in our result. That is, we would report our finding as an estimate of p_H, which I will term $p_{est} = H/N \pm \sqrt{(H/N)(1 - H/N)/N}$. The error in this approach can be seen easily by considering the case in which 0 heads appeared. Our 'most likely' estimate of p_H is indeed 0, but the formula given above would yield an uncertainty of 0. Since it is *possible* for N coin tosses to all come up tails, so long as $p_H \neq 1$, we would not wish to report $p_{est} = 0 \pm 0$. The mistake was to conflate the probability of getting H heads *if the true probability were* p_{est} and the probability *that* p_{est} *is equal to* [or, more carefully, within a specified confidence interval of] the true p_H given that we obtained H heads. Mathematically, using the notation $P(A \,|\, B)$ for the probability of A, *given* that we know B:

$$P(H = 0 \,|\, p_H = 0) = 100\% \,,$$

$$\text{yet} \quad P(p_H = 0 \,|\, H = 0) \neq 100\% \tag{7.1}$$

(because, for instance, $P(H = 0 \,|\, p_H = 0.0001) \neq 0$). In experimental science, we use observed data to test and constrain models. That is, based on our observations, we update our estimates of the probability that a given model is correct or incorrect; in many contexts (such as the present one), a 'model' may simply be the value of a parameter. (When Millikan reported a charge for the electron and an uncertainty, he was in essence reporting a probability distribution for a *parameter* in his theory.) The trick is that our models allow us to make (generally probabilistic) predictions about what we would expect to observe, *given* particular values for all relevant parameters, $P(\text{data} \,|\, \text{model})$—while what we wish to know is the probability of the model being correct (or the parameters having a given set of values), *given* our observed data, $P(\text{model} \,|\, \text{data})$. Fortunately, classical probability theory has a formula, namely Bayes's theorem, for carrying out just this inversion. Because a joint probability of two propositions A and B can be factorized in either of two ways—$P(A\&B) = P(B)P(A \,|\, B) = P(A)P(B \,|\, A)$—we have

$$P(B \,|\, A) = \frac{P(A \,|\, B)P(B)}{P(A)} \,. \tag{7.2}$$

Applied to the classical coin example, we can write the probability that p_H has some particular value, given that we observed H to be 0, as $P(p_H \mid H = 0) = P(H = 0 \mid p_H)P(p_H)/P(H = 0)$. Note that p_H here is a variable; the denominator on the right-hand side is independent of p_H (which is a good thing, since after having observed that $H = 0$, and without any other information about the model, how should we estimate what the prior probability was that we *would* have observed $H = 0$?). In other words, it merely serves as a normalisation constant. Now, we already know how to calculate the probability of H heads from the binomial distribution,

$$P(H \mid p_H) = p_H^H(1 - p_H)^{N-H} \binom{N}{H}, \tag{7.3}$$

and since $\binom{N}{H}$ is independent of p_H and we will normalize our distribution in any event, all that concerns us is that $P(H \mid p_H) \propto p_H^H(1 - p_H)^{N-H}$.

This function is termed the 'likelihood' $\mathcal{L}(p_H) \equiv P(H \mid p_H)$ of the model. Although in everyday speech, 'likelihood' and 'probability' may be used synonymously, it is essential to note that $\mathcal{L}$ is *not* the probability that your model p_H is correct in light of your observations, but rather the probability that you *would* have obtained those observations *if* p_H had been correct.

A common estimation technique—the one we implicitly used when we concluded that our best estimate of p_H was H/N—is known as 'maximum-likelihood estimation', for reasons which should be obvious. It is a simple exercise to show that this likelihood function $\mathcal{L}(p_H) \propto p_H^H(1-p_H)^{N-H}$ attains its maximum when $p_H = H/N$. But consider the case for $H = 0$, sketched in Fig. 7.1. Here the likelihood of the model is just the

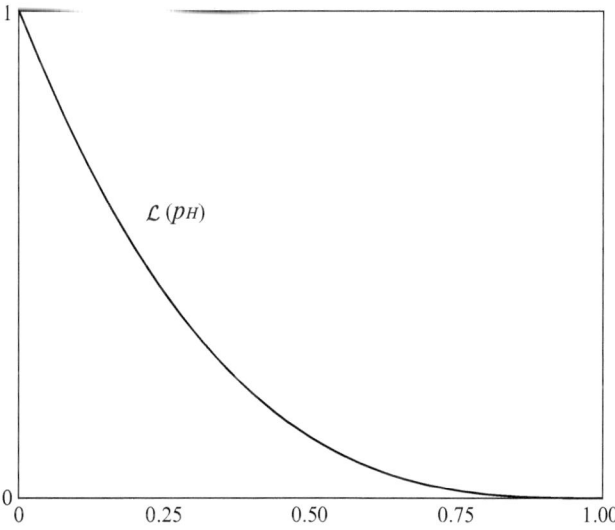

Fig. 7.1 The likelihood of a model p_H for the case where $H = 0$ heads are tossed (the example shown is for $N = 3$).

probability of flipping N tails in a row: $\mathcal{L} = p_T^N$, where $p_T = 1 - p_H$ is the probability of flipping tails. The point of maximum likelihood is $p_H = H/N = 0$, but there are two issues to address. The first is whether or not we can even conclude that $p_H = 0$ is the *most probable* value of p_H in the light of our observations. Recall that from (7.2), $P(p_H \mid H) \propto \mathcal{L}(p_H)P(p_H)$ and not just $\mathcal{L}(p_H)$ on its own. This $P(p_H)$ on the right-hand side is the *prior probability* (or probability density, in the case of a continuous parameter like p_H) that p_H was true to begin with. Being careless with these 'priors' is one of the most common errors made with statistics.

A simple famous example is the following. Suppose that you are tested for a rare disease (incidence 10^{-6}). The test is excellent; the probability of a positive result is 99% if you have the disease, and only 0.01% if you do not. Suppose you take the test and it comes back positive for the disease—then which 'theory' ('d', that you have the disease, or 'h', that you are healthy) has the maximum likelihood?

$$\mathcal{L}(d) \equiv P(+ \mid d) = 99\% \,,$$
$$\mathcal{L}(h) \equiv P(+ \mid h) = 0.01\% \,. \tag{7.4}$$

Clearly, the likelihood of 'd' is nearly 10^4 times larger than that of 'h'. But which is more *probable*? Well, since the test detects almost all true cases of the disease, roughly 10^{-6} of the population gets a '+' because they have the disease. On the other hand, since almost all the population is healthy, and the probability of a false positive is 10^-4, roughly 10^{-4} of the population gets a '+' because of simple error. In other words, 100 times more people get '+'s by error than because of the presence of the disease; there is still a 99% chance that you are healthy. Of course, the 1% chance that you have the disease is 10^4 times larger than the 10^{-6} estimate you would have made before taking the test; in fact, the *increase* in this probability (more strictly, in the ratio of the two probabilities) is given precisely by the ratio of the likelihoods:

$$\left[\frac{P(d \mid +)}{P(h \mid +)} \right] = \frac{\mathcal{L}(d)}{\mathcal{L}(h)} \left[\frac{P(d)}{P(h)} \right] . \tag{7.5}$$

What this shows us is that our experimental observations on their own (from which we calculate the likelihoods of the various models) cannot tell us the probability of one model or another being correct. The likelihoods serve only to *update* our estimates of these probabilities. Sadly, in survey after survey, a majority of doctors do not solve these problems correctly. Less importantly, but perhaps more shockingly, the same is true of a majority of physicists.

This interpretation of the likelihood based on data as an 'update rule' is crucial when one wishes to combine information from multiple experiments or observations. It is also important when other constraints exist for theoretical reasons—for instance, early measurements of the square m_ν^2 of the neutrino mass consistently turned up negative numbers, which were of course inconsistent with our understanding of physics. Had these results occurred simply because of experimental uncertainties,[1] the correct interpretation would have been to say that even though the point of maximum

---

[1] In fact, systematic errors were eventually identified.

likelihood occurred for $m_\nu^2 < 0$, the prior probability $P(m_\nu^2)$ vanished for negative squared masses, and no amount of data would change this. A reasonable candidate for $P(m_\nu^2 \,|\, \text{data})$ would be truncated below $m_\nu^2 = 0$, leaving an asymmetric curve not unlike that in Fig. 7.1.

But what should we do when we have no prior information, as in the case of the coin? The natural assumption might be to assume a 'flat' prior. This in itself can be tricky; for instance, making $P(m^2)$ flat is different from making $P(m)$ flat. Making $P(\log m)$ flat is even more different. This is the problem of choosing a measure. Many therefore advocate using a scale-invariant prior, the 'Bures prior'. For most applications in physics, this question is secondary. The fact is that we never prove or disprove a theory based on a single observation, but rather on a large set of observations. Every new observation multiplies the probability ratio in (7.5) by another data-based ratio of likelihoods; in the end, one hopes to be relatively insensitive to the initial choice of prior.

So, let us for simplicity suppose a flat prior, $P(p_H)\,dp_H = dp_H$ for p_H between 0 and 1. After flipping N tails, we *can* now conclude that $P(p_H \,|\, \text{data}) \propto \mathcal{L}(p_H) \propto p_T^N$. But would we really wish to report as our best estimate the *peak* of this asymmetric function, at $p_H = 0$, given that we know with certainty that the true value cannot possibly lie below that? Or wouldn't we rather report the mean of the distribution? You are all familiar with least-squares fits; typically, our goal is to minimize the squared error. Which value of p_H would we report in order to minimize the expectation value of the squared error from the true value of p_H? The answer is precisely the mean, $\langle p_H \rangle = \int_0^\infty dp_H\, p_H P(p_H \,|\, \text{data})$. Interestingly, in this case, the result is not the maximum-likelihood solution H/N, but rather $(H+1)/(N+2)$, a 'correction' originally proposed by Laplace.[2] Note that instead of being undefined for $N = 0$, this formula returns the mean of the (flat) prior distribution, 0.5, and that it can never reach 0 or 1. The subtlety here is again one of inversion. Adjusting p_H to minimize the χ^2 (essentially, the mean squared deviation of our observations from the model prediction) would yield the maximum-likelihood result (in the above example, $p_H=0$); this is not the same thing as minimizing the mean squared error of our estimate for p_H from its true value.

7.1.2 A quantum example

The relation of this formalism to quantum measurement theory arises because the quantum state (by which I mean either a wavevector $|\psi\rangle$ or a density matrix ρ) is fundamentally a way of making predictions about future observations: a description of our knowledge, or 'prior'. When we gain information by performing a measurement, our knowledge of the system (and our best estimates of future probabilities) changes. At the simplest level, this is the origin of 'collapse'. Let us consider a simple example central to quantum optics. Consider an atom initially in the state $|+\rangle \equiv (|g\rangle + |e\rangle)/\sqrt{2}$. Imagine letting this atom evolve for one half life, all the while observing it with a 100%-efficient photodetector. If you detect a photon, the atom must have been in $|e\rangle$,

---

[2] If instead of a flat prior, one uses the Bures prior, one arrives at $(H+0.5)/(N+1)$.

so you have that new information about the original state—but of course, since it has decayed, it must be in $|g\rangle$: such a measurement both provides information about the initial condition and simultaneously modifies the state.

On the other hand, if you wait one half life and observe no photons—what state should you conclude the atom is in? Since it hasn't decayed, you might suspect that it is still in $|+\rangle$. This can clearly be seen to be incorrect by going to extremes: if you waited for 100 years and still observed no photon, you would presumably reason that any excited atom would have decayed by now, so the atom must have been in the ground state all along. Even though no photon was emitted, your description of the quantum state should change from $|+\rangle$ to $|g\rangle$. *Observing nothing is also an observation.* So how should we apply this intuition at intermediate times, such as after a single half life? It turns out that the Bayesian methods introduced above are perfectly adequate. Our prior probabilities (given by the initial quantum state) are $P(g) = P(e) = \frac{1}{2}$, and if we consider 'g' and 'e' to be two different models of the atom and a single photon 'γ' to be the observation, then we can write our conditional probabilities $P(\gamma \,|\, e) = \frac{1}{2}$ and $P(\gamma \,|\, g) = 0$. The calculation is straightforward:

$$P(g \,|\, \gamma) \propto P(g)P(\gamma \,|\, g) = \tfrac{1}{2} \cdot 0 \,,$$
$$P(e \,|\, \gamma) \propto P(e)P(\gamma \,|\, e) = \tfrac{1}{2} \cdot \tfrac{1}{2} \,,$$

(7.6)

from which we see that $P(g \,|\, \gamma) = 0$ as expected and $P(e \,|\, \gamma) = 1$. Similarly,

$$P(g \,|\, \text{no } \gamma) \propto P(g)P(\text{no } \gamma \,|\, g) = \tfrac{1}{2} \cdot 1 \,,$$
$$P(e \,|\, \text{no } \gamma) \propto P(e)P(\text{no } \gamma \,|\, e) = \tfrac{1}{2} \cdot \tfrac{1}{2} \,,$$

(7.7)

from which we conclude that after *no photon* is observed, the probability of g is twice that of e, and hence must be $\frac{2}{3}$. Indeed, the result of this non-observation of a photon is to 'collapse' the state to a new state $\sqrt{\frac{2}{3}} \,|g\rangle + \sqrt{\frac{1}{3}} \,|e\rangle$, as suggested by this classical analysis (though of course the Bayesian reasoning does not tell us anything about the phase relationship between g and e; to understand that it is preserved, we will need the full quantum formalism, to be discussed in Section 7.3).

This example can also be seen as a simple case of a more general observation: *not all measurements are projections.* By observing no photon to be emitted for some length of time, we have acquired some information about the state—but only a limited amount. This modifies the state vector, but it does not project it onto one of two alternative eigenstates; in fact, the final states that result from the two possible measurement outcomes are not even orthogonal to one another.

7.2 Projective measurement, density matrices, and decoherence

7.2.1 Review of projective measurement

Before treating the modern approaches to measurement that can properly describe situations such as the 'non-observation' we have just discussed, we should briefly review the 'standard' measurement postulate treated in most textbooks, also referred to as 'von Neumann' or 'projective' measurement. By this postulate, objects that can be measured are 'observables' corresponding to Hermitian operators. Such an operator $\hat{A}$ has a set of orthonormal eigenstates $|n\rangle$ such that $\hat{A}|n\rangle = a_n|n\rangle$, where the (real) eigenvalues a_n are the only possible results of a measurement of the observable corresponding to $\hat{A}$. From the orthonormality relation $\langle m|n\rangle = \delta_{mn}$ and the completeness relation $\sum_m |m\rangle\langle m| = I$, one can show that any state $|\psi\rangle$ may be decomposed in this eigenbasis, as $|\psi\rangle = \sum_i c_i|i\rangle$, with the coefficients c_i given by the inner product $\langle i|\psi\rangle$. The first part of the measurement postulate is that when $\hat{A}$ is measured, a result a_i will be found with probability $P_i = |c_i|^2 = |\langle i|\psi\rangle|^2$. The second part of the measurement postulate is based on the requirement that measurements be repeatable, essentially related to the idea of updating our probability distribution based on our observations. If we have already measured $\hat{A}$ to be a_i, then we expect that a second measurement will also yield a_i, with 100% probability. This requires that all the coefficients $c_{j\neq i}$ vanish: the state thus 'collapses' into the corresponding eigenstate $|i\rangle$.

This statement neglects the possibility of degenerate eigenvalues (with corresponding eigen*spaces* rather than eigenstates). This is of particular importance once we wish to think about multipartite systems; since an observable of one subsystem does not depend directly on the state of another subsystem, it is automatically degenerate. So what is the full state of the whole system after a measurement? Consider, for example, an entangled state. For the sake of argument, imagine the coherent superposition of my holding a black chess pawn in my left hand and a white pawn in my right, or the reverse: $(|B_L\rangle|W_R\rangle + |W_L\rangle|B_R\rangle)/\sqrt{2}$. If I open my right hand, the probability of finding a white pawn there is $p(W_R) = \frac{1}{2}$. Of course, if I find this white pawn, the probability of finding a black pawn there drops instantly from $\frac{1}{2}$ to 0; this is the classical 'collapse' of probabilities. But it is equally clear that the probability of finding a white pawn in my *left* hand immediately jumps to 1, even though I have only made a measurement on the right hand. (Note that such a 'non-local' collapse is entirely classical, and arises because of correlations we can understand through a common cause; the non-locality inherent in quantum entanglement is of course deeper.) Classically, the analysis is simple; there are four possible models: $B_L B_R$, $B_L W_R$, $W_L B_R$, and $W_L W_R$. When I find a white pawn in my right hand, the second and fourth models have likelihoods of 1, while the first and third have likelihoods of 0. I thus update my prior by multiplying the individual model probabilities by these likelihoods and renormalizing. Since my prior state was a 50/50 superposition of the second and third models, only the second model ($B_L W_R$) survives. The quantum mechanical projection postulate is the natural extension of this classical idea: each *subspace* corresponding to a particular outcome gets scaled up in amplitude according to its likelihood. Of course, if the event in question is observation of an eigenvalue a_i, then the likelihoods are 1 for any eigenstate with this eigenvalue and 0 for any other eigenstate. We therefore *project* onto the subspace with eigenvalue a_i, and renormalize:

$$|\psi_f\rangle = \frac{\mathrm{Proj}(i)\,|\psi_i\rangle}{\sqrt{\langle\psi_i|\,\mathrm{Proj}(i)\,|\psi_i\rangle}}. \tag{7.8}$$

7.2.2   Density matrices

We introduce density matrices as a more general description of a quantum state, because not all states of knowledge can be described by wavevectors. Specifically, consider again the behaviour of an atom that can spontaneously emit. Let us begin with an atom in the excited state and 0 photons, writing this state as $|e, 0\gamma\rangle$. If we imagine, as before, that we have a collection system able to capture 100% of the emitted photons, then we can describe the state of the composite system after one half life by

$$|e, 0\gamma\rangle \rightarrow \frac{|e, 0\gamma\rangle}{\sqrt{2}} + \frac{|g, 1\gamma\rangle}{\sqrt{2}}. \tag{7.9}$$

By detecting 0 or 1 photons, we could 'collapse' the atomic state into e or g, respectively. But in the more realistic case where the hypothetical photon flies off to infinity with no possibility of us observing it again, how shall we describe the resulting situation? There is a 50/50 probability that the atom is in g or e, *certes*. But what phase should we choose? $(|g\rangle + |e\rangle)/\sqrt{2}$? $(|g\rangle - |e\rangle)/\sqrt{2}$? Any such superposition would possess some dipole moment;[3] what would break the symmetry and determine whether this dipole moment should be positive or negative (point to the right or the left)? Clearly, spontaneous emission should not define any preferential direction in space, so we need a way to describe a 50/50 *mixture* of ground and excited states that nevertheless has a vanishing expectation value of the dipole moment (which is to say, of the operator $|g\rangle\langle e|$ and its Hermitian adjoint).

Such *probabilistic* mixtures are called 'mixed states', to distinguish them from 'pure states', which are any states that can be written as state vectors. Note that the concept of purity is basis-independent—thus *superposition states* are still pure. We seek a mathematical description of situations where, beyond the intrinsic quantum mechanical uncertainties, we have some ignorance about the state of the system. There is a chance P_A that the system is in pure state A and a chance P_B that the system is in pure state B, for instance (where A and B could be the ground and excited states, as above). In such a case, the rule for an expectation value is clear; the expectation value for any observable in the mixed state should be the average, weighted by the probabilities P_A and P_B, of its expectation values in A and B individually. The same weighted-average rule should hold for the probability of any given event (the probability of finding a given state being, after all, the expectation value of the projector onto that state). This is the classical law of total probability: $P(x) = P(A)P(x\,|\,A) + P(B)P(x\,|\,B) + \ldots$, where $\{A, B, \ldots\}$ are any exhaustive set of mutually exclusive possibilities. The problem with drawing a direct quantum analogue to this expression is that our description of the state is not a direct list of probabilities $P(x\,|\,A)$ but rather a state vector (or wavefunction, or list of probability *amplitudes*, if you prefer). In particular, since $P(x) = |\psi(x)|^2$ is not linear, we cannot take $\psi(x) = \psi_A(x)P(A) + \psi_B(x)P(B)$, or even $\psi(x) = \psi_A(x)\sqrt{P(A)} + \psi_B(x)\sqrt{P(B)}$; any such expression would supplement the desired ('classical') weighted-average law with cross terms. The situation would be

---

[3] If not at $t = 0$ then anyway after some time evolution.

simpler if, instead of describing a physical situation with a state function that appears quadratically in expectation values $\langle\psi|\,X\,|\psi\rangle$ and in probabilities

$$P(i) = |\langle i|\psi\rangle|^2 = \langle i|\psi\rangle\langle\psi|i\rangle\,, \tag{7.10}$$

we could identify a mathematical object that uniquely and completely described the situation, while appearing *linearly* in such expressions. And this object leaps out at us from the right-hand side of (7.10). The *projector* $|\psi\rangle\langle\psi|$ has a one-to-one correspondence with the state vector ψ and hence describes the physical situation equally well, but it appears linearly in the expression for probability. It follows that if for a pure state we *define* the 'density matrix' $\rho_{\text{pure}} \equiv |\psi\rangle\langle\psi|$, we can use the same rule $P(i) \equiv \langle i|\,\rho\,|i\rangle$ for mixed states, simply by defining $\rho_{\text{mixed}} \equiv \sum_m P_m\rho_m$, where P_m are the probabilities to be in pure states described by the ρ_m, and the sum over m is over any number of states that may be mixed, with no requirement that this set be complete, orthonormal, or anything else. This construction allows us to use a single representation for any state—pure or mixed—and a single formula for probabilities or expectation values, the linearity of these formulas ensuring that the proper description of a mixed state is indeed the probability-weighted average of the pure states entering into the mixture. It is important to note that this decomposition is not unique—while ρ is a Hermitian operator and therefore possesses a unique decomposition into an eigenbasis (aside from degeneracy), it will in general possess an infinite number of indistinguishable expansions of this more general form. I will not discuss the mathematical properties of density matrices at more length, since they can be found in standard textbooks.

7.2.3   Update rule for density matrices

Having seen (7.8) for the update rule we apply to a state vector upon obtaining a measurement result, we now need an analogous expression to update our information when it is written as a density matrix. To do this, recall that the density matrix is a weighted average of projectors onto the different pure states the system 'might have been in'. Upon observation of a measurement result $|j\rangle$, each one of these component pure states will simply be projected onto $|j\rangle$ (to within normalization), such that

$$\sum_m P_m|\psi_m\rangle\langle\psi_m|$$
$$\implies \frac{1}{N}\sum_m P_m\,|j\rangle\,\langle j|\psi_m\rangle\langle\psi_m|j\rangle\,\langle j| = \frac{1}{N}\sum_m [P_m P(j\,|\,m)]\,|j\rangle\,\langle j|\,, \tag{7.11}$$

where N is a normalization constant. The expression on the right is straightforward to interpret in terms of Bayesian probabilities: the expression in square brackets is (again modulo normalization) the updated weighting probability $P(m\,|\,j)$; and the projector onto $|j\rangle$ appears because each component $|\psi_m\rangle$ collapsed onto that state when the measurement occurred. The constant N must be equal to $\sum_m P_m P(j\,|\,m)$, which is of course just P_j. This expression can easily be generalized to the situation where $|j\rangle$

is replaced by a subspace, by replacing $|j\rangle\langle j|$ with a projector $\mathrm{Proj}(j)$, which need not be rank 1:

$$\rho_f = \frac{\mathrm{Proj}(j)\,\rho\,\mathrm{Proj}(j)}{P_j}. \tag{7.12}$$

7.2.4 Losing information

Now we are ready to address the important question of lost information. Unitary (Schrödinger) evolution conserves information—that is to say, the entropy of a state remains constant, and the overlap of any two states also remains constant. Irreversible loss of information only occurs for 'open systems', as is the case for classical irreversibility as well. More rigorously, what this means is that when one subsystem is 'discarded', the evolution of the retained subsystem may be irreversible. This is what happened in the example of Section 7.2.2, of spontaneous emission. When we hypothesized that a photon might be lost to infinity and never retrieved, the consequent behaviour of the atom was not unitary, and this necessitated the invention of the density matrix formalism.

Let us for convenience split the universe into two parts or subsystems: the part that we are interested in studying we will term the 'system', while everything else (beyond our ability to measure) we will term the 'environment'. Then we will replace the basis states $|i\rangle$ by system–environment product states $|i\rangle_{\mathrm{sys}}|j\rangle_{\mathrm{env}}$. (In the earlier example, $|i\rangle_{\mathrm{sys}}$ might be the ground and excited states of the atom, while $|j\rangle_{\mathrm{env}}$ could be the 0- and 1-photon states of the field.) Now for some state ρ of the combined system and environment, let us calculate the expectation value of an operator A. This is done by calculating $\mathrm{Tr}\rho A$, which involves summing all the diagonal elements of ρA in any complete basis, for example $|i\rangle|j\rangle$:

$$\langle A\rangle \equiv \mathrm{Tr}\,\rho A = \sum_i \sum_j \langle i_{\mathrm{sys}}|\langle j_{\mathrm{env}}|\,\rho A\,|i_{\mathrm{sys}}\rangle|j_{\mathrm{env}}\rangle$$

$$= \sum_i \langle i_{\mathrm{sys}}|\left\{\sum_j \langle j_{\mathrm{env}}|\,\rho A\,|j_{\mathrm{env}}\rangle\right\}|i_{\mathrm{sys}}\rangle. \tag{7.13}$$

Now, we have assumed that we can carry out measurements only on the *system*, and not on the environment. Thus, A acts only on the system part of the state, and commutes with $|j_{\mathrm{env}}\rangle$:

$$\langle A\rangle = \sum_i \langle i_{\mathrm{sys}}|\left\{\sum_j \langle j_{\mathrm{env}}|\,\rho\,|j_{\mathrm{env}}\rangle\right\}A\,|i_{\mathrm{sys}}\rangle$$

$$= \mathrm{Tr}_{\mathrm{sys}}\,\rho_{\mathrm{red}}A, \tag{7.14}$$

where

$$\rho_{\mathrm{red}} \equiv \mathrm{Tr}_{\mathrm{env}}\,\rho \tag{7.15}$$

is termed the 'reduced density matrix', and represents the density matrix for the system, once the environment has been 'discarded' (or 'traced over', in mathematical jargon). As can be seen clearly in the above sums, the partial traces $\mathrm{Tr_{sys}}$ and $\mathrm{Tr_{env}}$ are computed by taking the diagonal matrix elements between basis vectors for the system and environment individually. Physically, tracing over the environment reflects ignoring the state of the environment. One way to conceive of doing this is to imagine an observer measuring the environment in some basis ($|k\rangle$). For any outcome he obtains, the system will be left in a density matrix $\langle k| \rho |k\rangle$ (with probability given by the trace of that—unnormalized—density matrix, as I leave as an exercise). If we have no access to this observer's result, our system will nonetheless be left in a weighted average of all possible $\langle k| \rho |k\rangle$'s. Summing these unnormalized matrices over a complete set of k includes this weighting automatically—and since this sum is a trace ($\mathrm{Tr_{env}}$), it is basis-independent: as one should perhaps have expected, the (average, mixed) state of our system cannot depend on which basis some unknown observer chooses to measure the environment in.

Mathematically, taking this trace over the environment means that we retain only terms diagonal in the environment variables—no cross terms (coherences) are retained between terms corresponding to different states of the environment. This is illustrated in Fig. 7.2 for the case of the spontaneously decaying atom in state $(|e0\rangle + |g1\rangle)/\sqrt{2}$.

The full state of the entangled atom and field is given by this 4×4 matrix. The state of the system when no photon is observed is given by the upper-left 2×2 block, while the state when 1 photon is observed is given by the lower-right 2×2 block. The final state of the system if we do not know which of these occurred is given by

$$\rho_{\mathrm{red}} = \langle 0_{\mathrm{env}}| \rho |0_{\mathrm{env}}\rangle + \langle 1_{\mathrm{env}}| \rho |1_{\mathrm{env}}\rangle$$

$$= \begin{pmatrix} \frac{1}{2} & 0 \\ 0 & 0 \end{pmatrix} + \begin{pmatrix} 0 & 0 \\ 0 & \frac{1}{2} \end{pmatrix}$$

$$= \begin{pmatrix} \frac{1}{2} & 0 \\ 0 & \frac{1}{2} \end{pmatrix}. \tag{7.16}$$

$$\rho = \begin{pmatrix} \begin{pmatrix} 1/2 & 0 \\ 0 & 0 \\ 0 & 0 \\ 1/2 & 0 \end{pmatrix} & \begin{pmatrix} 0 & 1/2 \\ 0 & 0 \\ 0 & 0 \\ 0 & 1/2 \end{pmatrix} \end{pmatrix}$$

Coherence lost

$$\mathrm{Tr_{env}}\rho = \begin{pmatrix} \frac{1}{2} & 0 \\ 0 & \frac{1}{2} \end{pmatrix}$$

There is still coherence between ↓↓ and ↑↑, but if the environment is not part of your interferometer, you may as well consider it to have 'collapsed' to ↑ or ↓ . This means there is no effective coherence if you look only at the system.

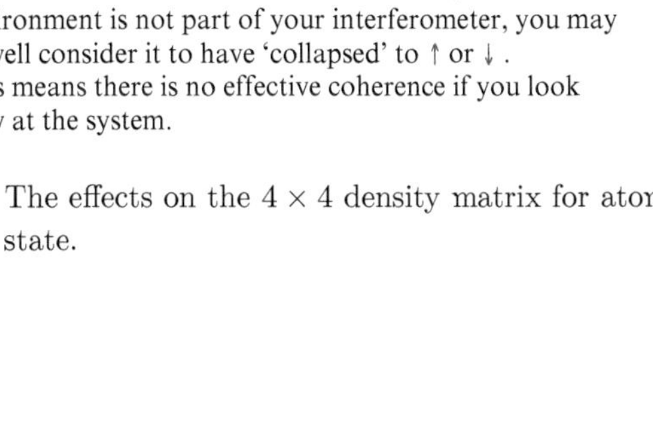

Fig. 7.2 [Colour online] The effects on the 4×4 density matrix for atom and field of tracing over the unobserved field state.

This expression is at the heart of the effective decoherence that is characteristic of open systems, but also of measurement; there is, of course, no difference between an 'environment' treated as above (which acquires information about the state of the system through an entangling interaction) and a measuring device, save for intent. As Feynman's rules for interference teach us, if there is any way—even in principle— to tell which of two histories (Feynman paths, if you like) was followed, then no interference may occur between these paths: their coherence is lost (see Section 7.4 for further discussion of this). Clearly, by observing the environment, one could determine whether the atom was in $|g\rangle$ or $|e\rangle$; this is why the reduced density matrix has 0's on the off-diagonals, representing vanishing coherence between the two. There can be no dipole moment, since a non-zero dipole arises from interference between states of different parity. Naturally, the full system still has perfect coherence between $|g1\rangle$ and $|e0\rangle$, but this coherence can only be studied by manipulating the two systems together; once we limit ourselves to studying a subsystem (or once the environment has such a large number of degrees of freedom that, as in statistical thermodynamics, it is inconceivable in practice that it could be manipulated sufficiently coherently), this coherence becomes practically unobservable. This is true, incidentally, even if there is no additional 'collapse' process, and the evolution of the universe is described by the Schrödinger equation alone ... In sum, effective 'collapse' (really decoherence: the impossibility of observing any effects that depend simultaneously on the amplitudes for Schrödinger's cat to have been alive and to have been dead) arises in two steps: (1) the entanglement of two systems and (2) the discarding (or neglecting) of one of these systems.

Importantly, the trace is basis-independent. It is not necessary that some demon in the environment choose to measure whether there was 1 photon or 0, and acquire this information; the mere fact that this is *conceivable* guarantees that there is no interference. If such a demon measured the field to be in the state $(|0\rangle + |1\rangle)/\sqrt{2}$, he would in fact collapse the atom into a state with a positive dipole moment. This is the content of the 'quantum eraser' idea. However, if the demon attempted such a measurement, he would be equally likely to find $(|0\rangle - |1\rangle)/\sqrt{2}$, collapsing the atom into a state with negative dipole moment; if we *ignore* the demon's subsystem, the density matrix is the same regardless of what basis he picks for a measurement. This is a reflection of the fact that no action that is performed on one subsystem can affect the statistics of another subsystem, once the two are no longer directly interacting. This is one way to see that EPR correlations satisfy a 'no-signalling' theorem, and do not violate relativistic causality.

7.3   Generalized measurement (POVMs)

All measurement is indirect. By this, I wish to say that when we talk about studying a system A, we really mean that the system interacts with a meter B, and we look at the meter. Really, this chain typically has many more steps, culminating whenever you feel like truncating it—when the information is written on your hard drive, or displayed on your computer screen, or when your eyes absorb the photons from the screen, or after your brain processes the information? But to understand the effect on the system and the maximum amount of information extractable, it is enough to consider the interaction of A with B. Now, the traditional (von Neumann) view of measurement is in terms of projectors onto complete sets of basis states for Hermitian observables; obviously, when in the real world we carry out a Stern–Gerlach experiment and a silver atom causes a spot to appear on a screen, this spot does not live in a two-dimensional Hilbert space and spit out a '$+\hbar/2$' or a '$-\hbar/2$' (Fig. 7.3). Only given some knowledge of the system do we make the *approximation* that spots in certain regions *nearly* guarantee one eigenstate and spots in others nearly guarantee the other, recognizing that some experimental uncertainty is always unavoidable. Any real observation is thus viewed as an approximation to the Platonic ideal of a measurement. A more appropriate view, I would maintain, is that any correlation can provide information, and thus constitutes a measurement. Projection operators are merely one idealization of this idea—and interestingly, they are not even always the optimal strategy, depending on one's goal. This is the motivation for discussing 'generalized measurement', a formalism that is intended to encompass both von Neumann-style measurement and the wide range of real-world techniques for extracting information from a system.

Let us explicitly consider the interaction between a system of interest and a second, 'meter' system. To accomplish an ideal von Neumann measurement, we would like an interaction to create perfect entanglement between the system and the meter, which should have the same dimensionality as the system. For instance, if we wish to observe whether an atom is in $|g\rangle$ or $|e\rangle$, we need some two-level measuring device, such as a spin that can be in state $|\downarrow\rangle$ or $|\uparrow\rangle$. Initially, the atom is in an unknown state, and the spin is 'initialized', for example to $|\downarrow\rangle$. Our goal would be to design an interaction that would lead to the evolution

$$\begin{aligned} |g\rangle\,|\downarrow\rangle &\rightarrow |g\rangle\,|\downarrow\rangle\,, \\ |e\rangle\,|\downarrow\rangle &\rightarrow |e\rangle\,|\uparrow\rangle\,, \end{aligned} \tag{7.17}$$

for instance, so that later observation of the spin would inform us as to the state of the atom. (The astute reader with some familiarity with quantum information will

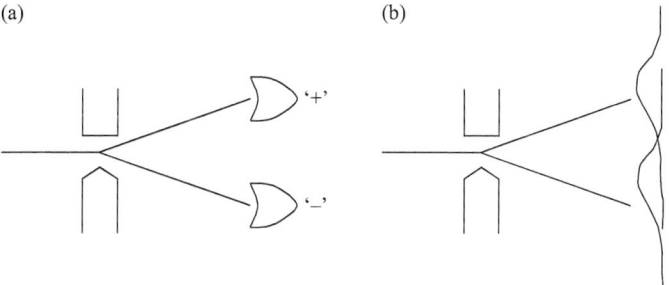

Fig. 7.3 A Stern–Gerlach measurement. (a) The idealized 'two-dimensional' picture of a projective measurement. (b) The reality: infinite-dimensional, and with finite uncertainties.

have observed that this interaction is essentially a controlled-NOT.) Meanwhile, since the spin has complete information about the state of the atom, this measurement has destroyed all coherence between e and g.

But recall our earlier discussion of an 'incomplete' measurement in Section 7.1.2, where only giving the atom one half life to interact with the field failed to maximally entangle the two systems:

$$|g\rangle |0\rangle \rightarrow |g\rangle |0\rangle \,,$$

$$|e\rangle |0\rangle \rightarrow \frac{|e\rangle |0\rangle + |g\rangle |1\rangle}{\sqrt{2}} \,. \tag{7.18}$$

Here, observing the field to be in '0' or '1' gives some information about the state of the atom, but not complete information; it is a therefore a reasonable model for at least one class of real-world measurements with finite uncertainty. Another example of a measurement with finite uncertainty is shown in Fig. 7.3(b). In the Stern–Gerlach effect, the (two-dimensional) spin of the atom is coupled to the (continuous) momentum of the atom, which is what is eventually read off (ideally), in the far field. In this case, the infinite dimensionality of the momentum 'pointer' is merely a technical detail, perhaps a weakness, but any true description of practical measurements should take into account the fact that the number of possible outcomes may be larger than the dimensionality of the system Hilbert space alone—perhaps even infinite. In the present situation, this may be construed as a minor technical nuisance, but, as we shall see later, there are cases in which there is an actual advantage to performing such measurements.

In the case of projective measurements, the probability of outcome i was given by the expectation value of the projector onto i: $P_i = \langle \mathrm{Proj}(i) \rangle = \mathrm{Tr}[\rho \mathrm{Proj}(i)]$. If the i's come from a complete set of orthonormal states, then the completeness relation $\sum_i \mathrm{Proj}(i) = I$ ensures that $\sum_i P_i = 1$, as desired. For a more general measurement, we still expect a basis-independent expression linear in ρ to describe the probability of a given outcome, and a small generalization suffices:

$$P_i = \mathrm{Tr}(E_i \rho) \,, \tag{7.19}$$

where the E_i need not be projectors, but must be *positive* (to guarantee that P_i is always positive) and must sum to the identity: $\sum_i E_i = I$. These operators can be written in terms of 'measurement operators' M_i as follows:

$$E_i = M_i^\dagger M_i \,. \tag{7.20}$$

Note that while the M_i determine the E_i, the reverse is not true. Any $(M_i^\dagger U^\dagger)(U M_i) = M_i^\dagger M_i$, after all. What is the meaning of the measurement operators? If an outcome i is found, then M_i describes the effect on the state: $|\psi\rangle \rightarrow M_i |\psi\rangle$. The simplest example is given by $M_i = |i\rangle \langle i|$, which leads automatically to $E_i = |i\rangle \langle i|$, and recovers the usual projective measurements, including both the probability formula and the projection postulate. But consider the case of spontaneous emission once more: the probability of detecting a photon depends on the initial probability of being in the excited state (multiplied by some 'efficiency' η), but upon detection of this photon, the atom is left in the ground state. The measurement operator for such a decay observation is $M_1 = \sqrt{\eta} \, |g\rangle \langle e|$. (This is stated here without proof, but it can be derived directly from the form of the interaction Hamiltonian that coupled the atom to the field, creating a photon through the term $a^\dagger |g\rangle \langle e|$.) It is easy to see that $E_1 = M_1^\dagger M_1 = \eta |e\rangle \langle e|$, so that the probability of detecting a photon is given by the probability for the atom to be in e, multiplied by η. The corresponding operator for the detection of 0 photons must be $E_0 = I - E_1$, since there are only two possible outcomes in this example:

$$E_0 = I - E_1 = |g\rangle \langle g| + (1 - \eta)|e\rangle \langle e| . \tag{7.21}$$

The probability of detecting no photon is the sum of two terms,: one for an atom in the ground state and the other for an atom in the excited state that failed to emit a photon. As we saw before, such a non-detection event makes it more likely—but not certain—that the atom is in g. How shall we see this in the update rule? Again, there are multiple 'square roots' of E_1, and the physically relevant one depends on the details of the interaction Hamiltonian, but the simplest case is easily seen to be

$$M_0 = |g\rangle \langle g| + \sqrt{1 - \eta} \, |e\rangle \langle e| . \tag{7.22}$$

An atom initially in $c_g |g\rangle + c_e |e\rangle$ is left in a state $c_g |g\rangle + \sqrt{1 - \eta} \, c_e |e\rangle$ upon non-observation of a photon. In the example we previously considered of $c_g = c_e = \frac{1}{\sqrt{2}}$ and $\eta = \frac{1}{2}$,

$$M_0 |\psi\rangle = \frac{|g\rangle + \frac{1}{\sqrt{2}} |e\rangle}{\sqrt{2}} , \tag{7.23}$$

confirming our classical conclusion that e is half as likely as g, implying a 33% probability to be in e and a 67% probability to be in g. Note that the state in (7.23) is not normalized; just as in the case of projection operators, the norm-squared of this resultant state ($\langle \psi | M_i^\dagger M_i |\psi\rangle = \langle E_I \rangle$) gives the probability of the result—here 50% (for an atom starting in g) plus 25% (for an atom starting in e but failing to decay), for a total of 75%. But when that event occurs, we are left in a reweighted *coherent* superposition of e and g. The reweightings of the probabilities can be seen to be simply $\langle g | M_i^\dagger M_i |g\rangle$ and $\langle e | M_i^\dagger M_i |e\rangle$, which are nothing but the Bayesian likelihoods of the models g and e, respectively (the conditional probability of outcome i *given* an initial state of g or e).

The update rule written for pure states can easily be rewritten in terms of density matrices, and proves again to be linear, which guarantees that the following update rule is generally applicable to any ρ, pure or mixed:

$$\rho \to \frac{M_i \rho M_i^\dagger}{P_i} , \tag{7.24}$$

where P_i is still given by $\langle E_i \rangle = \mathrm{Tr}(\rho M_i^\dagger M_i)$. Again, it is easy to verify that this reduces to the update rule for projective measurements when the M_i's are taken to be a complete set of orthonormal projection operators. These generalized measurements are also often referred to as 'POVMs', for 'positive operator-valued measurements', although this is such an uninformative mouthful that some authors have been known to resist even including it in their publications, arguing that 'POVM' should by now stand on its own as a synonym for generalized measurement. Naimark's theorem shows that *any* POVM can be accomplished in the manner I have been outlining here, that is, by coupling the system of interest to some (potentially higher-dimensional) pointer system, and then making a projective measurement on the pointer.

Quantum optics students should be familiar with the treatment of spontaneous emission on the Bloch sphere, where the inversion decays with a timescale T_1 while the coherences decay with a timescale T_2, where for pure radiative decay with no broadening, $T_2 = 2T_1$ (which can be understood at one level by arguing that if amplitudes decay as e^{-t/T_2}, then the excited-state probability, being proportional to the square of an amplitude, will decay as $e^{-2t/T_2} \equiv e^{-t/T_1}$). This means that as an atom initially in a pure state on the surface of the Bloch sphere decays, it does not follow a straight trajectory towards the South pole (g), but rather curves around, outside this naive path. From the perspective of generalized measurements, it is easy to derive this result, which most of us were taught to put in by hand when learning to write down the optical Bloch equations. Either a photon was emitted or not; if we do not know the result of this hypothetical measurement, then we must 'trace over' all outcomes—here, this simply means summing over all M_i with the appropriate weightings:

$$\rho \to \sum_i P_i \frac{M_i \rho M_i^\dagger}{P_i} = \sum_i M_i \rho M_i^\dagger . \tag{7.25}$$

Since this describes decoherence (the evolution from a pure state into a mixed state), there is no corresponding form in terms of state vectors.

7.3.1 An example: unambiguous state discrimination

A prototypical case in which one can see that coupling to a higher-dimensional system may actually be advantageous is the problem of unambiguous state discrimination. The problem is as follows: suppose that you are provided with a single quantum system, which is guaranteed to be prepared in either state $|a\rangle$ or state $|b\rangle$. How well can you tell which of the two states you have? Clearly, if $\langle a|b \rangle = 0$, then a single projective measurement gives you the answer with certainty. But if the overlap does not vanish, it is not possible to unambiguously discriminate the two states all of the time. For simplicity, suppose the two states lie in a two-dimensional Hilbert space; they could, for instance, refer to a horizontally polarized photon ('H') and a 45°-polarized photon ('45'). No projection that is guaranteed to detect every H can reject every 45. At least two different goals are conceivable; one might wish to simply minimize the error rate ('minimum-error discrimination') or one might wish to answer *with certainty* as often as possible ('unambiguous state discrimination'). The former case was solved by Helstrom, and the solution is more or less what one might expect: one chooses a basis symmetrically positioned about $|a\rangle$ and $|b\rangle$—in our $0/45$ example, one would project onto $-22.5°$ or $+67.5°$. It is easy to see that (for instance) a $+67.5°$ detector is much more likely to fire if the photon is polarized at $45°$ than if it is polarized at $0°$ (roughly 5.8 times so); this leads to an error rate of $\frac{1}{2}\left(1 - \sqrt{1 - |\langle a|b\rangle|^2}\right)$ if the two possibilities were equally likely to begin with: about 15% in our simple example.

But how could one ever be *certain* that the state was $|a\rangle$? The only way is to be certain that it is *not* $|b\rangle$, which suggests performing a projective measurement onto the orthogonal state $|\bar{b}\rangle$. Unfortunately, to be certain that the state *is* $|b\rangle$, one would have to project onto $|\bar{a}\rangle$, which is of course not orthogonal to $|\bar{b}\rangle$. Thus on any individual case, one must choose to do one or the other. For instance, we could project onto the $0/90$ (H/V) basis for the photon, but when an H is observed, we have no way of being certain whether the photon was polarized along H or along $45°$, and must simply report 'Don't know'. Only when we observe a V can we conclude that the photon *could not* have been H, and thus must have been along $45°$. Since only half the photons were presumed to be along $45°$, and only half of these are transmitted by a V polarizer, this strategy succeeds only one-quarter of the time, or, more generally $\frac{1}{2}\left(1 - |\langle a|b\rangle|^2\right)$. Interestingly, information-theoretic arguments show that the optimum success rate is $1 - |\langle a|b\rangle|$, or about 29% for our example. There is no way to achieve this optimum with projective measurements, and the reason can be understood in the following simple way. We are studying a two-dimensional system, and wish to be able to report either a or b when we find the appropriate result—but, aware that we will be unable to do so correctly 100% of the time, we must also report 'DK' ('Don't know') on occasion. In other words, we need to perform a measurement with three possible outcomes on this two-dimensional system, in violation of the assumptions of projective measurement. We hence need a generalized measurement, with three 'POVM elements' E_i, corresponding to the three results a, b, and DK. As mentioned earlier, the strategy for doing this is to 'expand the Hilbert space', for instance by allowing the system to interact with a three-dimensional pointer system.

Another way to think about this is to recognize that one would like the two non-orthogonal states $|a\rangle$ and $|b\rangle$ to evolve into perfectly distinguishable (i.e. orthogonal) pointer states $|A\rangle$ and $|B\rangle$ that we can observe and announce. However, since unitary evolution preserves the overlap, there is no purely unitary process which can accomplish this. Something non-unitary must be done; one non-unitary operation is projection. We can allow the system to interact with a pointer so that the overall state expands into a three-dimensional state, but then *project* onto the AB subspace, so that the two states are orthogonal *after projection*. Mathematically, the unitary evolution could be written

$$|a\rangle \rightarrow u\,|A\rangle + v\,|\mathrm{DK}\rangle \,,$$
$$|b\rangle \rightarrow w\,|B\rangle + x\,|\mathrm{DK}\rangle \,, \tag{7.26}$$

where A, B, and DK are three orthogonal pointer states indicating 'we are sure the state is a', 'we are sure the state is b', and 'we don't know', respectively. Unitarity guarantees that $|vx| = |\langle a|b\rangle|$, so the 'failure probability', that is, the probability of being required to admit we don't know, is

$$P_{DK} = \frac{|v|^2 + |x|^2}{2} \geq |vx| = |\langle a|b\rangle|. \tag{7.27}$$

The maximum success rate, achieved when v and x are chosen to saturate the inequality, is, as promised, $1 - |\langle a|b\rangle|$, or about 29% for the example we have discussed. This transformation can be pictured as a rotation of the two-dimensional vectors into a three-dimensional space, as in Fig. 7.4, where the z-axis represents the DK state,

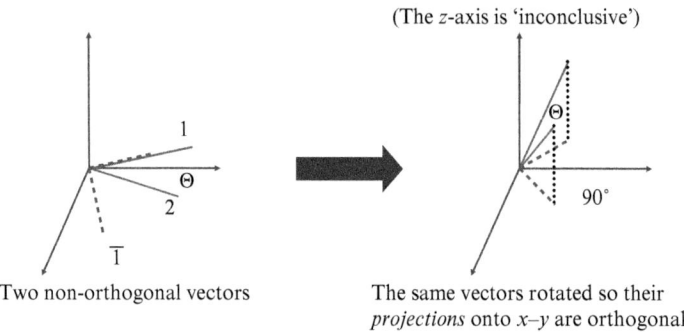

Two non-orthogonal vectors

The same vectors rotated so their
projections onto *x–y* are orthogonal

Fig. 7.4 [Colour online] The POVM for unambiguous discrimination of two non-orthogonal states in a two-dimensional Hilbert space, viewed geometrically as a rotation into a third dimension.

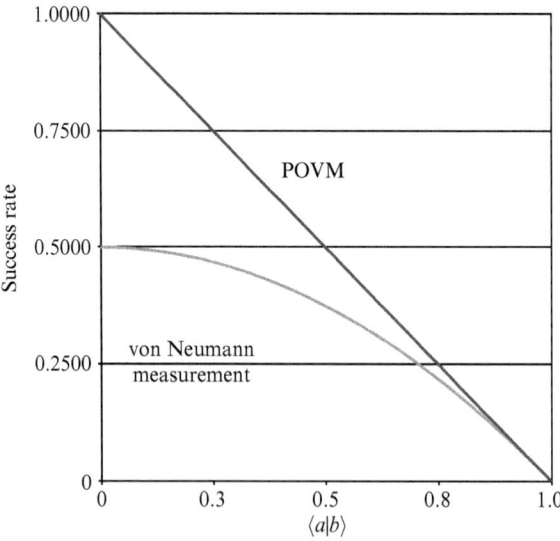

Fig. 7.5 [Colour online] The maximum success probabilities for unambiguous discrimination of two states, as a function of their overlap, for projective measurements and POVMs.

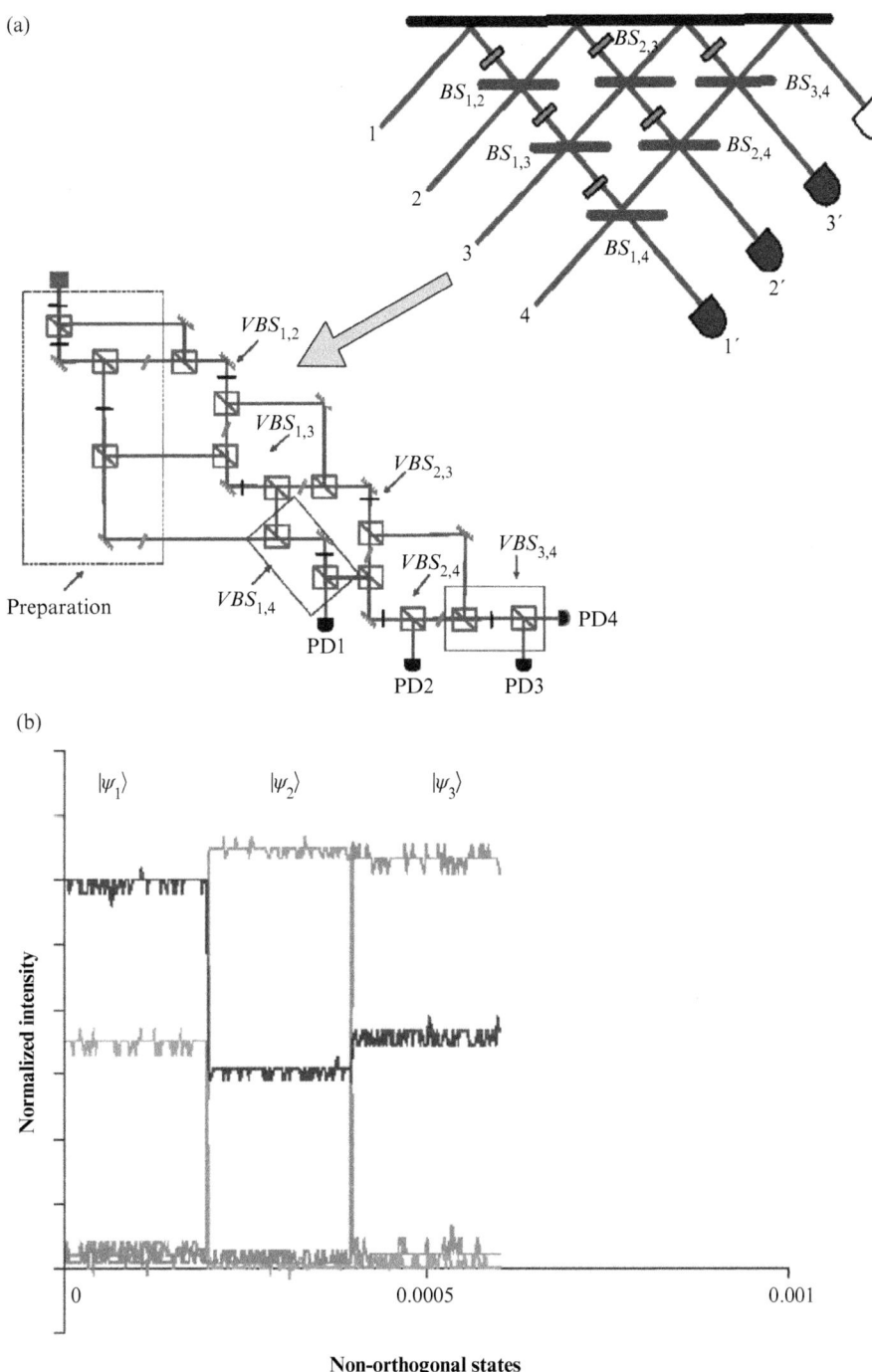

Fig. 7.6 [Colour online] Our group carried out an optical experiment to distinguish among three qutrit states, using the experimental setup shown in (a). In our results, shown in (b), the correct state was identified 55% of the time, which is much better than the < 33% maximum for projective measurements.

and the rotation angle is chosen to make the projections of a and b on the xy plane orthogonal (these projections are then the A and B vectors).

Expressed in the POVM formalism, the operators E_A and E_B are not full projectors onto states a and b, since these would not sum to unity. Instead, three operators are required, such that $I = E_A + E_B + E_{DK}$. In the optimal solution, $E_A = \kappa |\bar{b}\rangle \langle \bar{b}|$, $E_B = \kappa |\bar{a}\rangle \langle \bar{a}|$, and therefore $E_{DK} = I - \kappa \left(|\bar{b}\rangle \langle \bar{b}| + |\bar{a}\rangle \langle \bar{a}| \right)$. The constant κ must be chosen so that E_{DK} remains a positive operator, and when it is maximized under this constraint, one achieves the best-case unambiguous discrimination. Figure 7.5 shows how the success rate of this POVM-based solution compares with the maximum achievable using projective measurements. Note that since no projective strategy can work for more than one state at a time, the success rate with projective measurements is strictly less than 50% except when a and b are orthogonal; on the other hand, the POVM success rate grows smoothly to 100%. In higher dimensions, the advantage is even greater, because, in the most general case, one can distinguish up to d states in a d-dimensional space (the states must be linearly independent for unambiguous discrimination to be possible), and there will usually be no projective measurement basis that can unambiguously identify more than a single one of these d candidate states; an example is shown in Fig. 7.6.

7.4 Complementarity: Feynman's rules and the quantum eraser

We are all familiar with Bohr's complementarity principle, and the conclusion of the Bohr–Einstein debates, that measuring which-path information destroys interference. In Section 7.2.4, we saw how coupling to an environment and *ignoring* the final state of the environment accomplishes this, by erasing the off-diagonal terms in ρ. The implication is that actually measuring 'which path' a particle takes is not required: the mere possibility *in principle* of measuring this—that is to say, the existence of anything in the environmental state from which such information could conceivably be drawn—is already enough. This is the content of 'Feynman's rules for interference':

  I. If two or more fundamentally *indistinguishable* processes can lead to the same final event (e.g. a particle appearing at a given point on a screen), then add the complex amplitudes for these processes, and take the absolute square of the result to find the probability of the event.
 II. If processes are distinguishable, even *in principle*, then take the absolute squares of their amplitudes individually, and add the resulting probabilities to find the total probability of the event.

This language can still confuse people—what exactly do we mean by 'in principle'? For instance, as Einstein repeatedly pointed out, it is certainly possible to measure which slit a particle passes through. The question is whether that possibility exists *after* the particle has reached the screen. If the experiment was built in such a way that once the particle reaches the screen, there is *no longer* any conceivable way to measure which slit it followed, then interference occurs; otherwise, it does not. A more modern perspective would be to say that one should calculate the probability of a given final state of the whole universe, not just of a single particle. If two different processes can lead to the same final state of *everything*, then there is no way to tell from that final state which process occurred. If, on the other hand, some demon has written down which slit the photon went through, then there are two different final states for the universe: one with a photon on the screen and the words 'upper slit' recorded in the demon's lab book, and one with the same photon at the same point on the screen but the words 'lower slit' written in the lab book.

Let us see how this applies to our earlier description of measurement as based on an interaction of the system with a 'pointer' or environment. Let us, without much loss of generality, consider the state of the system in a two-path interferometer, which I will write $\psi_s = \psi_a + e^{i\phi}\psi_b$, where ψ_a and ψ_b are meant to represent the wavefunctions corresponding to particles traversing slits (or generalized 'paths') a and b, respectively. The final probabilities (or intensity pattern) are given by

$$|\psi_s|^2 = |\psi_a|^2 + |\psi_b|^2 + e^{i\phi}\psi_a^*\psi_b + e^{-i\phi}\psi_a\psi_b^*\,, \tag{7.28}$$

where the two cross terms on the right are of course the interference terms. But now let the system interact with a measuring apparatus 'MA', such that $\psi_s \to \psi_s\,|\mathrm{MA}\rangle$. In particular, we imagine that

$$\begin{aligned}
\psi_a &\to \psi_a\,|A\rangle\,, \\
\psi_b &\to \psi_b\,|B\rangle\,.
\end{aligned} \tag{7.29}$$

Repeating the calculation of the intensity pattern, we now have

$$|\psi_s|^2 \to |\psi_a|^2\langle A|A\rangle + |\psi_b|^2\langle B|B\rangle + e^{i\phi}\psi_a^*\psi_b\langle A|B\rangle + e^{-i\phi}\psi_a\psi_b^*\langle B|A\rangle\,. \tag{7.30}$$

A 'good' measurement is normally taken to be one in which a and b can be distinguished with certainty, i.e. $\langle A|B\rangle = 0$; clearly, in this case, the interference terms vanish. The first two terms are unchanged (since A and B are normalized), and the total probability becomes the incoherent sum of $|\psi_a|^2$ and $|\psi_b|^2$, as per rule II.

But this formalism allows us to treat the more general situation, in which our measurement provides *some* information about which path was followed, but not perfect information. In this case, the visibility is reduced, to a maximum value of $|\langle A|B\rangle|$, which runs from 1 in the case of no information (identical final environment states) to 0 in the case of perfect information. This is the origin of the 'duality relations' studied by Greenberger and Yasin, by Jaeger, Shimony, and Vaidman, by Englert, and by others, which can be written

$$D^2 + V^2 \leq 1\,, \tag{7.31}$$

where D is the 'distinguishability', defined as $|P_A - P_B|$, the absolute value of the probability 'bias' achieved between a and b by making the measurement, and V is the visibility as described above. To relate D to $\langle A|B\rangle$, it is sufficient to harken back to Section 7.3.1. What is the minimum error probability for distinguishing states A and B? We saw that it was $\frac{1}{2}\left(1 - \sqrt{1 - |\langle A|B\rangle|^2}\right)$. It can be shown that this is equivalent to saying $D \equiv |P_A - P_B| = \sqrt{1 - |\langle A|B\rangle|^2}$, from which (7.31) follows directly.

7.4.1   The quantum eraser

During the development of quantum mechanics, it was taken for granted—in practice, if not always in principle—that a measurement had to involve amplification up to the macroscopic realm, and what Bohr described as an 'uncontrollable, irreversible disturbance'. With the development of technologies such as cavity QED, however, it became clear that the sort of 'measurement apparatus' we have just described could be a single photon, or other individual quantum particle. Scully and various co-authors (notably Druhl, Hillery, Englert, and Walther) asked whether storing which-path information in a single quantum must necessarily disturb the system in the way Bohr imagined, and in particular, whether it might be possible to practically reverse it. (While the Schrödinger equation is time-reversal symmetric, it is difficult to even conceive practically of reversing the motions of all the electrons, holes, and phonons excited in a slab of Silicon when a photon is detected by a classical avalanche photodiode: this is the same problem of practical irreversibility we come up against in classical statistical mechanics. But if a measurement is effected by allowing a single atom to pass through a high-Q cavity for half a Rabi period, thus exchanging energy with a single mode of the radiation field, it is easy to imagine 'undoing' this interaction in precisely the same manner.)

As a toy example (whose connection to some of the classic experiments on entangled photons will easily become clear), imagine a source (Fig. 7.7) which emits a single 'signal' photon at a time, in a coherent superposition of paths s_1 and s_2 (this could simply be a beam splitter illuminated by a weak laser beam). Interference is of course observed at the two detectors as the path lengths are changed, for example by displacing the beam splitter. Now suppose that, unbeknownst to us, the NSA had convinced the manufacturer of the source to build in a 'back door', surreptitiously sending them an 'information photon' i_1 whenever s_1 was emitted, or i_2 whenever s_2 was emitted. This information photon of course plays the role of measuring apparatus, and if, as in Fig. 7.8, $\langle i_1 | i_2 \rangle = 0$, our interference will be destroyed (and we will learn of the presence of the eavesdropper).

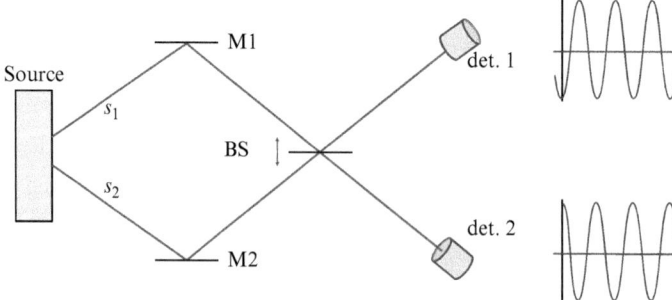

Fig. 7.7 [Colour online] A cartoon version of a single-photon interferometer: the source emits a 'signal' photon in a superposition of paths s_1 and s_2, and the probability of each detector firing varies sinuosoidally (or cosinusoidally) with the phase difference between the two paths.

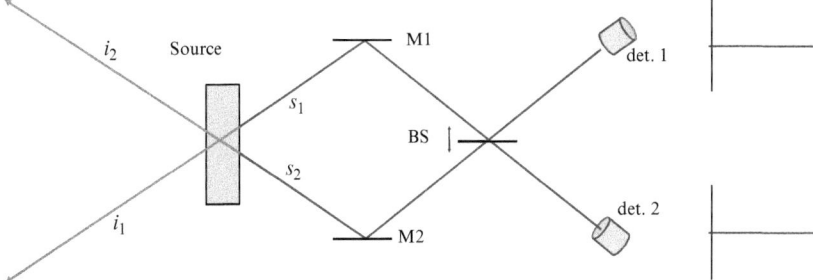

Fig. 7.8 [Colour online] Which-path information can be stored in an 'information photon' entangled with our signal, such that the emitted state is $c_1 |s_1\rangle |i_1\rangle + c_2 |s_2\rangle |i_2\rangle$, as of course is the case in spontaneous parametric down-conversion.

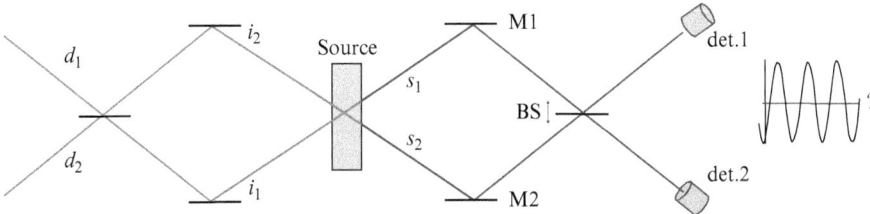

Fig. 7.9 [Colour online] Can the information carried by the 'i' photons be easily erased?

The question posed by Scully et al. was: if one could 'erase' this information, making it impossible for any one to determine which path our signal had taken, would interference be restored? To address this, let us think about how such erasure could take place. A simple proposal is shown in Fig. 7.9. If the two 'i' paths are combined at a 50/50 beam splitter, then detecting a photon exiting either output port of the beam splitter would provide no information at all about whether it originated along i_1 or i_2. Would this do the job? If so, consider the consequences. The 'erasure' beam splitter could be added at any point in time; hypothetically, long after I had detected my signal photon, after the information photon had been propagating along one of its two paths for years. An alien residing in the vicinity of α−Centauri could decide at the last minute whether to keep the information (destroying the interference) or to erase it (restoring the interference). And somehow, my dusty lab book from 4 years in the past would now have a record of interference or its absence, depending on what this alien had just done. Obviously, this is impossible, and it is one way to see that *no* unitary evolution on the measuring apparatus can ever change the status of the signal interferometer. Mathematically, this is because unitary evolution preserves inner products ($\langle A| U^\dagger U |B\rangle = \langle A|B\rangle$), so adding the beam splitter (which of course only leads to a unitary evolution of the photon modes) cannot change the quantum distinguishability of the paths, or therefore the visibility.

Specifically, if the beam splitter mixes i_1 and i_2 to create two new modes d_1 and d_2, it will in fact map each of them onto a different superposition:

$$|i_1\rangle \rightarrow t\,|d_1\rangle + r\,|d_2\rangle\,,$$
$$|i_2\rangle \rightarrow t\,|d_2\rangle + r\,|d_1\rangle\,,$$

(7.32)

where unitarity demands that $r^*t + t^*r = 0$, and hence that the two final states be orthogonal. This means that they are distinguishable in principle. And in fact, it is easy to see that they are distinguishable in practice—it suffices to add yet *another* beam splitter to recombine d_1 and d_2, such that they form two paths in a balanced Mach–Zehnder interferometer. Then i_1 will be guaranteed to exit one port of the resulting interferometer and i_2 the other; placing detectors at those two ports would have precisely the same effect as placing them just behind the source.

The missing element in the functional quantum eraser was hinted at twice above. On the one hand, no *unitary* evolution can change the overlap, but non-unitary evolution could (recall the solution to the unambiguous state discrimination problem). On the other, if the 'erasure beam-splitter' is inserted, then 'detecting a photon exiting [along path d_1 or d_2] would provide no information at all'. If a photon is actually *detected* in one of d_1 or d_2, then in some sense it is too late to add the extra beam splitter and distinguish between $(|d_1\rangle \pm |d_2\rangle)/\sqrt{2}$. Well, strictly speaking, even this is not true. If by 'detected' we simply mean that the field modes have interacted with some measuring apparatus, then remember that in the state of the whole universe (including the measuring apparatus), no coherence has been lost, and it is *mathematically* conceivable that Schrödinger's cat could be interfered back together again. The only way to guarantee that this does not happen is to *project* the system onto d_1 or d_2. This is the non-unitary operation that allows the overlap to be modified. Physically, it corresponds to the act of 'post-selection', i.e. studying only events where a given detector fired on the left, and discarding the others.

Let us analyse this in the framework of (7.29):

$$\frac{|s_1\rangle + e^{i\phi}\,|s_2\rangle}{\sqrt{2}} \longrightarrow \frac{|s_1\rangle\,|i1\rangle + e^{i\phi}\,|s_2\rangle\,|i2\rangle}{\sqrt{2}}$$

(7.33)

exhibited no interference terms, because $\langle i1|i2\rangle = 0$. But now let us *project* this state onto a particular equal superposition of i_1 and i_2, for example $\langle d1| \equiv (\langle i1| - i\,\langle i2|)/\sqrt{2}$ for a symmetric 50/50 beam splitter:

$$\frac{\langle i1| - i\,\langle i2|}{\sqrt{2}} \left(\frac{|s_1\rangle\,|i1\rangle + e^{i\phi}\,|s_2\rangle\,|i2\rangle}{\sqrt{2}} \right) = \frac{|s_1\rangle - ie^{i\phi}\,|s_2\rangle}{2}\,,$$

(7.34)

a resulting state that will of course exhibit interference (albeit with a $-\pi/2$ phase shift). The non-normalized nature of this state reflects the fact that the post-selection only succeeds 50% of the time.

Note that if one instead post-selected d_2, by projecting onto $\langle d2| \equiv (\langle i1| + i\,\langle i2|)/\sqrt{2}$, then the resulting system state would be $\propto (|s_1\rangle + ie^{i\phi}\,|s_2\rangle)/2$, with a

$+i$ in place of the $-i$; interference would still occur, but $180°$ out of phase with the pattern observed for a d_1 post-selection. This comes back to the point established earlier: even after the 'erasure beam splitter', the full ensemble displays no interference. It is necessary to project out a *subensemble*, and only after this non-unitary projection can it be possible for interference to be restored.

The symmetry of the situation should be apparent. What is really observed are *correlations* between the detectors for the information (or 'idler') photon and the detectors for the signal photon. For a given value of ϕ, detecting a photon at d_1 will increase the probability of detecting a photon at the upper signal detector; for a different value of ϕ, it will do the reverse. But one can equally well argue that in the former case, detection at the upper signal detector increased the probability of detecting a photon at d_1. The interference occurs not for subsystems but for the entire composite system, determining the probabilities of the four possible outcomes s_1d_1, s_1d_2, s_2d_1, and s_2d_2. We are free to treat one subsystem or the other as a 'measuring apparatus', but there is no distinct physical role for this in the theory.

7.5 'Interaction-free' measurement and the trouble with retrodiction

Perhaps the most remarkable example of a case of 'indirect' measurement is what was colloquially known as the 'Elitzur–Vaidman bomb' problem until political correctness (perhaps related to the fact that some of the experiments on this topic were carried out at Los Alamos when it was trying to shed its association with bombs) led to the seemingly more neutral name "'interaction-free' measurement" (IFM), which in turn proved so controversial that I shy away from typing it without extra scare quotes. Of course, we have already recognized that every measurement begins with an interaction, but the phrase 'interaction-free' here is meant to have a very specific sense. To understand it, it is helpful (although by no means essential) to return to the original setup of the problem. Imagine that you have purchased a supply of bombs that come equipped with triggers so sensitive that the passage of a single particle of *anything* would be guaranteed to set off an explosion—a single molecule of air, a single electron, a single photon would be enough. Now suppose that you are told that some of your stockpiled bombs have defective triggers, and will never blow up at all—but that the *only physical difference* between the working bombs and the defective bombs is that the former will blow up, and the latter will not. If this is the only physical difference, then it should seem intuitively clear that the only way to establish which kind of bomb one has is to try to set it off, and observe whether or not it explodes; this would make it impossible, of course, to establish that a given bomb was functional without destroying that bomb. (Analogous peacetime applications could be considered, such as detecting the presence of a piece of unexposed film without exposing the film, or taking an X-ray of a subject without running the risk that the subject will be harmed by the absorption of the X-rays, but for the present, we will content ourselves with the theoretical ramifications of the problem.) The goal Elitzur and Vaidman set themselves was to determine whether or not a bomb was operational—at least some of

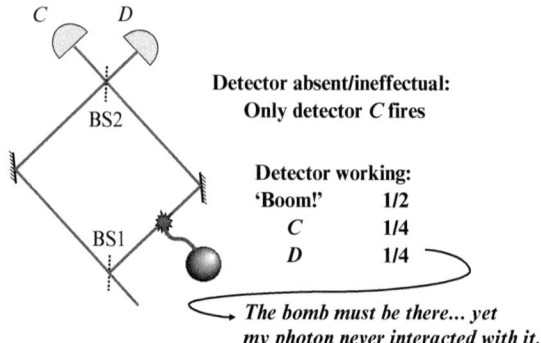

Fig. 7.10 [Colour online] Interaction-free measurement proceeds by placing the object to be studied in one path of a balanced Mach–Zehnder interferometer, which in the absence of any disturbance exhibits constructive interference at detector C and destructive at detector D, making the latter a 'dark port'.

the time—without destroying it. And since interaction with even a single photon was hypothesized to trigger any working bomb, this task can be considered in that limited sense 'interaction-free'.

While the classical intuition is faultless classically, and the task is indeed impossible, in quantum physics the situation is different. The setup can be seen in Fig. 7.10. A balanced (equal-path-length) Mach–Zehnder interferometer is built, and illuminated with one photon at a time. Owing to constructive interference, every photon reaches detector C, making detector D a 'dark port'. Now the trigger of a bomb is placed in one path of the interferometer. For simplicity, let us suppose the trigger detector is of the QND (quantum non-demolition) sort, that is, that the photon passes through unscathed, whether or not the trigger works. If this is a working trigger, then the bomb serves as a 'measuring apparatus' for the photon; if the photon follows the right-hand path, the bomb explodes, and if it follows the left-hand path, it does not. Presumably, these two final states have vanishing overlap, and if my lab is still in one piece, I can safely conclude that the photon took the left-hand path. But of course, if I have which-path information, no interference is possible: therefore, one-half of the photons that avoid the bomb (25% overall) reach detector D. On the other hand, if the trigger had been defective, no which-path information would have been obtained, and interference would persist: no photons would have reached D at all. Armed with this knowledge, I conclude that whenever a photon reaches D, this indicates the presence of a working bomb (one with the capability of establishing which-path information), *even* when (because I did not blow up) I am certain that the photon did not follow the path containing the bomb.

This is one of those straightforward consequences of quantum interference that are so striking that quantum optics researchers, when first told about it, seemed to split fairly evenly into a camp that called it impossible and a camp that called it trivial. It is of course related to the fact that quantum mechanics is not a theory of events, or even of probabilities, but of probability *amplitudes*. In some sense, Feynman's injunction that we must add the amplitudes for every path that could possibly have been followed means that it is not a theory of 'what happened', but of 'all the things that could have happened'. Thus, the question of whether or not the photon *could* have taken the other path without exploding a bomb is a salient one. To those who simplistically feel that quantum mechanics proves the old chestnut that 'a tree falling in a forest when there is no-one there makes no sound', I would counter that this effect demonstrates that, sometimes, we can tell whether or not the tree *would* have made a sound, even without felling it.

Of course, only 25% of the time do we confirm that the bomb is working, without destroying it—25% of the time, we get no information; and 50% of the time, we need to replace our destroyed laboratory. Well, this is not quite true. First of all, what can we conclude if detector C fires? We know that

$$P(C\,|\mathrm{working}) = \tfrac{1}{2} \quad \mathrm{and} \quad P(C\,|\mathrm{defective}) = 1\,, \tag{7.35}$$

and we remember from (7.2) that an observation of C will therefore double the ratio of $P(\mathrm{defective})$ to $P(\mathrm{working})$; if the two possibilities were equally likely, then a single photon at C makes the probability of a defective bomb $\tfrac{2}{3}$, two photons make it $\tfrac{4}{5}$, three photons make it $\tfrac{8}{9}$, and so on. In the case of a defective bomb, we can send as many photons through the interferometer as we like, to reach a desired level of certainty that it is in fact defective. For a working bomb, on the other hand, we can also send another photon through every time one lands at C, but, each time, the bomb will be twice as likely to blow up as to divert the photon to D. Thus, if we continue sending photons through the system until one of the (more or less) conclusive results is reached, $\tfrac{2}{3}$ of the working bombs will blow up and $\tfrac{1}{3}$ will be 'detected'. By modifying the reflectivities of the beam splitters appropriately, Elitzur and Vaidman showed that this success probability could be raised arbitrarily close to $\tfrac{1}{2}$—and Kwiat et al. later showed that by using a 'Zeno-effect'-based strategy (in practice, a multipass rather than a single-pass interferometer), one could in principle identify arbitrarily close to 100% of the bombs without detonating them!

7.5.1 Hardy's 'retrodiction paradox'

Clearly, there is still an interaction *Hamiltonian* at work in the IFM, even if, in specific cases, our final state may escape the nefarious influences of this interaction. But there is a deeper concern here. It is fundamental to quantum mechanics that measurement (at least of a quantity that was not already in a definite eigenstate) changes the state of the system. Elitzur and Vaidman and many others tried to study the case of a *quantum* bomb, and how this indirect measurement would affect it, if the measurement could proceed without blowing it up. The analysis due to Hardy has had the greatest long-standing impact. Instead of considering a bomb which could be functional or defective, Hardy considered a single electron in a superposition of being in the interferometer and being elsewhere. In his Gedankenexperiment, the photon interferometer was replaced with a positron interferometer (Fig. 7.11), and it was assumed to be arranged such that if the electron met the positron in region W, they would be certain to annihilate.

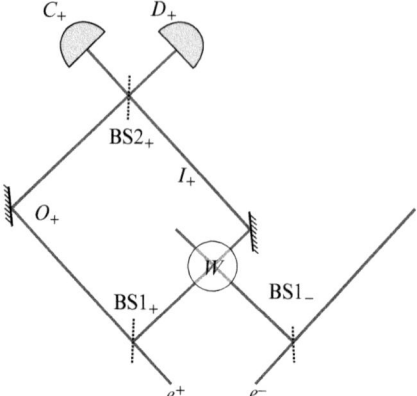

Fig. 7.11 [Colour online] Hardy's version of IFM for a quantum object: an electron that is in a superposition of being in 'interaction region' W and being outside. If an electron and positron meeting at W are certain to annihilate, then a positron reaching detector D_+ might lead us to conclude that the electron was in fact 'in'.

In this way, the electron in W functions as a working bomb, while the electron in an outside path plays the role of the defective bomb. According to the logic of IFM, the positron can only reach detector D_+ if the electron is in fact in W. But this should 'collapse' (or at least decohere) the state of the electron—something which could be studied by adding an additional electron beam splitter, to 'close the interferometer' for the electron, as in Fig. 7.12.

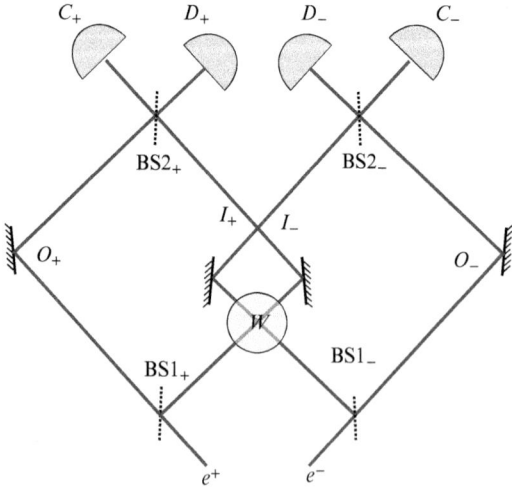

Fig. 7.12 [Colour online] If detection at D_+ tells us the electron was in W while detection at D_- tells us the positron was in W, yet the electron and positron annihilate whenever they meet in W, can the two D detectors ever fire simultaneously?

A glance at Fig. 7.12 should immediately convince you that the IFM discussion suggests that D_+ can only fire if the electron is in W, but since the *electron* is in its own balanced interferometer, D_- can only fire if the *positron* is in W. At first, this seems borne out by direct calculation. Let us start with an initial state

$$|\psi_i\rangle = \frac{|O_+\rangle + |I_+\rangle}{\sqrt{2}} \otimes \frac{|O_-\rangle + |I_-\rangle}{\sqrt{2}} \,. \tag{7.36}$$

After the interaction in W, this state has become

$$|\psi_f\rangle = \frac{|O_+\rangle\,|O_-\rangle + |I_+\rangle\,|O_-\rangle + |O_+\rangle\,|I_-\rangle}{2} + \frac{|\text{boom}\rangle}{2} \,. \tag{7.37}$$

Since $|C+\rangle$ is meant to be the port with fully constructive interference (prior to any electron–positron interaction), we define $|C+\rangle \equiv (|O_+\rangle + |I_+\rangle)/\sqrt{2}$, and $|D_+\rangle$ must then be the orthogonal state $(|O_+\rangle - |I_+\rangle)/\sqrt{2}$. If we detect the positron at detector D_+, the state of the electron can be determined by projecting onto this:

$$\langle D_+|\psi_f\rangle = \frac{|O_-\rangle - |O_-\rangle + |I_-\rangle}{2\sqrt{2}} = \frac{|I_-\rangle}{2\sqrt{2}} \,. \tag{7.38}$$

What do we conclude? First, that the probability of D_+ firing is $\left|\frac{1}{2\sqrt{2}}\right|^2$ or $\frac{1}{8}$—this is the 50% chance that the electron was in W, times the 25% chance of a 'successful interaction-free measurement' on the first try. Second, we conclude that whenever D_+ fires, the electron is indeed in the state I_-. So far, so good. But now let us ask what the probability of $D-$ firing on the same occasion is. Intuitively, an electron in I_- has a 50% chance of reaching detector $D-$, so we see immediately that the overall probability of D_+ *and* D_- firing on the same occasion is $\frac{1}{16}$, or

$$|\langle D_-|\,\langle D_+|\psi_f\rangle|^2 = \left|-\tfrac{1}{4}\right|^2 \,. \tag{7.39}$$

But wait. $D+$ could only fire if the electron was in W; $D-$ could only fire if the positron was in W; and if the electron and positron were both in W, they are sure to annihilate, and therefore no detector at all should fire. Yet a few-line calculation shows us that 1 time in 16, this paradoxical event occurs. This leads many to the conclusion that there is something fundamentally wrong with 'interaction-free' measurements: in essence, that although there may be cases in which you reach the correct conclusions, by avoiding the physical effects of the interaction, you have side-stepped one of the necessary conditions for a true measurement, and cannot be sure that your conclusions will always be valid. Hardy attributes this to the feature of quantum mechanics known as *contextuality*. Because of the theorem of Kochen and Specker, it is known to be impossible to assign 'non-contextual' values to observables. That is, if you wished to construct a 'hidden-variable theory' consistent with quantum mechanics, you could not assume that even on individual events, the value you would get if you measured some observable A would be independent of whether you measured A along with some other operators B and C or whether you measured it along with B' and C'—even if all the operators in either set (A, B, C) or (A, B', C') are known to commute and hence be compatible. The result of a measurement of A depends on the *context*. Thus, the question of whether the electron was in W or not could have a definite answer if you measured it directly—but a different answer if you looked for the electron at D_- instead.

To many, this paradox reinforced the traditional view of quantum measurement. You cannot measure something without disturbing it; and if we did not directly measure which path the electron took through its interferometer, we cannot assign a particular history to it based on observations of the positron; we cannot 'retrodict' its behaviour in this indirect fashion. In the following section, I will introduce another modern measurement paradigm that takes precisely the opposite view, and then examine how they may be reconciled.

7.6   Strong and weak measurements: from von Neumann to Aharonov

7.6.1   von Neumann's measurement interaction

Throughout these lectures, we have followed the view that a measurement proceeds by allowing a system to interact with some 'probe', after which we can examine the state of the probe. This is precisely how von Neumann himself modelled the physical process of measurement. He imagined a 'pointer', like the needle on a dial, whose position after a measurement interaction could depend on some observable of the system. In order to treat this interaction quantum mechanically, his pointer of course had to be a quantum system. Let us consider it to be described by some wavefunction $\psi(x_p)$, where x_p represents the pointer position. This position has some associated uncertainty Δx_p, but for the measurement to be useful, we would assume that Δx_p is not too big; an ideal 'classical' measurement would arise in the limit $\Delta x_p \to 0$.

In order to measure a system observable A_s, we would like to design an interaction such that the pointer position—initially assumed to be $\langle x_p \rangle = 0$—becomes proportional to the value of A_s, as shown schematically in Fig. 7.13. Since the pointer momentum operator P_p is the generator of translations, this can be accomplished by using an interaction Hamiltonian $\mathcal{H}_{\text{int}} = g(t) A_s P_p$, where $g(t)$ is some time-dependent interaction strength. This leads to

$$\frac{d}{dt} \langle x_p \rangle = \frac{1}{i\hbar} \langle [x_p, \mathcal{H}] \rangle$$

$$= \frac{g(t)}{i\hbar} \langle A_s \rangle [x_p, P_p]$$

$$= g(t) \langle A_s \rangle . \tag{7.40}$$

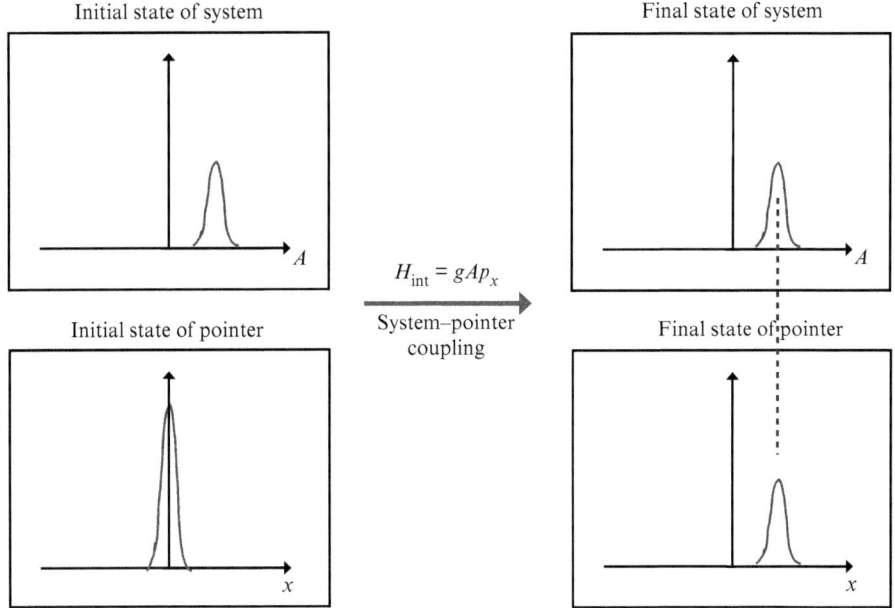

Fig. 7.13 [Colour online] von Neumann's hypothetical measurement interaction couples the system observable of interest, A_s, to the momentum p_x of the pointer. Since the momentum is the generator of translations, the net effect is to translate the pointer wavefunction from $\langle x_p \rangle = 0$ to $\langle x_p \rangle \propto \langle A_s \rangle$.

If we assume that the interaction is brief enough that no time evolution of A_s itself need be considered, we find immediately for the pointer shift

$$\Delta \langle x_p \rangle = \int dt\, g(t) \langle A_s \rangle \equiv G \langle A_s \rangle , \qquad (7.41)$$

where we have defined G as the integrated interaction strength.

The bottom line is that if system and pointer are coupled through a product of their respective operators, then the system operator is the observable *being measured*, while the pointer operator is the *conjugate* ('pointer momentum') to the observable that will serve as a readout ('pointer position'). Some find the situation more intuitive by flipping the roles. One could measure A_s by having the system exert a force on the pointer proportional to the value of A_s. Since a constant force is a potential $U(x_p) = -F x_p$, this would be accomplished by a Hamiltonian $\mathcal{H}_{\text{int}} = -(dF/da)A x_p$. Here the proportionality constant dF/da is the interaction strength, and x_p now plays the role of 'pointer momentum'. How can we see this? If this Hamiltonian acts on the system and pointer for some time, the applied force will impart a momentum shift to the pointer, which is proportional to the value of A_s. Thus, it is the *momentum* of the physical pointer that can be used to read out the system observable, and that we would typically refer to as the 'pointer position'.

Note that if A_s is in an eigenstate, then the operator in $\mathcal{H}_{\text{int}}$ can be replaced by the corresponding eigenvalue a, and the Hamiltonian becomes a pure displacement operator. The pointer wavefunction is shifted, without distortion, by an amount Ga. But if the system is initially in a superposition of states with different eigenvalues a_i, then, by the superposition principle, system + pointer evolve into a superposition in which each A_s-eigenstate is correlated with an appropriately displaced pointer:

$$\{c_1 \left| a_1 \right\rangle + c_2 \left| a_2 \right\rangle + c_3 \left| a_3 \right\rangle + \ldots\} \psi(x_p)$$

$$\rightarrow c_1 \left| a_1 \right\rangle \psi(x_p - Ga_1) + c_2 \left| a_2 \right\rangle \psi(x_p - Ga_2) + c_3 \left| a_3 \right\rangle \psi(x_p - Ga_3) + \ldots . \quad (7.42)$$

The typical assumption, following von Neumann, is that the scale of the position shifts (Ga_i) is much larger than the initial uncertainty of the pointer position Δx_p; otherwise, it would not be possible to determine the value of A_s from the pointer. If this is the case, then the pointer wavefunctions in the above expression are nearly orthogonal, making it a highly entangled state of system and pointer, represented schematically in Fig. 7.14. As in the discussion of (7.30), such a measurement completely destroys the coherence between difference eigenstates of A_s. The measurement has disturbed the system; in particular, by changing the phase relationship between A-eigenstates, it has disturbed the observable conjugate to A_s (as changing the phase

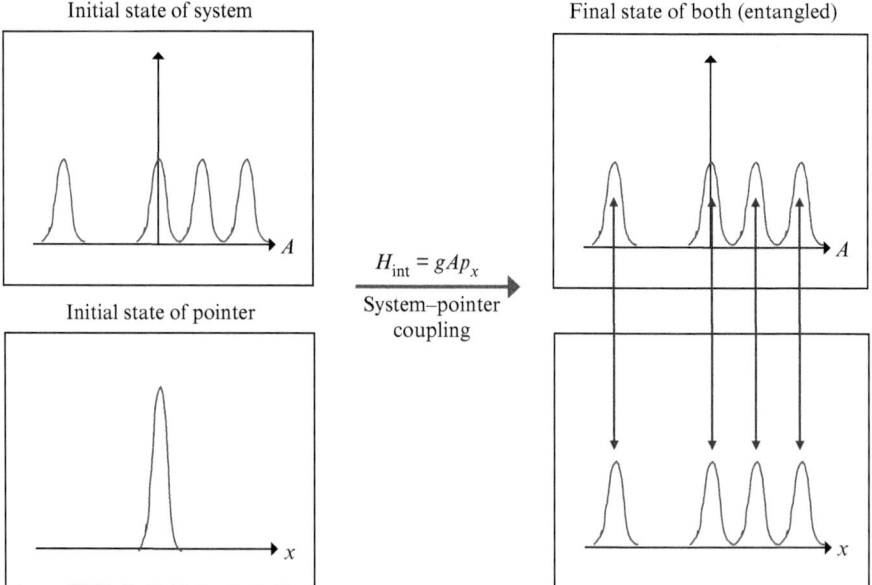

Fig. 7.14 [Colour online] If the characteristic pointer shifts are large compared with the initial width of the pointer wavefunction, then in general the final state of system and pointer will be entangled, each value of A_s being correlated with an appropriately shifted pointer state.

relationship between different positions disturbs the momentum, and changing the phase relationship between different momenta disturbs the position).

There is a complementary way to understand the origin of this disturbance. In the limit of a 'good' measurement, we assumed Δx_p had to be small. This means, of course, that ΔP_p must be large. But from the point of view of the system, the interaction Hamiltonian that acts on it through A_s is proportional to P_p; if this latter quantity is uncertain, then the system feels an effect of uncertain magnitude. Since A_s is the displacement operator for its own conjugate variable, the disturbance manifests itself on this conjugate. (A_s itself commutes with $\mathcal{H}_{\text{int}}$ by construction, and therefore cannot be disturbed by this interaction.) For instance, if A_s were the position of some particle, then the Hamiltonian would correspond to an actual force, proportional to P_p, acting on the particle. The more uncertain P_p, the more uncertain the force, and the greater the random disturbance to the particle's final momentum (the variable conjugate to A_s). The conclusion is already well known to you. The more accurately you measure A_s (e.g. particle position), the more disturbance you create for the conjugate variable (e.g. particle momentum). Note that in von Neumann's formalism, this result falls out directly of the coherent Schrödinger interaction between system and pointer, and no assumption of any 'state reduction' is required. Mathematically, the 'randomness' arises when we trace over the pointer with which the system has become entangled, leaving the latter in a mixed state with reduced coherence.

7.6.2 Weak measurement

Aharonov and his co-workers realized in the late 1980s that this formalism opened the door to a new possibility. If, contrary to von Neumann, we take the opposite limit, of large Δx_p, then it is possible to have a state with small ΔP_p, and hence little disturbance arising in the interaction Hamiltonian. Seen from the perspective of (7.42), the large position uncertainty means that the overlap between $\psi(x - Ga_i)$ and $\psi(x - Ga_j)$ may be nearly unity, so that very little information is gained from an individual measurement and there is very little concomitant loss of coherence (recall Section 7.4). (The state very nearly factors into a product of a system state and a pointer state, so that tracing over the latter leaves the former unchanged.) Note that the *average* shift of the pointer is still given by $G\langle A_s \rangle$, as we saw in (7.41), which did not require any assumptions about the width of the pointer wavepacket.

Part of the motivation for thinking about such 'weak' measurements, where the disturbance could be reduced at the expense of the amount of information extractable, was to enable one to discuss measurements on post-selected systems, such as those we encountered in the discussion of the quantum eraser, or indeed Hardy's paradox. Could we imagine measuring which path the electron had been in, but only on those occasions where it eventually reached a particular detector? Of course, the standard view of quantum mechanics would have us say no; we must choose whether to place a detector inside the interferometer or after the final beam splitter. But by now we realize that the choice between 'a detector' and none is too limiting; a measurement can be any interaction between the system and some second object, and this interaction may be stronger or weaker. The time-reversibility of the Schrödinger equation led Aharonov to argue that one should be able to get as much information about present observables from final conditions (post-selection) as from initial conditions (state preparation). In the usual paradigm, the symmetry is apparently broken by the 'uncontrollable, irreversible' nature of traditional measurement. I cannot really talk about the electron that was following a trajectory from a given input to a given output if at some intermediate time I measure where it is, violently disturbing its momentum. But, argued Aharonov et al., in the limit of an arbitrarily weak measurement, the disturbance to my system could be arbitrarily small. Of course, the deeper one moves into this limit, the smaller the effect on the pointer. However, one could repeat identical measurements on a large 'subensemble' of identically prepared *and* identically post-selected systems, and the average pointer shift would still provide information. (As observed above, in the absence of post-selection, the pointer shift remains proportional to $\langle A \rangle$, no matter how small the coupling G or how large the pointer uncertainty Δx_p.)

Figure 7.15 shows a particular example. I show you three boxes, labelled A, B, and C, and offer to play a game. I will hide a \$20 bill in the boxes, and if you can find the bill, you get to keep it. But to make the game more interesting, instead of placing the bill in a particular box, I place it in a coherent superposition $(|A\rangle + |B\rangle)/\sqrt{2}$. To make the game more fair, on the other hand, I give you an additional piece of information: I promise that I will find it in *another* superposition, $(|B\rangle + |C\rangle)/\sqrt{2}$. (Of course, I cannot guarantee that this will occur on every try; it is merely a post-selection. I promise you that if I *fail* to find the bill in that superposition, we will simply set up the game again and you will have a second chance, and so on until the post-selection succeeds.) Which box do you bet on?

Most physicists when asked this question refuse to answer; we have been trained to believe that the bill does not reside in any particular box, but really is in a superposition of A and B until I choose to measure whether or not it is in $(|B\rangle + |C\rangle)/\sqrt{2}$, at

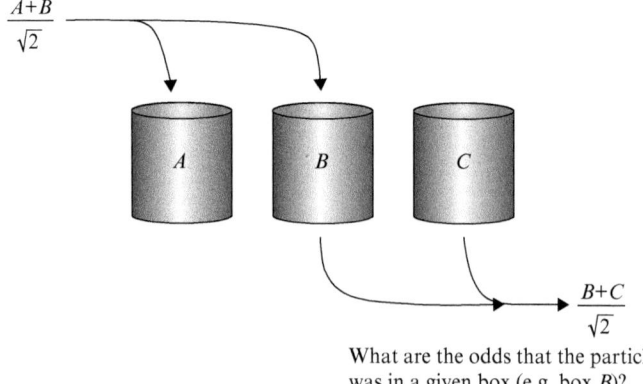

What are the odds that the particle
was in a given box (e.g. box *B*)?

Fig. 7.15 The quantum three-box problem. If I promise you that I will prepare a \$20 bill in the state $(|A\rangle + |B\rangle)/\sqrt{2}$ and that I will find it in the state $(|B\rangle + |C\rangle)/\sqrt{2}$, how much can you conclude about where it is likely to have been between the preparation and the post-selection? Would it be possible to verify this?

which instant it makes a 'quantum jump' into that state, or an orthogonal one. Most non-physicists immediately give the right answer, which is B. There is no way the bill could have been in C, given the promised preparation (we neglect the possibility of tunnelling between boxes); yet there is no way the bill could have been in A, in light of the state I found it in. According to this rather classical reasoning, there is a 100% probability that the bill was in B. But are we allowed to do this quantum mechanically or not? Is '$|A\rangle + |B\rangle$' tantamount to saying 'A or B', or not? The very fact that physicists hesitate when presented with this problem seems to me to point to a gap in our real intuition for the quantum world, and this is at least one important role for discussions of weak measurements.

For indeed, if—before the post-selection was attempted—you made a projective measurement of whether or not the bill was in one of the boxes, you would greatly disturb the state of the bill. There would still be correlations with the success of the post-selection, but you might not wish to think of the bill you observed as being in the process of following its natural trajectory from preparation to post-selection. Nevertheless, the correlations are easy to reason through: if you found the bill in box A, the post-selection could never succeed. And naturally, you could never find the bill in box C in the first place. Therefore, if you looked in A, B, and C on various occasions, the only place you could find the particle *and* have the post-selection succeed would be box B. Not finding the particle in B would collapse it into A and guarantee that the post-selection would fail. Thus, even for projective measurements, in this case the classical intuition holds. More generally, the results of strong measurements conditioned on later post-selection are given by the ABL (Aharonov–Bergmann–Lebowitz) rule, which is essentially the straightforward application of Bayesian reasoning to these inferences, along the lines that we have applied several times now in these lectures. In what follows, though, we will consider the results of *weak* measurements on such systems, asking what the average shift of a weakly coupled pointer would be, on those occasions when the post-selection succeeded.

This is of interest because there are more general situations in which one can really not tolerate the disturbance of these intermediate strong measurements, unlike the case of the three-box problem. For instance, consider a particle that is at $t = 0$ in a well-localized, real-valued wavepacket centred at $x = 0$. Its momentum is of course very uncertain, and has zero expectation value because of the real value of the wavefunction. But suppose that at $t = \tau$ you make a projective measurement, finding the particle in a well-localized wavepacket centred at $x = x_0$ (also with a very uncertain momentum). What would you conclude about the average value of the momentum at times $0 < t < \tau$? The strict textbook approach rejects the question—or, rather, suggests that $\langle P \rangle$ was 0, and remained 0 until the instant when the final measurement disturbed the system (after which $\langle P \rangle$ was in fact 0 again anyway). Common sense, on the other hand, suggests that a particle that travelled a distance x_0 in a time τ should have had a momentum somewhere around mx_0/τ. But of course, if a strong measurement of momentum were actually made between times 0 and τ, any momentum within the initial uncertainty could be found; and since any momentum eigenstate is delocalized, the post-selection would be equally likely regardless of which was obtained. It turns out in such cases that if a *weak* measurement is performed, the average result conforms with intuition; this is a result of a number of mathematical properties of weak measurements that make some researchers wish to ascribe some deep significance to the values they reveal. Such an attribution remains controversial, however, so we will focus on the incontrovertible experimental predictions of the formalism, which rely only on standard quantum theory.

Suppose a system begins in some initial state $|i\rangle$, and a pointer in some state $|\psi_p\rangle$. We once more assume an interaction Hamiltonian $\mathcal{H} = g(t)A_s P_p$, and an integrated interaction strength $G \equiv \int g(t)\,dt$, but we will now make the additional assumption that the measurement is weak, i.e. that the characteristic pointer shifts —GA— are small compared with the smallest lengthscale set by the pointer momentum distribution, $\hbar/|P|$. Schrödinger evolution yields

$$U\,|\psi_p\rangle\,|i\rangle = e^{-(i/\hbar)\int dt\,\mathcal{H}(t)}\,|\psi_p\rangle\,|i\rangle$$

$$\approx \left(1 - \frac{i}{\hbar}G A_s P_p\right)|\psi_p\rangle\,|i\rangle\,. \tag{7.43}$$

Now suppose we successfully post-select the particle to be in final state $|f\rangle$. The state of the pointer can be determined by projecting (7.43) onto $\langle f|$:

$$|\psi_p'\rangle = \langle f|\,U\,|\psi_p\rangle\,|i\rangle$$

$$\approx \langle f|i\rangle\,|\psi_p\rangle - \frac{iG}{\hbar}\,\langle f|\,A_s\,|i\rangle\,P_p\,|\psi_p\rangle$$

$$= \langle f|i\rangle\left[1 - \frac{iG}{\hbar}\frac{\langle f|\,A_s\,|i\rangle}{\langle f|i\rangle}P_p\right]|\psi_p\rangle$$

$$\approx \langle f|i\rangle e^{-iG a_w P_p/\hbar}\,|\psi_p\rangle\,, \tag{7.44}$$

where we have defined

$$a_w \equiv \frac{\langle f|\,A_s\,|i\rangle}{\langle f|i\rangle} \tag{7.45}$$

and have made the weakness assumption that $|GA_s P_p| \ll \hbar$.

Recalling that P_p is the generator of translations, we see that the action of the exponential in (7.44) is to translate $|\psi_p\rangle$ by the amount Ga_w. Recalling that (in the absence of post-selection) the pointer shift was $G\langle A_s\rangle$, we see that this shift corresponds to the result we would expect if the value of A were given by a_w, which is termed the 'weak value', and is given by (7.45).

7.6.3   Some implications of the weak-value formula

The question of whether or not to ascribe 'reality' to this weak value between the preparation and the post-selection is a difficult philosophical one. But it has a rigorously defined operational meaning, in that it tells us how strong the effect of the system would be on any other object coupled to A_s, in the limit where this coupling was not so large as to significantly disturb the evolution of the system between preparation and post-selection.

Some interesting properties are immediately evident from (7.45). If either the initial or the final state is an eigenstate of A_s, then the weak value is guaranteed to be the corresponding eigenvalue: we get information as reliably from final conditions as from initial conditions. Furthermore, the formula is linear; the weak value of $A + B$, generally denoted by $\langle A + B \rangle_{\mathrm{wk}}$, is equal to $\langle A \rangle_{\mathrm{wk}} + \langle B \rangle_{\mathrm{wk}}$, regardless of whether or not A and B commute. (Roughly speaking, this is why from information about $\langle X(0) \rangle_{\mathrm{wk}}$ and $\langle X(\tau) = X(0) + P\tau/m \rangle_{\mathrm{wk}}$—the localization of a particle at two instants in time—we can draw conclusions about $\langle P/m \rangle_{\mathrm{wk}}$, the velocity of the particle.) In a sense, this points to one of the most dramatic features of weak measurements, which is their non-contextuality. While the Kochen–Specker theorem tells us that we cannot imagine assigning definite values to particular observables on a shot-by-shot basis without knowledge of the 'context' of the measurement, the weak-value formula tells us that the weak value of a given observable in some post-selected ensemble is entirely independent of any other weak values that might be measured simultaneously. As we shall see, this makes weak measurement a particularly interesting approach from which to reanalyse Hardy's paradox, and its suggestion that retrodiction (the stuff of IFM, but really the stuff of weak measurement as well) is inherently untrustworthy.

It is also simple to show that if post-selection is done onto a complete, orthonormal set of final states, and the weak value is averaged over all these post-selections, weighted by the success frequency, one recovers the usual expectation value. This has to be the case, since if one sums up all the possible post-selections, one is back to measuring the pointer shift irrespective of the final state of the particle, as in (7.41). On the other hand, (7.45) has some disquieting features as well. For instance, as $\langle f | i \rangle \to 0$, the weak value may diverge. In fact, the weak value is free to take essentially arbitrary values, even ones entirely outside the eigenvalue spectrum of A_s. Of course, given the large pointer uncertainty assumed at the outset, there is never an individual event on which one can say that one observed a value of A_s beyond its spectrum; but the average shift could be extremely large. Since this only occurs when the post-selection probability $|\langle f | i \rangle|^2$ is very small, there is no contradiction with the behaviour of the full ensemble, for which the pointer shifts by an amount $G\langle A_s \rangle$, which is of course constrained by the eigenvalues of A_s.

Worse, a_w is not even guaranteed to be real, even for Hermitian operators A_s. But there is a straightforward interpretation of an imaginary part to the weak value. The effect of

$$e^{-iG(i\,\mathrm{Im}\,a_w)P_p/\hbar}\,|\psi_p\rangle \tag{7.46}$$

is to multiply the momentum-space wavefunction $\tilde{\psi}(p) \equiv \langle p | \psi_p \rangle$ by a weighting factor $e^{(G\,\mathrm{Im}\,a_w)p/\hbar}$. Unlike the real part of a_w, which indicates a position shift for the pointer (the expected result of a measurement), this imaginary part occasions a *momentum* shift of the pointer. For the simple case of an initial Gaussian pointer wavefunction, this is easy to calculate:

$$e^{-x^2/4\sigma^2} \rightarrow e^{-(x-Ga_w)^2/4\sigma^2}$$

$$\propto e^{-(x-G\,\mathrm{Re}\,a_w)^2/4\sigma^2} e^{(ixG\,\mathrm{Im}\,a_w)/2\sigma^2}, \tag{7.47}$$

where the first exponential indicates a position shift for the pointer of $\Delta x = G\,\mathrm{Re}\,a_w$ and the second is a momentum boost of $\Delta p = G\,\mathrm{Im}\,a_w/2\sigma^2$. Note that, unlike the position shift, the momentum shift depends on the width of the initial pointer wavepacket; as the measurement is made infinitely weak and $\sigma \rightarrow \infty$, this effect vanishes. It is a reflection of the *back-action* of the measurement on the system. Recall that when a_w is summed over all possible post-selections, it recovers the expectation value. Since $\langle A_s \rangle$ is real, this means that the imaginary parts cancel out; only when a particular post-selection is made can this change the expectation value of the pointer momentum, and for the simple reason that the presence of the pointer momentum in the interaction Hamiltonian could disturb the particle, making the post-selection more or less likely to succeed for different values of momentum (within the initial distribution). As σ gets larger and larger, the pointer momentum distribution may be made narrower and narrower, until, in the limit, the back-action goes away.

What are the implications for the three-box problem? Measuring whether or not the bill is in box B is simply measuring the projection operator $|B\rangle\langle B|$; the eigenvalues of this operator are 1 (for 'yes, the bill is in B') and 0 (for 'no, the bill is elsewhere'), and its expectation value represents the probability that the bill is in B. Given

$$|i\rangle = \frac{|A\rangle + |B\rangle}{\sqrt{2}} \tag{7.48}$$

and

$$\langle f| = \frac{\langle B| + \langle C|}{\sqrt{2}}, \tag{7.49}$$

it can be seen at once that $\langle f|\,\mathrm{Proj}(B)\,|i\rangle = \frac{1}{2}$ and $\langle f|i\rangle = \frac{1}{2}$, implying that $\langle \mathrm{Proj}(B)\rangle_{\mathrm{wk}} = 1$: the weak-valued probability is 100%, meaning on an operational level that when the post-selection succeeds, the effect of an interaction on a pointer would be *exactly* as strong as if the bill had simply been prepared entirely in box B.

The three-box problem discussed by Aharonov et al. is more surprising even than this. Recall that the only property essential to proving that we could not find the particle in A or C was the orthogonality of states A and C. But let us then make three new boxes, A', B, and C', and replace A with $A'+C'$ and C with the orthogonal state $A' - C'$. Specifically, suppose the bill is prepared in

$$|i\rangle = \frac{|A'\rangle + |B\rangle + |C'\rangle}{\sqrt{3}} \tag{7.50}$$

and post-selected in

$$|f\rangle = \frac{|A'\rangle + |B\rangle - |C'\rangle}{\sqrt{3}} . \qquad (7.51)$$

Now, the particle cannot be found in $|A'\rangle - |C'\rangle$, since that is orthogonal to $|i\rangle$, nor can it be found in $|A'\rangle + |C'\rangle$, since that is orthogonal to $|f\rangle$. Once more, the only remaining state is $|B\rangle$, and we conclude that it must be found there with certainty. But a moment's inspection will alert you to the fact that both $|i\rangle$ and $|f\rangle$ are symmetric under exchange of A' and B. The same argument then applies to A'. The particle cannot be found in $|B\rangle + |C'\rangle$, nor can it be found in $|B\rangle - |C'\rangle$, so it must be found in $|A'\rangle$. Indeed, it is easy to calculate $\langle P_{A'}\rangle_{\mathrm{wk}} = \langle P_B\rangle_{\mathrm{wk}} = 1$. But can a particle be in two places at once? More to the point, we know by completeness that $\mathrm{Proj}(A') + \mathrm{Proj}(B) + \mathrm{Proj}(C')$ equals the identity I, with expectation value 1. So, by linearity of weak values, $\langle P_{A'}\rangle_{\mathrm{wk}} + \langle P_B\rangle_{\mathrm{wk}} + \langle P_{C'}\rangle_{\mathrm{wk}} = 1$. This relation can only be satisfied because $\langle P_{C'}\rangle_{\mathrm{wk}} = -1$. While these negative weak values for positive-valued operators may seem especially disturbing, their operational meaning is precisely the same as before. If a pointer is allowed to interact with box C in such a way that if the particle was in box C, the pointer would get attracted to the box, this negative weak value means that upon successful post-selection of the system, the *average* pointer would exhibit a momentum shift away from the box. This is a real physical consequence of the weak values, and although I will not delve into it here, all such effects can be understood in terms of quantum interference.

This brings us back to the question of Hardy's paradox, and whether or not one can use 'post-selections' (detections of the positron at C_+ or D_+, for instance) to draw conclusions about the past whereabouts of the electron. A few minutes' thought should suffice to convince you that, indeed, when D_+ fires, the weak value of the projector onto $|I_-\rangle$ (the state in which the electron is in the interaction region W) is 1. In the *measurable* sense of the average effect on a weakly coupled pointer, the retrodiction is completely valid. And yet quantum mechanics tells us that D_+ and D_- may fire together—how is this possible? When D_+ and D_- both fire, does that mean that the weak-valued 'probability' for the electron and the positron to be in W simultaneously is 100%? As a matter of fact, no. Since the annihilation removed the term $|I_+\rangle |I_-\rangle$ from the superposition in (7.37), the weak value of the probability for both particles to be in W is strictly 0. How, then, can the weak-valued probability for *each* particle to be in W be 100%? The resolution is of precisely the same spirit as the three-box problem: there is 100% weak-valued probability for the electron to be in W and the positron to be outside, but there is *also* a 100% weak-valued probability for the positron to be in W and the electron to be outside; and since the complete set of projectors must add up to unity, there is a -100% weak-valued probability for *neither* particle to be in W. The situation is summarized in Fig. 7.16.

These results were recently confirmed in two experiments (Fig. 7.17). In conclusion, if one is willing to accept the statement that the value of an observable can only be determined in practice by measuring the size of its effect on some other physical system, and that if the interactions disturb the system, it is reasonable to use the

But what can we say about where the particles
were or weren't, once D_+ and D_- fire?

Probabilities	e^- in	e^- out	
e^+ in	**0**	1	1
e^+ out	1	–1	0
	1	0	

Fig. 7.16 [Colour online] The weak-value 'resolution' to Hardy's paradox. The weak-valued probability for both particles to be in W simultaneously vanishes, even while the weak-valued probability for *each* of them to be in W is 100%; the two statements can only be reconciled owing to the possibility of negative weak-valued probabilities.

(a)

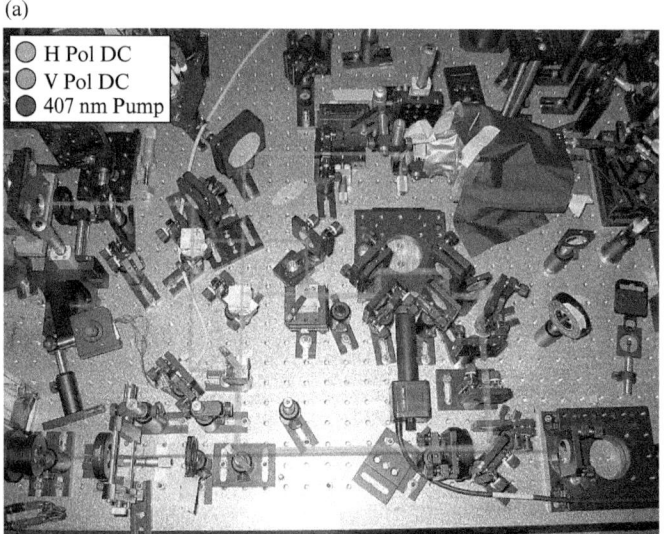

(b)

Experimental weak values ('probabilities'?)

	$N(I_-)$	$N(O_-)$	
$N(I_+)$	0.243±0.068	0.663±0.083	0.882±0.015
$N(O_+)$	0.721±0.074	−0.758±0.083	0.887±0.021
	0.925±0.024	**−0.039±0.023**	

Fig. 7.17 [Colour online] The Toronto experiment on Hardy's paradox. (a) The two-photon interferometer that replaced the electron and positron interferometers of the original proposal, quantum-enhanced up-conversion in a $\chi^{(2)}$ crystal playing the role of the $e^- e^+$ annihilation. (b) Experimentally extracted weak values, to be compared with the table in Fig. 7.16.

weakest possible interaction strengths to minimize the disturbance, while measuring the *ratio* of the effect to the interaction strength; then these weak values provide a natural way to talk about properties of systems at times intermediate to preparation and post-selection. Given the importance of post-selection in many aspects of quantum information, this should make them of particularly wide applicability. And, despite the seemingly paradoxical nature of Hardy's thought experiment and its implications about contextuality, once one follows this set of definitions, one finds that there are no paradoxes, but rather a fully self-consistent and operational framework within which one can visualize the time evolution of a quantum system.

7.6.4 A Bayesian rederivation of the weak-value formula

Given the way we introduced measurement, in terms of Bayesian inference, it may be interesting to see an alternative construction of the weak-value formula, based purely on conditional probabilities. Starting from a reasonable proposal for what quantum mechanical expression to write down to replace classical conditional probabilities, (7.45) can be derived in a few lines without any reference to the form of the von Neumann interaction, or Schrödinger evolution. Along with the other tantalizing mathematical properties of weak values, this close connection to probability theory is one more hint that perhaps these expressions do have some fundamental significance in quantum mechanics. Recall that a probability is given by the expectation value of a corresponding projection operator, with its eigenvalues of 0 and 1. Let us make the assumption that the *joint* probability of two propositions j and f is given by $\langle \text{Proj}(f)\text{Proj}(j) \rangle$, where, since projectors onto different bases do not in general commute, we will choose the time-ordered product, with the earlier measurements on the right and the later ones on the left.

The expectation value of A can be written

$$\langle A \rangle = \sum_j a_j P(j) \,, \tag{7.52}$$

where a_j are the eigenvalues of A and $P(j)$ are their respective probabilities. If by a 'weak value' what we really mean is simply the 'conditional expectation value', i.e. the average value of A *given* that a desired final state $|f\rangle$ is observed, then it is natural to write by analogy

$$\langle A \rangle_{\text{wk}} = \sum_j a_j P(j \,|\, f) \,. \tag{7.53}$$

We have already seen that

$$P(j \,|\, f) = \frac{P(j \& f)}{P(f)} \,. \tag{7.54}$$

Given our joint-probability hypothesis, we can evaluate this as

$$P(j \,|\, f) = \frac{\langle \text{Proj}(f)\text{Prof}(j) \rangle}{\langle \text{Proj}(f) \rangle} \tag{7.55}$$

and plug this into (7.53) to calculate

$$
\begin{aligned}
\langle A \rangle_{\mathrm{wk}} &= \sum_j a_j \frac{\langle \mathrm{Proj}(f)\mathrm{Prof}(j)\rangle}{\langle \mathrm{Proj}(f)\rangle} \\
&= \frac{\langle \mathrm{Proj}(f) \sum_j a_j \mathrm{Prof}(j)\rangle}{\langle \mathrm{Proj}(f)\rangle} \\
&= \frac{\langle \mathrm{Proj}(f) A \rangle}{\mathrm{Proj}(f)} \\
&= \frac{\langle i|f\rangle \langle f| A |i\rangle}{\langle i|f\rangle \langle f|i\rangle} \\
&= \frac{\langle f| A |i\rangle}{\langle f|i\rangle}
\end{aligned}
\tag{7.56}
$$

Acknowledgements

I would like to acknowledge all members of my group, present and past, for the collaborations that have informed these notes, along with the many people who have influenced the development of my thinking about the topics treated in these lectures, notably Yakir Aharonov, Lev Vaidman, Sandu Popescu, János Bergou, Robin Blume-Kohout, and Howard Wiseman.

Further reading

More detailed references related to quantum measurement

K. Jacobs, *Quantum Measurement Theory and its Applications*. Cambridge University Press, Cambridge, 2014.

M. A. Nielsen and I. L. Chuang, *Quantum Computation and Quantum Information*. Cambridge University Press, Cambridge, 2000.

C. W. Helstrom,. *Quantum Detection and Estimation Theory*. Academic Press, New York, 1976.

H. M. Wiseman and G. J. Milburn, *Quantum Measurement and Control*. Cambridge University Press, 2009.

J. von Neumann, *Mathematical Foundations of Quantum Mechanics*. Princeton University Press, Princeton, NJ, 1955.

V. B. Braginsky, F. Ya. Khalili, and K. S. Thorne, *Quantum Measurement*. Cambridge University Press, Cambridge, 1992.

W. Zurek, Phys. Today **44**, 36 (October 1991).

A.M. Steinberg, Lecture Notes on Experimental Quantum Measurement. Available online at http://www.physics.utoronto.ca/~aephraim/2206/#notes.

State discrimination

I. D. Ivanovic, Phys. Lett. A **23**, 257 (1987).

A. Chefles and S. M. Barnett, J. Mod. Opt. **45**, 1295 (1998).

S. M. Barnett and E. Riis, J. Mod. Opt. **44**, 1061 (1997).

B. Huttner, A. Muller, J. D. Gautier, H. Zbinden, and N. Gisin, Phys. Rev. A **54**, 3783 (1996).

R. B. M. Clarke, A. Chefles, S. M. Barnett, and E. Riis, Phys Rev. A **63**, 040305 (2001).

R. B. M. Clarke, V. M. Kendon, A. Chefles, S. M. Barnett, E. Riis, and M. Sasaki, Phys Rev. A **64**, 012303 (2001).

T. Rudolph, R. W. Spekkens, and P. S. Turner, Phys. Rev. A **68**, 0101301 (2003).

M. Takeoka, M. Ban, and M. Sasaki, Phys. Rev. A **68**, 012307 (2003).

A. Chefles, Phys. Lett. A **239**, 339 (1998).

D. Dieks, Phys. Lett. A **126**, 303 (1998).

A. Peres, Phys. Lett. A **128**, 19 (1988).

A. Chefles and S. M. Barnett, Phys. Lett. A **250**, 223 (1998).

Y. Sun, M. Hillery, and J. A. Bergou, Phys. Rev. A **64**, 022311 (2001).

J. A. Bergou, M. Hillery, and Y. Sun, J. Mod. Opt. **47**, 487 (2000).

Y. Sun, J. A. Bergou, and M. Hillery, Phys. Rev. A **66**, 032315 (2002).

J. A. Bergou, U. Herzog, and M. Hillery, Phys. Rev. Lett. **90**, 257901 (2003).

M. Mohseni, A.M. Steinberg, and J. Bergou, Phys. Rev. Lett. **93**, 200403 (2004).

M. A. P. Touzel, R. B. A. Adamson, and A. M. Steinberg, Phys. Rev. A **76**, 002314 (2007).

Complementarity and quantum erasers

Z. Y. Ou, L. J. Wang, X. Y. Zou, and L. Mandel, Phys. Rev. A **41**, 566 (1990).

M. Hillery and M. O. Scully, in *Quantum Optics, Experimental Gravitation, and Measurement Theory*, ed. P. Meystre et al. Plenum, New York, 1983, pp. 65–85.

M. O. Scully, B.-G. Englert, and H. Walther, Nature **351**, 111 (1991).

D. M. Greenberger and A. Yasin, Phys. Lett. A **128**, 391 (1988).

G. Jaeger, A. Shimony, and L. Vaidman, Phys. Rev. A.**51**, 54 (1995).

B.-G. Englert, Phys. Rev. Lett. **77**, 2154 (1996).

P. G. Kwiat, A. M. Steinberg, and R. Y. Chiao, Phys. Rev. A **45**, 7729 (1992).

Y.-H. Kim, R. Yu, S. P. Kulik, Y. H. Shih, and M. O. Scully, Phys. Rev. Lett. **84**, 1 (2000).

For a review, see A. M. Steinberg, P. G. Kwiat, and R. Y. Chiao, in *Atomic, Molecular, & Physics Handbook*, ed. G. W. F. Drake. AIP Press, New York, 1996, pp. 901–918. Available online at http://www.physics.utoronto.ca/~steinber/Quantum_Optical.pdf.

Interaction-free measurements

A. C. Elitzur and L. Vaidman, Found. Phys. **23**, 987 (1993).

P. G. Kwiat, H. Weinfurter, T. Herzog, A. Zeilinger, and M. A. Kasevich, Phys. Rev. Lett. **74**, 4763 (1995).

L. Hardy, Phys. Rev. Lett. **68**, 2981 (1992).

Y. Aharonov, A. Botero, S. Popescu, B. Reznik, and J. Tollaksen, Phys. Lett. A **301**, 130 (2002).

J. S. Lundeen and A. M. Steinberg, Phys. Rev. Lett **102**, 020404 (2009).

K. Yokota, T. Yamamoto, M. Koashi, and N. Imoto, New J. Phys. **11**, 033011 (2009).

Cloning

W. K. Wootters and W. H. Zurek, Nature **299**, 802 (1982).

A. Peres, 'How the no-cloning theorem got its name', arXiv:quant-ph/0205076 (2002).

N. Herbert. Found. Phys. **12**, 117 (1982)

P. W. Milonni and M. L. Hardies. Phys. Lett. A **92**, 321 (1982).

A. Garuccio, in *The Present Status of the Quantum Theory of Light*, ed. S. Jeffers et al. Kluwer, Dordrecht, 1997.

K. Furuya, P.W. Milonni, A.M. Steinberg, and M. Wolinsky, Phys. Lett. A **251**, 294 (1999).

E. Nagali, T. de Angelis, F. Sciarrino, and F. de Martini, Phys. Rev. A **76**, 042126 (2007).

J. Fiurášek and N. J. Cerf, Phys. Rev. A **77**, 052308 (2008).

J.-S. Xu, C.-F. Li, L. Chen, X.-B. Zou, and G.-C. Guo, Phys. Rev. A **78**, 032322 (2008).

Dense coding and teleportation

C. H. Bennett and S. J. Wiesner, Phys. Rev. Lett. **69**, 2881 (1992).

K. Mattle, H. Weinfurter, P. G. Kwiat, and A. Zeilinger, Phys. Rev. Lett. **76**, 4656 (1996).

C. H. Bennett, G. Brassard, C. Crépeau, R. Jozsa, A. Peres, and W. K. Wootters, Phys. Rev. Lett. **70**, 1895 (1993).

D. Bouwmeester, J.-W. Pan, K. Mattle, M. Eibl, H. Weinfurter, and A. Zeilinger, Nature **390**, 575 (1997).

A. Furusawa, J. L. Sørensen, S. L. Braunstein, C.A. Fuchs, H. J. Kimble, and E. S. Polzik, Science **282**, 706 (1998).

Error-correcting codes

A. Steane, Proc. R. Soc. Lond. A **452**, 2551 (1996).

P. W. Shor, Phys. Rev. A **52**, R2493 (1995).

E. Knill, R. Laflamme, A. Ashikhmin, H. Barnum, L. Viola, and W. H. Zurek, 'Introduction to quantum error correction', arXiv:quant-ph/0207170 (2002).

Single-photon tomography

A. G. White, D. F. V. James, W. J. Munro, and P. G. Kwiat, Phys. Rev. A **65**, 012301 (2001).

D. F. V. James, P. G. Kwiat, W. J. Munro, and A. G. White, Phys. Rev. A **64**, 052312 (2001).

Ancilla-assisted photon-polarization tomography

J. Altepeter, D. Branning, E. Jeffrey, T. C. Wei, P. G. Kwiat, R. T. Thew, J. L. O'Brien, M. A. Nielsen, and A. G. White, Phys. Rev. Lett. **90**, 193601 (2003).

Phase-space tomography on single-photon fields

A. I. Lvovsky, H. Hansen, T. Aichele, O. Benson, J. Mlynek, and S. Schiller, Phys. Rev. Lett. **87**, 050402 (2001).

Two-photon process tomography

M. W. Mitchell, C. W. Ellenor, S. Schneider, and A. M. Steinberg, Phys. Rev. Lett. **91**, 120402 (2003).

Applications of process tomography

Y. S. Weinstein, M. A. Pravia, E. M. Fortunato, S. Lloyd, and D. G. Cory, Phys. Rev. Lett. **86**, 1889 (2001).

N. Boulant, M. A. Pravia, E. M. Fortunato, T. F. Havel, and D. G. Cory, Quantum Inf. Process. **1**, 135 (2002) [see also Phys. Rev. A **67**, 042322 (2003)].

A. G. White, A. Gilchrist, G. J. Pryde, J. L. O'Brien, M. J. Bremner, and N. K. Langford, J. Opt. Soc. Am. B **24**, 172 (2007).

Linear-optics quantum computation

E. Knill, R. Laflamme, and G. J. Milburn, Nature **409**, 46 (2001).

D. Gottesmann and I. L. Chuang, Nature **402**, 390 (1999).

T. C. Ralph, N. K. Langford, T. B. Bell, and A. G. White, Phys. Rev. A **65**, 062324 (2002).

T. B. Pittman, B. C. Jacobs, and J. D. Franson, Phys. Rev. Lett. **88**, 257902 (2002).

J. L. O'Brien, G. J. Pryde, A. G. White, T. C. Ralph, and D. Branning, Nature **426**, 264 (2003).

N. K. Langford, T. J. Weinhold, R. Prevedel, K. J. Resch, A. Gilchrist, J. L. O'Brien, G. J. Pryde, and A. G. White, Phys. Rev. Lett. **95**, 210504 (2005).

Measurement-based quantum computation and related approaches

M. A. Nielsen, Phys. Lett. A **308**, 96 (2003).

R. Raussendorf and H. J. Briegel, Phys. Rev. Lett. **86**, 5188 (2001).

R. Raussendorf and H. J. Briegel, Phys. Rev. A **68**, 022312 (2003).

P. Aliferis and D. W. Leung, Phys. Rev. A **70**, 062314 (2004).

M. A. Nielsen, Rep. Math. Phys. **57**, 147 (2006).

P. Walther, K. J. Resch, T. Rudolph, E. Schenck1, H. Weinfurter, V. Vedral, M. Aspelmeyer, and A. Zeilinger, Nature **434**, 169 (2005).

H. J. Briegel, D. E. Browne, W. Dür, R. Raussendorf, and M. Van den Nest, Nat. Phys. **5**, 19 (2009).

M. Anderlini, P. J. Lee, B. L. Brown, J. Sebby-Strabley, W. D. Phillips, and J. V. Porto, Nature **448**, 452 (2007).

K. Nemoto and W. J. Munro, Phys. Rev. Lett **93**, 250502 (2004).

J. D. Franson, B. C. Jacobs, and T. B. Pittman, Phys. Rev. A **70**, 062302 (2004).

N00N states and generation

H. Lee, P. Kok, N. J. Cerf, and J. P. Dowling, Phys. Rev. A **65**, 030101 (2002).

J. Fiurášek, Phys. Rev. A **65**, 053818 (2002).

P. Walther, J.-W. Pan, M. Aspelmeyer, R. Ursin, S. Gasparoni, and A. Zeilinger, Nature **429**, 158 (2004).

M. W. Mitchell, J. S. Lundeen, and A. M. Steinberg, Nature **429**, 161 (2004).

K. J. Resch, K. L. Pregnell, R. Prevedel, A. Gilchrist, G. J. Pryde, J. L. O'Brien, and A. G. White, Phys. Rev. Lett. **98**, 223601 (2007).

Weak measurements

Y. Aharonov and L. Vaidman, Phys. Rev. A **41**, 11 (1991).

Y. Aharonov, D. Z. Albert, and L. Vaidman, Phys. Rev. Lett. **60**, 1351 (1988).

N. W. M. Ritchie, J. G. Story, and R. G. Hulet, Phys. Rev. Lett. **66**, 1107 (1991).

A. M. Steinberg, Phys. Rev. A **52**, 32 (1995).

H. M. Wiseman, Phys. Rev. A **65**, 032111 (2002).

N. Brunner, A. Acín, D. Collins, N. Gisin, and V. Scarani, Phys. Rev. Lett. **91**, 180402 (2003).

K. J. Resch and A. M. Steinberg, Phys. Rev. Lett. **92**, 130402 (2004).

R. Mir, J. S. Lundeen, M. W. Mitchell, A. M. Steinberg, J. L. Garretson, and H. M. Wiseman, New J. Phys. **9**, 287 (2007).

O. Hosten and P. G. Kwiat, Science **319**, 787 (2008).

J.S. Lundeen and A. M. Steinberg, Phys. Rev. Lett. **102**, 020404 (2009).

K. Yokota, T. Yamamoto, M. Koashi, and N. Imoto, New. J. Phys. **11**, 033011 (2009).

P. Ben Dixon, D. J. Starling, A. N. Jordan, and J. C. Howell, Phys. Rev. Lett. **102**, 173601 (2009).

S. Kocsis, B. Braverman, S. Ravets, M. J. Stevens, R. P. Mirin, L. K. Shalm, and A. M. Steinberg, Science **332**, 1170 (2011).

A. Feizpour, X. Xing, and A. M. Steinberg, Phys. Rev. Lett. **107**, 133603 (2011).

8
Quantum optics and cavity QED with quantum dots in photonic crystals

Jelena VUČKOVIĆ

Department of Electrical Engineering and Applied Physics
Stanford University, California, USA

Jelena Vučković: *Quantum optics and cavity QED with quantum dots in photonic crystals*. In:
Quantum Optics and Nanophotonics. Edited by: Claude Fabre, Vahid Sandoghdar, Nicolas Treps,
and Leticia F. Cugliandolo. Oxford University Press (2017), pp. 365–406. © Oxford University
Press 2017. DOI: 10.1093/acprof:oso/9780198768609.003.0008

Chapter Contents

This chapter will focus primarily on studies of quantum optics with epitaxially grown semiconductor quantum dots embedded in photonic crystal cavities. Therefore, we will start by giving a brief introduction to photonic crystals and quantum dots, although the reader is advised to refer to other references [19, 32] for more detailed studies of these topics. We will then proceed with an introduction to cavity quantum electro-dynamics (QED) effects [18, 21], with particular emphasis on the demonstration of these effects on the quantum dot–photonic crystal platform. Finally, we will focus on applications of such cavity QED effects.

8.1 Photonic crystals and microcavities

8.1.1 Photonic crystals

Photonic crystals are media with periodic modulation of the dielectric constant in up to three dimensions [19, 20, 48]. In the literature, 'photonic crystal' usually refers to a structure with a dielectric constant that is periodic in two or three dimensions, while one-dimensional periodic media are referred to as distributed Bragg reflectors.

Let us consider a non-magnetic periodic medium (with permeability μ_0), described by a periodic dielectric constant $\varepsilon(\mathbf{r}) = \varepsilon(\mathbf{r} + \mathbf{a})$, where $\mathbf{a}$ is an arbitrary lattice vector. The allowed electromagnetic modes of frequency ω in such a medium can be obtained as solutions of the wave equation

$$\nabla \times \nabla \times \mathbf{E} = \omega^2 \varepsilon(\mathbf{r}) \mu_0 \mathbf{E}, \tag{8.1}$$

where the dielectric constant is periodic, $\varepsilon(\mathbf{r}) = \varepsilon(\mathbf{r} + \mathbf{a})$, as described above.

Similarly to the wavefunction of an electron in a periodic potential imposed by the crystal lattice, it can be shown that the electromagnetic field within such a periodic dielectric medium can be expanded in terms of electromagnetic eigenmodes that satisfy Bloch's theorem and are therefore referred to as Bloch modes (Bloch states) [50]. The electric field of the three-dimensional Bloch modes can be described as

$$\mathbf{E_k}(\mathbf{r}) = e^{i\mathbf{k}\cdot\mathbf{r}} \mathbf{u_k}(\mathbf{r}),$$

$$\mathbf{u_k}(\mathbf{r} + \mathbf{a}) = \mathbf{u_k}(\mathbf{r}). \tag{8.2}$$

where $\mathbf{k}$ is the wavevector and $\mathbf{u_k}(\mathbf{r})$ is the periodic part of the Bloch state. Therefore, given the frequency ω, we can solve the wave equation in such a periodic medium to find the wavevector $\mathbf{k}$ and the allowed Bloch modes of the system at that frequency. The relation $\omega(\mathbf{k})$ is called the photonic band diagram (or dispersion relation).

If there is a band of frequencies, between $\omega_{\min}$ and $\omega_{\max}$, in which the wave equation inside such a periodic crystal has no solutions for any real $\mathbf{k}$, such a band is called the forbidden (stop) band, or the photonic bandgap. In practice, Bloch modes do not exist in this range of frequencies, implying that electromagnetic waves with frequencies within the photonic bandgap cannot propagate through the photonic crystal. In other words, the photonic crystal behaves as a mirror for electromagnetic waves with frequencies inside the photonic bandgap. By analogy with the case of one-dimensional (1D) periodic media, such reflection can also be viewed as distributed Bragg reflection of the electromagnetic wave from interfaces between regions with different dielectric constants. As in the calculation of band diagrams in solid state physics, in order to determine the size of the photonic bandgap, it is sufficient to solve for the photonic band diagram only inside the irreducible first Brillouin zone (this is a result of the band diagram's periodicity and symmetry in k-space).

In photonic crystals in which the dielectric constant is periodic in less than three dimensions, a photonic bandgap can occur, but only for waves propagating in a certain direction or set of directions in which the periodicity of the dielectric constant occurs. Three-dimensional (3D) photonic crystals, however, can have a complete photonic bandgap, meaning that in a certain frequency region, wave propagation is prohibited through the crystal in (1) any direction in space and (2) for any polarization.

Although 3D photonic crystals offer the opportunity for light manipulation in all three dimensions in space, they are very difficult to fabricate. For this reason, most research has focused on planar photonic crystals, i.e. 1D and two-dimensional (2D) photonic crystals of finite depth, which can be made using standard microfabrication methods. The light confinement in planar photonic crystals results from the combined action of distributed Bragg reflection in the photonic crystal and total internal reflection in the remaining dimensions. The imperfect confinement by total internal reflection produces some unwanted radiation loss, which is usually a limiting factor in the performance of these structures; however, most of the functionality of 3D photonic crystals can still be achieved by careful design.

There are five parameters of planar photonic crystals that one can control, as illustrated in Fig. 8.1(a): the refractive index n of the material, the type of photonic crystal lattice (hexagonal, square, ...), the thickness d of the slab, the lattice periodicity a, and the hole radius r. The refractive index is typically in the range between 3.4 and 3.6 for photonic crystals made out of semiconductors such as Si, GaAs, and InP and operating at near-infrared wavelengths. In vertically symmetric planar photonic crystals, modes can be classified either as vertically even, i.e. transverse electric (TE)-like, or vertically odd, i.e. transverse magnetic (TM)-like. At the centre of the photonic crystal slab (the centre in the z-direction, labeled here as $z = 0$), TE-like modes look exactly like the TE modes of 2D photonic crystals (infinite in the z-direction): the only non-zero field components that they have are H_z, E_x, and E_y. Similarly, the TM-like modes are the same as the TM modes at the centre of the slab, and have only E_z, H_x, and H_y non-zero field components. Away from the centre of the slab, both TE-like and TM-like modes contain all E and H field components, i.e. are not purely TE or TM; this is a result of the mode confinement in the vertical direction. It should also be pointed out that the TE-like and TM-like modes are sometimes referred to as even and odd, respectively, based on whether they exhibit even or odd mirror symmetry in the vertical direction, at the centre of the slab. Even mirror symmetry means that H_z, E_x, and E_y are even at $z = 0$, with the remaining three field components being odd, while odd mirror symmetry means that E_z, H_x, and H_y are even at $z = 0$, with the remaining three field components being odd. This is a result of the application of an even or odd mirror plane at $z = 0$, for which $M_z\mathbf{E} = \sigma_z\mathbf{E}$, with $\sigma_z = 1$ or $\sigma_z = -1$, respectively (the mirror operator M_z is defined as $M_z\mathbf{E} = M_z(E_x, E_y, E_z) = (E_x, E_y, -E_z)$).

The photonic band diagram (dispersion diagram) of TE-like modes of a planar photonic crystal with a hexagonal lattice of air holes and with parameters $n = 3.4$,

(a)

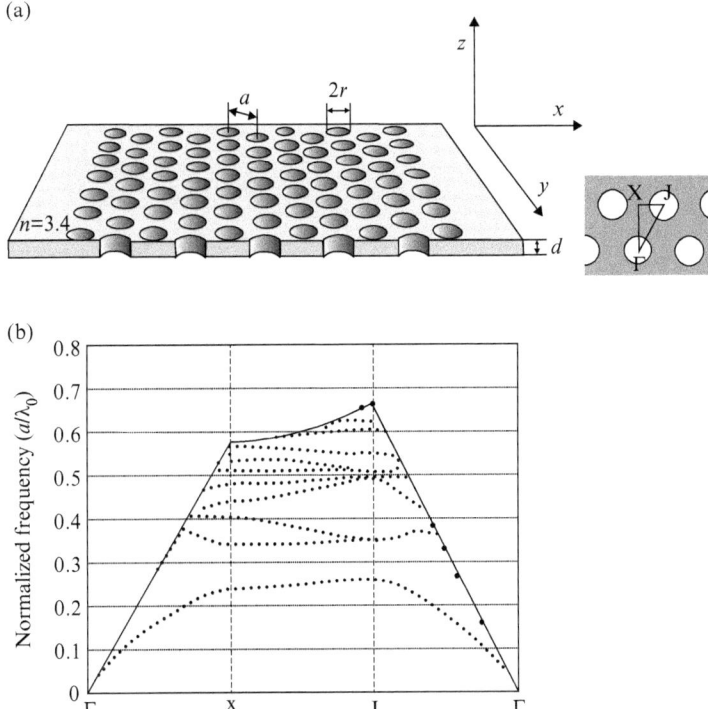

(b)

Fig. 8.1 (a) Optically thin membrane patterned with a hexagonal array of air holes. The inset illustrates the orientation of the principal directions in reciprocal space. (b) Band diagram for TE-like modes of a thin slab surrounded by air on both sides and patterned with a hexagonal array of air holes. The parameters of the photonic crystal are $n = 3.4$, $d/a = 0.65$, and $r/a = 0.3$. The solid line represents the light line.

$d/a = 0.65$, and $r/a = 0.3$ is shown in Fig. 8.1(b). On the horizontal axis, we plot the values of the wavevector in the plane, and on the vertical axis, we plot the normalized frequency of light in units of a/λ_0, where λ_0 is the wavelength of light in air. These calculations were performed using the first-principles 3D finite-difference time-domain (FDTD) method, which is based on discretization of Maxwell's equations in space and time. This band diagram is plotted only along the high-symmetry directions of the irreducible first Brillouin zone (ΓX, XJ and ΓJ). These directions are labelled along the x-axis of the band diagram in the figure.

In the band diagrams of 2D periodic structures like that in Fig. 8.1(b), we plot only the in-plane value of the wavevector $k_\parallel = (k_x, k_y)$ on the horizontal axis, where $|k_\parallel|^2 = k_x^2 + k_y^2$. However, the total wavevector also has a k_z component, and we can write $k_x^2 + k_y^2 + k_z^2 = k^2 = (\omega/c)^2$. A wave that is free to propagate in the z-direction (leaky mode) will have a real k_z, i.e. its z-dependence can be expressed as a plane wave $e^{ik_z z}$, where k_z is real. On the other hand, a wave that is confined by total internal reflection in the z-direction (a guided mode) will decay exponentially in that direction (i.e. it will be evanescent in the z-direction) and therefore have an imaginary k_z. If we plot the light line $\omega_{ll} = c|k_{\parallel}|$ (corresponding to a wave without any k_z) on the band diagram (see Fig. 8.1(b)), then for an evanescent mode, we will have $k_x^2 + k_y^2 + k_z^2 = (\omega/c)^2 < (\omega_{ll}/c)^2$, and for a leaky mode, we will have $k_x^2 + k_y^2 + k_z^2 = (\omega/c)^2 > (\omega_{ll}/c)^2$. Therefore, modes below the light line are confined by total internal reflection and can be guided in the slab, and are called guided modes (they are evanescent in the z-direction and are confined by total internal reflection), whereas modes above the light line (not plotted in Fig. 8.1(b)) are called leaky modes.

In 8.1(b), one can also observe the frequency region where guided modes do not exist for any value of the in-plane wavevector; this is the photonic bandgap. Guided modes are organized into modes above the bandgap (air band), and modes below the bandgap (dielectric band); the names 'dielectric' and 'air' band are based on where the electric field energy of a mode is mostly concentrated. For example, the dielectric band modes mostly concentrate their electric field energy in the high-refractive-index region (semiconductor), while the air band modes mostly concentrate their electric field energy in the air region (holes) [19]. The air band is sometimes also referred to as the conduction band, and the dielectric band as the valence band, by analogy with the energy band diagrams of semiconductors.

8.1.2   Planar photonic crystal microcavities

As already discussed, two-dimensional photonic crystals of finite depth can exhibit a photonic bandgap for electromagnetic waves propagating in the plane of the crystal. Inside the photonic bandgap, the density of optical states is zero; outside the bandgap, Bloch modes exist that can be classified based on their k-vector. However, by perturbing a photonic crystal lattice (i.e. by introducing lattice defects), one can permit localized modes that have frequencies within the photonic bandgap. Such modes have to be evanescent inside the photonic crystal, i.e. they have to decay exponentially away from the defect. In other words, the defect behaves as an optical cavity, and the surrounding photonic crystal represents mirrors surrounding the cavity. Therefore, the defects introduce peaks into the density of optical states inside the photonic bandgap. Moreover, the defects break the discrete translational symmetry of the photonic crystal, and one can no longer classify the modes based on their k-vector.

The simplest ways of forming a microcavity starting from the unperturbed hexagonal photonic crystal lattice of air holes, for example, are by changing the radius of a single hole or by changing its refractive index [46]. The former method is more interesting from the perspective of fabrication. By simply increasing the radius of a single hole, an acceptor defect state is excited, i.e. pulled into the bandgap from the dielectric band. On the other hand, by decreasing the radius of an individual hole (or by tuning its refractive index between 1 and the refractive index of the slab), a donor defect state is excited and pulled into the bandgap from the air band, as shown in Fig. 8.2. It is also important to note that in this section, we are focusing only on TE-like modes, for which a bandgap exists in such planar photonic crystals.

The two most important parameters characterizing the cavity mode (in addition to its wavelength or frequency) are its quality factor Q and the mode volume V_{mode}. The Q-factor describes the storage time of a photon (or, classically, of electromagnetic

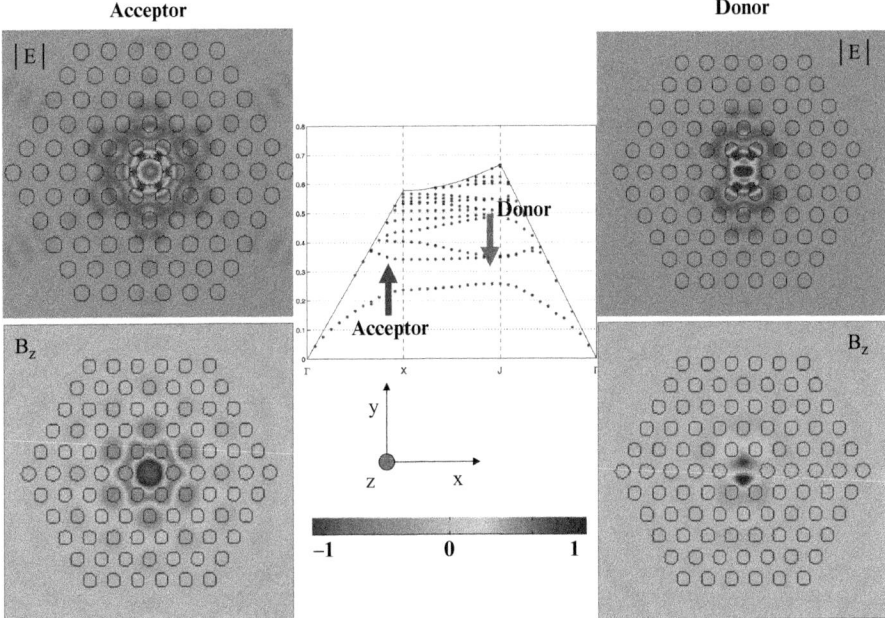

Fig. 8.2 [Colour online] Acceptor defect state and donor defect state excited by changing the radius of a single photonic crystal hole.

field energy) inside a cavity; the energy stored inside the mode with frequency ω and quality factor Q will decay in time t as

$$W(t) = W_0 e^{-\omega t/Q} = W_0 e^{-2\kappa t}, \tag{8.3}$$

where $\kappa = \omega/2Q$ is the cavity field decay rate. The volume of space in which the field is concentrated is described by the cavity mode volume

$$V_{\mathrm{mode}} = \frac{\iiint \varepsilon(\mathbf{r})|\mathbf{E}(\mathbf{r})|^2 \, d^3 \mathbf{r}}{\max\left\{\varepsilon|\mathbf{E}(\mathbf{r})|^2\right\}}. \tag{8.4}$$

Clearly, high Q increases the photon storage time, and small V_{mode} increases field localization, and this is usually the range in which we prefer our system to operate in order to increase the light–matter interaction, as described later.

Even these simple single-defect microcavities produced by changing the radius or refractive index of a single photonic crystal hole can localize light into volumes as small as one-half of a cubic wavelength in the material. However, more sophisticated designs can lead to much higher Q-factors while preserving such small mode volumes [40]. The highest Q-factors of photonic crystal cavities have been demonstrated in silicon at 1550 nm, where material absorption is minimal. Q-factors as high as 10^6 have been reported recently by passive, transmission measurements [3].

In this chapter, we focus primarily on the so-called L3 photonic crystal cavity produced by removing three holes in a hexagonal photonic crystal along a line and

(a)

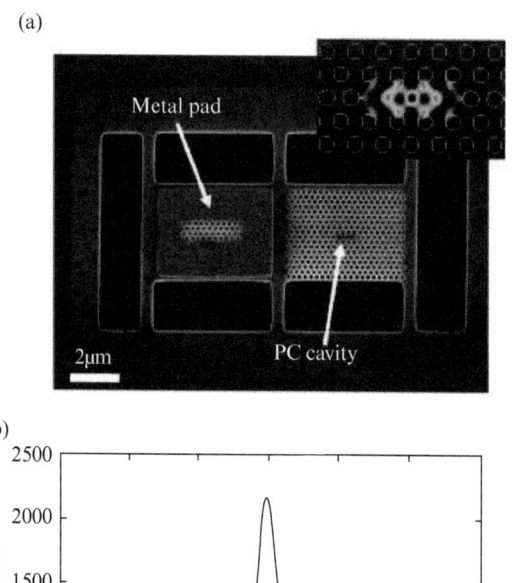

(b)

Fig. 8.3 [Colour online] (a) L3 photonic crystal cavity fabricated in GaAs. The inset shows the simulated electric field pattern inside such a cavity. The simulated optical mode volume is $V_{\mathrm{mode}} \sim 0.7(\lambda/n)^3$. (b) Measured spectrum of such a cavity with quality factor $Q = 25\,300$. (From Englund, D., Faraon, A., Fushman, I., Stoltz, N., Petroff, P., & Vučković, J. (2007). Controlling cavity reflectivity with a single quantum dot. Nature, 450(7171), 857–861. doi:10.1038/nature06234. Copyright 2007 Nature Publishing Group.)

perturbing some neighbouring holes [1]. A scanning electron microscope (SEM) image of such a microcavity fabricated in GaAs, the electric field pattern, and the measured spectrum for one of the fabricated cavities are shown in Fig. 8.3.

8.2   Quantum dots

8.2.1   Semiconductors

Let us consider a simple model of a solid in which the positive ions comprise a uniform array of fixed sites in a periodic structure referred to as the crystal lattice and the electrons are free to move through this lattice. This model can be used to describe semiconductors. To simplify the analysis, we will also make a so-called one-electron approximation: we neglect the interaction between electrons and the vibrations of the crystal, and we assume that an electron moves through the crystal as if it were alone and that it experiences a potential that has the same periodicity as the crystal lattice and results from the periodic arrangement of the crystal lattice ions. This is already a sufficient amount of information to set up the (time-independent) Schrödinger equation for such an electron:

$$-\frac{\hbar^2}{2m}\nabla^2\psi(\mathbf{r}) + V_p(\mathbf{r})\psi(\mathbf{r}) = E\psi(\mathbf{r})\,, \tag{8.5}$$

or, in the form of an energy eigenvalue equation,

$$H\psi(\mathbf{r}) = E\psi(\mathbf{r})\,,$$
$$H = -\frac{\hbar^2}{2m}\nabla^2 + V_p(\mathbf{r})\,. \tag{8.6}$$

The potential that an electron experiences is periodic, with the crystal lattice periodicity $\mathbf{R}_L$:

$$V_p(\mathbf{r} + \mathbf{R}_L) = V_p(\mathbf{r})\,, \tag{8.7}$$

where $m = 9.1 \times 10^{-31}$ kg is the mass of an electron. This is completely analogous to the wave equation for electromagnetic waves inside photonic crystals, and so the energy eigenstates for an electron are assumed to be in the form of the previously introduced Bloch states:

$$\psi_{\mathbf{k}}(\mathbf{r}) = e^{i\mathbf{k}\cdot\mathbf{r}}u_{\mathbf{k}}(\mathbf{r})$$
$$u_{\mathbf{k}}(\mathbf{r} + \mathbf{a}) = u_{\mathbf{k}}(\mathbf{r})\,. \tag{8.8}$$

The energy eigenstates of an electron inside a periodic potential (crystal lattice) are Bloch states, satisfying the Bloch boundary condition (8.8).

Therefore, by solving this Schrödinger equation, we can obtain all possible eigenenergies of an electron inside the crystal structure described with a periodic potential $V_p(\mathbf{r})$. In other words, we can find its energy band diagram, i.e. the allowed combinations of eigenenergy E and the Bloch-state wavevector k. In the case of semiconductors, such as GaAs, the band diagram exhibits an energy bandgap—a region of energies that an electron cannot have inside such a material (also called a semiconductor bandgap).

As in the case of photonic crystals, in order to determine the size of the energy bandgap, it is sufficient to solve the band diagram only along the edges of the irreducible first Brillouin zone.

8.2.2 Effective-mass approximation

Let us assume that we are interested only in the region of the band diagram around the minimum or maximum at $k = 0$ (e.g. the bottom of the conduction band or the top of the valence band) in direct band semiconductors, with the bandgap equal to V.

In this case, we can make a parabolic approximation, i.e. assume that conduction and valence bands around $k = 0$ can be approximated as parabolic:

$$E_c(k) = V + \frac{\hbar^2 k^2}{2m_{\text{eff},c}},$$
$$E_v(k) = \frac{\hbar^2 k^2}{2m_{\text{eff},v}},$$

(8.9)

where $m_{\text{eff},c}$ and $m_{\text{eff},v}$ are the effective masses of the conduction and valence bands, which are positive and negative, respectively.

A localized electron state can be expressed as a superposition of the Bloch states from one band in a narrow region around $k = 0$:

$$\psi(\mathbf{r}, t) = \sum_{\mathbf{k}} c_{\mathbf{k}} u_{\mathbf{k}}(\mathbf{r}) e^{i\mathbf{k}\cdot\mathbf{r}} e^{-iE_{\mathbf{k}}t/\hbar}.$$

(8.10)

For a narrow region of k in superposition, $u_{\mathbf{k}}(\mathbf{r}) \approx u_0(\mathbf{r})$, implying that

$$\psi(\mathbf{r}, t) \approx u_0(\mathbf{r}) \sum_{\mathbf{k}} c_{\mathbf{k}} e^{i\mathbf{k}\cdot\mathbf{r}} e^{-iE_{\mathbf{k}}t/\hbar} = u_0(\mathbf{r}) \psi_{\text{env}}(\mathbf{r}, t),$$

(8.11)

where $\psi_{\text{env}}(\mathbf{r}, t)$ is the envelope wavefunction. Using the parabolic approximation to the bands, it can then be shown that the envelope function satisfies

$$-\frac{\hbar^2}{2m_{\text{eff}}} \nabla^2 \psi_{\text{env}}(\mathbf{r}, t) + V(\mathbf{r}) \psi_{\text{env}}(\mathbf{r}, t) = i\hbar \frac{\partial \psi_{\text{env}}(\mathbf{r}, t)}{\partial t},$$

(8.12)

where $V(r)$ is the size of the bandgap as a function of position. Equation (8.12) looks like the Schrödinger equation for the envelope function. This result is also called the effective-mass approximation. The boundary conditions used for solving this equation are the continuity of $m_{\text{eff}}^{-1} \nabla \psi_{\text{env}}(\mathbf{r}, t)$ and $\psi_{\text{env}}(\mathbf{r}, t)$.

In the effective-mass approximation, an electron is treated as a particle of mass equal to the effective mass m_{eff} (instead of the free-electron mass $m_0 = 9.1 \times 10^{-31}$ kg). We forget about the periodic part of the Bloch state, and treat only the wavefunction envelope.

8.2.3 Quantum dots

The only way to completely localize the wavefunction and obtain discrete energy levels is by localizing it in more than one dimension, i.e. by surrounding the particle by a potential barrier in all directions. This is achieved in a quantum dot. Quantum dots are usually made by sandwiching a tiny chunk of one type of semiconductor with a smaller bandgap inside another semiconductor with a larger bandgap, thereby producing a potential barrier for electrons and holes in three dimensions.

Figure 8.4(a) shows an example of InAs quantum dots embedded inside GaAs ($1\,\mu\text{m}^2$ arrays). Such quantum dots are formed by self-assembly during the growth process called molecular-beam epitaxy (MBE), as a result of lattice mismatch between InAs and GaAs [32]. This process of self-assembly of quantum dots is called

Stransky–Krastanov growth. Since quantum dot islands are formed by self-assembly, their locations are random, and there is a distribution of sizes and shapes, leading to a variation in the energies of transitions of different quantum dots on the same wafer (inhomogeneous broadening).

It is interesting to note that usually not all InAs from the thin deposited layer is turned into such islands. A very thin sheet remains under the quantum dots, which is in fact a very thin quantum well, also referred to as the wetting layer. This layer somewhat affects the 3D confinement of carriers.

Quantum dots exhibit discrete atomic-like spectra (shown in Fig. 8.4(c)), and for this reason they are often referred to as artificial atoms. (As shown below, the discrete lines in the spectrum of quantum dots are produced by recombination of carriers (electrons and holes) occupying discrete energy levels in the conduction and valence bands.

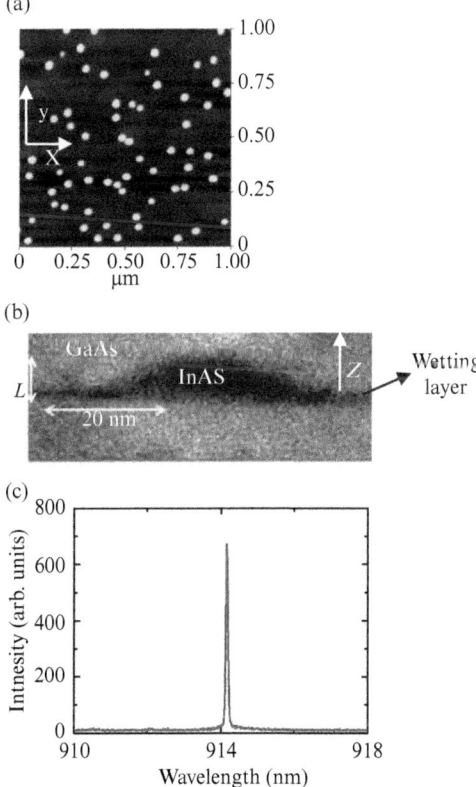

Fig. 8.4 (a) Atomic force microscope image of an InAs quantum dot array. The area shown is $1\,\mu\text{m} \times 1\,\mu\text{m}$. (b) Transmission electron microscope image of a cross section of an assembled InAs/GaAs quantum dot (courtesy of Professor Jonathan Finley, TU Munich). (c) Photoluminescence spectrum of a single InAs/GaAs quantum dot, showing a neutral exciton line.

The energy structure of the quantum dot can be found by solving the following Hamiltonian for an electron–hole pair in a quantum dot:

$$H = H_e + H_h + H_c \,, \tag{8.13}$$

where the subscripts e, h, and c refer to electron, hole, and Coulomb interaction, respectively:

$$H_e = -\frac{\hbar^2}{2} \nabla \left(\frac{1}{m_{\text{eff},e}} \nabla \right) + V_e(\mathbf{r}) \,,$$

$$H_h = -\frac{\hbar^2}{2} \nabla \left(\frac{1}{m_{\text{eff},h}} \nabla \right) + V_h(\mathbf{r}) \,, \tag{8.14}$$

$$H_c = -\frac{e^2}{4\pi\varepsilon} \frac{1}{|r_e - r_h|} \,.$$

Typically, we first solve single-particle problems for electron and hole (H_e and H_h) and then treat the Coulomb interaction (H_c) as a perturbation. The potential $V(r)$ is usually assumed to be harmonic in the x–y plane and an infinite potential well in the z-direction (along the quantum dot's growth axis; see Fig. 8.4(b)):

$$V(x, y, z) = \begin{cases} \dfrac{1}{2} m_{\text{eff}} \omega_0^2 (x^2 + y^2) & \text{for } 0 \le z \le L \,, \\ \infty & \text{for } z > L \text{ or } z < 0 \,. \end{cases} \tag{8.15}$$

This expression holds for both electrons and holes, but the appropriate parameters have to be substituted ($m_{\text{eff},e}$ and $m_{\text{eff},h}$). L is the height of the quantum dot in the z-direction (as shown in Fig. 8.4(b)) and $\hbar\omega_0 = 30$–80 meV.

The energy levels resulting from this Hamiltonian are

$$E = (n_x + n_y + 1)\hbar\omega_0 + \frac{1}{2m_{\text{eff}}} \left(\frac{\pi\hbar n_z}{L} \right)^2 \tag{8.16}$$

and the corresponding envelope wavefunctions are

$$\psi_{\text{env}}(x, y, z) \propto H_{n_x}(x) H_{n_y}(y) e^{-(m_{\text{eff}}\omega_0/2\hbar)(x^2+y^2)} \sin\left(\frac{\pi n_z z}{L} \right) \,, \tag{8.17}$$

where n_x and n_y are non-negative integers and n_z is a positive integer. In the case of InAs/GaAs quantum dots, the height L is small, so that usually only $n_z = 1$ is confined. H_{n_x} and H_{n_y} are Hermite polynomials. The second term in the sum for E and the last factor in the product for ψ describe the potential well in the z-direction, while the remaining terms describe two harmonic oscillators, in the x- and y-directions. As described in Section 8.2.2, the full wavefunction also contains a periodic part (in addition to the envelope).

It should also be noted that the expression (8.16) for the eigenenergies does not include the effect of the Coulomb interaction H_c. To include the effect of the Coulomb interaction on energy eigenstates, we can treat H_c as a perturbation Hamiltonian:

$$\Delta E_0 = \langle 0| H_c |0\rangle + \sum_i \frac{|\langle 0| H_c |i\rangle|^2}{E_i - E_0} \,. \tag{8.18}$$

Here $|i\rangle$ represent the excited eigenstates of the system without perturbation (including both electron and hole) and E_i are the corresponding eigenenergies:

$$|i\rangle \rightarrow \psi_i(r_e)\psi_i(r_h)\,,$$
$$E_i = E_{ie} + E_{ih}\,.$$

(8.19)

Similarly, $|0\rangle$ and E_0 correspond to the ground state and the corresponding eigenenergy:

$$|0\rangle \rightarrow \psi_0(r_e)\psi_0(r_h)\,,$$
$$E_0 = E_{0e} + E_{0h}\,.$$

(8.20)

By solving this perturbation Hamiltonian for a typical InAs/GaAs quantum dot, one can obtain corrections to the energy of a single electron–hole pair of the order of $\Delta E_0 = -20\,\text{meV}$ [17]. A single electron–hole pair in the quantum dot is referred to as a neutral exciton (X).

In a similar way, it can be shown that when adding charges of the same sign to a quantum dot, the energy required to add each new particle increases by 10–20 meV, due to electrostatic (Coulomb) repulsion (estimated through perturbation theory). An excitonic complex consisting of two electrons and one hole is referred to as a negatively

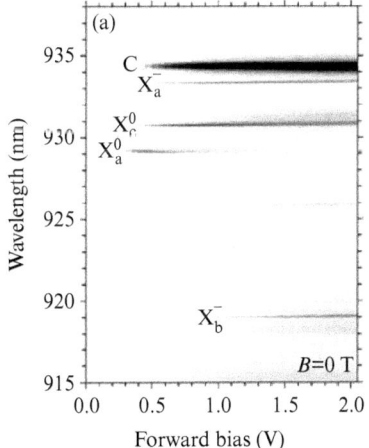

Fig. 8.5 Overview of spectral lines as a function of applied bias on a nanocavity containing a few quantum dots (labeled 'a', 'b', and 'c'). The lines are labeled as the nanocavity mode (C), neutral excitons (X_a^0 and X_c^0), and charged excitons (X_a^- and X_b^-. As the quantum dot gradually charges, the neutral exciton line disappears, while the trion line appears. This is most visible for quantum dot 'a' beyond 1.2 V (where the neutral exciton line X_a^0 disappears, and the charged exciton X_a^- line appears), implying that this quantum dot is then charged with maximal probability. (From Lagoudakis, K. G., Fischer, K., Sarmiento, T., Majumdar, A., Rundquist, A., Lu, J., Bajcsy, M., & Vučković, J. (2013). Deterministically charged quantum dots in photonic crystal nanoresonators for efficient spin–photon interfaces. New J. Phys., 15(11), 113056. doi:10.1088/1367-2630/15/11/113056.)

charged trion (X$^-$), etc. Following the discussion above, after recombination of various carrier complexes inside a quantum dot, different photon energies are expected at the output. This can also be seen in Fig. 8.5, where, by applying a voltage bias (electric field), the quantum dot goes through various charge configurations, and its photoluminescence spectrum at the output changes [24].

8.3 Introduction to cavity quantum electrodynamics

8.3.1 Quantum mechanical treatment of the atom–electromagnetic field interaction: Jaynes–Cummings Hamiltonian

We are considering a two-level emitter with excited state $|ex\rangle$ and ground state $|g\rangle$ and a transition frequency ($\nu = E/\hbar$) between two energy levels (Fig. 8.6). We also assume that the system is on resonance or close to resonance with the fundamental optical cavity mode frequency ω. Under these conditions, the excitation of other cavity modes can be neglected, and the system can be modelled as a single two-level system coupled to a single cavity mode. This coupled system can be described by the Hamiltonian

$$H = H_E + H_F + H_{\text{int}} \,. \tag{8.21}$$

The three terms of the Hamiltonian are the emitter Hamiltonian H_E, the field Hamiltonian H_F, and the emitter–field interaction Hamiltonian H_{int}.

The emitter Hamiltonian is $H_E = \frac{1}{2}\hbar\nu\sigma_z$, where $\sigma_z = |ex\rangle\langle ex| - |g\rangle\langle g|$ is the population operator. $|ex\rangle$ and $|g\rangle$ represent a complete set of energy eigenstates for the emitter, with corresponding eigenvalues $\frac{1}{2}\hbar\nu$ and $-\frac{1}{2}\hbar\nu$, respectively. They also represent a basis for the emitter, i.e. the closure relation holds: $1 = |ex\rangle\langle ex| + |g\rangle\langle g|$. The field Hamiltonian is $H_F = \hbar\omega\left(a^\dagger a + \frac{1}{2}\right)$, where a and $a^\dagger$ are photon annihilation and creation operators, respectively. The eigenstates of H_F are Fock (photon number) states $|n\rangle$, with corresponding eigenvalues $\hbar\omega\left(n + \frac{1}{2}\right)$.

The interaction Hamiltonian between the emitter and the cavity field can be described semiclassically in the dipole approximation:

$$H_{\text{int}} = -\mathbf{d} \cdot \mathbf{E} \,, \tag{8.22}$$

where $\mathbf{d} = \sum_j e\mathbf{r}_j$ is the classical dipole moment of the emitter, $\mathbf{r}_j$ is the position of the jth electron in this system, and $e = 1.6 \times 10^{-19}$ C. For example, for a single-electron atom, $\mathbf{d} = e\mathbf{r}$, where $\mathbf{r}$ is the position of the electron relative to the nucleus (and,

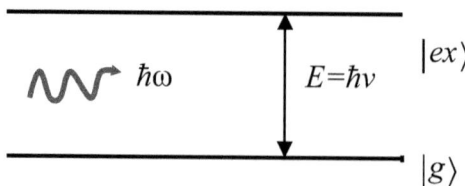

Fig. 8.6 A two-level system interacting with a single mode of the electromagnetic field.

similarly, for a neutral exciton in a quantum dot, $\mathbf{r}$ could be roughly interpreted as the 'distance' between electron and hole). This semiclassical Hamiltonian can be converted into a quantum mechanical representation (acting on atomic eigenstates $|ex\rangle$ and $|g\rangle$ and the field eigenstates $|n\rangle$). First, we apply the unit operator $1 = |ex\rangle\langle ex| + |g\rangle\langle g|$ to both sides of the dipole moment: $e\mathbf{r} = 1e\mathbf{r}1$. The resulting expression can be simplified by introducing the dipole moment matrix element $\boldsymbol{\mu}_{ij} = \langle i| e\mathbf{r} |j\rangle$ and atom transition operators, defined as $\sigma_{ij} = |i\rangle\langle j|$. The transition operators for two particular cases are also referred to as the raising and lowering operators and are denoted by $\sigma_+ = \sigma_{eg} = |ex\rangle\langle g|$ and $\sigma_- = \sigma_{ge} = |g\rangle\langle ex|$, respectively. Therefore, after this transformation, we can represent the dipole moment in terms of the operators acting on the atomic states:

$$e\mathbf{r} = \boldsymbol{\mu}_{ee}\sigma_{ee} + \boldsymbol{\mu}_{gg}\sigma_{gg} + \boldsymbol{\mu}_{eg}\sigma_{eg} + \boldsymbol{\mu}_{ge}\sigma_{ge} . \tag{8.23}$$

The final step in converting H_{int} into quantum mechanical form is representing the E-field quantum mechanically, using the electric field operator

$$\mathbf{E}(\mathbf{r}_E) = i\sqrt{\frac{\hbar\omega}{2V_{\text{mode}} \max\left\{\varepsilon(\mathbf{r})|\mathbf{E}(\mathbf{r})|^2\right\}}} \; \hat{\mathbf{e}}\left[E(\mathbf{r}_E)a - E^*(\mathbf{r}_E)a^\dagger\right] . \tag{8.24}$$

In this case, we assume that the system is single-mode, $\hat{\mathbf{e}}$ is the electric field orientation at the location $\mathbf{r}_E$ of the quantum emitter, a and $a^\dagger$ are the annihilation and creation operators for the cavity mode, $E(\mathbf{r}_E)$ is the classical value of the field at the location of the emitter, and V_{mode} is the cavity mode volume:

$$V_{\text{mode}} = \frac{\iiint \varepsilon(\mathbf{r})|\mathbf{E}(\mathbf{r})|^2 \, d^3\mathbf{r}}{\varepsilon_M |\mathbf{E}(\mathbf{r}_M)|^2} . \tag{8.25}$$

Here $\mathbf{r}_M$ denotes the point where the electric field energy density $\varepsilon(\mathbf{r})|E(\mathbf{r})|^2$ is maximum and ε_M is the dielectric constant at this point $\varepsilon_M = \varepsilon(\mathbf{r}_M)$, i.e.

$$V_{\text{mode}} = \frac{\iiint \varepsilon(\mathbf{r})|\mathbf{E}(\mathbf{r})|^2 \, d^3\mathbf{r}}{\max\left\{\varepsilon|\mathbf{E}(\mathbf{r})|^2\right\}} . \tag{8.26}$$

By combining the expressions (8.23) and (8.24) for the dipole moment and the electric field operator, we can express the interaction Hamiltonian as

$$H_{\text{int}} = -i\sqrt{\frac{\hbar\omega}{2V_{\text{mode}} \max\left\{\varepsilon(\mathbf{r})|\mathbf{E}(\mathbf{r})|^2\right\}}}$$

$$\times \left(\boldsymbol{\mu}_{ee} \cdot \hat{\mathbf{e}}\sigma_{ee} + \boldsymbol{\mu}_{gg} \cdot \hat{\mathbf{e}}\sigma_{gg} + \boldsymbol{\mu}_{eg} \cdot \hat{\mathbf{e}}\sigma_{eg} + \boldsymbol{\mu}_{ge} \cdot \hat{\mathbf{e}}\sigma_{ge}\right)\left[E(\mathbf{r}_E)a - E^*(\mathbf{r}_E)a^\dagger\right] . \tag{8.27}$$

This expression can be simplified further by noting that $\boldsymbol{\mu}_{eg} = \boldsymbol{\mu}_{ge}^*$ and $|\boldsymbol{\mu}_{ee}| = |\boldsymbol{\mu}_{gg}| = 0$ because of parity (for any state $|i\rangle$, $\boldsymbol{\mu}_{ii} = \langle i| e\mathbf{r} |i\rangle = e\langle i| \mathbf{r} |i\rangle = 0$, since the position operator is an odd function). After multiplying all of the remaining terms, we have

$$H_{\text{int}} = -i\sqrt{\frac{\hbar\omega}{2V_{\text{mode}} \max\{\varepsilon(\mathbf{r})|\mathbf{E}(\mathbf{r})|^2\}}}$$
$$\times \left[\boldsymbol{\mu}_{eg} \cdot \hat{\mathbf{e}} E(\mathbf{r}_E)\sigma_+ a + \boldsymbol{\mu}_{eg}^* \cdot \hat{\mathbf{e}} E(\mathbf{r}_E)\sigma_- a \right. \tag{8.28}$$
$$\left. - \boldsymbol{\mu}_{eg} \cdot \hat{\mathbf{e}} E^*(\mathbf{r}_E)\sigma_+ a^\dagger - \boldsymbol{\mu}_{eg}^* \cdot \hat{\mathbf{e}} E^*(\mathbf{r}_E)\sigma_- a^\dagger \right].$$

The terms proportional to $\sigma_- a$ and $\sigma_+ a^\dagger$ do not conserve energy and can be neglected. Therefore, we are left with the following interaction Hamiltonian:

$$H_{\text{int}} = -i\sqrt{\frac{\hbar\omega}{2V_{\text{mode}} \max\left\{\varepsilon(\mathbf{r})|\mathbf{E}(\mathbf{r})|^2\right\}}} \left[\boldsymbol{\mu}_{eg} \cdot \hat{\mathbf{e}} E(\mathbf{r}_E)\sigma_+ a - \boldsymbol{\mu}_{eg}^* \cdot \hat{\mathbf{e}} E^*(\mathbf{r}_E)\sigma_- a^\dagger \right]. \tag{8.29}$$

Let us introduce the emitter–field coupling parameter $g(\mathbf{r})$:

$$g(\mathbf{r}_E) = \frac{1}{\hbar}\sqrt{\frac{\hbar\omega}{2V_{\text{mode}} \max\left\{\varepsilon(\mathbf{r})|\mathbf{E}(\mathbf{r})|^2\right\}}}\, \boldsymbol{\mu}_{eg} \cdot \hat{\mathbf{e}} E(\mathbf{r}_E). \tag{8.30}$$

Then we can express the interaction Hamiltonian in the form

$$H_{\text{int}} = i\hbar\left[g^*(\mathbf{r}_E) a^\dagger \sigma_- - g(\mathbf{r}_E)\sigma_+ a \right]. \tag{8.31}$$

The coupling parameter $g(\mathbf{r}_E)$ is the product of the Rabi frequency g_0, a position-dependent part $\psi(\mathbf{r}_E)$, and a polarization-dependent part $\cos\xi$:

$$g(\mathbf{r}_E) = g_0 \psi(\mathbf{r}_E)\cos\xi, \tag{8.32}$$

where

$$g_0 = \frac{\mu_{eg}}{\hbar}\sqrt{\frac{\hbar\omega}{2\varepsilon_M V_{\text{mode}}}},$$

$$\psi(\mathbf{r}_E) = \frac{E(\mathbf{r}_E)}{|E(\mathbf{r}_M)|}, \tag{8.33}$$

$$\cos\xi = \frac{\boldsymbol{\mu}_{eg} \cdot \hat{\mathbf{e}}}{\mu_{eg}},$$

with $\mu_{eg} = |\boldsymbol{\mu}_{eg}|$. Therefore, the coupling strength $|g(\mathbf{r}_E)|$ reaches its maximum value of $|g_0|$ when the emitter is located at the point $\mathbf{r}_M$ at which the electric field energy density is maximum and when its dipole moment is aligned with the electric field, i.e., when $\psi(\mathbf{r}_E) = 1$ and $\cos\xi = 1$. To maximize $|g_0|$, we should pick the cavity mode with as small a mode volume V_{mode} as possible. This describes the motivation behind using photonic crystal cavities for cavity QED experiments.

To summarize, the total Hamiltonian of the coupled emitter–cavity field system is given by

$$H = H_A + H_F + H_{\text{int}}, \tag{8.34}$$

where

$$H_A = \tfrac{1}{2}\hbar\nu\sigma_z \,,$$

$$H_F = \hbar\omega\left(a^\dagger a + \tfrac{1}{2}\right),$$

$$(8.35)$$

$$H_{\text{int}} = i\hbar\left[g^*(\mathbf{r}_E)a^\dagger\sigma_- - g(\mathbf{r}_E)\sigma_+ a\right].$$

The Hamiltonian H is called the Jaynes–Cummings Hamiltonian, and describes a lossless emitter–field system [21].

8.3.2 Strong-coupling regime: Rabi oscillation and anharmonic dressed states ladder

Without the interaction turned on ($H_{\text{int}} = 0$), the total Hamiltonian of the emitter–field system is

$$H_0 = \tfrac{1}{2}\hbar\nu\sigma_z + \hbar\omega\left(a^\dagger a + \tfrac{1}{2}\right).$$

$$(8.36)$$

The eigenstates of this Hamiltonian are $|ex, n\rangle$ and $|g, n+1\rangle$, i.e. the products of atomic and field eigenstates. The corresponding eigenenergies are the sums of the atomic and field eigenenergies:

$$E_1 = \tfrac{1}{2}\hbar\nu + \hbar\omega\left(n + \tfrac{1}{2}\right) = \tfrac{1}{2}\hbar\delta + \hbar\omega(n+1)\,,$$

$$E_2 = -\tfrac{1}{2}\hbar\nu + \hbar\omega\left(n + \tfrac{3}{2}\right) = -\tfrac{1}{2}\hbar\delta + \hbar\omega(n+1)\,,$$

$$(8.37)$$

where $\delta = \nu - \omega$ is the detuning between the cavity field frequency and the atomic transition frequency. When detuning is not present ($\delta = 0$), it follows that the energy levels are degenerate and equal to $E_0 = \hbar\omega(n+1)$.

When the interaction Hamiltonian is turned on, these states are coupled, and the evolution of the state of the system can be found by solving the time-dependent Schrödinger equation:

$$H\left|\psi(t)\right\rangle = i\hbar\frac{d\left|\psi(t)\right\rangle}{dt}\,,$$

$$(8.38)$$

$$\left|\psi(t)\right\rangle = [C_{e,n}(t)\left|ex, n\right\rangle + C_{g,n+1}(t)\left|g, n+1\right\rangle]e^{-i(n+1)\omega t}\,,$$

When detuning is not present ($\delta = 0$), and assuming that the system is initially in the state $|ex, n\rangle$ at $t = 0$, the solution to the coupled system of equations is

$$C_{g,n+1}(t) \propto \sin\left(|g(\mathbf{r}_E)|\sqrt{n+1}\,t\right),$$

$$C_{e,n}(t) \propto \cos\left(|g(\mathbf{r}_E)|\sqrt{n+1}\,t\right).$$

$$(8.39)$$

The probabilities of the system being in the states $|g, n+1\rangle$ and $|ex, n\rangle$ are respectively

$$|C_{g,n+1}(t)|^2 = \tfrac{1}{2}\left\{1 - \cos\left[2|g(\mathbf{r}_E)|\sqrt{n+1}\,t\right]\right\},$$

$$|C_{e,n}(t)|^2 = \tfrac{1}{2}\left\{1 + \cos\left[2|g(\mathbf{r}_E)|\sqrt{n+1}\,t\right]\right\}.$$

$$(8.40)$$

Therefore, in the strong-coupling regime, the time evolution of the system can be described as an oscillation at the frequency $2|g(\mathbf{r}_E)|\sqrt{n+1}$ between the states $|ex, n\rangle$ and $|g, n+1\rangle$. This process is called Rabi oscillation. For the case when the system is initially in the state $|ex, 0\rangle$, i.e. no photons are present in the cavity and the atom is in the excited state, the frequency of the Rabi oscillation is equal to $2|g(\mathbf{r}_E)|$.

The coupled emitter–field system (with interaction turned on) has new eigenstates. These can be found by solving for the eigenvectors and eigenvalues of the Hamiltonian with the interaction included. The representation of the total Hamiltonian of the coupled emitter–cavity field system in the basis $|ex, n\rangle, |g, n+1\rangle$ is

$$H = \begin{bmatrix} \langle ex, n| \ H \ |ex, n\rangle & \langle ex, n| \ H \ |g, n+1\rangle \\ \langle g, n+1| \ H \ |ex, n\rangle & \langle g, n+1| \ H \ |g, n+1\rangle \end{bmatrix},$$

(8.41)

$$= \begin{bmatrix} \frac{1}{2}\hbar\delta + \hbar\omega(n+1) & -i\hbar g\sqrt{n+1} \\ i\hbar g^* \sqrt{n+1} & -\frac{1}{2}\hbar\delta + \hbar\omega(n+1) \end{bmatrix}$$

and the eigenenergies (eigenvalues of this Hamiltonian) are

$$E_\pm = \hbar\omega(n+1) \pm \sqrt{\left(\tfrac{1}{2}\hbar\delta\right)^2 + \hbar^2|g|^2(n+1)} .$$

(8.42)

When detuning between the emitter and the field is not present ($\delta = 0$), it follows that

$$E_\pm = \hbar\omega(n+1) \pm \hbar|g|\sqrt{n+1} ,$$

(8.43)

and the corresponding eigenstates are

$$E+ : \quad |n+1, +\rangle = \tfrac{1}{\sqrt{2}}(|ex, n\rangle + |g, n+1\rangle) \quad \text{(even mode)},$$

$$E- : \quad |n+1, -\rangle = \tfrac{1}{\sqrt{2}}(|ex, n\rangle - |g, n+1\rangle) \quad \text{(odd mode)}.$$

(8.44)

The states $|n+1, +\rangle$ and $|n+1, -\rangle$ are termed dressed states (normal modes) and the states $|ex, n\rangle$ and $|g, n+1\rangle$ are termed bare states. Dressed states are eigenstates of the total Hamiltonian, including interaction (Figs. 8.7 and 8.8). Bare states are eigenstates of the Hamiltonian without the interaction turned on. If the system is prepared in the bare state and the interaction Hamiltonian is turned on, then Rabi oscillation between the bare states occurs, as we have shown. Note that if the coupling coefficient g is real, then there will be an additional phase term in front of $|ex, n\rangle$ in the normal modes given by (8.44) ($-i$ in the even mode and $+i$ in the odd mode). On the other hand, if g is purely imaginary, then the eigenstates will be defined as above. This is a result of the direct solution of the eigenstates of the Hamiltonian. Since all of the observable quantities are proportional to $|g|$, the definitions of g as real or imaginary are equivalent, and this phase term is not important.

Therefore, if the atom–field interaction is turned on and detuning between the cavity field frequency and the atom transition frequency is not present, then the energy level E_0 splits into two energy levels E_+ and E_-, corresponding to the even and odd

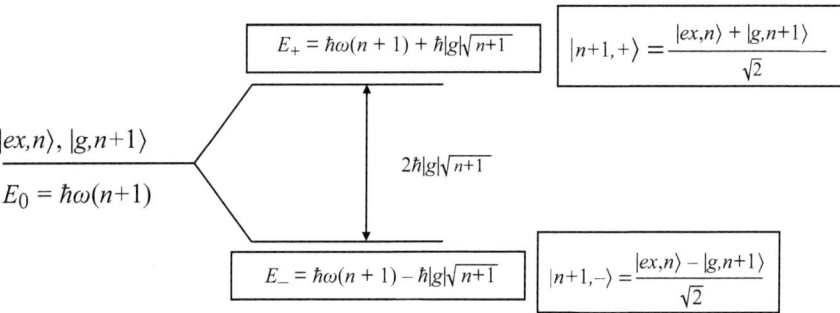

Fig. 8.7 Splitting of the eigenstates of the uncoupled emitter–field system into dressed states once the emitter–field interaction is turned on. n is the number of photons in the field mode, and $|ex\rangle$ and $|g\rangle$ are the excited and ground states of the emitter, as shown in Fig. 8.6.

modes, respectively. This splitting is usually used as the indication that the atom–cavity system has reached the strong-coupling regime. Such splitting happens for any excitation level (integer $n > 0$, as shown in Fig. 8.7), and therefore the new structure of eigenstates is represented by a ladder as shown in Fig. 8.8.

The dressed-states ladder in Fig. 8.8 is plotted only for zero detuning between the atom and the cavity. When the detuning δ is not zero, the eigenenergies as a function

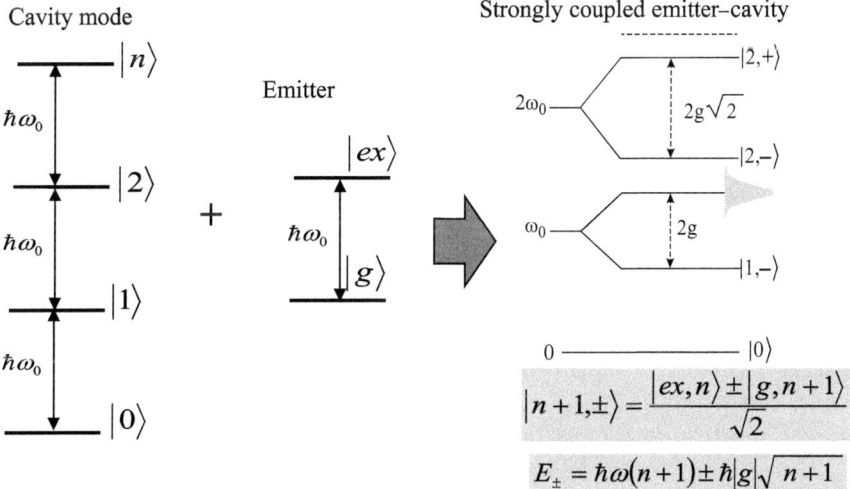

Fig. 8.8 The cavity mode has energy eigenstates described by a harmonic ladder (left), while a single two-level emitter has energy eigenstates $|ex\rangle$ and $|g\rangle$ (middle). Once they are strongly coupled, new energy eigenstates are formed called dressed states (right), and the energy eigenvalues are positioned on an anharmonic ladder called the dressed-states ladder.

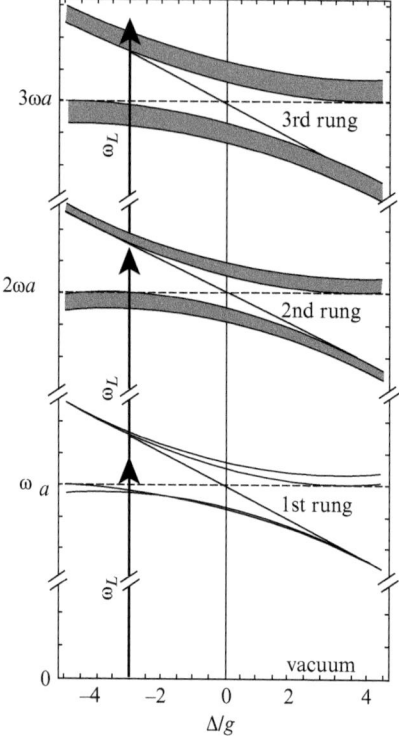

Fig. 8.9 [Colour online] Dressed-states ladder for a detuned emitter–cavity system. (Courtesy of Professor Fabrice Laussy, Universidad Autónoma de Madrid, reproduced from Laussy, F. P., Valle, E. D., Schrapp, M., Laucht, A., & Finley, J. J. (2012). Climbing the Jaynes–Cummings ladder by photon counting. J. Nanophotonics, 6(1), 061803. doi:10.1117/1.jnp.6.061803. Copyright 2012 SPIE.)

of detuning will be represented by plotting the expression for $E_{\pm}$ with δ given in (8.42), leading to the anticrossing curves shown in Fig. 8.9 [26].

As can be seen in Figs. 8.8 and 8.9, the ladder of dressed states is anharmonic. In other words, the splitting between dressed-state energy levels is not constant. This can be employed to demonstrate effects such as photon blockade, where the coupling of the second photon with frequency $\omega_0 \pm g$ into the cavity QED system is prohibited, since only one photon with that frequency can occupy one of the first two dressed states (Fig. 8.10). This effect was first demonstrated in an atomic cavity QED system [6] and then in a quantum dot cavity QED system [14, 36].

Similarly, higher-order manifolds in the ladder can be probed, leading to filtering of n-photon states via the effect called photon-induced tunnelling. In the example shown in Fig. 8.10, photon pairs are filtered by exciting the second manifold. This effect has been demonstrated in solid state physics (a quantum dot–cavity system) [14] and in atomic physics [38].

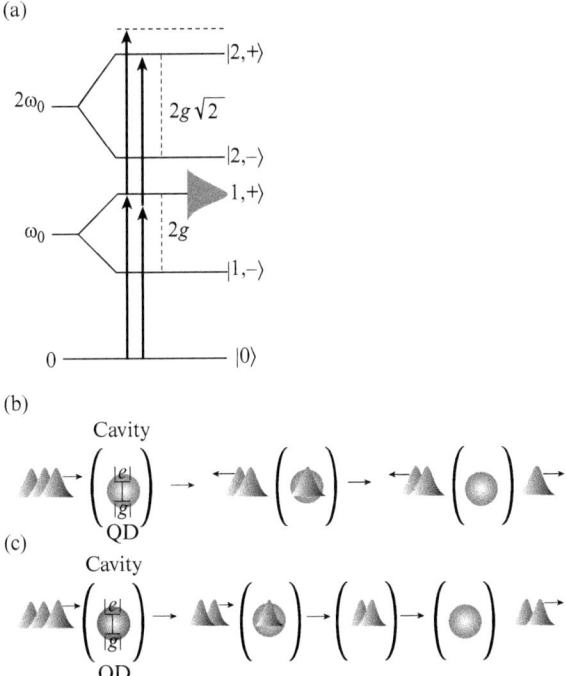

Fig. 8.10 [Colour online] Excitation of one of the eigenstates in the first manifold of the dressed-states ladder can be done with frequency $\omega_0 + g$ (in a tuned system), shown by the left [blue] vertical arrow in (a). Only one excitation (one photon) can be placed into this first manifold, since it is a result of mixing between one photon and one atom. However, the distance from this eigenstate to any eigenstates in the second manifold is different from $\omega_0 + g$ (because of the ladder anharmonicity). Therefore, only one photon from the excitation can be coupled to the system, and once it is coupled, all the others will be repelled (b). This effect is referred to as the photon blockade. Similarly, if the excitation has frequency given by the right [red] vertical arrow in (a), then an eigenstate in the second manifold is reached (where two photons can be placed). In this case, only pairs of photons can be coupled to the system, as shown in (c). This effect is referred to as photon-induced tunnelling.

Usually, eigenstates are probed by transmission of a probe laser through the system. For example, if we have an empty cavity (no strongly coupled emitter inside it), it will feature a transmission spectrum as in Fig. 8.11(a). However, if an emitter is loaded into the cavity and strongly coupled to the cavity field, the cavity transmission spectrum changes. If the excitation is weak and only the first manifold of the ladder can be accessed, the transmission spectrum exhibits two peaks, with the splitting equal to $2|g|$, as in Fig. 8.11(b). These two peaks correspond to the two eigenstates in the first manifold. With stronger excitation, additional peaks start appearing in the transmission spectrum (positioned between the two initial peaks), corresponding to higher-order manifolds [38].

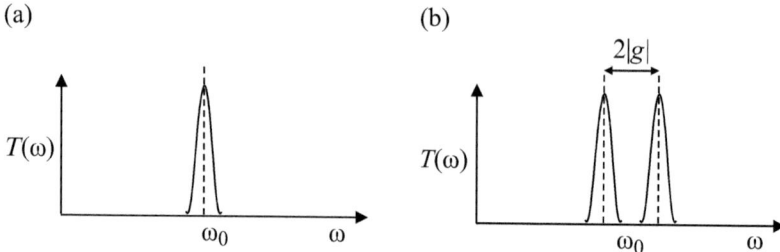

Fig. 8.11 Rabi splitting: transmission spectra of (a) an empty cavity mode and (b) a strongly coupled emitter–cavity system for weak excitation, when only eigenstates in the first manifold can be accessed.

8.3.3 Strongly coupled cavity QED system with losses

Here we will follow a very simple semiclassical model to include the effect of losses in a cavity QED system. Recall that the eigenfrequencies (eigenvalues of the Jaynes–Cummings Hamiltonian) for the first manifold of the system without losses are

$$E_\pm = \hbar\omega \pm \sqrt{\left(\tfrac{1}{2}\hbar\delta\right)^2 + \hbar^2|g|^2}\,, \tag{8.45}$$

where $\delta = \nu - \omega$ is the atom–cavity detuning. This expression was derived by setting the ground state for the field at $\frac{1}{2}\hbar\omega$ and for atom at $-\frac{1}{2}\hbar\nu$, as described in Section 8.3.1. If we set both to zero, the energy eigenvalues are

$$\frac{E_\pm}{\hbar} = \omega_\pm = \omega - \tfrac{1}{2}\omega + \tfrac{1}{2}\nu \pm \sqrt{\left(\tfrac{1}{2}\delta\right)^2 + |g|^2}$$
$$= \tfrac{1}{2}(\omega + \nu) \pm \sqrt{\left(\tfrac{1}{2}\delta\right)^2 + |g|^2}\,. \tag{8.46}$$

Let us now assume that the system has losses: the emitter can decay into other channels (e.g. spontaneous emission into other modes, or non-radiative decay) with rate γ and the cavity is imperfect, so the cavity mode can lose photons with cavity field decay rate κ. In other words, κ and γ are half width at half maxima of the cavity and emitter spectra, respectively. To account for losses, we will replace emitter and cavity mode frequencies with complex frequencies:

$$\nu \to \nu - i\gamma\,, \tag{8.47}$$

$$\omega \to \omega - i\kappa\,. \tag{8.48}$$

Now let us plug these complex frequencies back into the expression for eigenfrequencies of the strongly coupled system:

$$\omega_\pm = \tfrac{1}{2}(\omega + \nu) - \tfrac{1}{2}i(\kappa + \gamma) \pm \sqrt{\left\{\tfrac{1}{2}[\delta - i(\kappa - \gamma)]\right\}^2 + |g|^2}\,. \tag{8.49}$$

These are eigenfrequencies of the first manifold in the cavity QED system with losses [37]. The real part of $\omega_{\pm}$ gives frequencies of eigenmodes and the imaginary part describes their damping.

Let us assume that we have a good emitter, with linewidth γ much narrower than the cavity linewidth κ. We also assume that the emitter–cavity coupling strength g is larger than any losses in the system ($g > \frac{1}{2}\kappa$ and $\kappa \gg \gamma$) and that the emitter–cavity mode detuning $\delta = 0$. To reach this regime, one has to increase g relative to κ, which can be achieved by reducing the cavity mode volume V_{mode} (thereby increasing g) and by increasing the cavity Q factor (thereby decreasing κ). In this regime, the expression under the square root is positive and we have new eigenfrequencies

$$\omega_{\pm} = \omega \pm \sqrt{|g|^2 - \tfrac{1}{4}\kappa^2} - \tfrac{1}{2}i\kappa. \tag{8.50}$$

Therefore, we expect to see two distinct lines in the spectrum with frequencies $\omega_{\pm} = \omega \pm \sqrt{|g|^2 - \frac{1}{4}\kappa^2}$ that have full widths at half maxima of κ. The system is strongly coupled, exhibiting Rabi splitting, as shown in Fig. 8.11, but the splitting between the two peaks is not exactly $2g$ as in the lossless system, and the peaks are broadened by cavity decay. In the case when $g \gg \frac{1}{2}\kappa$, the splitting approaches $2g$, as in the lossless system:

$$\omega_{\pm} = \omega \pm |g| - \tfrac{1}{2}i\kappa. \tag{8.51}$$

The two peaks correspond to new eigenstates of the system, which are the entangled states of the emitter and the cavity field, as shown in Fig. 8.8.

8.3.4   Weakly coupled cavity QED system—Purcell regime

Let us now consider a different regime of emitter–cavity field interaction in which the cavity field decay rate dominates ($\frac{1}{2}\kappa \gg g \gg \gamma$) and there is detuning ($\delta = 0$). In the expression (8.49) for $\omega_{\pm}$, the expression under the square root is now negative. Therefore, both eigenstates are at the same frequency ω (since the real parts of $\omega_{\pm}$ are the same), but their dampings are different. The damping (i.e. the imaginary part of $\omega_{\pm}$) is

$$\tfrac{1}{2}\kappa \mp \sqrt{\tfrac{1}{4}\kappa^2 - |g|^2} = \frac{\kappa}{2}\left(1 \mp \sqrt{1 - \frac{4|g|^2}{\kappa^2}}\right)$$

$$\approx \frac{\kappa}{2}\left[1 \mp \left(1 - \frac{1}{2}\frac{4|g|^2}{\kappa^2}\right)\right] \approx \begin{cases} |g|^2/\kappa, \\ \kappa. \end{cases} \tag{8.52}$$

Therefore, in this regime, we expect two eigenstates with the same frequency (ω) but different damping; their energy will decay with rates $2|g|^2/\kappa$ and 2κ, respectively. This is the so-called *Purcell regime*—a special situation in the weak-coupling regime of cavity QED: there will be no Rabi splitting in the spectrum, and emitter and cavity will just cross, but the decay rate (spontaneous emission rate) of the emitter will be modified by the presence of the cavity. In this regime, one eigenstate (cavity mode) has an energy decay rate 2κ. The other eigenstate (at the same frequency ω) is the emitter, with modified radiative decay rate equal to $2|g|^2/\kappa$. This modified spontaneous emission rate is very different from the spontaneous emission rate of that same emitter without the cavity, in bulk (uniform material with refractive index n), where it is equal to $\Gamma_n = n\Gamma_0$ and $\Gamma_0 = \mu_e^2 \nu^3/3\pi\varepsilon_0\hbar c^3$ is the Einstein A-coefficient [39].

The ratio between the spontaneous emission rate of an emitter in a cavity and the same emitter without the cavity is the Purcell factor

$$F = \frac{2|g|^2/\kappa}{n\Gamma_0} . \tag{8.53}$$

Clearly, the Purcell factor reaches its maximum value when the emitter is on resonance with the cavity and spatially aligned with the cavity field maximum (when $g = g_0$). By plugging in the expressions for g_0 (see Section 8.3.1), κ, and Γ_0 into (8.53), we find that the maximum Purcell factor is

$$F = \frac{3}{4\pi^2} \left(\frac{\lambda}{n}\right)^3 \frac{Q}{V_{\text{mode}}} . \tag{8.54}$$

If the emitter is not spatially aligned to the cavity mode maximum, then $g = g_0\psi(\mathbf{r})\cos\xi$, as shown in Section 8.3.1, where ψ describes its spatial offset from the E-field energy density maximum and $\cos\xi$ describes the misalignment between its dipole moment and the E-field polarization at that location. The reduction in g also degrades the Purcell enhancement. Finally, the previous analysis was done for an emitter that is tuned to the cavity resonance, i.e. $\delta = \nu - \omega = 0$. If there is spectral detuning between the two, the Purcell factor F is also degraded, in proportion to the density of optical states,

$$D_c(\nu) = \frac{1}{\pi} \frac{\omega/2Q}{(\nu - \omega)^2 + (\omega/2Q)^2} . \tag{8.55}$$

Thus, the spontaneous emission is not an intrinsic property of a quantum emitter (e.g. an atom or exciton), but is rather a property of an emitter coupled to its electromagnetic environment [35]. The spontaneous emission rate is directly proportional to the density of electromagnetic states, and can be modified with respect to its value in free space by placing the emitter in a cavity. Inside a cavity, this rate can be significantly enhanced in proportion to Q/V_{mode}. Similarly, by decreasing the density of optical states (e.g. inside a photonic bandgap), one can reduce the emission rate [48].

8.3.5 Spontaneous emission coupling factor and relation to lasing threshold

In the previous subsection, we showed that the spontaneous emission rate of an emitter (e.g. an atom, or a quantum dot exciton) can be increased by increasing the Q/V_{mode} ratio of a cavity. We will now show that an increase in this ratio also leads to a reduction of the lasing threshold.

Consider an emitter inside a cavity and with an enhanced spontaneous emission rate $\Gamma_c = F\Gamma_0$ (as a result of coupling to a cavity mode), where F is the Purcell factor and Γ_0 is the spontaneous emission rate of that same emitter without a cavity (e.g. in free space or bulk). Let us denote by Γ_{other} the spontaneous emission rate of this emitter into all modes other than the cavity mode. We can express $\Gamma_{\text{other}} = f\Gamma_0$, where $f < 1$ in the photonic bandgap, but otherwise $f \approx 1$. The total spontaneous emission rate $\Gamma_{\text{total}} = \Gamma_{\text{other}} + \Gamma_c$. The fraction of the light emitted by an emitter that is coupled into one particular cavity mode is known as the spontaneous emission coupling factor β:

$$\beta = \frac{\Gamma_c}{\Gamma_{\text{total}}} = \frac{\Gamma_c}{\Gamma_c + \Gamma_{\text{other}}} = \frac{F}{F + f}. \tag{8.56}$$

Therefore, if the spontaneous emission rate of an emitter is strongly enhanced by its interaction with a cavity mode ($F \gg 1$), or if the emission to other modes is significantly suppressed ($f \ll 1$), then the spontaneous emission coupling factor $\beta \approx 1$. This means that nearly all spontaneous emission goes into the lasing mode. The fraction of spontaneous emission going into non-lasing modes is one of the fundamental losses in a laser, and by decreasing it one can lower the laser threshold. Therefore, by increasing β, we can decrease the lasing threshold.

Let us now prove this conclusion from the simple laser rate equations. We denote the number of photons in the cavity mode by p and the total number of emitters in the excited state by N. An excited emitter can decay into the ground state by three mechanisms: (a) emission of a photon to modes other than the cavity mode with rate Γ_{other} (spontaneous emission);(b) emission of a photon into the cavity mode with rate $\Gamma_{\text{cav}}(p + 1)$ (stimulated emission); and (c) decay without emission of a photon with rate $1/\tau_{\text{nr}}$ (non radiative decay). It is also assumed that the emitters are pumped back to the excited state with rate R_{pump}. Therefore,

$$\frac{dN}{dt} = R_{\text{pump}} - N\Gamma_{\text{other}} - N\Gamma_{\text{cavity}}(p + 1) - \frac{N}{\tau_{\text{nr}}}. \tag{8.57}$$

The last term can be neglected, under the assumption that the non-radiative decay rate is small. Similarly, the rate equation for the number of photons in the cavity mode is

$$\frac{dp}{dt} = -2\kappa p + N\Gamma_{\text{cavity}}(p + 1), \tag{8.58}$$

where $\kappa = \omega/2Q$ is the cavity field decay rate. Clearly, the number of photons increases as a result of the emission rate into the cavity mode, and decreases as a result of the cavity loss.

Equations (8.57) and (8.58) constitute the rate equations for this system. If we solve them in a steady state ($d/dt = 0$), express N in terms of p from (8.58), and plug that back into (8.57), we obtain

$$R_{\text{pump}} = \frac{2\kappa p}{\Gamma_{\text{cavity}}(p + 1)}[\Gamma_{\text{other}} + \Gamma_{\text{cavity}}(p + 1)]. \tag{8.59}$$

On the other hand, we know that

$$\beta = \frac{\Gamma_{\text{cavity}}}{\Gamma_{\text{cavity}} + \Gamma_{\text{other}}} \implies \frac{1}{\beta} - 1 = \frac{\Gamma_{\text{other}}}{\Gamma_{\text{cavity}}}. \tag{8.60}$$

By plugging (8.60) into (8.59),

$$R_{\text{pump}} = \frac{2\kappa p}{p+1} \left(\frac{1}{\beta} + p \right). \tag{8.61}$$

The lasing threshold is defined as the condition where the mean photon number in a cavity is equal to 1, i.e. $p = 1$ [49]. Therefore, above threshold, $p > 1$, i.e. stimulated emission occurs (since there is more than one photon to stimulate emission). From (8.61), it follows that the pump rate at the lasing threshold is

$$R_{\text{pump}} = \frac{\kappa}{\beta}(\beta + 1) = \frac{\omega}{2Q\beta}(\beta + 1) = \frac{\omega}{2Q}\left(1 + \frac{1}{\beta}\right). \tag{8.62}$$

For $\beta \ll 1$, as is the case in conventional lasers, $R_{\text{pump,th}} \approx \omega/2Q\beta$. Therefore, the lasing threshold pump rate decreases with an increase in the spontaneous emission coupling factor β and the cavity Q-factor. In other words, by employing smaller-volume and higher-Q cavities, one can achieve lasing at lower pumping powers.

8.4   Quantum optics and cavity QED experiments with quantum dots in photonic crystal cavities

8.4.1   Rabi splitting and anticrossing

As an illustration of the strong-coupling regime with a lossy cavity and a good emitter ($g > \frac{1}{2}\kappa$ and $\kappa \gg \gamma$), let us consider a single InAs/GaAs quantum dot exciton in a GaAs photonic crystal cavity, as shown in Fig. 8.3. [10, 12]. Figure 8.12(b) shows the result of cross-polarized reflectivity measurement on a strongly coupled quantum dot–cavity system that operates in this regime. Cross-polarized reflectivity measurement is described in detail in our earlier work [10], but here we only note that such a measurement leads to a result identical to that of transmission measurement through the system. Figure 8.12(a) shows the anticrossing curve (for another quantum dot–cavity system) where the quantum dot is tuned across the cavity resonance by changing the sample temperature. A clear Rabi splitting is visible in both cases. From the experimental results, we extract that a typical range of parameters in our strongly coupled quantum dot–cavity system is $g/2\pi = 10$–$25\,\text{GHz}$, $\kappa/2\pi = 8$–$16\,\text{GHz}$, and $\gamma/2\pi \sim 1\,\text{GHz}$. The use of photonic crystal cavities with very small mode volume enables operation in the ultrafast g regime, and allows the strong-coupling regime ($g > \frac{1}{2}\kappa$) to be reached even with moderate values of the cavity Q-factor of $Q \sim 10^4$.

It is worth noting that at higher probe laser powers (roughly exceeding one photon per cavity lifetime), the dip in the split spectrum (and Rabi splitting) disappears (Fig. 8.12(c)). This is the result of the system saturation. Classically, the probe laser

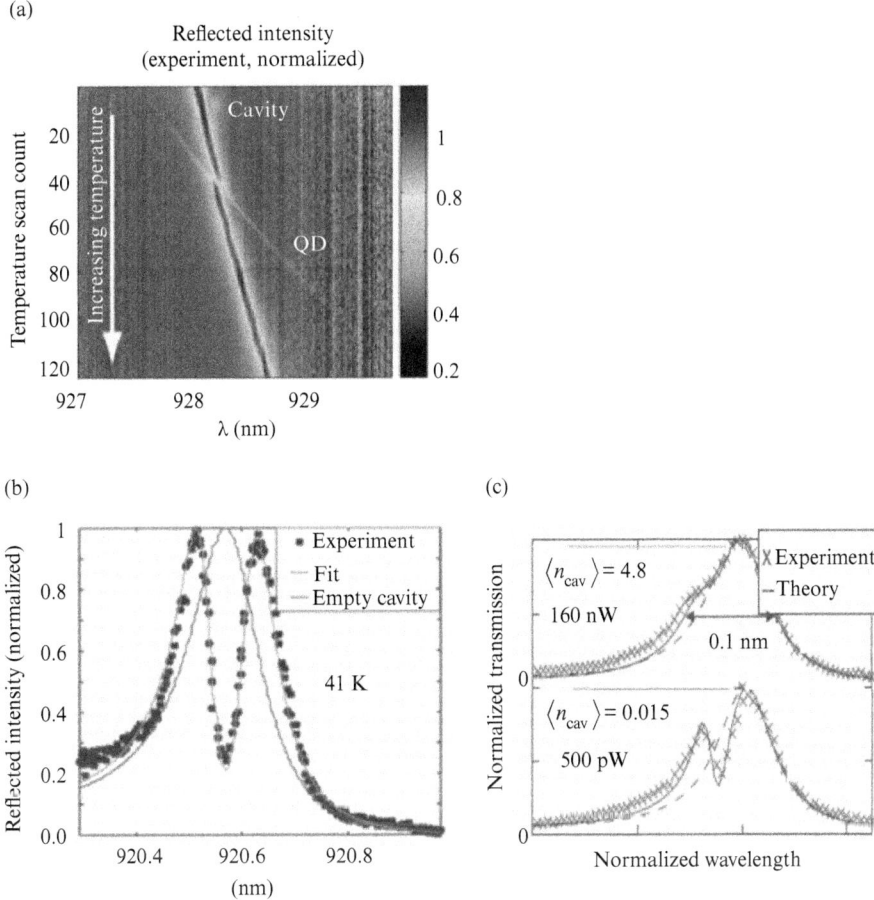

Fig. 8.12 [Colour online] (a) Mapping the anticrossing curve (the first manifold) of the strongly coupled quantum dot–photonic crystal cavity system by changing the sample temperature. (b) Rabi splitting of the photonic crystal cavity resonance after a quantum dot exciton is strongly coupled to it. (c) At higher power of the probe laser, the strongly coupled system saturates and the Rabi splitting disappears. (From Englund, D., Faraon, A., Fushman, I., Stoltz, N., Petroff, P., & Vučković, J. (2007). Controlling cavity reflectivity with a single quantum dot. Nature, 450(7171), 857–861. doi:10.1038/nature06234. Copyright 2007 Nature Publishing Group.)

sends enough photons to keep the emitter constantly excited, so for all other incoming photons the system behaves as an empty cavity. Quantum mechanically, at higher probe laser powers, one can excite higher manifolds of the dressed state ladder (Fig. 8.8), which appear in the spectrum as peaks at frequencies between the two peaks of the first manifold. Therefore, these higher-order peaks fill up the dip in the Rabi split spectrum and the system is saturated.

8.4.2 Controlled amplitude and phase shifts; switching with a strongly coupled system

It is possible to employ the described saturation effects to perform controlled amplitude and phase switching using the strongly coupled system. Let us shine two beams on the system (instead of one); they should be at the same frequency, or slightly detuned from each other, but both still within the cavity resonance. In this case, both beams can jointly saturate the system. We will refer to one of them as signal and the other as control. If only a signal beam is present, its power is insufficient to saturate the system. Since its frequency coincides with the dip of the strongly coupled system, the beam is just reflected back (Fig. 8.13(a)). On the other hand, if the signal beam is accompanied by a control beam (Fig. 8.13(b)), they jointly saturate the system, the Rabi splitting disappears, and both beams are transmitted to the other side. The structure thus behaves as an AND logic gate for two beams.

Figure 8.14 shows the results of such an experiment with a single quantum dot strongly coupled to a photonic crystal cavity, and with two continuous-wave (CW) beams that are slightly detuned from each other [16]. The signal beam is very weak and cannot saturate the system on its own. When the control beam photon number in the cavity approaches unity ($n_c \sim 1$), the system is saturated and the amplitude and phase of the transmitted signal beam look like those for an empty cavity. Therefore, with only a few nanowatts of power in the control beam ($n_c \sim 1$, when saturation happens), both the amplitude and phase of the transmitted signal beam are completely changed. For complete saturation, phase shifts of up to $\pi/4$ are achievable.

For practical applications, it is desirable to perform such switching with optical pulses. The results of such experiments with a single quantum dot strongly coupled to a photonic crystal cavity are shown in Fig. 8.15 [11]. Signal and control beams

(a)

(b)

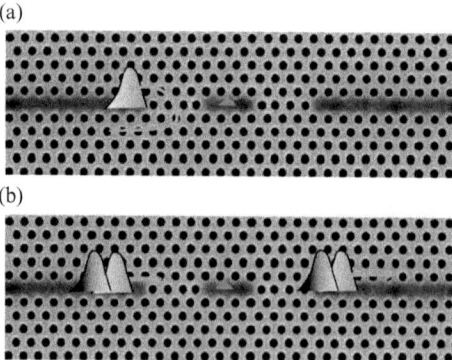

Fig. 8.13 [Colour online] Switching with a strongly coupled cavity QED system. The signal beam alone (a) has insufficient power to saturate the system and is reflected back (since its frequency is within the dip of the Rabi split system), but if it is accompanied by a control beam (b), they saturate the system and are transmitted together to the other side.

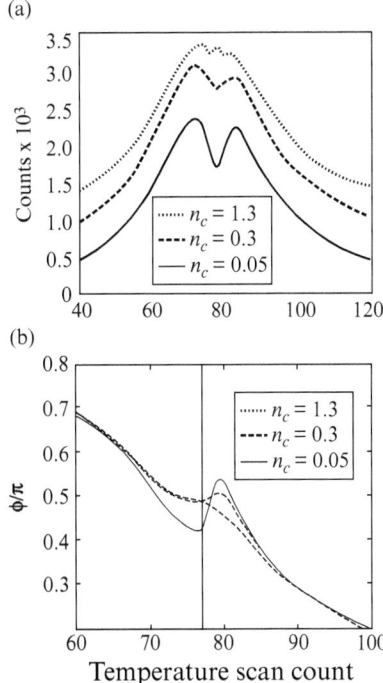

(a)

(b)

Temperature scan count

Fig. 8.14 [Colour online] Controlled amplitude (a) and phase (b) shifts with a single quantum dot strongly coupled to a photonic crystal cavity. The plots map the amplitude and phase of the probe beam at the output as a function of the control beam photon number n_c. At saturation, the amplitude and phase curves look like the corresponding curves for an empty cavity. Both control and probe beams are continuous-wave and are slightly detuned from each other. (From Fushman, I., Englund, D., Faraon, A., Stoltz, N., Petroff, P., & Vučković, J. (2008). Controlled phase shifts with a single quantum dot. Science, 320(5877), 769–772. doi:10.1126/science.1154643. Reprinted with permission from AAAS.)

consist of 40 ps pulses and are resonant with the cavity. Individually, they cannot saturate the system, but jointly they can. We map transmission through the system as a function of the delay between the signal and control pulses. When the delay is greater than the pulse duration, there is no saturation, and pulses bounce back (as in Fig. 8.13(a)). However, when the delay is comparable to or smaller than the pulse duration, the system is saturated and the pulses propagate together to the other side (as in Fig. 8.13(b)). As can be seen in Fig. 8.15(d), the transmission rises when the pulse delay becomes comparable to or smaller than the pulse duration. Therefore, with a strongly coupled quantum dot–cavity system, we can perform all optical switching at 40 GHz speed and with optical powers of the order of few nanowatts. Similar switching experiments have also been demonstrated in [7] and [44].

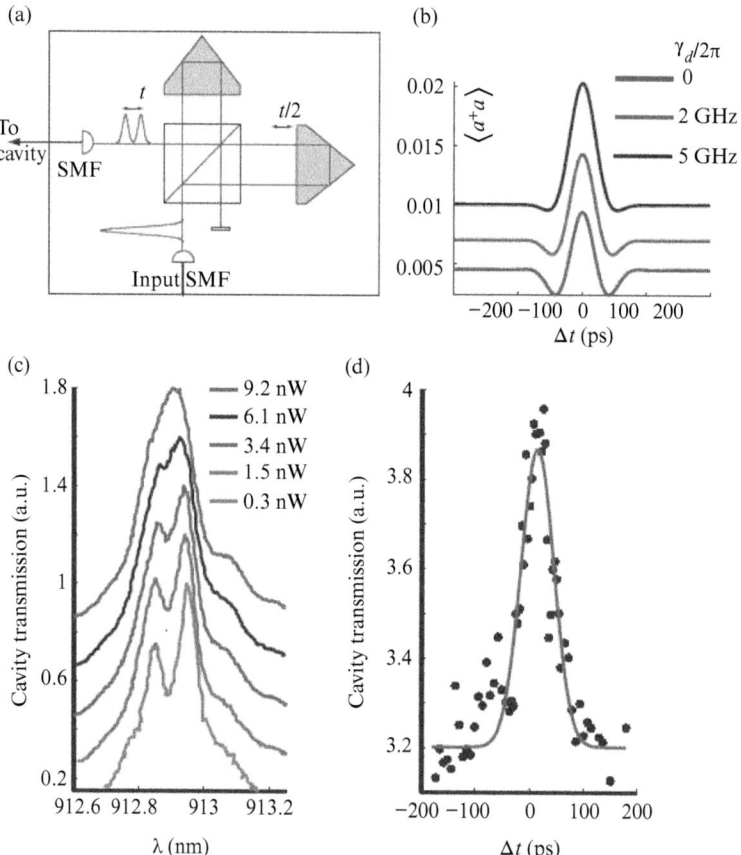

Fig. 8.15 [Colour online] Interaction of two weak laser pulses through a quantum dot–cavity system. (a) Time-delay setup for producing pulses at a separation Δt. (b) Simulated interaction of two laser pulses, represented by the instantaneous intracavity photon number $\langle a^\dagger a \rangle$ as a function of the time delay Δt between the two 40 ps long Gaussian pulses. Curves are calculated for a set of different rates of pure quantum dot dephasing γ_d, which causes a reduction of the transmission dips before and after the peak. Pure dephasing also causes a blurring of the spectral normal mode splitting, which in turn raises the transmission for increasing γ_d. (c) Pump-power dependence of the cavity transmission for coincident pulses repeating at 80 MHz. (d) Signal observed when the quantum dot–cavity system is probed with two 40 ps pulses as a function of their delay. When the two pulses have a temporal overlap inside the cavity, the quantum dot saturates and the overall cavity reflection increases. The power in each of the two pulses corresponds roughly to the 3.4 nW trace in (c). The best agreement with the theoretical plot is found for a pure dephasing rate $\gamma_d/2\pi \sim 5$ GHz. (Reprinted with permission from Englund, D., Majumdar, A., Bajcsy, M., Faraon, A., Petroff, P., & Vučković, J. (2012). Ultrafast photon–photon interaction in a strongly coupled quantum dot–cavity system. *Phys. Rev. Lett.*, 108(9). doi:10.1103/physrevlett.108.093604. Copyright 2012 by the American Physical Society.)

8.4.3 Generation of non-classical light with a strongly coupled quantum dot–cavity system: photon blockade and tunnelling

The effects of photon blockade and photon-induced tunnelling resulting from the anharmonicity of the dressed-states ladder were explained in Section 8.3.2. As shown in Fig. 8.10, such effects can be used to generate non-classical light (Fock states) by transmission of the laser beam at the appropriate frequency through a strongly coupled cavity QED system. Photon blockade was first demonstrated in an atomic cavity QED system [6] and higher-order manifolds were also probed in an atomic cavity QED system [38]. Here we present the results of photon blockade and photon-induced tunnelling in a strongly coupled quantum dot cavity QED system [14].

To show that in laser transmission we are generating a certain non-classical state of light, we measure the second-order photon correlation function $g^{(2)}(\tau)$ [39] using a Hanbury Brown and Twiss type setup shown in Fig. 8.16:

$$g^{(2)}(\tau) = \frac{\langle a^\dagger a^\dagger(\tau) a(\tau) a \rangle}{\langle a^\dagger a \rangle^2}, \tag{8.63}$$

where $a = a(0)$, $a^\dagger = a^\dagger(0)$, and the photon number operator is defined as $N = a^\dagger a$. Let us assume that the transmitted state through the system can be expressed as a superposition of the Fock states, $|\psi\rangle = \sum_n c_n |n\rangle$, where the probability of the nth Fock state is $P_n = |c_n|^2$. Then (by employing $a^\dagger |n\rangle = \sqrt{n+1} |n+1\rangle$ and $a |n\rangle = \sqrt{n} |n-1\rangle$), it follows that

$$g^{(2)}(\tau) = \frac{\langle \psi| a^\dagger a^\dagger a a |\psi\rangle}{\langle \psi| a^\dagger a |\psi\rangle^2} = \frac{\sum_n n(n-1)P_n}{\left(\sum_n nP_n\right)^2}. \tag{8.64}$$

Using this expression, let us evaluate $g^{(2)}(0)$ for several cases of interest:

1. *Perfect single-photon pulse train*:

$$P_n = \begin{cases} 1 & (n = 1), \\ 0 & (n \neq 1). \end{cases}$$

   It is simple to show that $g^{(2)}(0) = 0$.

2. *Sparse single-photon pulse train* (mostly vacuum with sporadic single-photon pulses):

$$P_n = \begin{cases} \varepsilon & (n = 1), \\ 1 - \varepsilon & (n = 0), \\ 0 & (\neq 0, 1), \end{cases}$$

   where $\varepsilon \ll 1$. By plugging this into the expression for $g^{(2)}(0)$, we can show that $g^{(2)}(0) = 0$ still holds.

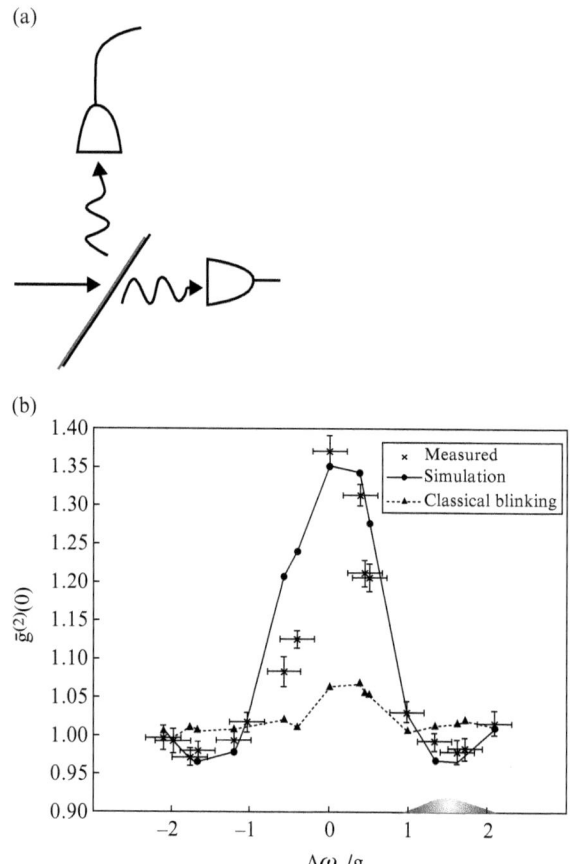

Fig. 8.16 [Colour online] (a) The Hanbury Brown and Twiss setup for $g^{(2)}(t)$ measurements consists of a beamsplitter and two single-photon counters. (b) Measurement of $g^{(2)}(0)$ for transmission of a laser beam through a strongly coupled quantum dot–photonic crystal cavity system. The horizontal axis corresponds to the tuning of the input laser frequency relative to the empty cavity frequency. Transmission through an empty cavity should exhibit Poissonian statistics (like a laser itself: $g^{(2)}(0) = 1$). However, in this case, we observe the regimes of antibunching ($g^{(2)}(0) < 1$) corresponding to sub-Poissonian statistics (imperfect single-photon stream in the regime of photon blockade), and super-Poissonian statistics (bunching $g^{(2)}(0) > 1$) in the regime of photon-induced tunnelling. (From Faraon, A., Fushman, I., Englund, D., Stoltz, N., Petroff, P. & Vučković, J. Coherent generation of nonclassical light on a chip via photon-induced tunnelling and blockade. Nat. Phys. 4(11), 859–863. doi:10.1038/nphys1078. Copyright 2008 Nature Publishing Group.)

3. *Perfect two-photon pulse train*:

In this case, $g^{(2)}(0) = \frac{1}{2}$.

$$P_n = \begin{cases} 1 & (n = 2), \\ 0 & n \neq 2. \end{cases}$$

In this case, $g^{(2)}(0) = \frac{1}{2}$.

4. *Perfect N-photon pulse train*:

$$P_n = \begin{cases} 1 & (n = N), \\ 0 & (n \neq N). \end{cases}$$

In this case, $g^{(2)}(0) = (N-1)/N$.

5. *Sparse train of two-photon pulses*:

$$P_n = \begin{cases} \varepsilon & (n = 2), \\ 1 - \varepsilon & (n = 0), \\ 0 & (n \neq 0, 2). \end{cases}$$

In this case, $g^{(2)}(0) = 1/(2\varepsilon)$. Since $\varepsilon \ll 1$ for a sparse train, it follows that $g^{(2)}(0) > 1$ in this case, i.e. we expect to observe bunching.

6. *Sparse train of three-photon pulses*:

$$P_n = \begin{cases} \varepsilon & (n = 3), \\ 1 - \varepsilon & (n = 0), \\ 0 & (n \neq 0, 3). \end{cases}$$

In this case, $g^{(2)}(0) = 2/(3\varepsilon)$. Again, we expect bunching ($g^{(2)}(0) > 1$), since $\varepsilon \ll 1$ for a sparse train.

Similarly, if the output state is a combination of these, it can be shown that $g^{(2)}(0)$ varies between antibunching (sub-Poissonian regime, with $g^{(2)}(0) < 1$) and bunching (super-Poissonian regime, with $g^{(2)}(0) > 1$). For a coherent state (at the output of the laser) that features Poissonian photon statistics, it can be shown that $g^{(2)}(0) = 1$ [39].

The measurement of $g^{(2)}(0)$ for a laser beam transmitted through a strongly coupled quantum dot–cavity system as a function of the laser detuning from the empty cavity frequency is shown in Fig. 8.16 [14]. Clearly, $g^{(2)}(0)$ goes between antibunching (in the photon blockade regime, where sub-Poissonian light is generated, i.e. the output is a superposition of Fock states with dominant single-photon component) and bunching (in the photon-induced tunnelling regime, where the output is mostly vacuum with sparse higher-order Fock states.

8.4.4 Strong coupling of a single quantum dot to a photonic molecule

As an extension of the single quantum dot–photonic crystal cavity QED experiments described in Section 8.4.1, we now consider strong coupling between a single quantum dot and a photonic crystal molecule (i.e. a system of two coupled cavities) [30].

A photonic molecule consists of two coupled photonic crystal cavities in this case, as shown in Fig. 8.17(a). The spectrum of a such a system is expected to exhibit two coupled modes that correspond to bonding and antibonding states and whose

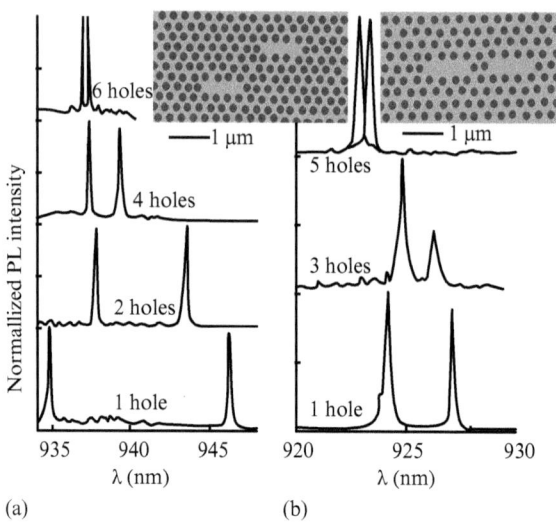

Normallized PL intensity

6 holes

4 holes

2 holes

1 hole

—1 μm

5 holes

3 holes

1 hole

—1 μm

935   940   945   920   925   930
λ (nm)   λ (nm)

(a)   (b)

Fig. 8.17 Spectra of photonic crystal molecules as a function of the separation of coupled cavities. As expected, the increase in separation between the cavities decreases their coupling strength and reduces the spectral separation between the coupled modes. (Reprinted with permission from Majumdar, A., Rundquist, A., Bajcsy, M., & Vučković, J. (2012). Cavity quantum electrodynamics with a single quantum dot coupled to a photonic molecule. Phys. Rev. B, 86(4). doi:10.1103/physrevb.86.045315. Copyright 2012 by the American Physical Society.)

separation increases as the coupling between the cavities increases (which in turn can be done by decreasing their spatial separation—see Fig. 8.17).

Let us consider a system of coupled cavities as in Fig. 8.17(a) with a single quantum dot inside it. If the system consists of three strongly coupled oscillators (two modes and one quantum dot), it is expected that the modes will form supermodes and the quantum dot will anticross both of them. Indeed, by tuning a quantum dot across both supermodes (by a combination of temperature and nitrogen condensation tuning), we observe its anticrossing with both supermode peaks, as shown in Fig. 8.18. Such a system has potential applications in non-classical light generation [27].

8.4.5 Weak coupling of a single quantum dot to a cavity: Purcell regime and spontaneous emission rate suppression

In the Purcell regime of cavity QED described in Section 8.3.4, we expect to observe enhancement of the spontaneous emission rate of a quantum emitter as a result of its coupling to a cavity resonance. This enhancement scales as Q/V_{mode} of a cavity, and is maximal for an emitter that is spectrally and spatially aligned with the cavity. If the emitter is spectrally detuned, the Purcell enhancement follows a Lorentzian lineshape given by the density of optical states in a cavity. The experimental results shown in Fig. 8.19 indeed demonstrate such a result for a single InAs/GaAs quantum

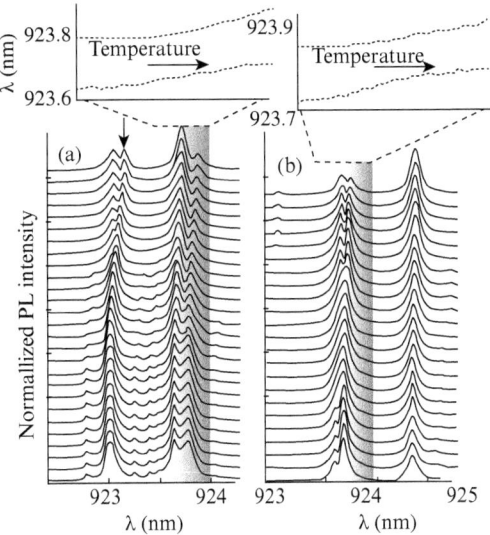

Fig. 8.18 [Colour online] Strong coupling of a single InAs/GaAs quantum dot to a photonic crystal molecule, like that shown in Fig. 8.17(a). Anticrossing of the quantum dot with both supermodes is observed as it is tuned across them. Thus, the system consists of three strongly coupled oscillators. (Reprinted with permission from Majumdar, A., Rundquist, A., Bajcsy, M., & Vučković, J. (2012). Cavity quantum electrodynamics with a single quantum dot coupled to a photonic molecule. Phys. Rev. B, 86(4). doi:10.1103/physrevb.86.045315. Copyright 2012 by the American Physical Society.)

dot in a distributed Bragg reflector micropost cavity (basically a 1D photonic crystal cavity) [45].

The results shown in Fig. 8.19 confirm significant Purcell enhancement for a quantum dot resonant with the cavity mode, and show that the Purcell factor as a function of quantum dot detuning from the cavity does indeed follow a Lorentzian lineshape given by the cavity spectrum.

Similarly, as described in Section 8.3.4, if the density of photon states $D_c(\nu)$ is zero or significantly suppressed, then the spontaneous emission rate should also be suppressed (as $\Gamma/\Gamma_0 \propto D_c(\nu)$). This is for example, possible in photonic crystals, as a result of photonic bandgap, as shown in Fig. 8.20 [9].

8.4.6 Ultralow threshold photonic crystal laser

In Section 8.3.5, we learned that Purcell enhancement also leads to enhancement of the spontaneous emission coupling factor β and a reduction in the lasing threshold. This has been employed in optically pumped photonic crystal lasers with InAs quantum dots as an active medium to demonstrate $\beta \sim 0.85$ and lasing threshold power of 124 nW [42]. However, for practical applications it is necessary to inject lasers electrically. This has recently been achieved in the same GaAs photonic crystal cavity lasers with an

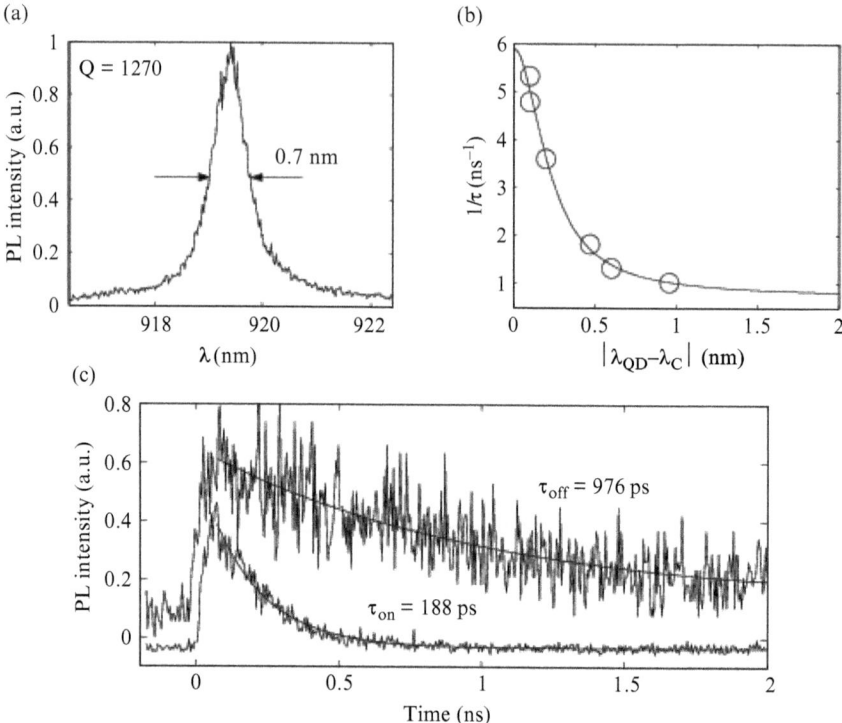

Fig. 8.19 Weak coupling of a single quantum dot to a distributed Bragg reflector micropost cavity. (a) Cavity Q factor. (b) The quantum dot spontaneous emission rate as it is tuned across the cavity resonance, following the Lorentzian lineshape described by the cavity spectrum. (c) Time-resolved lifetime measurements for this quantum dot, showing the maximum Purcell effect (shortest lifetime) and longest lifetime. (Reproduced from Vučković, J., Fattal, D., Santori, C., Solomon, G. S., & Yamamoto, Y. (2003). Enhanced single-photon emission from a quantum dot in a micropost microcavity. Appl. Phys. Lett., 82(21), 3596. doi:10.1063/1.1577828, with the permission of AIP Publishing.)

ensemble of quantum dots as an active medium, using the configuration of lateral pin junction shown in Fig. 8.21 [8]. A demonstration of an ultralow threshold electrically injected laser in this configuration is shown in Fig. 8.22, with lasing threshold current as low as 180 nA (with 1 V bias voltage).

8.5 Summary

This chapter has provided an introduction to quantum optics and solid state cavity QED with quantum dots in photonic crystal cavities. Although the focus has been on the physics of the quantum dot–photonic crystal cavity system, there are many applications that would greatly benefit from this platform. In addition to the already

(a)

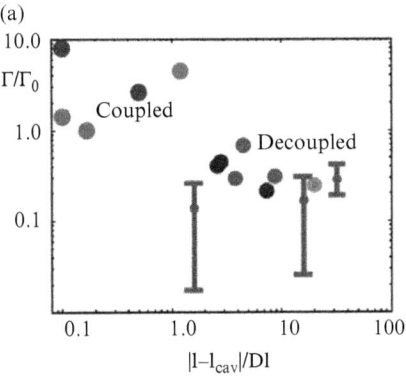

(b)

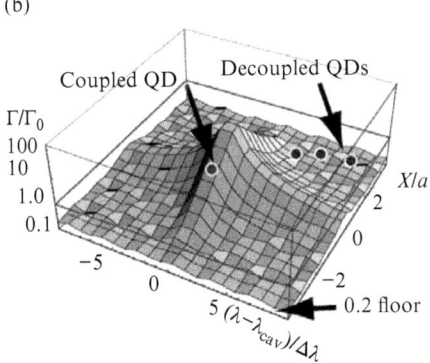

Fig. 8.20 [Colour online] (a) Spontaneous emission rate modification of individual quantum dots as a function of their detuning from 2D photonic crystal cavity resonance. Tuned quantum dots exhibit Purcell enhancement or not much enhancement (depending on whether they are spatially aligned with the cavity), while spectrally detuned quantum dots always exhibit suppression of the spontaneous emission, as a result of the reduction of the photon density of states in photonic bandgap. The [green] bars show simulation results. (b) Predicted spontaneous emission rate modification in the photonic crystal cavity as a function of normalized spatial and spectral misalignment from the cavity (a is the lattice periodicity). This plot assumes $Q = 1000$ and polarization matching between the emitter dipole and cavity field. (Reprinted with permission from Englund, D., Fattal, D., Waks, E., Solomon, G., Zhang, B., Nakaoka, T., Arakawa, Y., Yamamoto, Y. & Vučković, J. (2005). Controlling the spontaneous emission rate of single quantum dots in a two-dimensional photonic crystal. Phys. Rev. Lett., 95(1). doi:10.1103/physrevlett.95.013904. Copyright 2005 by the American Physical Society.)

mentioned all-optical switches at a few-photon level and ultralow-threshold lasers, combining such blocks into circuits could be used to construct all-optical computers, optical interconnects in computers, or quantum repeaters for long-distance quantum communication [22, 33]. Finally, although we have focused on a single InAs/GaAs quantum dot as a quantum emitter inside a planar GaAs photonic crystal cavity

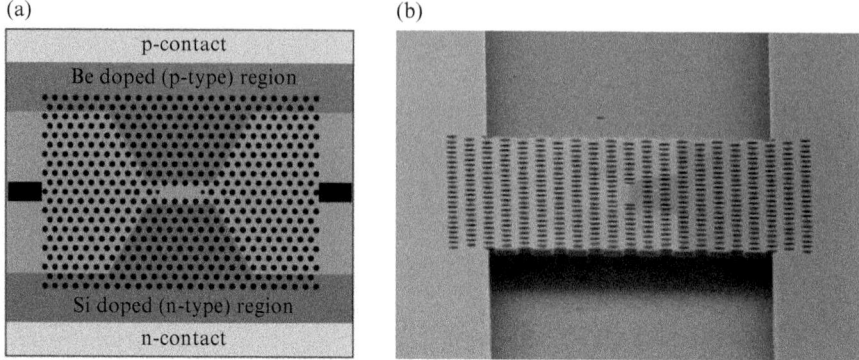

Fig. 8.21 [Colour online] (a) Lateral p–i–n junction embedded in a photonic crystal cavity for electrical injection into the cavity region. (b) Such a cavity fabricated in GaAs with InAs quantum dots as an active medium. (From Ellis, B., Mayer, M. A., Shambat, G., Sarmiento, T., Harris, J., Haller, E. E. & Vučković, J. (2011). Ultralow-threshold electrically pumped quantum-dot photonic-crystal nanocavity laser. Nat. Photonics, 5(5), 297–300. doi:10.1038/nphoton.2011.51. Copyright 2011 Nature Publishing Group.)

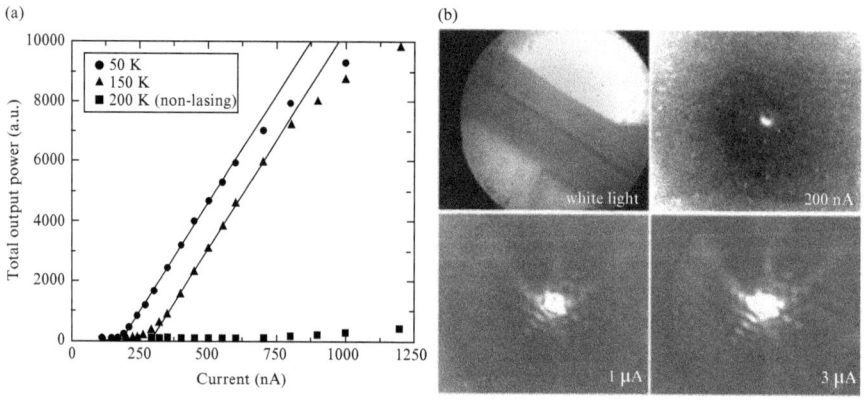

Fig. 8.22 [Colour online] (a) Laser optical output power–injection current characteristics at different temperatures. At 50 K, the lasing threshold is as low as 180 nW. (b) Optical image of the laser and lasing patterns at different injection currents. (From Ellis, B., Mayer, M. A., Shambat, G., Sarmiento, T., Harris, J., Haller, E. E. & Vučković, J. (2011). Ultralow-threshold electrically pumped quantum-dot photonic-crystal nanocavity laser. Nat. Photonics, 5(5), 297–300. doi:10.1038/nphoton.2011.51. Copyright 2011 Nature Publishing Group.)

[4, 9, 23, 51], the physics described here applies to other types of quantum emitters, including those in the solid state, such as nitrogen vacancy centres in diamond coupled to cavities [13, 15, 29], quantum dots in other types of structures [2, 5, 25, 28, 34, 37, 41], neutral atoms in microcavities [6, 31, 43], and even circuit QED systems [47].

Acknowledgements

I would like to thank all of the present and former members of my research group who have worked on the described cavity QED and quantum optics projects with quantum dots in photonic crystals: Professor Dirk Englund (MIT), Professor Andrei Faraon (Caltech), Dr Ilya Fushman (Index Ventures), Professor Edo Waks (University of Maryland and JQI), Professor Vanessa Sih (University of Michigan), Professor Arka Majumdar (University of Washington, Seattle), Dr Erik Kim (HGST), Professor Michal Bajcsy (IQC, Waterloo), Dr Bryan Ellis (Foveon), Dr Gary Shambat (Google), Dr Per Kaer (DTU), Dr Konstantinos Lagoudakis, Dr Tomas Sarmiento, Dr Kai Mueller, Dr Armand Rundquist (Neophotonics), Mr. Kevin Fisher, and Mr Jan Petykiewicz. I would also like to thank Professor Yoshi Yamamoto (Stanford) and my former co-workers from his research group, Dr Charlie Santori and Dr David Fattal (HP Labs), with whom I performed Purcell effect measurements on a single quantum dot in a micropost cavity, as well as Professor Pierre Petroff (emeritus, UCSB), Professor Jim Harris (Stanford) and Professor Seth Bank (UT Austin) for growth of quantum dots used in some of these experiments. In addition, I would like to thank ARO (Dr TR Govindan), AFOSR (Dr Tatjana Curcic and Dr Gernot Pomrenke), ONR (Dr Chagaan Baatar), DARPA (Dr Jag Shah), and NSF (Dr Wendy Fuller-Mora, Dr Dan Finotello, and Dr Paul Sokol) for funding these projects. I would also like to thank the organizers of this summer school, Professor Claude Fabre, Professor Nicolas Treps, and Professor Vahid Sandoghdar, for the invitation to Les Houches, and for many stimulating discussions there with them and all of the school participants. Finally, thanks to Claude for carefully proofreading this chapter.

References

[1] Y. Akahane, T. Asano, B.-S. Song, and S. Noda. High-Q photonic nanocavity in a two-dimensional photonic crystal. *Nature* **425** (2003) 944–947.

[2] K. Aoki, D. Guimard, M. Nishioka, M. Nomura, S. Iwamoto, and Y. Arakawa. Coupling of quantum-dot light emission with a three-dimensional photonic crystal nanocavity. *Nat. Photonics* **2** (2008) 688–692.

[3] T. Asano, B.-S. Song, and S. Noda. Analysis of the experimental Q factors (∼1 million) of photonic crystal nanocavities. *Opt. Express* **14** (2006) 1996–2002.

[4] A. Badolato, K. Hennessy, M. Atat, J. Dreiser, P. M. Petroff, and A. Imamoğlu. Deterministic coupling of single quantum dots to single nano cavity modes. *Science* textbf308 (2005) 1158–1161.

[5] O. Benson. Assembly of hybrid photonic architectures from nanophotonic constituents. *Nature* **480** (2011) 193–199.

[6] K. M. Birnbaum, A. Boca, R. Miller, A. D. Boozer, T. E. Northup, and H. J. Kimble. Photon blockade in an optical cavity with one trapped atom. *Nature* **436** (2005) 87–90.

[7] R. Bose, D. Sridharan, H. Kim, G. Solomon, and E. Waks. Low-photon-number optical switching with a single quantum dot coupled to a photonic crystal cavity. *Phys. Rev. Lett.* **108** (2012) 227402.

[8] B. Ellis, M. Mayer, G. Shambat, T. Sarmiento, E. Haller, J. S. Harris, and J. Vučković. Ultralow-threshold electrically pumped quantum-dot photonic-crystal nanocavity laser. *Nat. Photonics* **5** (2011) 297–300.

[9] D. Englund, D. Fattal, E. Waks, G. Solomon, B. Zhang, T. Nakaoka, Y. Arakawa, Y. Yamamoto, and J. Vučković. Controlling the spontaneous emission rate of single quantum dots in a 2D photonic crystal. *Phys. Rev. Lett.* **95** (2005) 013904.

[10] D. Englund, A. Faraon, I. Fushman, N. Stoltz, P. Petroff, and J. Vučković. Controlling cavity reflectivity with a single quantum dot. *Nature* (2007) 857–861.

[11] D. Englund, A. Majumdar, M. Bajcsy, A. Faraon, P. Petroff, and J. Vučković. Ultrafast photon–photon interaction in a strongly coupled quantum dot–cavity system. *Phys. Rev. Lett.* **108** (2012) 093604.

[12] D. Englund, A. Majumdar, A. Faraon, M. Toishi, N. Stoltz, P. Petroff, and J. Vučković. Resonant excitation of a quantum dot strongly coupled to a photonic crystal nanocavity. *Phys. Rev. Lett.* **104** (2010) 073904.

[13] D. Englund, B. Shields, K. Rivoire, F. Hatami, J. Vučković, H. Park, and M. D. Lukin. Deterministic coupling of a single nitrogen vacancy center to a photonic crystal cavity. *Nano Lett.* **10** (2010) 3922–3926.

[14] A. Faraon, I. Fushman, D. Englund, N. Stoltz, P. Petroff, and J. Vučković. Coherent generation of nonclassical light on a chip via photon-induced tunnelling and blockade. *Nat. Phys.* **4** (2008) 859–863.

[15] A. Faraon, P. E. Barclay, C. Santori, K.M.C. Fu, and R.G. Beausoleil. Resonant enhancement of the zero-phonon emission from a colour centre in a diamond cavity. *Nat Photonics* **5** (2011) 301–305.

[16] I. Fushman, D. Englund, A. Faraon, N. Stoltz, P. Petroff, and J. Vučković. Controlled phase shifts with a single quantum dot. *Science* **320** (2008) 769–772.

[17] M. Grundmann, O. Stier, and D. Bimberg. InAs/GaAs pyramidal quantum dots: strain distribution, optical phonons, and electronic structure. *Phys. Rev. B* **52** (1995) 11969–11981.

[18] S. Haroche. Nobel Lecture: Controlling photons in a box and exploring the quantum to classical boundary. *Rev. Mod. Phys.* **85** (2013) 1083–1102.

[19] J. D. Joannopoulos, S. G. Johnson, J. N. Winn, and R. D. Meade. *Photonic Crystals: Molding the Flow of Light*, 2nd edn. Princeton, NJ: Princeton University Press, 2008.

[20] S. John. Strong localization of photons in certain disordered dielectric superlattices. *Phys. Rev. Lett.* **58** (1987) 2486–2489.

[21] H. J. Kimble. Structure and dynamics in cavity quantum electrodynamics. In *Cavity Quantum Electrodynamics* (ed. P. R. Berman), pp. 203–266. San Diego: Academic Press, 1994.

[22] H. J. Kimble. The quantum internet. *Nature* **453** (2008) 1023–1030.

[23] A. Kress, F. Hofbauer, N. Reinelt, M. Kaniber, H. J. Krenner, R. Meyer, G. Böhm, and J. J. Finley. Manipulation of the spontaneous emission dynamics of quantum dots in two-dimensional photonic crystals. *Phys. Rev. B* **71** (2005) 241304.

[24] K. G. Lagoudakis, K. Fischer, T. Sarmiento, A. Majumdar, A. Rundquist, J. Lu, M. Bajcsy, and J. Vučković. Deterministically charged quantum dots in photonic

crystal nanoresonators for efficient spin–photon interfaces. *New J. Phys* **15** (2013) 113056.

[25] A. Laucht, T. G, S. P, R. Saive, S. Frederick, N. Hauke, M. Bichler, M. C. Amann, A. Holleitner, M. Kaniber, and J. Finley. Broadband Purcell enhanced emission dynamics of quantum dots in linear photonic crystal waveguides. *J. Appl. Phys.* **112** (2012) 093520.

[26] F. P. Laussy, E. del Valle, M. Schrapp, A. Laucht, and J. J. Finley. Climbing the Jaynes–Cummings ladder by photon counting. *J. Nanonphotonics* **6** (2012) 061803.

[27] T. C. H. Liew and V. Savona. Single photons from coupled quantum modes. *Phys. Rev. Lett.* **104** (2010) 183601.

[28] P. Lodahl. All-solid-state quantum optics employing quantum dots in photonic crystals. In *Quantum Optics with Semiconductor Nanostructures* (ed. F. Jahnke), pp. 395–420. Cambridge: Woodhead Publishing, 2012.

[29] M. Lončar and A. Faraon. Quantum photonic networks in diamond. *MRS Bull.* **38** (2013) 144–148.

[30] A. Majumdar, A. Rundquist, M. Bajcsy, and J. Vučković. Cavity quantum electrodynamics with a single quantum dot coupled to a photonic molecule. *Phys. Rev. B* **86** (2012) 045315.

[31] P. Maunz, T. Puppe, I. Schuster, N. Syassen, P. W. H. Pinkse, and G. Rempe. Normal-mode spectroscopy of a single bound atom–cavity system. *Phys. Rev. Lett.* **94** (2005) 033002.

[32] P. Michler. *Single Semiconductor Quantum Dots*. Berlin: Springer-Verlag, 2009.

[33] J. O'Brien, A. Furusawa, and J. Vučković. Photonic quantum technologies. *Nat. Photonics* **3** (2009) 687–695.

[34] D. Press, S. Goetzinger, S. Reitzenstein, C. Hofmann, A. Loeffler, M. Kamp, A. Forchel, and Y. Yamamoto. Photon antibunching from a single quantum dot–microcavity system in the strong coupling regime. *Phys. Rev. Lett.* **98** (2007) 117402.

[35] E. M. Purcell. Spontaneous emission probabilities at radio frequencies. *Phys. Rev.* **69** (1946) 681.

[36] A. Reinhard, T. Volz, M. Winger, A. Badolato, K. J. Hennessy, E. L. Hu, and A. Imamoğlu. Strongly correlated photons on a chip. *Nat. Photonics* **6** (2012) 93–96.

[37] J. P. Reithmaier, G. Sek, A. Löffler, C. Hofmann, S. Kuhn, S. Reitzenstein, L. V. Keldysh, V. D. Kulakovskii, T. L. Reinecke, and A. Forchel. Strong coupling in a single quantum dot–semiconductor microcavity system. *Nature* **432** (2004) 197–200.

[38] I. Schuster, A. Kubanek, A. Fuhrmanek, T. Puppe, P. W. H. Pinkse, K. Murr, and G. Rempe. Nonlinear spectroscopy of photons bound to one atom. *Nat. Phys.* **4** (2008) 382–385.

[39] M. O. Scully and M. S. Zubairy. *Quantum Optics*. Cambridge: Cambridge University Press, 1997.

[40] B.-S. Song, S. Noda, and T. Asano. Photonic devices based on in-plane hetero photonic crystals. *Science* **300** (2003) 1537–1540.

[41] K. Srinivasan and O. Painter. Linear and nonlinear optical spectroscopy of a strongly coupled microdisk–quantum dot system. *Nature* **450** (2007) 862–865.

[42] S. Strauf, K. Hennessy, M. T. Rakher, Y.-S. Choi, A. Badolato, L. C. Andreani, E. L. Hu, P. M. Petroff, and D. Bouwmeester. Self-tuned quantum dot gain in photonic crystal lasers. *Phys. Rev. Lett.* **96** (2006) 127404.

[43] J. D. Thompson, T. G. Tiecke, N. P. de Leon, J. Feist, A. V. Akimov, M. Gullans, A. S. Zibrov, V. Vuletić, and M. D. Lukin. Coupling a single trapped atom to a nanoscale optical cavity. *Science* **340** (2013) 1202–1205.

[44] T. Volz, A. Reinhard, M. Winger, A. Badolato, K. J. Hennessy, E. L. Hu, and A. Imamoğlu. Ultrafast all-optical switching by single photons. *Nat. Photonics* **6** (2012) 605–609.

[45] J. Vučković, D. Fattal, C. Santori, G. Solomon, and Y. Yamamoto. Enhanced single photon emission from a quantum dot in a micropost microcavity. *Appl. Phys. Lett.* **82** (2003) 3596–3598.

[46] J. Vučković, M. Loncar, H. Mabuchi, and A. Scherer. Design of photonic crystal microcavities for cavity QED. *Phys. Rev. E* **65** (2002) 016608.

[47] A. Wallraff, D. I. Schuster, A. Blais, L. Frunzio, R.-S. Huang, J. Majer, S. Kumar, S. M. Girvin, and R. J. Schoelkopf. Circuit quantum electrodynamics: coherent coupling of a single photon to a Cooper pair box. *Nature* **431** (2004) 162–165.

[48] E. Yablonovitch. Inhibited spontaneous emission in solid-state physics and electronics. *Phys. Rev. Lett.* **58** (1987) 2059–2062.

[49] Y. Yamamoto and A. Imamoğlu. *Mesoscopic Quantum Optics*. New York: Wiley, 1999.

[50] A. Yariv and P. Yeh. *Optical Waves in Crystals: Propagation and Control of Laser Radiation*. Hoboken, NJ: Wiley, 2002.

[51] T. Yoshie, A. Scherer, J. Hendrickson, G. Khitrova, H. M. Gibbs, G. Rupper, C. Ell, O. B. Shchekin, and D. G. Deppe. Vacuum Rabi splitting with a single quantum dot in a photonic crystal nanocavity. *Nature* **432** (2004) 200–203.

9
Casimir forces and vacuum energy

Serge REYNAUD and Astrid LAMBRECHT

Laboratoire Kastler Brossel, CNRS, ENS, UPMC
Campus Jussieu, Paris, France

Serge Reynaud and Astrid Lambrecht: *Casimir forces and vacuum energy*. In: *Quantum Optics and Nanophotonics*. Edited by: Claude Fabre, Vahid Sandoghdar, Nicolas Treps, and Leticia F. Cugliandolo. Oxford University Press (2017), pp. 407–455. © Oxford University Press 2017.
DOI: 10.1093/oso/9780198768609.003.0009

Chapter Contents

9.1   The Casimir force

9.1.1   Introduction

The emergence of quantum theory has profoundly altered our conception of space by forcing us to consider it as permanently filled by vacuum field fluctuations [162, 163]. These vacuum fluctuations are electromagnetic fields propagating with the speed of light, like any free field, and corresponding to an energy of half a photon per mode. They have a number of observable consequences in microscopic physics, for example radiative corrections in subatomic physics [107], spontaneous emission processes, the Casimir–Polder interaction, and the Lamb shift in atomic physics [58].

Vacuum fluctuations also have observable mechanical effects in macroscopic physics, the archetype of these effects being the Casimir force between two motionless mirrors. This force was predicted in 1948 by H. B. G. Casimir [48] and was soon observed experimentally [142]. Experiments have been improved after years of development and now reach a good level of accuracy [67, 68, 70, 71] (more references are given below). However, comparison with theoretical predictions has raised a number of difficulties, which have been discussed in a large number of papers (references can be found in [16, 64, 127]).

This comparison is particularly interesting because of its fascinating relation to some open questions in fundamental physics. The Casimir effect is connected with the puzzles of gravitational physics through the problem of vacuum energy as well as with the principle of relativity of motion [115]. Effects beyond the proximity force approximation also make apparent the rich interplay of vacuum and geometry [10]. Casimir physics also plays an important role in tests of gravity at submillimetre ranges [1, 2, 5, 89]. For scales of the order of the micrometre, such gravity tests are performed by comparing Casimir force measurements with theory, and the comparison has to take into account the many differences between real experiments and the idealized case considered initially by Casimir [135].

At the end of this short general introduction, it has also to be stressed that the Casimir and closely related Casimir–Polder forces [49, 88, 104, 166, 182] have strong connections with various active domains and interfaces of physics, such as atomic and molecular physics, condensed matter and surface physics, chemical and biological physics, and micro- and nanotechnology [91, 173].

9.1.2   Outline

Section 9.2 contains an introduction and a description of the current state of the art for experiments and their comparison with theory, while Sections 9.3 and 9.4 provide a pedagogical presentation of the main features of the theory of Casimir forces.

Section 9.2 begins with a short history of quantum field fluctuations in vacuum. We then review various arguments involved in the comparison with theory of the experiments devoted to the measurement of the Casimir force. As experiments are performed with gold-covered plates, the force depends on non-universal properties of the real plates used in the experiments. As they are performed at room temperature, the effect of thermal field fluctuations has to be added to that of vacuum fluctuations.

The most precise experiments are performed in the plane–sphere geometry rather than the geometry of two parallel planes, although the latter is theoretically easier to handle. Finally, surfaces are non-ideal, and effects such as roughness, electrostatic patches, and contamination affect the comparison between theory and experiment.

Section 9.3 contains a simple derivation of the Casimir effect in a model of scalar fields on the one-dimensional line. This model allows one to introduce the *quantum optics* approach to the Casimir effect. This approach is based on the existence of field fluctuations that pervade empty space and exert *radiation pressure* on mirrors at rest in vacuum. The force is thus calculated as the result of different pressures acting on the inner and outer sides of the two mirrors that form a cavity. This approach is often also called the *scattering formalism* because all properties of the Casimir force are determined by the reflection amplitudes of the fields on the two mirrors [108].

Section 9.4 then treats the case of two plane and parallel mirrors at rest in electromagnetic vacuum in three-dimensional space. It describes the models generally used for the metallic mirrors used in the experiments [95, 134] and discusses the results obtained in this manner. For mirrors characterized by Fresnel reflection amplitudes deduced from a linear and local dielectric function, the scattering formalism leads to the same results as Lifshitz's method [79, 145, 146]. The section ends with a presentation of the general scattering formalism that allows one to deal with non-specular reflection and arbitrary geometries [82, 136, 139, 165, 184, 197].

9.2   Vacuum fluctuations and Casimir forces

9.2.1   The birth of the quantum vacuum

The classical idealization of absolute empty space was affected by the discovery of black body radiation, which is present everywhere at non-zero temperature and exerts a pressure on the boundaries of any cavity. It is precisely for explaining the properties of black body radiation that Planck introduced the first quantum law in 1900 [178] (for discussions, see [65]).

In modern terms, the first Planck law gives the energy per electromagnetic mode characterized by its frequency $\omega = 2\pi\nu$ as the product of the energy of a single photon $\hbar\omega \equiv h\nu$ by a mean number of photons $\overline{n}$ per mode:

$$\overline{E}_{1900} = \overline{n}\hbar\omega\,, \qquad \overline{n} \equiv \frac{1}{e^{\hbar\omega/k_\mathrm{B}T} - 1}\,. \tag{9.1}$$

Unsatisfied with his first derivation, Planck [179] wrote in 1911 a new expression for the mean energy per mode:

$$\overline{E}_{1911} = \left(\tfrac{1}{2} + \overline{n}\right)\hbar\omega\,. \tag{9.2}$$

The difference between the two Planck laws corresponds to the zero-point energy. Whereas the first law describes a cavity entirely emptied of radiation at the limit of zero temperature ($\overline{n} \to 0$ when $T \to 0$), the zero-point energy added in the second law persists at zero temperature.

The story of the two Planck laws and of the discussions they raised is related for example in [161]. The arguments used by Planck in 1911 are no longer considered satisfactory, but an argument that is still valid was proposed by Einstein and Stern in 1913 [81]. In order to state this argument, let us consider the limit for $\overline{E}$ at high temperatures ($k_B T \gg \hbar\omega$):

$$\overline{E}_{1900} = \frac{\hbar\omega}{e^{\hbar\omega/k_B T} - 1} = k_B T - \frac{\hbar\omega}{2} + \frac{(\hbar\omega)^2}{4k_B T} + \cdots,$$

$$\overline{E}_{1911} = \frac{\hbar\omega}{2} + \overline{E}_{1900} = k_B T + \mathcal{O}\left(\frac{(\hbar\omega)^2}{k_B T}\right). \tag{9.3}$$

In contrast to the first Planck law (9.1), which falls off the correct classical limit by a constant offset $\frac{1}{2}\hbar\omega$, the second Planck law (9.2) matches the correct classical limit at high temperatures. This feature is emphasized by the modern writing of this law, which clearly has no term linear in $\hbar\omega$ because the right-hand side is an even function of ω (note that $\coth x \equiv 1/\tanh x$):

$$\overline{E}_{1911} = \left(\tfrac{1}{2} + \overline{n}\right)\hbar\omega = \frac{\hbar\omega}{2}\coth\frac{\hbar\omega}{2k_B T}. \tag{9.4}$$

Debye was the first to insist on observable consequences of zero-point fluctuations in atomic motion, by discussing their effect on the intensities of diffraction peaks [66], whereas Mulliken gave the first experimental proof of these consequences by studying vibrational spectra of molecules [169]. At this point, we should emphasize that these discussions took place before the existence of these fluctuations was confirmed by fully consistent quantum theoretical calculations. Nowadays, vacuum fluctuations are just an immediate consequence of Heisenberg inequalities (see e.g. references in [188–190]).

9.2.2 The puzzle of vacuum energy

Nernst is credited for having been the first to emphasize in 1916 that zero-point fluctuations should also exist for free electromagnetic fields [171] (for discussions, see [36]), thus discovering what physicists now call *quantum vacuum*. He also noticed that the associated energy constituted a challenge for gravitational theory. When summing the zero-point energies over all field modes, a finite energy density is obtained for the first Planck law—this is the solution of the *ultraviolet catastrophe*—but an infinite value is produced from the second law. When introducing a high-frequency cutoff, the calculated energy density remains finite, but it is still much larger than the mean energy observed in the world around us through gravitational phenomena [217]. Pedagogical derivations and numbers illustrating this major problem are given in [3].

This puzzle has led famous physicists to deny the reality of vacuum fluctuations. In particular, Pauli stated in his textbook on Wave Mechanics [174] (translation from [85]):

At this point it should be noted that it is more consistent here, in contrast to the material oscillator, not to introduce a zero-point energy of $\frac{1}{2}\hbar\omega$ per degree of freedom. For, on the one hand, the latter would give rise to an infinitely large energy per unit volume due to the infinite number of degrees of freedom, on the other hand, it would be principally unobservable since nor can it be emitted, absorbed or scattered and hence, cannot be contained within walls and, as is evident from experience, neither does it produce any gravitational field.

A part of these statements is simply inescapable: it is just a matter of evidence that the mean value of vacuum energy as predicted by quantum theory does not contribute to gravitation as an ordinary energy. However, we know nowadays that vacuum fluctuations are *scattered* by matter, as shown by the numerous effects of the associated scattering in subatomic [107] and atomic [58] physics. And the Casimir effect discussed subsequently in this chapter may be seen as the manifestation of vacuum fluctuations when they are *contained within walls*. We should note at this point that different points of view coexist about the significance of vacuum fluctuations [4, 85, 117, 129, 199, 202, 206, 208].

The puzzle of vacuum energy was discovered nearly a century ago and it is not yet solved. It has led and still leads to many ideas: for example, a *dark energy length-scale* λ=85 μm can be defined by setting the cutoff used to calculate the vacuum energy so that it fits the now measured cosmic vacuum energy density [21, 175, 192]. It is thus natural to check if gravity could be affected below this dark energy length-scale [17, 78, 211]. It has been shown in torsion-balance experiments [120] that Yukawa modifications of the gravitational inverse-square law can have a magnitude equal to that of gravity only if their range has a value smaller than 56 μm, which means that these kinds of ideas must be rejected, at least in their simplest forms. More possibilities, corresponding for example to power-law modifications associated with compact extra dimensions [6], are discussed in [1, 2, 5].

The search for scale-dependent modifications of the gravitational force law have now been pushed down to even smaller ranges, approaching the micrometre distance range where Casimir forces are predominant. For these *Casimir tests* of the law of gravity to make sense, the accuracy and reliability of theoretical and experimental values have to be assessed cautiously and independently of each other. In particular, systematic effects have to be identified and eliminated, whenever this is possible. This implies in particular the need to deal carefully with the many differences between the idealized situation studied by Casimir and the configurations of real experiments [140].

9.2.3   The Casimir force between ideal and real mirrors

In his initial calculation [48], Casimir considered an idealized configuration, with perfectly smooth, flat and parallel plates (Fig. 9.1) in the limit of zero temperature and perfect reflection. L denotes the distance between the two plates and A their area, supposed to be large enough ($A \gg L^2$). He thus obtained expressions for the energy

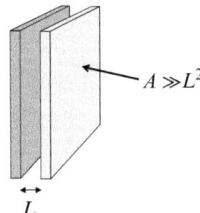

Fig. 9.1 Configuration considered by Casimir: two perfectly parallel planes placed in vacuum experience an attractive pressure given by (9.6) under the assumptions of perfect reflection and zero temperature.

E_{Cas} and force $F_{\mathrm{Cas}} \equiv -\mathrm{d}E_{\mathrm{Cas}}/\mathrm{d}L$ which exhibit a universal behaviour associated with the confinement of vacuum energy

$$E_{\mathrm{Cas}} = -\frac{\hbar c \pi^2 A}{720 L^3}\,,$$

$$F_{\mathrm{Cas}} = -\frac{\hbar c \pi^2 A}{240 L^4}\,,$$

(9.5)

where c is the speed of light and $\hbar = h/2\pi$ the reduced Planck constant. The signs have been chosen according to the standard thermodynamic conventions (the relation with the thermodynamics of the Casimir effect will be discussed later on). The negative energy corresponds to a binding energy and the negative force to an attractive force, which is also a negative pressure

$$P_{\mathrm{Cas}} \equiv \frac{F_{\mathrm{Cas}}}{A} = -\frac{\hbar c \pi^2}{240 L^4}\,.$$

(9.6)

The order of magnitude of the pressure is $|P_{\mathrm{Cas}}| \sim 1\,\mathrm{mPa}$ for two mirrors at the distance $L = 1\,\mu\mathrm{m}$ typical of Casimir force measurements (see below). The formula (9.6) describes an extremely rapid increase of the pressure when the distance is decreased, and it would lead to a value $\sim 1\,\mathrm{TPa}$ typical of strong molecular cohesion when it is extrapolated down to atomic distances $\sim 0.1\,\mathrm{nm}$. This means that the Casimir force is a quantum force like molecular cohesion forces, which has a weaker magnitude only because it is measured at distances much larger than typical atomic distances. Note, however, that the formula (9.6) cannot be used at atomic distances, where the ideal assumptions used to derive it are no longer valid, as we now explain.

9.2.4 The effect of imperfect reflection

Indeed, perfectly reflecting mirrors do not exist, except as idealizations giving fair descriptions of reality in limiting cases only. The mirrors used in the experiments are made of metal and they have a good reflection only at frequencies below the plasma frequency. Accounting for their imperfect reflection and its frequency dependence is thus essential for obtaining a reliable theoretical expectation of the Casimir pressure [135]. In other words, the real Casimir pressure depends on the non-universal optical properties of the material plates used in the experiments. It can be written as the product of the ideal result (9.6) by a dimensionless factor that accounts for these optical properties:

$$P = P_{\text{Cas}}\eta_{\text{P}} \, . \tag{9.7}$$

The expression of η_{P} in terms of the optical properties of the mirrors will be given in Section 9.4.

Most descriptions of the metallic mirrors used in the experiments are based on Fresnel reflection laws at the two interfaces between vacuum and metallic bulks with optical properties described by a linear and local dielectric response function. This dielectric function ε is a sum of contributions corresponding respectively to bound ($\bar{\varepsilon}$) and conduction electrons, the latter being directly related to the conductivity σ:

$$\varepsilon[\omega] = \bar{\varepsilon}[\omega] + \frac{\sigma[\omega]}{-\mathrm{i}\omega} \, . \tag{9.8}$$

Note that the functions ε, $\bar{\varepsilon}$, and σ are all defined as reduced quantities, with their SI counterparts being $\varepsilon_0\varepsilon$, $\varepsilon_0\bar{\varepsilon}$, and $\varepsilon_0\sigma$ ($\varepsilon_0 = 1/\mu_0 c^2$ is the vacuum permittivity). With these conventions, ε and $\bar{\varepsilon}$ are dimensionless, while σ has the dimension of a frequency.

The dielectric function (9.8) has to be obtained from optical data [134, 214] as they are tabulated for example for gold in [172]. At low frequencies, $\bar{\varepsilon}$ tends to a constant, while the contribution of conduction electrons diverges and σ tends to a constant σ_0. Optical data have then to be extrapolated at low frequencies by using the dissipative Drude model for the conductivity of the metal [7]:

$$\sigma_{\text{D}}[\omega] = \frac{\omega_{\text{P}}^2}{\gamma - \mathrm{i}\omega} \, . \tag{9.9}$$

Here ω_{P} is the plasma frequency and γ the relaxation parameter for conduction electrons. This model meets the well-known fact that gold has a finite static conductivity

$$\sigma_0 = \frac{\omega_{\text{P}}^2}{\gamma} \, . \tag{9.10}$$

For reasons which will become clear in the following, the limiting case of a lossless plasma of conduction electrons ($\gamma = 0$ in (9.9)) is often considered:

$$\sigma_{\text{P}}[\omega] = \frac{\omega_{\text{P}}^2}{-\mathrm{i}\omega} \, . \tag{9.11}$$

This so-called *plasma model* cannot be an accurate description of metallic mirrors. As a matter of fact, it contradicts the fact that gold has a finite static conductivity (9.10) while also leading to a poorer fit of tabulated optical data than the more general Drude model [134]. However, γ is much smaller than ω_P for a good metal (e.g., $\gamma \simeq 0.004\omega_P$ for gold). As the difference between (9.9) and (9.11) is appreciable only at low frequencies $\omega \lesssim \gamma$, where ε is very large for both models, one might expect that it does not affect too much the value of the Casimir force. This naive expectation is met at small distances or low temperatures, but not in the general case of arbitrary distances and temperatures, as explained in the following.

9.2.5 The effect of temperature

Most experiments are performed at room temperature, so the effect of thermal fluctuations has to be added to that of vacuum fields [35, 94, 158, 207]. This important point will be discussed in a detailed manner in Sections 9.3 and 9.4. At this point, we focus on the strong correlation effect obtained between the effects of temperature and dissipation, which has given rise to a large number of contradictory papers (see e.g. the references in [30, 31, 103, 126, 164]).

Boström and Sernelius [28] were the first to remark that, in spite of the naive expectation described at the end of Section 9.2.4, dissipation has a large effect on the value of the Casimir force at distances accessible in experiments and at room temperature. Their result is illustrated in Fig. 9.2, where the ratio η_P giving the real Casimir pressure (9.7) is drawn for the Drude and plasma models at room temperature ($T = 300\,\text{K}$), using the formulas in [103]. These two models correspond to the dielectric function (9.8) with $\bar{\varepsilon}=1$ and σ substituted by (9.9) and (9.11), respectively. The simplification $\bar{\varepsilon} = 1$ does not change the difference between the predictions of the

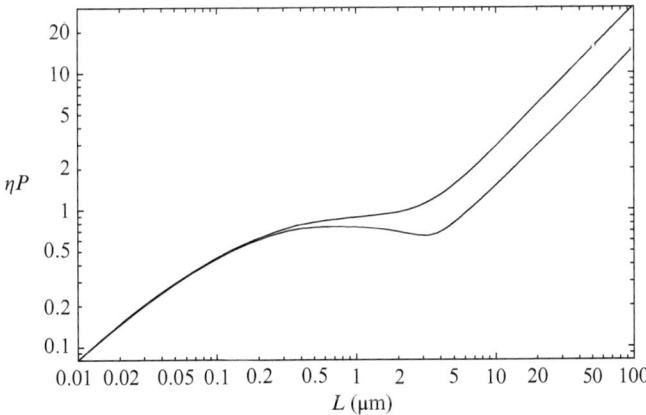

Fig. 9.2 Variation with distance L of the Casimir pressure shown as the ratio η_P of the real (P) to the ideal (P_{Cas}) Casimir pressure (see (9.7)). P is smaller than P_{Cas} ($\eta_P < 1$) at small distances ($L \lesssim 1\,\mu\text{m}$), owing to imperfect reflection, whereas it is larger ($\eta_P > 1$) at large distances ($1\,\mu\text{m} \lesssim L$), owing to the contribution of thermal photons. In the latter large-distance domain, there is a significant difference between the Drude (lower curve) and plasma (upper curve) models, both drawn here for room temperature.

two models and it can also be dropped, by having $\bar{\varepsilon}$ deduced from the optical data. The parameters of the optical models are chosen to match values typical for thick layers of gold ($\omega_P = 2\pi c/\lambda_P$, with $\lambda_P = 136\,\text{nm}$ and $\gamma/\omega_P = 0.004$).

A striking difference appears between the predictions of the two models [28]. These predictions, which are close to each other at short distances, exhibit an increasing difference for distances of the order or larger than $1\,\mu\text{m}$. In particular, the ratio of the plasma to Drude prediction for the Casimir pressure goes to a factor 2 in the limit of large distances. In fact, the result of the plasma model coincides in this limit with that obtained for perfect mirrors, whereas the result of the Drude model reaches only half that value. It is worth recalling here that this last result is reproduced by the derivations of Casimir pressures from microscopic models of the metallic mirrors [22, 37, 118].

As a matter of principle, there should be no doubt that the Drude model is a better representation of the optical properties of real plates at low frequencies than the plasma model. At this point, however, we have to face discrepancies in the comparison between experimental results and theoretical predictions, and, unexpectedly, some experimental results appear to lie closer to the predictions of the plasma model than to those of the Drude model [127]. Before coming to this point, we have still to discuss another important feature of the recent precise experiments, which are performed in the plane–sphere geometry rather than in the plane–plane geometry in which most calculations are done.

9.2.6   The effect of geometry

The configuration used for most Casimir experiments corresponds to a plane and a sphere, with L the distance of closest approach and R the radius of the sphere, supposed to be large (Fig. 9.3). The force in this plane–sphere geometry is usually calculated by using the so-called *proximity force approximation* (PFA) [72, 75]. Let us emphasize here that this approximation is valid only in the limit of a very large radius ($R \gg L$) and that the question of its accuracy for a finite value of R/L remains an open question (there will be more discussion of this topic later in the chapter).

We assume here provisionally that the PFA is precise enough for the purpose of theory–experiment comparison. It follows that the force between the plane and the sphere can be obtained by integrating over the distribution of local inter-plate

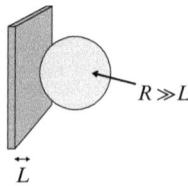

Fig. 9.3   Configuration for most Casimir experiments: a plane and a spherical mirror placed in vacuum experience an attractive force.

distances the pressure P calculated in the geometry with two parallel planes. In the plane–sphere geometry, this gives

$$F_{\mathrm{PFA}}(L) = \int_L^\infty \mathrm{d}A\, P(\mathcal{L})\,, \qquad \mathrm{d}A = 2\pi R\,\mathrm{d}\mathcal{L}\,, \tag{9.12}$$

where $P(\mathcal{L})$ is the pressure evaluated between two planes at at a distance $\mathcal{L}$ from each other; $\mathcal{L}$ runs over distances larger than the distance of closest approach L and $\mathrm{d}A$ is the corresponding surface element.

In the experiment, the gradient G of the Casimir force in the plane–sphere geometry is measured (see Section 9.2.7). This quantity is deduced from (9.12) as

$$G_{\mathrm{PFA}}(L) \equiv \frac{\partial F_{\mathrm{PFA}}}{\partial L} = -2\pi R P(L)\,. \tag{9.13}$$

Within the PFA, this measurement gives the Casimir pressure evaluated at distance L between two planes, which can then be substituted by the expression (9.7) (with η_{P} to be given in Section 9.4).

9.2.7 Casimir experiments

We now present briefly the experimental methods and results. With this aim, we focus our attention on a few experiments: the experiment at IUPUI [68, 70, 71], which has been run for ten years, with results pointing to an unexpected conclusion later confirmed at UCR [53], and the experiment in Yale [212], which points to a different conclusion. We also give here a list of other Casimir measurements that have produced information of interest on the topics discussed in this paper: [11, 12, 29, 50, 51, 69, 75, 80, 99, 119, 123, 141, 148, 154, 156, 168, 170, 200, 209, 215].

The experiment at IUPUI is described in [68, 70, 71]. A summary and update can be found in the slides associated with a talk given recently by R. S. Decca at a Pan-American Advanced Study Institute school [74]. The experiment uses dynamic measurements of the resonance frequency of a torsion micro-oscillator. For the free micro-oscillator, i.e. in the absence of the Casimir force, the resonance frequency is determined by the stiffness coefficient K_0 and the moment of inertia I:

$$\omega_0^2 = \frac{K_0}{I}\,. \tag{9.14}$$

When a gold-covered sphere is approached from the gold-covered plane of the micro-oscillator plate, the effective stiffness is modified as the gradient of the Casimir force G. The resonance frequency is thus shifted to a new value:

$$\omega^2 = \frac{K}{I}\,, \qquad K = K_0 - b^2 G\,, \tag{9.15}$$

where b is the lever arm. As the radius of the sphere is $R \simeq 150\,\mu\mathrm{m}$ and the range of distances $0.16\,\mu\mathrm{m} < L < 0.75\,\mu\mathrm{m}$, the condition $R \gg L$ is met. Using the expression (9.13) that gives the gradient G within the PFA, and measuring b, I, and R accurately, the shift of the squared frequency is then transformed into a reading of the Casimir pressure $P(L)$ as it would be between two planes at distance L:

$$\omega^2 - \omega_0^2 = \frac{b^2}{I} 2\pi R P(L). \tag{9.16}$$

P is given by (9.7) and η_P will be discussed in Section 9.4. Note that the separation L between bodies is measured separately through two-colour interferometry, up to a global offset adjusted in the data analysis process [70, 71].

When compared with the theoretical prediction, this measurement leads to unexpected conclusions [70, 71]: the measurements appear to agree with the predictions obtained from the lossless plasma model $\gamma = 0$ but to deviate significantly from those deduced from the Drude model that accounts for dissipation (see Fig. 1 in [70]). These experiments are performed in a range of distances $0.16\,\mu m < L < 0.75\,\mu m$ where the difference between the predictions of the two models is small. This entails that the problem of accuracy, which is also the control of systematic errors, is a critical issue. However, the deviation of experimental results from theoretical expectations (based on the Drude model) is clearly larger than the statistical dispersion of these results (the bars on Fig. 1 in [70]). More details of statistical and systematic errors in this experiment can be found in [70, 71].

Different conclusions are reached from the results of a more recent experiment performed at Yale [212]. This experiment aims at measurements at larger distances $0.7\,\mu m < L < 7\,\mu m$, where the force is smaller while the thermal contribution and the effect of dissipation are larger (see Fig. 9.2). The experimental technique is based on a torsion balance and uses a much larger sphere, with $R = 156$ mm, which allows for measurements of weaker pressures. This experiment clearly sees the thermal effect, and its results fit the predictions drawn from the dissipative Drude model, after the contribution of the electrostatic patch effect has been subtracted [212]. These new results have to be confirmed by further studies [167]. The main issue in this experiment is that the pressure due to electrostatic patches is larger than that due to the Casimir effect, so a proper modelling of this contribution is critical, whereas the patch pattern has not been characterized independently. This is in fact a more general problem, since the patch properties have not been measured in other experiments either (see the further discussion below).

9.2.8   Discussion

The conclusion at this point is that the Casimir effect is measured with good precision in several experiments, with a persisting problem, however, in terms of accuracy. The results of the most precise experiment, improved over a decade at IUPUI and confirmed recently at UCR, appear to favour theoretical predictions obtained with the lossless plasma model and to deviate from the predictions obtained with the best-motivated model, namely the dissipative Drude model. The Yale experiment fits predictions drawn from this Drude model, after the subtraction of a large contribution of the electrostatic patch effect. For the IUPUI experiment, the pressure difference goes up

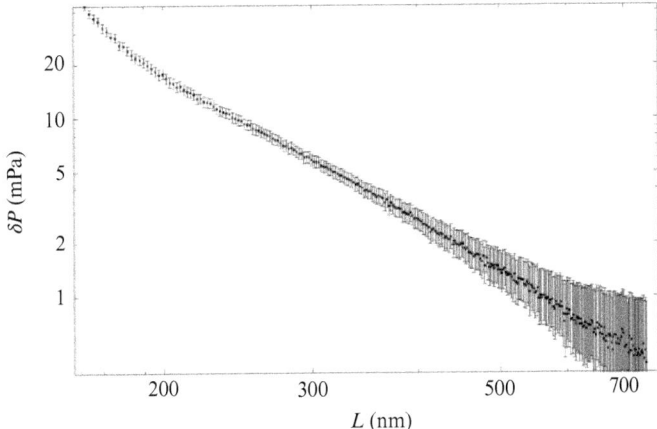

Fig. 9.4 Difference $\delta P = P_{\text{exp}} - P_{\text{th}}$ between the experimental and theoretical values of the Casimir pressure as a function of the distance L. Experimental values were kindly provided by R. S. Decca and theoretical values were calculated by R. O. Behunin et al., with the Drude model and at room temperature.

to $\sim$50 mPa at the smallest distances $\sim$160 nm where the pressure itself is $\sim$1000 mPa, which entails that the accuracy is certainly not at the 1% level, as has been occasionally claimed.

The difference $\delta P = P_{\text{exp}} - P_{\text{th}}$ between the experimentally measured (P_{exp}) and theoretically predicted (P_{th}) values of the Casimir pressure is drawn in Fig. 9.4 as a function of the distance L. Experimental values and error bars correspond to data kindly provided by R. S. Decca [70, 71, 74]. Theoretical values were calculated by R. O. Behunin et al. [18], using the optical data for gold extrapolated at low frequencies to a Drude model, and at room temperature. Systematic corrections were done in [18] as in [70, 71] and similar results were obtained. The discrepancy clearly appears on Fig. 9.4 and it is of particular importance in the context of gravity tests at submillimetre ranges [2, 5]. The deviation seen on Fig. 9.4 does not look like a Yukawa law, but it certainly looks like a combination of power laws!

This discrepancy between theory and experiment may have various origins, in particular artefacts in the experiments or inaccuracies in the theoretical evaluations. They may also come from yet unmastered systematic effects in the comparison between experimental data and theoretical predictions. They could in principle be the first hint of the existence of new forces beyond the Standard Model—though such a strong statement should only be considered after a cautious examination of the more mundane explanations associated in particular with systematic effects.

The theoretical formula used to calculate the Casimir pressure between real plates will be derived in Section 9.4. It will reproduce the ideal Casimir expression at the limits of perfect reflection and null temperature while being valid at any temperature for any model of mirrors obeying well-motivated physical properties [108, 134], including the case of dissipative mirrors [95]. When the reflection amplitudes are deduced from the Fresnel laws, and semi-infinite bulk mirrors are characterized by a linear and local dielectric response function, the results reproduce those of I. E. Dzyaloshinskii, E. M. Lifshitz and L. P. Pitaevskii [79, 145, 146]. It then remains to specify this dielectric function and its low-frequency behaviour. Here, it is worth emphasizing that the Drude model, though obviously much better motivated than the lossless plasma model, is not a very accurate description of conduction phenomena in real metals. More detailed descriptions can be considered, which can for example be determined from microscopic models of conduction in metals. Attempts in this direction and discussions can be found for example in [62, 63, 73, 96, 176, 177, 213]. To date, they have been unable to explain the discrepancy.

A possible source of systematic error is the use of the PFA to derive expressions for the plane–sphere geometry from those known from the plane–plane geometry. This approximation is expected to be valid in the limit where the aspect ratio $x \equiv L/R$ goes to zero. Even in this case, the accuracy of the PFA for a finite value of x remains an open question after the remarkable advances made recently on this topic [45], which will be described in Section 9.4. The question remains open, even though most specialists would probably bet that the deviation from PFA is not able to bridge the gap between experiment and theory. Other possible sources of systematic error involve the effects of surface physics on Casimir experiments. The problem of surface roughness has been studied in a thorough manner [32–34, 151, 152, 220].

9.2.9 Electrostatic patches and contamination

Electrostatic patches and contamination, already alluded to, are a worrying source of such systematic effects. Electrostatic patches have been known for a long time to be a source of worries for a large number of high-precision measurements [39, 47, 76, 77, 86, 87, 100, 101, 181, 185, 193, 201, 216, 219], and in particular for Casimir experiments [56, 124, 125, 155, 210] and short-range gravity tests [2]. The patch effect is due to the fact that the surface of a metallic plate cannot be an equipotential, since it is made of microcrystallites with different work functions. For clean metallic surfaces studied by the techniques of surface physics, the resulting *voltage roughness* is correlated with the *topographic roughness* as well as with the orientation of microcrystallites [92]. For surfaces exposed to air, the situation is changed owing to the unavoidable contamination by adsorbents, which spread out the electrostatic patches, enlarge correlation lengths, and reduce voltage dispersions [198].

The pressure due to electrostatic patches between two planes can be computed by solving the Poisson equation [210]. Its evaluation depends on the spectra describing the correlations of the patch voltages or, equivalently on the associated noise spectrum $C(k)$, with k a patch wavevector. In an analysis of the patch pressure devoted to the results of Casimir experiments, including those performed recently, the spectrum was assumed to be flat between two *sharp cutoffs* at a minimum wavevector k_{min} and a maximum one k_{max}. Assuming furthermore that k_{min} and k_{max} were given by the grain size distribution measured with an atomic force microscope (AFM), it was concluded that the patch pressure was much smaller than the discrepancy between experiment and theory [68, 70, 71].

A *quasilocal* model has recently been proposed with the aim of proposing a much better motivated representation of patches [18]. The model is based on a tessellation of the sample surface and a random assignment of the voltage on each patch. It produces a smooth spectrum different from the sharp-cutoff model used in previous analyses, since there are now contributions to the patch pressure coming from arbitrarily low values of k, even if the patch size distribution has an upper bound. When the patch effect is estimated with the parameters deduced from the grain size distribution as in [68, 70], a much larger contribution of patches is obtained. In fact, the calculated patch pressure is now larger than the residuals between experimental data and theoretical predictions, which means that patches could be a crucial systematic effect for Casimir force measurements [18].

As the computed patch pressure is model-dependent, it seems natural to try to find a model between the two cases presented above that would reproduce at least qualitatively the residuals. By varying the parameters of the quasilocal model, it was found in [18] that the output of the model depended mainly on two parameters, namely the size of the largest patches $\ell_{\mathrm{patch}}^{\mathrm{max}}$ and the root mean square (rms) voltage dispersion V_{rms}, and that a best fit on these two parameters produced a qualitative agreement between the residuals and the patch pressure. The best-fit values for the parameters $\ell_{\mathrm{patch}}^{\mathrm{max}}$ and V_{rms} are quite different from those obtained by identifying patch and grain sizes. With $\ell_{\mathrm{patch}}^{\mathrm{max}}$ larger than the maximum grain size and V_{rms} smaller than the rms voltage that would be associated with random orientations of clean microcrystallites, these values are, however, compatible with contamination of metallic surfaces, which had to be expected anyway.

It follows that the difference between the IUPUI experimental data and theoretical predictions can be fitted, at least qualitatively, by a simple model of electrostatic patches. This conclusion is, however, only the result of a fit, with the parameters of the patch model not being measured independently. In order to reach a firm conclusion, the patch spectrum has to be measured independently, by using the dedicated technique of Kelvin probe force microscopy (KPFM), which is able to achieve the necessary size and voltage resolutions [149, 150]. When these characteristics are available, the contribution of the patches to the Casimir measurements can be evaluated and unambiguously subtracted when comparing theory and experiments.

Preliminary results of such characterizations have recently been published [20]. Note that the evaluation of the force was done for the plane–sphere geometry [19].

9.3 A simple derivation of the Casimir effect in one dimension

This section contains a derivation of the Casimir effect in a model of scalar fields propagating along the two directions on the one-dimensional (1D) line. In this simple model, which lays the basis for more complicated calculations to appear in Section 9.4, we introduce the *quantum optics* approach to the Casimir effect. The approach is based on the scattering of vacuum field fluctuations obtained in the ground state of the associated *quantum field theory*. Each mirror is described by a scattering operator [93, 147], which is reduced here to a 2×2 matrix containing reflection and transmission amplitudes. Two mirrors form a Fabry–Perot cavity, with all field transformations being deduced from the two elementary scattering matrices. The Casimir force then results from the difference between the radiation pressures exerted on the inner and outer sides of the mirrors by the vacuum field fluctuations. Equivalently, the Casimir free energy can be written as the shift of field energy due to the presence of the Fabry–Perot cavity [180, 206]. The formula obtained in this manner is valid and regular at thermal equilibrium at any temperature and for any optical model of mirrors obeying causality and high-frequency transparency properties.

The radiation pressure interpretation of the Casimir force was presented for perfect mirrors in [160] and extended to the case of real mirrors in [108]. The calculations were then systematically expanded in particular for applications to the problem of the *dynamical Casimir effect* [109–114, 132, 133]. It has also served as a basis for the scattering formalism for the static Casimir effect [95, 136], of which we will give a pedagogical presentation in the following.

9.3.1   Quantum field theory on the 1D line

We consider here quantum field theory on the 1D line, which is also quantum field theory in two-dimensional space–time (one time coordinate t and one space coordinate x). The field propagation is thus described by d'Alembert's wave equation, originally written for the propagation of transverse vibrations of a string, and which also describes many wave phenomena such as electrical propagation in a transmission line, acoustic waves, and so on.

9.3.1.1   *Propagation equation on the 1D line*

We write it here for a single vibration described by a scalar potential $\Phi(x, t)$:

$$\frac{\partial^2 \Phi}{\partial t^2} - c^2 \frac{\partial^2 \Phi}{\partial x^2} = 0. \tag{9.17}$$

The general solution of this equation is given by d'Alembert's formula, that is, the superposition of rightward (φ^+) and leftward (φ^-) traveling waves propagating at the velocity c in opposite directions along the x-axis:

$$\Phi(t, x) = \varphi^+(u_+) + \varphi^-(u_-), \tag{9.18}$$

where $u_\pm$ are today called the light cone variables,

$$
\begin{aligned}
u_+ &\equiv t - \frac{x}{c}, \\
u_- &\equiv t + \frac{x}{c}.
\end{aligned}
\tag{9.19}
$$

In this simplest version of field theory, there is one normal mode for each frequency $\omega \in [0, \infty]$ and each propagation direction $\eta = \pm 1$. The standard methods of quantum field theory [58, 107] then allow one to write the rightward (φ^+) and leftward (φ^-) traveling waves as Fourier decompositions over canonical mode operators:

$$\varphi^\eta(u) = \int_0^\infty \frac{d\omega}{2\pi} \sqrt{\frac{\hbar}{2\omega}} \left(a_{\omega,\eta} e^{-i\omega u} + a_{\omega,\eta}^\dagger e^{i\omega u} \right). \tag{9.20}$$

The annihilation and creation operators $a_{\omega,\eta}$ and $a_{\omega,\eta}^\dagger$ correspond respectively to positive and negative frequencies in the decompositions (9.20). Note that the fields in space–time $\varphi^\eta(u)$ are real-valued, so that annihilation and creation operators are Hermitian conjugate of each other. They obey the following canonical commutation relations:

$$\left[a_{\omega,\eta}, a_{\omega',\eta'}^\dagger \right] = 2\pi\delta(\omega - \omega')\delta_{\eta,\eta'},$$

$$\left[a_{\omega,\eta}, a_{\omega',\eta'} \right] = \left[a_{\omega,\eta}^\dagger, a_{\omega',\eta'}^\dagger \right] = 0. \tag{9.21}$$

The Hamiltonian $\mathcal{H}$ for d'Alembert's wave equation (9.17) is the integral over space of the energy density $e(t, x)$:

$$\mathcal{H} = \int dx \, e(t, x), \qquad e \equiv \frac{1}{2}\left(\frac{\partial\Phi}{\partial t} \right)^2 + \frac{c^2}{2}\left(\frac{\partial\Phi}{\partial x} \right)^2. \tag{9.22}$$

Using d'Alembert's formula (9.18) for fields, one derives another d'Alembert's formula describing the general energy density as the superposition of rightward- and leftward-traveling flows:

$$e(t, x) = e^+(u_+) + e^-(u_-), \qquad e^\eta(u_\eta) = \left(\frac{\partial\varphi^\eta}{\partial u_\eta} \right)^2. \tag{9.23}$$

9.3.1.2 *Vacuum and thermal fluctuations*

The vacuum $|\mathrm{vac}\rangle$ is the fundamental state of the quantum field, with an infinite number of modes each containing no photons. This means that all annihilation operators vanish in this state, while creation operators have their action determined by the commutation relations (9.21).

In the following, we use the correlation functions of vacuum fields, which are deduced from these elementary properties:

$$\langle a_{\omega,\eta} \rangle_\mathrm{vac} = \langle a_{\omega,\eta}^\dagger \rangle_\mathrm{vac} = 0,$$

$$\langle a_{\omega,\eta} a_{\omega',\eta'}^\dagger \rangle_\mathrm{vac} = 2\pi\delta(\omega - \omega')\delta_{\eta,\eta'},$$

$$\langle a_{\omega,\eta}^\dagger a_{\omega',\eta'} \rangle_\mathrm{vac} = 0,$$

$$\langle a_{\omega,\eta} a_{\omega',\eta'} \rangle_\mathrm{vac} = \langle a_{\omega,\eta}^\dagger a_{\omega',\eta'}^\dagger \rangle_\mathrm{vac} = 0, \tag{9.24}$$

where $\langle\cdots\rangle_{\mathrm{vac}} \equiv \langle\mathrm{vac}|\cdots|\mathrm{vac}\rangle$. Higher-order correlation functions are deduced from the fact that vacuum fields may be dealt with as Gaussian random variables.

In a state at thermal equilibrium at temperature T, the first and last lines in (9.24) are unchanged, whereas the second and third lines become

$$\langle a_{\omega,\eta} a^{\dagger}_{\omega',\eta'}\rangle_{\mathrm{therm}} = 2\pi\delta(\omega - \omega')\delta_{\eta,\eta'}(1+\overline{n}),$$

$$\langle a^{\dagger}_{\omega,\eta} a_{\omega',\eta'}\rangle_{\mathrm{therm}} = 2\pi\delta(\omega - \omega')\delta_{\eta,\eta'}\,\overline{n}, \tag{9.25}$$

where $\langle\cdots\rangle_{\mathrm{therm}} \equiv \langle\mathrm{therm}|\cdots|\mathrm{therm}\rangle$ and $\overline{n}$ is the mean photon number in the Planck law (9.2). Note that (9.25) is reduced to (9.24) when $T \to 0$. Note also that all field commutators are unchanged, which is consistent with the fact that they are directly connected to the propagators.

Using the Fourier decompositions (9.20) of the fields and the correlation functions (9.25), we deduce that the mean values of the energy densities (9.23) have the following spectral decompositions in the general case $(T \neq 0)$:

$$\langle e^{\eta}(u)\rangle_{\mathrm{therm}} = \int_0^{\infty} \frac{\mathrm{d}\omega}{2\pi} \int_0^{\infty} \frac{\mathrm{d}\omega'}{2\pi} \sqrt{\frac{\hbar\omega}{2}} \sqrt{\frac{\hbar\omega'}{2}} \langle a_{\omega,\eta} a^{\dagger}_{\omega',\eta'} + a^{\dagger}_{\omega,\eta} a_{\omega',\eta'}\rangle_{\mathrm{therm}}$$

$$= \int_0^{\infty} \frac{\mathrm{d}\omega}{2\pi} \frac{\hbar\omega}{2}(1+2\overline{n}). \tag{9.26}$$

This is the modern expression of the second Planck law [179] for the energy per mode given as the sum of vacuum and thermal contributions (compare it with (9.2)). The limit of zero temperature corresponds to $\overline{n} = 0$ in this expression.

9.3.2   One mirror on the 1D line

We now consider the situation where one mirror is placed into vacuum at a position q_1 (Fig. 9.5).

9.3.2.1   *Scattering by one mirror on the 1D line*

The general solution is in this case a generalized d'Alembert's formula with different expressions for fields on the left (Φ_{L}) and right (Φ_{R}) sides of the mirror:

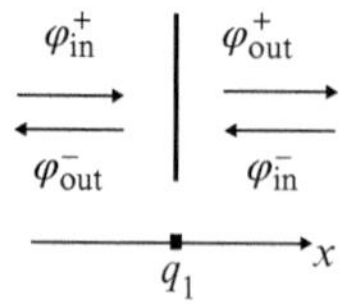

Fig. 9.5 Schematic representation of a mirror at position q_1 on a 1D line: this mirror is a point-like scatterer that couples rightward- and leftward-propagating waves.

$$\Phi(t, x) = \Phi_{\rm L}(t, x) + \Phi_{\rm R}(t, x),$$

$$\Phi_{\rm L}(t, x) = \varphi_{\rm in}^+(u_+) + \varphi_{\rm out}^-(u_-), \quad x < q_1, \tag{9.27}$$

$$\Phi_{\rm R}(t, x) = \varphi_{\rm out}^+(u_+) + \varphi_{\rm in}^-(u_-), \quad q_1 < x.$$

The indices $+$ and $-$ represent, as previously, rightward and leftward propagation, respectively, while subscripts 'in' and 'out' correspond to incoming and outgoing traveling waves. These waves are coupled by scattering on the mirror.

For the limiting case of perfectly reflecting mirrors, the field Φ vanishes at the left and right sides of the mirrors. Output fields are then easily deduced from input ones as $\varphi_{\rm out}^\pm(t, q_1) = -\varphi_{\rm in}^\mp(t, q_1)$. In Fourier space, this is written as a scattering process with reflection amplitudes having a unit modulus and a phase determined by the position of the mirror, $\varphi_{\rm out}^\pm[\omega] = -e^{\mp 2ikq_1}\varphi_{\rm in}^\mp[\omega]$, with $k \equiv \omega/c$. As the scattering process preserves the frequency for a motionless mirror, these equations could as well have been written for annihilation ($\omega > 0$) and creation ($\omega < 0$) operators.

9.3.2.2 Real mirrors on the 1D line

Real mirrors cannot be perfectly reflecting at all frequencies. They are described by a more general scattering matrix containing transmission as well as reflection amplitudes (with $k \equiv \omega/c$):

$$\begin{pmatrix} \varphi_{\rm out}^+[\omega] \\ \varphi_{\rm out}^-[\omega] \end{pmatrix} = S_1[\omega] \begin{pmatrix} \varphi_{\rm in}^+[\omega] \\ \varphi_{\rm in}^-[\omega] \end{pmatrix},$$

$$\tag{9.28}$$

$$S_1[\omega] = \begin{pmatrix} t_1[\omega] & r_1[\omega]e^{-2ikq_1} \\ r_1[\omega]e^{2ikq_1} & t_1[\omega] \end{pmatrix}.$$

The scattering amplitudes r_1 and t_1 have been defined for a mirror located at $x = 0$ and the general case then obtained by introducing the phases $e^{\mp 2ikq_1}$ determined by the position of the mirror. We have considered the particular case of a symmetrical scattering matrix for a mirror located at $x = 0$. A more general treatment would not change any important result in the following.

The scattering matrix preserves frequency since energy is conserved for a stationary scattering, but it depends on frequency as a consequence of fundamental physical properties [108]. The scattering process considered here obeys the following properties (the unitarity assumption, valid only for lossless mirrors, will be relaxed later on):

1. Fields are real in the time domain, so that $S[\omega]^\dagger = S[-\omega]$; that is, also $t_1[\omega]^* = t_1[-\omega]$ and $r_1[\omega]^* = r_1[-\omega]$.
2. The scattering process obeys causality, so that the amplitudes can be prolongated as analytical functions in the upper half of the complex plane (more details are given later).

3. The scattering process obeys unitarity, so that $S[\omega]S[\omega]^\dagger = \mathcal{I}$; that is, also $|r_1[\omega]|^2 + |t_1[\omega]|^2 = 1$ and $t_1[\omega]r_1[\omega]^* + r_1[\omega]t_1[\omega]^* = 0$.
4. Reflection tends to vanish in the high-frequency limit $\lim_{\omega\to\infty} S_1[\omega] \to \mathcal{I}$, so $\lim_{\omega\to\infty} t_1[\omega] \to 1$ and $\lim_{\omega\to\infty} r_1[\omega] \to 0$.

These general properties may be illustrated with an example, which corresponds in particular to a transmission line with a localized impedance mismatch [114]. For this simple example, d'Alembert's wave equation (9.17) is changed to

$$\frac{\partial^2 \Phi}{\partial t^2} - c^2 \frac{\partial^2 \Phi}{\partial x^2} + 2c\Omega\delta(x - q_1)\Phi(t, q) = 0, \tag{9.29}$$

and the solution obtained as (9.28) with

$$r_1[\omega] = \frac{\Omega}{i\omega - \Omega}, \qquad t_1[\omega] = \frac{i\omega}{i\omega - \Omega}. \tag{9.30}$$

The general properties 1–4 enumerated in the preceding paragraph can easily be checked. The parameter Ω appears as the physical cutoff above which reflection tends to vanish. The limit of perfect reflection can be defined by the case of large values of Ω. It has been shown that this definition allows one to escape the difficulties encountered when studying perfectly reflecting mirrors [111, 112].

9.3.2.3 *Force on one mirror on the 1D line*

We come now to the evaluation of the force acting on the mirror represented in Fig. 9.5. With this aim, we first study the energy flows in the same situation, as sketched in Fig. 9.6.

The general solution is now a d'Alembert formula (9.23) for energy densities with different expressions on the left (e_L) and right (e_R) sides of the mirror:

$$e(t, x) = e_L(t, x) + e_R(t, x),$$

$$e_L(t, x) = e_{in}^+(u_+) + e_{out}^-(u_-), \quad x < q_1, \tag{9.31}$$

$$e_R(t, x) = e_{out}^+(u_+) + e_{in}^-(u_-), \quad q_1 < x.$$

Similar expressions can be written for the momentum densities by just putting a sign η in front of the energy densities e^η. The force on the mirror is then deduced

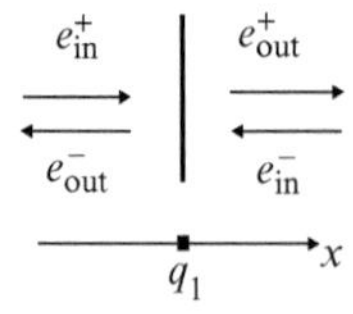

Fig. 9.6 Schematic representation of energy flows on the 1D line due to scattering on a mirror at position q_1 (see Fig. 9.5).

from a momentum balance upon scattering and it is found to be proportional to the difference of the energy densities $e_L(t, q)$ and $e_R(t, q)$ on the left and right sides of the mirror [109].

In the limiting case of perfectly reflecting mirrors, the field Φ vanishes at the left and right sides of the mirrors. The output fields are then easily deduced from the input ones as $e_{\text{out}}^{\pm}(t, q_1) = e_{\text{in}}^{\mp}(t, q_1)$, and the force thus obtained as $F(t) = 2[e_{\text{in}}^{+}(t, q) - e_{\text{in}}^{-}(t, q)]/c$. The mean values $\langle e_{\text{in}}^{\pm}(t, q)\rangle_{\text{therm}}$ of leftward and rightward energy densities are infinite (see (9.26)). But these mean values are also equal for leftward and rightward densities, so the force on the mirror is zero: $\langle F\rangle_{\text{therm}} = 0$.

This result depends crucially on the fact that we have evaluated the mean force on a mirror at rest. There exist non-vanishing fluctuations of force on the mirror, as well as a non-null mean force for a moving mirror [109].

9.3.2.4 *Force on one real mirror*

In the general case of a non-perfect mirror, the same result is proven by the following reasoning. First, the force is deduced from the momentum balance upon scattering and the expressions (9.31) for the energy densities:

$$
\begin{aligned}
F(t) &= \frac{e_L(t, q) - e_R(t, q)}{c} \\
&= \frac{e_{\text{in}}^{+}(t, q) + e_{\text{out}}^{-}(t, q) - e_{\text{out}}^{+}(t, q) - e_{\text{in}}^{-}(t, q)}{c}.
\end{aligned}
\tag{9.32}
$$

Then, the mean energy densities are calculated using the expressions (9.23) for the energy densities and the description (9.28) of the scattering process. From the unitarity of the S-matrix, one deduces that mean radiation pressures are still equal on the two sides (the equation written at thermal equilibrium at temperature T):

$$
\langle e_{\text{out}}^{+}\rangle = \langle e_{\text{out}}^{-}\rangle = \langle e_{\text{in}}^{+}\rangle = \langle e_{\text{in}}^{-}\rangle
$$

$$
= \int_0^\infty \frac{d\omega}{2\pi} \hbar\omega \left(\tfrac{1}{2} + \overline{n}\right),
\tag{9.33}
$$

so that the mean value of the force still vanishes:

$$
\langle F\rangle = 0.
\tag{9.34}
$$

9.3.3 Two mirrors on the 1D line

The situation changes fundamentally as soon as we consider that there are two mirrors present on the 1D line, which form a Fabry–Perot cavity. There is now a difference between the inner and outer sides of the mirrors, as sketched in Fig. 9.7.

9.3.3.1 *Scattering by a cavity on the 1D line*

The spatial positions of the two mirrors are denoted by q_1 and q_2, with the length of the cavity $L \equiv q_2 - q_1 > 0$. Each mirror couples rightward- and leftward-traveling

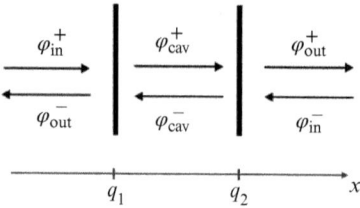

Fig. 9.7 Schematic representation of the scattering of fields by two mirrors at positions q_1 and q_2 on the 1D line.

waves. In contrast with the case of one mirror, the fields undergo multiple scattering and there appears an intracavity region. The general solution for the fields is now written with different expressions for fields on the left (Φ_L) and right (Φ_R) sides of the cavity and those (Φ_C) within the cavity:

$$
\begin{aligned}
\Phi(t,x) &= \Phi_L(t,x) + \Phi_C(t,x) + \Phi_R(t,x)\,, \\
\Phi_L(t,x) &= \varphi_{in}^+(u_+) + \varphi_{out}^-(u_-)\,, & x < q_1\,, \\
\Phi_C(t,x) &= \varphi_{cav}^+(u_+) + \varphi_{cav}^-(u_-)\,, & q_1 < x < q_2\,, \\
\Phi_R(t,x) &= \varphi_{out}^+(u_+) + \varphi_{in}^-(u_-)\,, & q_2 < x\,.
\end{aligned}
\tag{9.35}
$$

The scattering effect of the cavity now has to be described by two matrices, instead of one in the one-mirror case, namely a global scattering matrix S that gives the output fields in terms of the input ones and a resonance matrix R that gives the intracavity fields:

$$
\begin{pmatrix} \varphi_{out}^+ \\ \varphi_{out}^- \end{pmatrix} = S \begin{pmatrix} \varphi_{in}^+ \\ \varphi_{in}^- \end{pmatrix},
$$

$$
\begin{pmatrix} \varphi_{cav}^+ \\ \varphi_{cav}^- \end{pmatrix} = R \begin{pmatrix} \varphi_{in}^+ \\ \varphi_{in}^- \end{pmatrix}.
\tag{9.36}
$$

9.3.3.2 Scattering and resonance matrices

The global S-matrix can be evaluated from the elementary matrices S_1 and S_2 associated with the two mirrors and the free-propagation phase shifts. The results can be written as follows (all amplitudes depend on $\omega \equiv ck$):

$$
S = \frac{1}{d} \begin{pmatrix} t_1 t_2 & dr_2 e^{-ikL} + t_2^2 r_1 e^{ikL} \\ dr_1 e^{-ikL} + t_1^2 r_2 e^{ikL} & t_1 t_2 \end{pmatrix}.
\tag{9.37}
$$

The denominator d is an important function, with its zeros corresponding to the resonances of the cavity:

$$
d[\omega] = 1 - r[\omega] e^{2ikL}\,, \qquad r[\omega] \equiv r_1[\omega] r_2[\omega]\,.
\tag{9.38}
$$

For the problem under consideration, the global S-matrix obeys the same properties as the elementary matrices S_1 and S_2. In particular, it is unitary:

$$S[\omega]S[\omega]^\dagger = \mathcal{I}. \tag{9.39}$$

The resonance matrix can be deduced from the elementary matrices S_1 and S_2 associated with the two mirrors and the free-propagation phase shifts. The results are obtained as (amplitudes depend on ω and expressions are simplified by assuming $q_2 = -q_1 = L/2$)

$$R = \frac{1}{d}\begin{pmatrix} t_1 & t_2^2 r_1 e^{ikL} \\ t_1^2 r_2 e^{ikL} & t_2 \end{pmatrix}. \tag{9.40}$$

The resonance matrix shares some of the properties listed above with the S-matrix, but it is not unitary. It turns out that $RR^\dagger$ can be written as [110]

$$R[\omega]R[\omega]^\dagger = \mathcal{I} + Q[\omega] + Q[\omega]^\dagger,$$

$$Q = \frac{1}{d}\begin{pmatrix} r_1 r_2 e^{2ikL} & r_1 e^{ikL} \\ r_2 e^{ikL} & r_1 r_2 e^{2ikL} \end{pmatrix}. \tag{9.41}$$

9.3.4   Casimir force on the 1D line

We now evaluate the energy densities and the forces in the case of two mirrors on the 1D line.

9.3.4.1   *Energy densities*

The general solution for the energy densities is now written as in (9.31), considering the configuration sketched in Fig. 9.7 for the cavity:

$$
\begin{aligned}
e(t,x) &= e_\mathrm{L}(t,x) + e_\mathrm{C}(t,x) + e_\mathrm{R}(t,x), \\
e_\mathrm{L}(t,x) &= e_\mathrm{in}^+(u_+) + e_\mathrm{out}^-(u_-), & x &< q_1, \\
e_\mathrm{C}(t,x) &= e_\mathrm{cav}^+(u_+) + e_\mathrm{cav}^-(u_-), & q_1 &< x < q_2, \\
e_\mathrm{R}(t,x) &= e_\mathrm{out}^+(u_+) + e_\mathrm{in}^-(u_-), & q_2 &< x.
\end{aligned}
\tag{9.42}
$$

We deduce expressions for the forces acting on each mirror 1 or 2, obtained as differences of the radiation pressures on the left and right sides of the mirror (compare with (9.32)):

$$
\begin{aligned}
F_1(t) &= e_\mathrm{L}(t,q_1) - e_\mathrm{C}(t,q_1), \\
F_2(t) &= e_\mathrm{C}(t,q_2) - e_\mathrm{R}(t,q_2).
\end{aligned}
\tag{9.43}
$$

The energy densities on the outer sides of the cavity have the same expressions as in the one-mirror case, provided that the scattering matrix (9.37) of the cavity is used. As this global S-matrix is unitary (see (9.39)), one deduces that the mean energy densities on the outer sides are still given by (9.33). Hence, they are infinite but equal on the two sides of the cavity, so the mean value of the global force on the cavity vanishes:

$$\langle F_1 + F_2 \rangle = 0 \,. \tag{9.44}$$

9.3.4.2   *Intracavity energy densities*

The situation is different for the energy densities on the intracavity sides of the mirrors, since their calculation is now determined by the properties of the resonance matrix. Using the property (9.41), one deduces that the mean values of these energy densities are given by [110]

$$\langle e_{\text{cav}}^+ \rangle = \langle e_{\text{cav}}^- \rangle = \int_0^\infty \frac{d\omega}{2\pi}\, \hbar\omega \left(\tfrac{1}{2} + \overline{n}\right) g[\omega] \,, \tag{9.45}$$

where $g[\omega]$ is the common value of the diagonal elements in the matrix $RR^\dagger$,

$$g[\omega] \equiv 1 + f[\omega] + f[\omega]^* = \frac{1 - |r|^2}{|1 - r\,e^{2ikL}|^2} \,,$$

$$f[\omega] \equiv \frac{r e^{2ikL}}{1 - r e^{2ikL}} \,. \tag{9.46}$$

The real function $g[\omega]$ represents the modification of the energy density inside the cavity with respect to that outside the cavity, which is in fact identical for input and output fields. The same result would have been obtained for a classical calculation with an input field at frequency ω. Here this function $g[\omega]$ describes the change of energy densities for vacuum as well as thermal fluctuations. Its relation (9.46) to the *closed-loop function* $f[\omega]$ will be used in the following to transform the expression of the Casimir force.

9.3.4.3   *Casimir force as a result of radiation pressures*

Collecting these results, one deduces the following expression for the Casimir force, defined as the mean force on the right-hand mirror or the opposite of that on the left-hand mirror (see (9.44)):

$$F \equiv \langle F_2 \rangle = - \langle F_1 \rangle$$

$$= \int_0^\infty \frac{d\omega}{2\pi c}\, \hbar\omega (1 + 2\overline{n})(g[\omega] - 1) \,. \tag{9.47}$$

This expression is obtained as the difference between the radiation pressures (9.45) and (9.33) on the inner and outer sides of the mirrors.

Resonant frequencies correspond to an increase of energy in the cavity ($g > 1$) and they produce repulsive contributions to the Casimir force. In contrast, frequencies out of resonance correspond to a decrease of energy in the cavity ($g < 1$) and they produce attractive contributions to the Casimir force. The net force is the integral of these contributions over all modes. This interpretation of the Casimir force as a result of radiation pressures of vacuum and thermal fluctuations produces a final expression that is finite for any properly defined model of mirrors [108]. This is seen more easily by using causality properties to rewrite (9.47) as an integral over imaginary frequencies. Let us stress at this point that this rewriting is just a mathematical transformation that does not affect the physical content of (9.47). The rewriting will, however, spoil its direct intelligibility, since imaginary frequencies do not correspond to physical modes.

9.3.4.4   Casimir force as an integral over imaginary frequencies

We now rewrite the Casimir force as an integral over imaginary frequencies by using the causality properties of the scattering amplitudes. We give here a simplified description (a more general derivation can be found in [98]).

We first write the Casimir force (9.47) as the real part of a complex integral F_r defined over the positive part $\mathbb{R}^+$ of the real axis:

$$F = F_r + F_r^* \, ,$$

$$F_r \equiv \int_0^\infty \frac{\mathrm{d}z}{2\pi c} \, \hbar z f[z] \coth \frac{\hbar z}{2k_{\mathrm{B}}T} \, . \tag{9.48}$$

We have used the relation (9.46) between g and f, and have replaced the frequency ω by a complex variable z running over $\mathbb{R}^+$. We have also replaced $1 + 2\bar{n}[\omega]$ by its explicit form (9.4). Now the closed-loop function $f[z]$ is defined from causal reflection amplitudes and propagation phases. Considered as a function of $z \in \mathbb{C}$, it has poles in the lower half of the complex plane, which correspond to resonances of the Fabry–Perot cavity, while it is analytical in the upper half of the complex plane $\mathrm{Im}\, z \geq 0$. Meanwhile, the function $\coth(\hbar z/2k_{\mathrm{B}}T)$ is analytical in the right half of the complex plane $\mathrm{Re}\, z > 0$ but has poles at the Matsubara frequencies, which are regularly spaced on the imaginary axis [157]:

$$z_n = \mathrm{i}\xi_n \, , \qquad \xi_n = n\xi_1 \, , \qquad \xi_1 = \frac{2\pi k_{\mathrm{B}}T}{\hbar} \, . \tag{9.49}$$

It follows that the integral F_r can be transformed using Cauchy's theorem. Specifically, we apply Cauchy's theorem to the integral of the integrand appearing in (9.48) over a closed contour consisting of $\mathbb{R}^+$, the positive part of the imaginary axis shifted by a small positive real number, and a quarter of a circle with a very large radius. This last part vanishes as a consequence of the high-frequency transparency of the mirrors. The integral over the whole contour also vanishes, since the integrand is an analytical function in the domain enclosed by the contour. As a consequence, the integral F_r may be written in the equivalent form F_i ($\varepsilon \to 0^+$):

$$F_r = F_i = \int_\varepsilon^{\varepsilon + \mathrm{i}\infty} \frac{\mathrm{d}z}{2\pi c} \, \hbar z f[z] \coth \frac{\hbar z}{2k_{\mathrm{B}}T} \, . \tag{9.50}$$

The same transformation is then performed for F_r^*, which is the integral of the same function over the negative part $\mathbb{R}^-$ of the real axis, running from $-\infty$ to 0. F_r^* is equal to F_i^*, the integral of the same integrand over the positive part of the imaginary axis shifted by a small negative real number $-\varepsilon$ and running from $-\varepsilon - i\infty$ to $-\varepsilon$.

In the end, the Casimir force (9.48) is the integral over a contour that encircles the imaginary axis, and it is thus found to be a discrete sum of the values of the function $zf[z]$ at the Matsubara poles z_n:

$$F = -2k_{\mathrm{B}}T \sideset{}{'}\sum_n \frac{\kappa_n r[ic\kappa_n]e^{-2\kappa_n L}}{1 - r[ic\kappa_n]e^{-2\kappa_n L}}, \tag{9.51}$$

$$\kappa_n \equiv \frac{\xi_n}{c} = n\kappa_1, \qquad \kappa_1 = \frac{2\pi k_{\mathrm{B}}T}{\hbar c}.$$

The primed summation symbol indicates that the contribution of the zeroth Matsubara pole at $n = 0$ is counted for only one half (the symbol is written here for a function $\varphi(n)$):

$$\sideset{}{'}\sum_n \varphi(n) \equiv \tfrac{1}{2}\varphi(0) + \sum_{n=1}^{\infty} \varphi(n). \tag{9.52}$$

This final expression is always finite for any properly defined model of mirrors.

9.3.4.5 *Limiting cases*

In the limit $T \to 0$, the ensemble of Matsubara poles becomes a cut along the imaginary axis (the function $\coth(\hbar z/2k_{\mathrm{B}}T)$ thus goes to $+1$ for $\mathrm{Re}\, z > 0$ and to -1 for $\mathrm{Re}\, z < 0$). The discrete sum (9.51) is then written as an integral over the positive part of the imaginary axis:

$$F_0 = -\frac{\hbar c}{\pi} \int_0^\infty \mathrm{d}\kappa \, \frac{\kappa r[i\xi]}{e^{2\kappa L} - r[i\xi]}, \qquad \kappa \equiv \frac{\xi}{c}. \tag{9.53}$$

For perfect mirrors, i.e. when r may be taken as having unit value at all frequencies contributing to the integral (9.53), a universal result is obtained, which no longer depends on the specific properties of the mirrors:

$$F_0[r \to 1] = -\frac{\hbar c}{\pi} \int_0^\infty \mathrm{d}\kappa \, \frac{\kappa}{e^{2\kappa L} - 1} = -\frac{\hbar c\pi}{24L^2}. \tag{9.54}$$

We have used the fact that the Riemann zeta function

$$\zeta(s) = \frac{1}{\Gamma(s)} \int_0^\infty \frac{x^{s-1}}{e^x - 1} \, \mathrm{d}x = \sum_{n=1}^{\infty} \frac{1}{n^s} \tag{9.55}$$

takes the value $\zeta(2) = \pi^2/6$ for $s = 2$. Other values of the same function will appear in various places in Casimir force calculations with perfectly reflecting mirrors.

For real mirrors, the product of the reflection amplitudes $r = r_1 r_2$ always has a modulus smaller than unity. It follows that the integral is always regular, with a smaller modulus than for perfect mirrors. If the two mirrors are identical ($r_1 = r_2$), then their product r is positive and the force is attractive as in the case of perfect mirrors. The expression of the Casimir force as the integral (9.54) over imaginary frequencies is convenient when discussing the meaning of the limit of perfect mirrors. Taking as an example the model (9.30), we indeed see that (9.53) tends to (9.54) as soon as the characteristic frequency Ω at which the reflection falls down is larger than the typical frequencies $\omega \sim c/L$ contributing to the integral.

9.3.5   The Casimir free energy and phase-shift interpretation

We now write an expression for the Casimir free energy and show that it can be given an interpretation in terms of scattering phase shifts. We also obtain expressions for the Casimir entropy and Casimir internal energy.

9.3.5.1   *The Casimir free energy*

We come back to the expression (9.48) for the force as an integral over real frequencies and write it as the differential of a free energy $\mathcal{F}$ with respect to L:

$$F = -\frac{\partial \mathcal{F}(L, T)}{\partial L},$$

$$\mathcal{F} = \int_0^\infty \frac{\mathrm{d}\omega}{2\pi} \left(1 + 2\overline{n}_\omega\right) \frac{\hbar}{2\mathrm{i}} \ln \frac{d}{d^*}, \tag{9.56}$$

$$d \equiv 1 - r e^{2\mathrm{i}kL}.$$

We show in the following that this formula can be given a nice interpretation in terms of phase shifts [108]. Note that though $\mathcal{F}$ could be changed by an L-independent contribution without changing the result for F, the form of $\mathcal{F}$ given in (9.56) is fixed by the phase-shift interpretation discussed below.

9.3.5.2   *The phase-shift interpretation*

With the aim of introducing the phase-shift interpretation, we put the cavity of Fig. 9.7 inside a quantization box with a much larger size $\mathcal{L} \gg L$ and periodic conditions, as sketched in Fig. 9.8. In the absence of the scatterer, the modes in the large box would have their wavelengths determined by the box size through $k_n^\pm \mathcal{L} = 2n\pi$, where $\pm$ labels

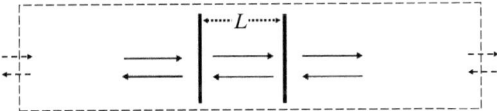

Fig. 9.8   The scattering system, a cavity of length L, is put in a much larger quantization box, and the Casimir effect is calculated as the change of free energy of the large box.

the rightward and leftward propagation directions, which correspond to degenerate solutions. The eigenmodes in the presence of the scatterer are thus determined by the eigenvalues $e^{i\delta_n^{\pm}}$ of the unitary S-matrix through $k_n^{\pm}\mathcal{L} + \delta_n^{\pm} = 2n\pi$. The change of global energy of the modes at a given frequency ω is then determined by the quantity $-\hbar c(\delta_n^+ + \delta_n^-)$ proportional to the sum of the two phase shifts at this frequency. As a consequence, we do not need to solve the full eigenvalue problem for the S-matrix, but we need only calculate the logarithm of its determinant, $\ln \det S = i(\delta_n^+ + \delta_n^-)$.

The relation between the expression (9.56) for the free energy and the phase-shift interpretation is then fixed by noting that the expression (9.37) for the S-matrix associated with the cavity leads to the following relation between the determinants:

$$\det S_{12} = (\det S_1)(\det S_2)\frac{d^*}{d}, \tag{9.57}$$

where we have denoted by S_{12} the scattering matrix for the compound system consisting of the two mirrors 1 and 2 (it was denoted simply by S in (9.37)). We thus deduce

$$\ln \det S_{12} = \ln \det S_1 + \ln \det S_2 + \ln \frac{d^*}{d}. \tag{9.58}$$

It is now clear that the Casimir free energy (9.56) is given by a difference between changes in free energies $\Delta\mathcal{F}$ calculated in three different scattering configurations:

$$\mathcal{F} = \Delta\mathcal{F}_{12} - \Delta\mathcal{F}_1 - \Delta\mathcal{F}_2, \tag{9.59}$$

with each of these quantities determined by the phase shifts for the associated S-matrix:

$$\Delta\mathcal{F}_{12} = \int_0^{\infty} \frac{d\omega}{2\pi}(1 + 2\overline{n}_{\omega})\frac{i\hbar}{2}\ln \det S_{12}. \tag{9.60}$$

Similar expressions hold for the changes in free energies $\Delta\mathcal{F}_1$ and $\Delta\mathcal{F}_2$ associated respectively with mirrors 1 and 2 placed in the large box in Fig. 9.8.

In fact, each of the changes is itself a difference of free energies calculated in the presence and in the absence of the scatterer:

$$\Delta\mathcal{F}_{12} \equiv \mathcal{F}_{12} - \mathcal{F}_0, \qquad \Delta\mathcal{F}_i \equiv \mathcal{F}_i - \mathcal{F}_0, \quad i = 1, 2, \tag{9.61}$$

with the subscript 0 labelling the configuration with no scatterer. In the end, the Casimir free energy is a difference involving four different configurations

$$\mathcal{F} = (\mathcal{F}_{12} - \mathcal{F}_0) - (\mathcal{F}_1 - \mathcal{F}_0) - (\mathcal{F}_2 - \mathcal{F}_0)$$

$$= \mathcal{F}_{12} - \mathcal{F}_1 - \mathcal{F}_2 + \mathcal{F}_0. \tag{9.62}$$

It is worth emphasizing at this point that the definitions of $\mathcal{F}_{12}$, $\mathcal{F}_1$, $\mathcal{F}_2$, and $\mathcal{F}_0$ are affected by the problem of vacuum energy discussed above. Meanwhile, $\Delta\mathcal{F}_{12}$, $\Delta\mathcal{F}_1$, and $\Delta\mathcal{F}_2$ are self-energies, and their proper evaluation should involve some renormalization procedure. In contrast, the Casimir free energy is always a finite quantity, the evaluation of which does not require renormalization [108].

9.3.5.3 Casimir thermodynamics

We can then write other thermodynamic functions associated with the Casimir effect. In particular, we may define an entropy $\mathcal{S}$ from the free energy $\mathcal{F}$,

$$\mathcal{S} = -\frac{\partial \mathcal{F}(L,T)}{\partial T}, \tag{9.63}$$

and then an internal energy $\mathcal{E}$,

$$\mathcal{E} = \mathcal{F} + T\mathcal{S} = \mathcal{F} - T\frac{\partial \mathcal{F}(L,T)}{\partial T}. \tag{9.64}$$

These thermodynamic functions obey the usual thermodynamic relations such as

$$\begin{aligned}
\mathrm{d}\mathcal{F} &= -F\,\mathrm{d}L - \mathcal{S}\,\mathrm{d}T, \\
\mathrm{d}\mathcal{E} &= -F\,\mathrm{d}L + T\,\mathrm{d}\mathcal{S}.
\end{aligned} \tag{9.65}$$

The last relation just means that the change $\mathrm{d}\mathcal{E}$ of internal energy under a transformation is the sum of a mechanical work $-F\,\mathrm{d}L$ and a heat term $T\mathrm{d}\mathcal{S}$.

These thermodynamic functions can be deduced from the free energy $\mathcal{F}(L,T)$ written previously as an integral over real frequencies or, equivalently, from its form as a Matsubara sum:

$$\mathcal{F}(L,T) = k_{\mathrm{B}}T \sum_{n}{}' \ln d[\mathrm{i}\xi_n]. \tag{9.66}$$

This expression can be obtained from (9.56) by using anew Cauchy's theorem. Equivalently, it can be obtained from the Casimir force (9.51) written above as a sum over Matsubara poles.

9.4 The Casimir force in three-dimensional space

We come now to the last section of these lecture notes, which is devoted to the discussion of the Casimir force in three-dimensional (3D) space.

We will mainly discuss the geometry of two plane and parallel mirrors, described by specular scattering amplitudes. This case constitutes a trivial extension of the derivation in 1D space, though a few points have to be treated with greater care. As for the 1D case, we will write a formula valid and regular at thermal equilibrium at any temperature and for any optical model of mirrors obeying causality and high-frequency transparency properties [108]. It reproduces the Casimir ideal formula in the limits of perfect reflection and zero temperature, but can also be used for calculating the Casimir force between arbitrary mirrors, as soon as the reflection amplitudes are specified. For mirrors characterized by Fresnel reflection amplitudes deduced from a linear and local dielectric function, the scattering formalism leads to the same results as Lifshitz's method [79, 145, 146].

The section ends with a short presentation of the general scattering formalism that allows one to deal with non-specular reflection and arbitrary geometries [139].

9.4.1 Free electromagnetic fields in 3D space

9.4.1.1 Modes for electromagnetic fields

The free modes for electromagnetic fields in 3D space are the free solutions of Maxwell's equations in vacuum. They correspond to frequency and wavevector related through the dispersion relation

$$\frac{\omega^2}{c^2} = k_x^2 + k_y^2 + k_z^2 = \mathbf{k}^2 + k_z^2, \qquad \mathbf{k}^2 \equiv k_x^2 + k_y^2. \tag{9.67}$$

As mirrors will be specified later to have their surfaces parallel to the (x, y) plane, k_z is the longitudinal part of the wavevector and $\mathbf{k} \equiv (k_x, k_y)$ its transverse part. We introduce a direction of propagation as in 1D calculations: $\eta = +1$ corresponds to rightward propagation and $\eta = -1$ to leftward propagation. η is the sign of k_z ($\eta = \text{sign}(k_z) = \text{sign}(\cos\theta)$) and it appears in the expression for k_z in terms of frequency and transverse wavevector:

$$k_z = \eta\sqrt{\frac{\omega^2}{c^2} - \mathbf{k}^2}. \tag{9.68}$$

The propagation direction is defined by the incidence angle θ and azimuth φ:

$$k = \frac{\omega}{c}\widehat{k}, \qquad \widehat{k} = \begin{pmatrix} \widehat{k}_x \\ \widehat{k}_y \\ \widehat{k}_z \end{pmatrix} = \begin{pmatrix} \sin\theta\cos\varphi \\ \sin\theta\sin\varphi \\ \cos\theta \end{pmatrix}. \tag{9.69}$$

Unit wavevector and polarization vectors for electric ($\widehat{\alpha}$) and magnetic ($\widehat{\beta}$) fields form an orthonormal basis for each polarization $p =$ TE, TM, with $\widehat{\beta}^p = \widehat{k} \times \widehat{\alpha}^p$ in each case:

$$\widehat{\alpha}^{\text{TE}} = \widehat{\beta}^{\text{TM}} = \begin{pmatrix} -\sin\varphi \\ \cos\varphi \\ 0 \end{pmatrix},$$

$$\widehat{\alpha}^{\text{TM}} = -\widehat{\beta}^{\text{TE}} = \begin{pmatrix} \cos\theta\cos\varphi \\ \cos\theta\sin\varphi \\ -\sin\theta \end{pmatrix}. \tag{9.70}$$

9.4.1.2 Electric and magnetic fields

Electric and magnetic fields are then obtained as linear superpositions of modes labelled by m and η

$$E = \sum_{m,\eta}\sqrt{\frac{\hbar\omega}{2\varepsilon_0}}\,\widehat{\alpha}_{m,\eta}\left(-ia_{m,\eta}e^{-i\omega t + i\mathbf{k}\cdot\mathbf{r} + ik_z z} + ia_{m,\eta}^\dagger e^{i\omega t - i\mathbf{k}\cdot\mathbf{r} - ik_z z}\right),$$

$$B = \sum_{m,\eta}\sqrt{\frac{\hbar\omega\mu_0}{2}}\,\widehat{\beta}_{m,\eta}\left(-ia_{m,\eta}e^{-i\omega t + i\mathbf{k}\cdot\mathbf{r} + ik_z z} + ia_{m,\eta}^\dagger e^{i\omega t - i\mathbf{k}\cdot\mathbf{r} - ik_z z}\right). \tag{9.71}$$

Here, m gathers all quantum numbers except η and can be written in two alternative forms:

$$m \equiv (\mathbf{k}, |k_z|, p) \quad \text{or} \quad m \equiv (\mathbf{k}, \omega, p) . \tag{9.72}$$

η is treated separately because the two modes $\eta = \pm 1$ corresponding to the same m will be coupled by the scattering processes studied later. The sum over modes has the following definition (A is an area introduced for defining integrals over $\mathbf{k}$):

$$\sum_m \equiv \sum_p \iint \frac{A \, d^2\mathbf{k}}{4\pi^2} \int \frac{d|k_z|}{2\pi}$$

$$= \sum_p \iint \frac{A \, d^2\mathbf{k}}{4\pi^2} \int \frac{\omega \, d\omega}{2\pi c^2 |k_z|} . \tag{9.73}$$

Finally, the symbols $a_{m,\eta}$ and $a^\dagger_{m,\eta}$ appearing in the positive- and negative-frequency components are the annihilation and creation operators of quantum electrodynamics [57]. Their commutation relations and correlations are immediate generalizations of those, (9.21) and (9.25), written for the 1D case.

Note that we have used the electromagnetic constants in vacuum ε_0 and μ_0. In the following, the symbol ε will be used as a relative permittivity, with its value in vacuum being unity, and the relative permittivity will keep its unit value in vacuum.

9.4.1.3 *Energy density and radiation pressure*

Energy–momentum densities and radiation pressures are components of the Maxwell stress tensor [107]. In particular, the energy per unit volume e is the component T_{00}:

$$e = \frac{\varepsilon_0 E^{\mathrm{T}} E}{2} + \frac{D^{\mathrm{T}} B}{2\mu_0} , \tag{9.74}$$

where T indicates the matrix transpose. Substituting the expressions for free fields and calculating the mean value of the energy density in a thermal state, we obtain a simple generalization of the expression (9.26) written for the 1D case:

$$\langle e(u) \rangle_{\text{therm}} = \sum_{m,\eta} \frac{\hbar \omega_m}{2} (1 + 2\bar{n}_m) . \tag{9.75}$$

We come then to the radiation pressure on plane mirrors parallel to the (x, y) plane, which corresponds to the component T_{zz} of the Maxwell stress tensor:

$$\frac{\varepsilon_0 (E_x E_x + E_y E_y - E_z E_z)}{2} + \frac{B_x B_x + B_y B_y - B_z B_z}{2\mu_0} . \tag{9.76}$$

The quantum average of this quantity in a thermal state leads to

$$\sum_{m,\eta} \frac{\hbar \omega_m}{2} \cos^2 \theta_m (1 + 2\bar{n}_m) . \tag{9.77}$$

The sums over modes written in the preceding expressions are infinite, but the force on mirrors will be obtained as a finite difference between the radiation pressure acting on their two sides.

9.4.2   Scattering on plane mirrors in 3D space

We consider plane mirrors parallel to the (x, y) plane, which are surfaces of constant z in 3D space. Scattering on such mirrors is a simple extension of that described in the 1D case because of symmetry considerations. Invariance under time translations and lateral space translations implies that the frequency ω, transverse wavevector $\mathbf{k}$, and polarization p are preserved. The scattering process is *specular* and it only affects the parameter η. We first consider lossless mirrors and then address the problem of losses.

9.4.2.1   A lossless mirror in 3D space

This specular scattering on a lossless mirror is schematically represented in Fig. 9.9. In (a), the angles of incidence are shown for the input and output fields. In (b), however, they are only implicit. It thus follows that the sketch of the scattering process is like that drawn in Fig. 9.5 for the 1D case.

As in the 1D case, the scattering matrix is just a 2×2 matrix giving the two output fields in terms of the two input ones:

$$\begin{pmatrix} E_{\text{out}}^+ \\ E_{\text{out}}^- \end{pmatrix} = S_1 \begin{pmatrix} E_{\text{in}}^+ \\ E_{\text{in}}^- \end{pmatrix},$$

$$S_1[m] = \begin{pmatrix} t_1[m] & r_1[m]\mathrm{e}^{-2\mathrm{i}|k_z|q_1} \\ r_1[m]\mathrm{e}^{2\mathrm{i}|k_z|q_1} & t_1[m] \end{pmatrix}.$$

(9.78)

Scattering amplitudes have been written as in (9.28): r_1 and t_1 are defined for a mirror located at $x = 0$ and depend on the quantum number m for real mirrors; the general case is then described by phases determined by the position q_1 of the mirror and the longitudinal wavevector k_z.

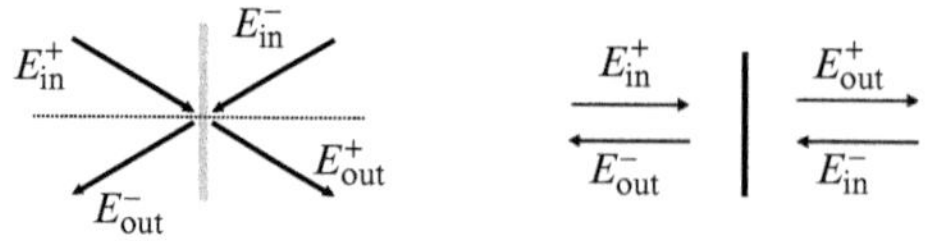

Fig. 9.9 Schematic representation of specular scattering on a lossless mirror in 3D space, with angles of incidence shown explicitly in (a) but only implicitly in (b).

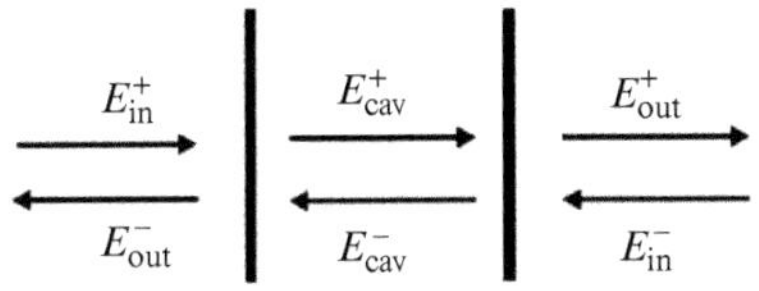

Fig. 9.10 Schematic representation of specular scattering on a cavity in 3D space, with the same conventions as in Fig. 9.9(b).

9.4.2.2 A lossless cavity in 3D space

Specular scattering on a cavity made of two mirrors is represented schematically in Fig. 9.10.

Most calculations are the same as for the 1D case, with the effect of the cavity described by a global scattering matrix and a resonance matrix as in (9.36):

$$\begin{pmatrix} E_{\text{out}}^+ \\ E_{\text{out}}^- \end{pmatrix} = S \begin{pmatrix} E_{\text{in}}^+ \\ E_{\text{in}}^- \end{pmatrix},$$

$$\begin{pmatrix} E_{\text{cav}}^+ \\ E_{\text{cav}}^- \end{pmatrix} = R \begin{pmatrix} E_{\text{in}}^+ \\ E_{\text{in}}^- \end{pmatrix}. \tag{9.79}$$

The matrices S and R are obtained from the scattering amplitudes associated with the mirrors 1 and 2 as in the 1D case, that is to say from (9.37) and (9.40) with the phase factors $e^{\pm ikL}$ replaced by $e^{\pm i|k_z|L}$.

As the global S-matrix is unitary, radiation pressures on the outer sides of the mirrors are given by the expression (9.77). Meanwhile, radiation pressures on the inner sides of the mirrors are modified by a factor given by the common value of the diagonal elements in $RR^\dagger$:

$$g_m \equiv 1 + f_m + f_m^* = \frac{1 - |r_m|^2}{|1 - r_m e^{2i|k_z|L}|^2},$$

$$f_m \equiv \frac{r_m e^{2i|k_z|L}}{1 - r_m e^{2i|k_z|L}}. \tag{9.80}$$

The effective pressure on the mirror is then given by the difference between the pressures on its two sides:

$$\frac{1}{A} \sum_m \hbar \omega_m \cos^2 \theta_m \left(1 + 2\overline{n}_m\right)(g_m - 1) = \frac{1}{A} \sum_m \hbar \omega_m \cos^2 \theta_m \left(1 + 2\overline{n}_m\right)(f_m + f_m^*). \tag{9.81}$$

The left-hand side corresponds to the radiation-pressure interpretation already discussed in the 1D case, while the right-hand side, written in terms of the causal function f, is the basis for transformations using Cauchy's theorem. However, some elements are now to be treated with more care before the Casimir pressure is obtained.

9.4.3   The Casimir pressure between two plates in 3D space

We now extend the derivations to take into account two effects that play an important role in calculations in 3D space. First, we discuss the effect of dissipation inside the mirror and the associated fluctuations. Then, we address the contributions of evanescent waves, which have to be added to those of ordinary propagation waves [95].

9.4.3.1   *Generalization for lossy mirrors*

As already explained, the mirrors used in the experiments are made of metal and they have losses associated with dissipation inside their material. As a consequence, the scattering matrices defined above cannot be considered to be unitary. In other words, there are additional fluctuations accompanying losses that should be taken into account in a large scattering matrix describing the coupling of the modes of interest and the noise modes [14, 15, 60].

At this point, it is worth using a theorem that gives the commutators of the intracavity fields as the product of those well known for fields outside the cavity and the factors g_m introduced in (9.80). This theorem, which has been demonstrated with an increasing range of validity in [15, 95, 108], proves that the expression written in (9.80) for lossless mirrors is still true for lossy mirrors. We then assume thermal equilibrium for the whole system, which means that input fields as well as fluctuations associated with electrons, phonons, and any loss mechanism inside the mirrors have to correspond to the same temperature T, whatever their microscopic origin may be. Then, the radiation pressures are given by the expressions written above, and (9.81) becomes the Casimir pressure, valid for lossy as well as lossless mirrors:

$$P = \frac{1}{A} \sum_m \hbar \omega_m \cos^2 \theta_m \left(1 + 2\bar{n}_m\right)(f_m + f_m^*). \tag{9.82}$$

9.4.3.2   *Generalization for evanescent waves*

Up to now, we have discussed the contributions of ordinary waves freely propagating outside and inside the cavity, which correspond to real wavevectors with k_z real, i.e. also $\omega > c|\mathbf{k}|$. Equation (9.82) thus reflects the intuitive picture of radiation pressure of vacuum and thermal fluctuations, as discussed in the 1D case. In the 3D case, we have also to take into account the contribution of evanescent waves (see Section 1.5.4 in [27]), which correspond to imaginary values of k_z, i.e. also frequencies in the interval $0 < \omega < c|\mathbf{k}|$.

Those waves propagate inside the mirrors with an incidence angle larger than the limit angle and they also exert a radiation pressure on the mirrors, due to the frustrated reflection phenomenon. In fact, the expression (9.82) for the Casimir pressure has to be understood as including these contributions [95]. They are obtained through an analytical continuation of those of ordinary waves, using the well-defined analytic behaviour of the causal function f_m. This analytical continuation has to be defined for fixed values of the lateral wavevector $\mathbf{k}$ and polarization p.

Using the definition (9.73) of the sum over modes and the relation $\cos \theta = ck_z/\omega$, one finally gets the Casimir pressure (9.82) as the real part of an integral over the positive part $\mathbb{R}^+$ of the real axis (to be compared with (9.48)):

$$P = P_r + P_r^* ,$$

$$P_r \equiv \sum_p \iint \frac{\mathrm{d}^2\mathbf{k}}{4\pi^2} \int_0^\infty \frac{\mathrm{d}\omega}{2\pi} \, \hbar |k_z| f_m \coth \frac{\hbar\omega}{2k_\mathrm{B}T} \cdot \tag{9.83}$$

We stress again that the integral over $\mathbb{R}^+$ includes the contributions of evanescent waves $0 < \omega < c|\mathbf{k}|$ besides those of ordinary waves $c|\mathbf{k}| < \omega$.

9.4.3.3 Casimir pressure as an integral over imaginary frequencies

We now transform the expression (9.83) for the Casimir pressure by using Cauchy's theorem as in the 1D case. In the end of the derivation, the Casimir pressure is the integral over a contour that encircles the imaginary axis, and it is thus found to be a discrete sum of the values of the function $|k_z| f_m$ at the Matsubara poles:

$$P = -k_\mathrm{B}T \sum_p \int \frac{\mathrm{d}^2\mathbf{k}}{(2\pi)^2} \sum_n{}' \frac{2\kappa_n r_\mathbf{k}^p[i\xi_n] e^{-2\kappa_n L}}{1 - r_\mathbf{k}^p[i\xi_n] e^{-2\kappa_n L}} , \tag{9.84}$$

where κ is obtained from $|k_z|$ through an analytical continuation for imaginary frequencies $\omega_n = i\xi_n$:

$$\kappa_n = \sqrt{\mathbf{k}^2 + \frac{\xi_n^2}{c^2}}, \qquad \xi_n = n\frac{2\pi k_\mathrm{B}T}{\hbar} \cdot \tag{9.85}$$

As in the 1D case, this expression is finite for any properly defined model of mirrors.

9.4.3.4 Zero-temperature limit

In the limit $T \to 0$, the ensemble of Matsubara poles becomes a cut along the imaginary axis and the Matsubara sum (9.84) becomes an integral over the positive part of the imaginary axis ($P_0 \equiv P_{T=0}$):

$$P_0 = -2\sum_p \iint \frac{\mathrm{d}^2\mathbf{k}}{4\pi^2} \int_0^\infty \frac{\mathrm{d}\xi}{2\pi} \, \hbar\kappa \frac{r_\mathbf{k}^p[i\xi] e^{-2\kappa L}}{1 - r_\mathbf{k}^p[i\xi] e^{-2\kappa L}} ,$$

$$\kappa = \sqrt{\mathbf{k}^2 + \frac{\xi^2}{c^2}} \cdot \tag{9.86}$$

For perfect mirrors, i.e. when r may be taken as having unit value at all frequencies contributing to the integral (9.86), a universal result is obtained, which no longer depends on the specific properties of the mirrors:

$$P_0[r \to 1] = -\frac{\hbar c}{\pi^2} \int_0^\infty \frac{\kappa^3 \, \mathrm{d}\kappa}{e^{2\kappa L} - 1} = -\frac{\hbar c\pi^2}{240 L^4} \cdot \tag{9.87}$$

We have used the fact that the Riemann zeta function (9.55) takes the value $\zeta(4) = \pi^4/90$ for $s = 4$.

9.4.4 Models for metallic mirrors

We now come back to the discussion presented in Section 9.2 for the models of metallic mirrors used in the experiments. The common model is that of bulk mirrors, in fact thick slabs, made with metals such as gold. The optical response of the metal is described by a local dielectric response function and the reflection amplitudes on each mirror are then deduced using the Fresnel laws.

9.4.4.1 Fresnel reflection amplitudes

Reflection amplitudes on a thick slab described by a local dielectric response ε are given by the well-known Fresnel laws for the TE and TM polarizations (see Section 1.5.1 in [27] or Section 86 in [143]):

$$r_1^{\mathrm{TE}}[\omega] = \frac{k_z - K_z}{k_z + K_z}, \qquad r_1^{\mathrm{TM}}[\omega] = \frac{K_z - \varepsilon k_z}{K_z + \varepsilon k_z}, \tag{9.88}$$

where K_z and k_z are the longitudinal wavevectors in matter and vacuum, respectively,

$$K_z = \sqrt{\varepsilon \frac{\omega^2}{c^2} - \mathbf{k}^2}, \qquad k_z = \sqrt{\frac{\omega^2}{c^2} - \mathbf{k}^2}. \tag{9.89}$$

When the frequency is continued to imaginary values, these expressions become

$$r_1^{\mathrm{TE}}[i\xi] = \frac{\kappa - K}{\kappa + K}, \qquad r_1^{\mathrm{TM}}[i\xi] = \frac{K - \varepsilon[i\xi]\kappa}{K + \varepsilon[i\xi]\kappa},$$

$$K = \sqrt{\varepsilon[i\xi]\frac{\xi^2}{c^2} + k^2}, \qquad \kappa = \sqrt{\frac{\xi^2}{c^2} + k^2}. \tag{9.90}$$

When the reflection amplitudes (9.90) are inserted in the expression (9.84) for the pressure, the formula obtained in 1956 by E. M. Lifshitz [145] and derived again in 1961 by I. E. Dzyaloshinskii, E. M. Lifshitz, and L. P. Pitaevskii [79] is recovered (see also Chapter VIII in [146]). We may stress at this point that this formula was obtained through a different derivation, with all fluctuations originating from matter. In fact, this expression was written in terms of the dielectric function and not of reflection amplitudes. To the best of our knowledge, Kats was the first to notice that it could be written in terms of the reflection amplitudes [121]. We may quote at this point a few alternative derivations of this formula through different methods [46, 105, 122, 131, 144, 183, 186, 187, 204].

We may remark that the optical response of the bulk material cannot always be described by a local dielectric function. In this case, the description in terms of reflection amplitudes is still valid, though the reflection amplitudes cannot be written in the specific forms (9.90). More detailed descriptions can be determined for example from microscopic models of metallic conduction. Attempts in this direction and discussions can be found for example in [62, 63, 73, 96, 176, 177, 213].

9.4.4.2 Models for dielectric functions

As already discussed in Section 9.2, the dielectric function ε is usually obtained from optical data [134, 214]. These data are then extrapolated to low frequencies by using the dissipative Drude model (9.9) for the conductivity of the metal [7]. In contrast to the plasma model (9.11), which corresponds to the lossless limit $\gamma = 0$, the dissipative Drude model meets the well-known fact that gold has a finite static conductivity (9.10).

The two models lead to different predictions for the Casimir pressure, in particular at the limits of large distances or large temperatures. This can be attributed to different limiting cases at zero frequency. The Drude model indeed corresponds to

$$\lim_{\omega \to 0} r_{\mathbf{k}}^{\text{TE}}[\omega] = 0 \qquad \lim_{\omega \to 0} r_{\mathbf{k}}^{\text{TM}}[\omega] = 1 \,, \qquad \gamma \neq 0 \,, \tag{9.91}$$

whereas the plasma model leads to

$$\lim_{\omega \to 0} r_{\mathbf{k}}^{\text{TE}}[\omega] = 1 \,, \qquad \lim_{\omega \to 0} r_{\mathbf{k}}^{\text{TM}}[\omega] = 1 \,, \qquad \gamma = 0 \,. \tag{9.92}$$

9.4.4.3 High-temperature limit

In the limit of high temperatures $\kappa_1 L \gg 1$, the Matsubara poles with $n > 0$ give an exponentially small contribution to the Casimir pressure (9.84). The result is thus dominated by the contribution of the zeroth pole $n = 0$ ($P_\infty \equiv P_{T \to \infty}$):

$$P_\infty = -k_{\text{B}} T \sum_p \int \frac{d^2 \mathbf{k}}{(2\pi)^2} \frac{|\mathbf{k}| r_{\mathbf{k}}^p[0] e^{-2|\mathbf{k}|L}}{1 - r_{\mathbf{k}}^p[0] e^{-2|\mathbf{k}|L}} \,. \tag{9.93}$$

It follows that the asymptotic Casimir pressure (9.93) has different values for a dissipative model,

$$P_\infty = -\frac{k_{\text{B}} T}{2\pi} \int_0^\infty \frac{k^2 \, dk}{e^{2kL} - 1} = -\frac{k_{\text{B}} T}{8\pi L^3} \zeta(3) \,, \qquad \gamma \neq 0 \,, \tag{9.94}$$

and a lossless model,

$$P_\infty = -\frac{k_{\text{B}} T}{\pi} \int_0^\infty \frac{k^2 \, dk}{e^{2kL} - 1} = -\frac{k_{\text{B}} T}{4\pi L^3} \zeta(3) \,, \qquad \gamma = 0 \,. \tag{9.95}$$

$\zeta(3) \simeq 1.202$ is the value of the Riemann zeta function (9.55) for $s = 3$. As already discussed in Section 9.2, there is a large factor 2 between the two predictions because the two polarizations contribute equally for the plasma model, as well as for perfect mirrors, whereas only one polarization contributes for the Drude model.

9.4.5 The Casimir free energy in 3D space

Using the results already obtained in the 1D case, we write an expression for the Casimir free energy and show that it can be given an interpretation in terms of scattering phase shifts. We also obtain expressions for the Casimir entropy and Casimir internal energy.

9.4.5.1 The Casimir free energy and phase-shift interpretation

The expression (9.83) for the Casimir pressure can be written as the differential $P = -\partial \mathcal{F}/\partial V$ of a free energy $\mathcal{F}(V,T)$ with respect to the volume $V \equiv AL$:

$$\mathcal{F} = \sum_p A \iint \frac{\mathrm{d}^2 \mathbf{k}}{4\pi^2} \int_0^\infty \frac{\mathrm{d}\omega}{2\pi} (1 + 2\bar{n}_\omega) \frac{\hbar}{2\mathrm{i}} \ln \frac{1 - r e^{2\mathrm{i}|k_z|L}}{1 - r^* e^{-2\mathrm{i}|k_z|L}} . \tag{9.96}$$

The interpretation of this formula in terms of phase shifts is the same as in the 1D case [108] and we do not repeat here the derivation presented in Section 9.3. At the end of this derivation, the Casimir free energy (9.96) is given by a difference (9.59) of changes of free energies calculated in three different scattering configurations, with each of these quantities determined by an integral (9.60) over all modes. As already emphasized, the Casimir free energy is always a finite quantity, the evaluation of which does not require a renormalization [108].

9.4.5.2 Casimir thermodynamics

We can then write other thermodynamic functions associated with the Casimir effect. In particular, we may define an entropy $\mathcal{S} = \partial \mathcal{F}/\partial T$ and then an internal energy $\mathcal{E} = \mathcal{F} + T\mathcal{S}$. These obey the usual thermodynamic relations ($V \equiv AL$)

$$\mathrm{d}\mathcal{F} = -P\,\mathrm{d}V - \mathcal{S}\,\mathrm{d}T , \qquad \mathrm{d}\mathcal{E} = -P\,\mathrm{d}V + T\,\mathrm{d}\mathcal{S} . \tag{9.97}$$

These thermodynamic functions can be deduced from the free energy (9.96) written previously as an integral over real frequencies or, equivalently, from its form as a Matsubara sum

$$\mathcal{F} = k_\mathrm{B} T \sum_n{}' \mathrm{Tr} \ln d[\mathrm{i}\xi_n] . \tag{9.98}$$

We have used the denominator of the S-matrix to write the free energy,

$$d_\mathbf{k}^p[\mathrm{i}\xi_n] = 1 - r_\mathbf{k}^p[\mathrm{i}\xi_n] e^{-2\kappa_n L} , \qquad \kappa_n = \sqrt{\frac{\xi_n^2}{c^2} + k^2} , \tag{9.99}$$

and we have also introduced a trace over polarizations and transverse wavevectors in order to simplify the expression,

$$\mathrm{Tr}\, d \equiv \sum_p \int \frac{A\,\mathrm{d}^2 \mathbf{k}}{(2\pi)^2} d_\mathbf{k}^p . \tag{9.100}$$

The expression (9.98) for the free energy is completely equivalent to the expression (9.84) written above for the pressure.

9.4.6 General scattering formula

We now present a more general scattering formula allowing us to calculate the Casimir force between stationary disjoint objects with arbitrary shapes. We also review briefly some of the applications of this new method that have been dedicated in recent years to the study of non-trivial geometries. More exhaustive reviews of the topic can be found in [82, 136, 139, 165, 184, 197].

9.4.6.1 The non-specular scattering formula

We consider a geometrical configuration with two disjoint scatterers O_1 and O_2 at rest in electromagnetic vacuum (Fig. 9.11). The scattering matrix is now a large matrix that accounts for non-specular reflection mixing different wavevectors and polarizations while preserving frequency if the scatterers are at rest. Of course, the non-specular scattering formula is the generic one, while specular reflection is an idealization for perfectly plane and flat plates.

The reasoning presented above leads to an expression of the Casimir free energy as the sum of all phase-shifts contained in the large S-matrix [136]. Causality and high-frequency transparency then allow one to write it as a Matsubara sum (to be compared with (9.98)):

$$\mathcal{F} = k_\mathrm{B}T \sum_m{}' \mathrm{Tr}\ln\mathcal{D}(\mathrm{i}\xi_m)\,, \qquad \xi_m \equiv \frac{2\pi m k_\mathrm{B}T}{\hbar}\,. \tag{9.101}$$

Each term in this sum is the trace of a large matrix that describes all couplings at a given frequency. The matrix $\mathcal{D}$ (here written at Matsubara frequencies $\omega_m = \mathrm{i}\xi_m$) is the denominator of the scattering matrix; it describes the resonance properties of the cavity formed by the two objects (to be compared with (9.99)):

$$\mathcal{D} = 1 - \mathcal{R}_1\mathrm{e}^{-\mathcal{K}L}\mathcal{R}_2\mathrm{e}^{-\mathcal{K}L}\,. \tag{9.102}$$

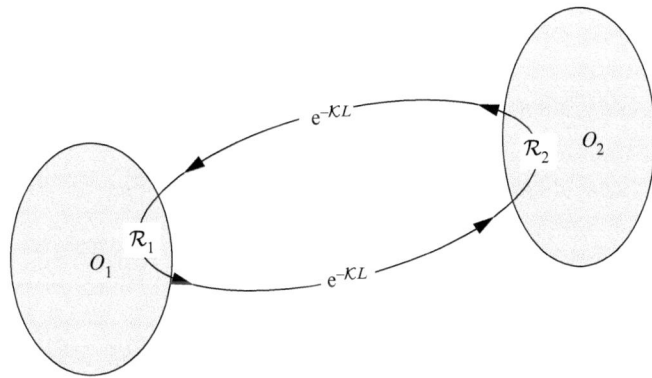

Fig. 9.11 General scattering configuration with two objects O_1 and O_2 in vacuum: the Casimir free energy can be calculated from the denominator (9.102) describing the resonance properties of the cavity formed by the two objects and given by the reflection matrices $\mathcal{R}_1$ and $\mathcal{R}_2$ and the propagation matrices $\mathrm{e}^{-\mathcal{K}L}$.

The matrices $\mathcal{R}_1$ and $\mathcal{R}_2$ represent reflection on the two objects O_1 and O_2, respectively, while $e^{-\mathcal{K}L}$ describes propagation. Note that the matrices $\mathcal{D}$, $\mathcal{R}_1$, and $\mathcal{R}_2$, which were diagonal in the plane-wave basis for specular scattering, are no longer diagonal in the general case of non-specular scattering. The propagation factors remain diagonal in this basis, with their eigenvalues κ written as in (9.99). Note that the expression (9.101) does not depend on the choice of a specific basis.

9.4.6.2 Applications to non-trivial geometries

The fact that geometry plays a non-trivial role in the context of vacuum energy and Casimir forces has been noticed for a long time [8, 9]. In the plane–sphere geometry, for example, the PFA can only be valid when the radius is much larger than the separation [116, 203, 205]. Recently, calculations have used the general non-specular scattering formula in order to push the theory beyond the PFA and thus open the way to a new domain [82, 136, 139, 165, 184, 197].

The first application of the non-specular scattering formula to calculations beyond the PFA was developed in [151, 152] to study the roughness correction to the Casimir force between two planes, in a perturbative expansion with respect to the roughness amplitude. Another geometry studied with this method is that of surfaces with periodic corrugations. As lateral translation symmetry is broken, the Casimir force contains a lateral component that is smaller than the normal one, but has nevertheless been measured in dedicated experiments [54]. Calculations beyond the PFA were first performed with the simplifying assumptions of perfect reflection [38] or shallow corrugations [195, 196]. As was to be expected, the PFA was found to be accurate only in the limit of large corrugation wavelengths.

In recent years, experiments have been able to probe the beyond-PFA regime [52, 55, 106]. Exact calculations of the forces between real mirrors with deep corrugations have been performed [137, 138] and comparisons between theory and experiments presented [13, 97]. The same kinds of calculations have been performed for studying the Casimir torque that appears between two corrugated surfaces with non-aligned corrugations [194].

9.4.6.3 The plane–sphere geometry

Another important application concerns the plane–sphere geometry used in most Casimir force experiments and for which explicit *exact calculations* (see the discussion of the meaning of this expression below) have recently been developed (see the discussions of the various methods devoted to this problem in [26, 83, 128, 191, 218]).

Indeed, exact multipolar expansions have been written for the force in the plane–sphere geometry [84, 153]. These calculations have now been performed for metallic surfaces coupled to electromagnetic vacuum, at zero [40] or non-zero [41, 42] temperature, and this has opened the way to a comparison with theory of the only experimental study devoted to a test of PFA in the plane–sphere geometry [130]. The results of these calculations are often presented in terms of the ratio ρ of the exact result to the PFA approximation, which is close to unity when the aspect ratio $x \equiv L/R$ is small. An alternative representation is in terms of the *slope* β defined by $\rho \equiv 1 + x\beta$.

In fact, the *exact result* is an infinite series corresponding to a multipolar expansion over the numbers (ℓ, m) that label the basis of spherical waves (with $|m| \leq \ell$). Sums have to be truncated to $\ell \leq \ell_{\max}$ for the numerics and the precision is thus limited for small values of the aspect ratio. The minimum aspect ratio for precise calculations scales as the inverse of the maximum multipolar index $\ell_{\max}$. For some time, extrapolations at low values of x were used to obtain bounds useful for the values of experimental interest. This procedure used the idea that the slope β could be represented as a Taylor expansion of the aspect ratio [23, 25, 90]. However, this idea was finally shown to be wrong in the high-temperature limit, which corresponds to a classical Casimir effect with an entropic origin [24, 44]. As a consequence, the question of accurate evaluations for the deviation from PFA remains open.

9.4.6.4 *Nanospheres and atoms close to surfaces*

We will end this chapter by considering another case of interest, that of dielectric nanospheres above a plane, which, in the limit of a very small sphere, tends to the Casimir–Polder problem.

Calculations are easier for nanospheres than for large spheres because a small number of multipoles is sufficient to achieve good accuracy. An example of such calculations is given in [43] for nano-diamonds above a copper plate. Of course, the Casimir–Polder expression between an atom and a plane is recovered in the limit of large distances $R \gg L$, where the dipole approximation is sufficient. The Casimir–Polder limit is written in terms of a polarizability for the atom (an expression is given in [159]). The polarizability of a small sphere is that of a large atom, since it is proportional to the volume of the sphere [43].

9.4.6.5 *Conclusion*

The general scattering formula allows one to describe all long-range interactions in a unified manner in terms of the change in vacuum energy due to the presence of scatterers in vacuum. Atoms as well as mirrors are completely characterized by scattering amplitudes, although, of course, these have different properties in the two cases. In particular, atomic scattering is weak, so that perturbation theory is in general sufficient, whereas scattering by mirrors may be saturated (with up to 100% reflection). Atoms are local probes of the vacuum, whereas mirrors are not.

The same general scattering formula thus allows one to treat different cases leading to a variety of phenomena [136], for example atom–atom, atom–plate, or plate–plate interactions, with plane or spherical plates as well as nanostructured surfaces. For example, it has been possible to study the Casimir–Polder forces or torques for atoms near corrugated surfaces [59, 61, 102].

Acknowledgements

The authors thank R. O. Behunin, A. Canaguier-Durand, I. Cavero-Pelaez, H. B. Chan, D. A. R. Dalvit, R. S. Decca, G. Dufour, T. Ebbesen, E. Fischbach, C. Genet, R. Guérout, G.-L. Ingold, F. Intravaia, M.-T. Jaekel, A. Liscio, J. Lussange, P. A.

Maia Neto, K. A. Milton, V. V. Nesvizhevsky, G. Palasantzas, P. Samori, C. Speake, A. Voronin, and Y. Zeng for contributions or discussions related to this chapter, and the ESF Research Networking Programme CASIMIR (www.casimir-network.com) for providing excellent opportunities for discussions of the Casimir effect and related topics.

References

[1] E. G. Adelberger, B. R. Heckel, and A. E. Nelson, *Annu. Rev. Nucl. Part. Sci.* **53**, 77 (2003).

[2] E. G. Adelberger, J. H. Gundlach, B. R. Heckel, S. Hoedl, and S. Schlamminger, *Prog. Part. Nucl. Phys.* **62**, 102 (2009).

[3] R. J. Adler, B. Casey, and O. C. Jacob, *Am. J. Phys.* **63**, 620 (1995)

[4] I. J. R. Aitchinson, *Contemp. Phys.* **26**, 331 (1985).

[5] I. Antoniadis, S. Baessler, M. Büchner et al., *C. R. Phys.* **12**, 755 (2011).

[6] N. Arkani-Hamed, S. Dimopoulis, and G. R. Dvali, *Phys. Lett. B* **436**, 257 (1998).

[7] N. W. Ashcroft and N. D. Mermin, *Solid State Physics* (HRW International, 1976).

[8] R. Balian and B. Duplantier, *Ann. Phys. (NY)* **104**, 300 (1977).

[9] R. Balian and B. Duplantier, *Ann. Phys. (NY)* **112**, 165 (1978)

[10] R. Balian, *Séminaire Poincaré* **1** (Birkhäuser, 2003), p. 55.

[11] A. A. Banishev, C.-C. Chang, G. L. Klimchitskaya, V. M. Mostepanenko, and U. Mohideen, *Phys. Rev. B* **85**, 195422 (2012).

[12] A. A. Banishev, G. L. Klimchitskaya, V. M. Mostepanenko, and U. Mohideen, *Phys. Rev. Lett.* **110**, 137401 (2013).

[13] Y. Bao, R. Guérout, J. Lussange, et al., *Phys. Rev. Lett.* **105**, 250402 (2010).

[14] S. M. Barnett, C. R. Gilson, B. Huttner, and N. Imoto, *Phys. Rev. Lett.* **77**, 1739 (1996).

[15] S. M. Barnett, J. Jeffers, A. Gatti, and R. Loudon, *Phys. Rev. A* **57**, 2134 (1998).

[16] R. G. Barrera and S. Reynaud, eds., *Focus on Casimir Forces, New J. Phys.* **8** (2006).

[17] S. R. Beane, *Gen. Relativ. Gravit.* **29** 945 (1997).

[18] R. O. Behunin, F. Intravaia, D. A. R. Dalvit, P. A. Maia Neto, and S. Reynaud, *Phys. Rev. A* **85**, 012504 (2012).

[19] R. O. Behunin, Y. Zeng, D. A. R. Dalvit, and S. Reynaud, *Phys. Rev. A* **86**, 052509 (2012).

[20] R. O. Behunin, D. A. R. Dalvit, R. S. Decca, et al., *Phys. Rev. A* **90**, 062115 (2014).

[21] C. L. Bennett, M. Halpern, G. Hinshaw, et al., *Astrophys. J. Suppl.* **148**, 1 (2003).

[22] G. Bimonte, *Phys. Rev. A* **79**, 042107 (2009).

[23] G. Bimonte, T. Emig and M. Kardar, *Appl. Phys. Lett.* **100**, 074110 (2012).

[24] G. Bimonte and T. Emig, *Phys. Rev. Lett.* **109**, 160403 (2012).

[25] G. Bimonte, T. Emig, R. L. Jaffe, and M. Kardar, *EPL* **97**, 50001 (2012).

[26] M. Bordag, and V. Nikolaev, *J. Phys. A* **41**, 164002 (2008)

[27] M. Born and E. Wolf, *Principles of Optics*, 7th edn (Cambridge University Press, 1999).

[28] M. Boström and B. E. Sernelius, *Phys. Rev. Lett.* **84**, 4757 (2000).

[29] G. Bressi, G. Carugno, R. Onofrio, and G. Ruoso, *Phys. Rev. Lett.* **88**, 041804 (2002).

[30] I. Brevik, S. A. Ellingsen, and K. Milton, *New J. Phys.* **8**, 236 (2006).

[31] I. Brevik and J. S. Hoye, *Eur. J. Phys.* **35**, 015012 (2014).

[32] W. Broer, G. Palasantzas, J. Knoester and V. B. Svetovoy, *EPL* **95**, 30001 (2011).

[33] W. Broer, G. Palasantzas, J. Knoester, and V. B. Svetovoy, *Phys. Rev. B* **85**, 155410 (2012).

[34] W. Broer, G. Palasantzas, J. Knoester, and V. B. Svetovoy, *Phys. Rev. B* **87**, 125413 (2013).

[35] L. S. Brown and G. J. Maclay, *Phys. Rev.* **184**, 1272 (1969).

[36] P. F. Browne, *Apeiron* **2**, 72 (1995).

[37] P. R. Buenzli and P. A. Martin, *Europhys. Lett.* **72**, 42 (2005).

[38] R. Büscher and T. Emig, *Phys. Rev. Lett.* **94**, 133901 (2005).

[39] J. B. Camp, T. W. Darling, and R. E. Brown, *J. Appl. Phys.* **69**, 7126 (1991).

[40] A. Canaguier-Durand, P. A. Maia Neto, I. Cavero-Pelaez, et al., *Phys. Rev. Lett.* **102**, 230404 (2009).

[41] A. Canaguier-Durand, P. A. Maia Neto, A. Lambrecht, and S. Reynaud, *Phys. Rev. Lett.* **104**, 040403 (2010).

[42] A. Canaguier-Durand, P. A. Maia Neto, A. Lambrecht, and S. Reynaud, *Phys. Rev. A* **82**, 012511 (2010); **83**, 039905(E) (2011).

[43] A. Canaguier-Durand, A. Gérardin, R. Guérout, et al., *Phys. Rev. A* **83** 032508 (2011).

[44] A. Canaguier Durand, G.-L. Ingold, M.-T. Jaekel, et al., *Phys. Rev. A* **85** 052501 (2012).

[45] A. Canaguier-Durand, R. Guérout, P. A. Maia Neto, A. Lambrecht, and S. Reynaud, *Int. J. Mod. Phys. Conf. Ser.* **14**, 250 (2012).

[46] P. Candelas, *Ann. Phys. (NY)* **143**, 241 (1982).

[47] J. D. Carter and J. D. D. Martin, *Phys. Rev. A* **83**, 032902 (2011).

[48] H. B. G. Casimir, *Proc. K. Ned. Akad. Wet. (Phys.)* **51**, 79 (1948).

[49] H. B. G. Casimir and D. Polder, *Phys. Rev.* **73**, 360 (1948).

[50] R. Castillo-Garza, J. Xu, G. L. Klimchitskaya, V. M. Mostepanenko, and U. Mohideen, *Phys. Rev. B* **85**, 075402 (2013).

[51] H. B. Chan, V. A. Aksyuk, R. N. Kleiman, D. J. Bishop, and F. Capasso, *Phys. Rev. Lett.* **87**, 211801 (2001).

[52] H. B. Chan, Y. Bao, J. Zou, et al., *Phys. Rev. Lett.* **101**, 030401 (2008).

[53] C.-C. Chang, A. A. Banishev, R. Castillo-Garza, et al., *Phys. Rev. B* **85**, 165443 (2012).

[54] F. Chen, U. Mohideen, G. L. Klimchitskaya, and V. M. Mostepanenko, *Phys. Rev. Lett.* **88**, 101801 (2002).

[55] H. C. Chiu, G. L. Klimchitskaya, V. N. Marachevsky, et al., *Phys. Rev. A* **80**, 121402 (2009).

[56] A. A. Chumak, P. W. Milonni, and G. P. Berman, *Phys. Rev. B* **70**, 085407 (2004).

[57] C. Cohen-Tannoudji, J. Dupont-Roc, and G. Grynberg, *Photons and Atoms* (Wiley, 1989).

[58] C. Cohen-Tannoudji, J. Dupont-Roc, and G. Grynberg, *Atom–Photon Interactions* (Wiley, 1992).

[59] A. M. Contreras-Reyes, R. Guérout, P. A. Maia Neto, et al., *Phys. Rev. A* **82**, 052517 (2010).

[60] J. M. Courty, F. Grassia and S. Reynaud, in *Noise, Oscillators and Algebraic Randomness*, ed. M. Planat (Springer-Verlag, 2000), p. 71.

[61] D. A. R. Dalvit, P. A. Maia Neto, A. Lambrecht, and S. Reynaud, *Phys. Rev. Lett.* **100**, 040405 (2008).

[62] D. A. R. Dalvit and S. K. Lamoreaux, *Phys. Rev. Lett.* **101**, 163203 (2008).

[63] D. A. R. Dalvit and S. K. Lamoreaux, *Phys. Rev. Lett.* **102**, 189304 (2009).

[64] D. A. R. Dalvit, P. W. Milonni, D. C. Roberts, and F. S. S. Rosa, eds., *Casimir Physics* (Lecture Notes in Physics **834**, Springer-Verlag, 2011).

[65] O. Darrigol, *Ann. Physik* **9**, 951 (2000).

[66] P. Debye, *Ann. Physik* **43**, 49 (1914).

[67] R. S. Decca, E. Fischbach, G. L. Klimchitskaya, et al., *Phys. Rev. D* **68** 116003 (2003).

[68] R. S. Decca, D. López, E. Fischbach, et al., *Ann. Phys. (NY)* **318**, 37 (2005).

[69] R. S. Decca, D. López, H. B. Chan, et al., *Phys. Rev. Lett.* **94**, 240401 (2005).

[70] R. S. Decca, D. López, E. Fischbach, et al., *Phys. Rev. D* **75**, 077101 (2007).

[71] R. S. Decca, D. López, E. Fischbach, et al., *Eur. Phys. J. C* **51**, 963 (2007).

[72] R. S. Decca, E. Fischbach, G. L. Klimchitskaya, et al., *Phys. Rev. D* **79**, 124021 (2009).

[73] R. S. Decca, E. Fischbach, B. Geyer, et al., *Phys. Rev. Lett.* **102**, 189303 (2009).

[74] R. S. Decca, Slides of a talk at the PASI School on Frontiers of Casimir Physics, Ushuaia (2012); http://physics.iupui.edu/graduate/program.

[75] B. V. Derjaguin, I. I. Abrikosova, and E. M. Lifshitz, *Q. Rev. Chem. Soc.* **10**, 295 (1956).

[76] L. Deslauriers, S. Olmschenk, D. Stick, et al., *Phys. Rev. Lett.* **97**, 103007 (2006).

[77] R. Dubessy, T. Coudreau, and L. Guidoni, *Phys. Rev. A* **80**, 031402 (2009).

[78] G. Dvali, G. Gabadadze, M. Kolanovic, and F. Nitti, *Phys. Rev. D* **65**, 024031 (2001).

[79] I. E. Dzyaloshinskii, E. M. Lifshitz and L. P. Pitaevskii, *Sov. Phys. Usp.* **4**, 153 (1961).

[80] T. Ederth, *Phys. Rev. A* **62** 062104 (2000).

[81] A. Einstein and O. Stern, *Ann. Physik* **40**, 551 (1913).

[82] T. Emig, N. Graham, R. L. Jaffe, and M. Kardar, *Phys. Rev. Lett.* **99**, 170403 (2007).

[83] T. Emig and R. L. Jaffe, *J. Phys. A* **41**, 164001 (2008)

[84] T. Emig, *J. Stat. Mech.: Theory Exp.* P04007 (2008).

[85] C. P. Enz, in *Physical Reality and Mathematical Description*, C. P. Enz and J. Mehra, eds. (Reidel, 1974), p. 124.

[86] R. J. Epstein, S. Seidelin, D. Leibfried, et al., *Phys. Rev. A* **76**, 033411 (2007).

[87] C. W. F. Everitt, D. B. DeBra, B. W. Parkinson, et al., *Phys. Rev. Lett.* **106**, 221101 (2011).

[88] G. Feinberg and J. Sucher, *Phys. Rev. A* **2**, 2395 (1970).

[89] E. Fischbach and C. Talmadge, *The Search for Non Newtonian Gravity* (AIP Press/Springer-Verlag, 1998).

[90] C. D. Fosco, F. C. Lombardo, and F. D. Mazzitelli, *Phys. Rev. D* **84**, 105031 (2011).

[91] R. H. French, V. A. Parsegian, R. Podgornik, et al., *Rev. Mod. Phys.* **82**, 1887 (2010).

[92] N. Gaillard, M. Gros-Jean, D. Mariolle, F. Bertin, and A. Bsiesy, *Appl. Phys. Lett.* **89**, 154101 (2006).

[93] M. Gell-Mann and M. L. Goldberger, *Phys. Rev.* **91**, 398 (1953).

[94] C. Genet, A. Lambrecht, and S. Reynaud, *Phys. Rev. A* **62**, 012110 (2000).

[95] C. Genet, A. Lambrecht, and S. Reynaud, *Phys. Rev. A* **67**, 043811 (2003).

[96] B. Geyer, G.L. Klimchitskaya, U. Mohideen, and V. M. Mostepanenko, *Phys. Rev. Lett.* **102**, 189301 (2009).

[97] R. Guérout, J. Lussange, H. B. Chan, A. Lambrecht, and S. Reynaud, *Phys. Rev. A* **87**, 052514 (2013).

[98] R. Guérout, A. Lambrecht, K. Milton, and S. Reynaud, *Phys. Rev. E* **90**, 042125 (2014).

[99] B. W. Harris, F. Chen, and U. Mohideen, *Phys. Rev. A* **62**, 052109 (2000).

[100] D. A. Hite, Y. Colombe, A. C. Wilson ,et al., *Phys. Rev. Lett.* **109**, 103001 (2012).

[101] D. A. Hite, Y. Colombe, A. C. Wilson, et al., *MRS Bull.* **38**, 826 (2013).

[102] F. Impens, A. M. Contreras-Reyes, P. A. Maia Neto, et al., *EPL* **92**, 40010 (2010).

[103] G.-L. Ingold, A. Lambrecht, and S. Reynaud, *Phys. Rev. E* **80**, 041113 (2009).

[104] F. Intravaia, C. Henkel, and M. Antezza, in *Casimir Physics*, D. A. R. Dalvit, P. W. Milonni, D. C. Roberts, and F. S. S. Rosa, eds. (Lecture Notes in Physics **834**, Springer-Verlag, 2011), p. 345.

[105] F. Intravaia and R. Behunin, *Phys. Rev. A* **86**, 062517 (2012).

[106] F. Intravaia, S. Koev, I. Woong Jung, et al., *Nat. Commun.* **4**, 2515 (2013).

[107] C. Itzykson and J.-B. Zuber, *Quantum Field Theory* (McGraw-Hill, 1980).

[108] M.-T. Jaekel and S. Reynaud, *J. Physique I* **1**, 1395 (1991).

[109] M.-T. Jaekel and S. Reynaud, *Quantum Opt.* **4**, 39 (1992).

[110] M.-T. Jaekel and S. Reynaud, *J. Physique I* **2**, 149 (1992).

[111] M.-T. Jaekel and S. Reynaud, *Phys. Lett. A* **167**, 227 (1992).

[112] M.-T. Jaekel and S. Reynaud, *J. Physique I* **3**, 1 (1993).

[113] M.-T. Jaekel and S. Reynaud, *J. Physique I* **3**, 1093 (1993).

[114] M.-T. Jaekel and S. Reynaud, *Phys. Lett. A* **172**, 319 (1993).

[115] M.-T. Jaekel and S. Reynaud, *Rep. Prog. Phys.* **60**, 863 (1997).

[116] R. L. Jaffe and A. Scardicchio, *Phys. Rev. Lett.* **92**, 070402 (2004).

[117] R. L. Jaffe, *Phys. Rev. D* **72**, 021301(R) (2005) .

[118] B. Jancovici and L. Šamaj, *Europhys. Lett.* **72**, 35 (2005).

[119] G. Jourdan, A. Lambrecht, F. Comin and J. Chevrier, *Euro. Phys. Lett.* **85**, 31001 (2009).

[120] D. J. Kapner, T. S. Cook, E.G. Adelberger, et al., *Phys. Rev. Lett.* **98**, 021101 (2007).

[121] E. I. Kats, *Sov. Phys. JETP* **46**, 109 (1977).

[122] O. Kenneth and I. Klich, *Phys. Rev. B* **78**, 014103 (2008).

[123] W. J. Kim, M. Brown-Hayes, D. A. R Dalvit, J. H. Brownell, and R. Onofrio, *Phys. Rev. A* **78**, 020101(R) (2008).

[124] W. J. Kim, A. O. Sushkov, D. A. R. Dalvit, and S. K. Lamoreaux, *Phys. Rev. A* **81**, 022505 (2010).

[125] W. J. Kim and U. D. Schwarz, *J. Vac. Sci. Technol. B* **28**, C4A1 (2010).

[126] G. L. Klimchitskaya and V. M. Mostepanenko, *Contemp. Phys.* **47**, 131 (2006).

[127] G. L. Klimchitskaya, U. Mohideen, and V. M. Mostepanenko, *Rev. Mod. Phys.* **81**, 1827 (2009).

[128] K. Klingmüller and H. Gies, *J. Phys. A* **41**, 164042 (2008)

[129] H. Kragh, *Arch. Hist. Exact Sci.* **66**, 199 (2012); *Astron. Geophys.* **53**, 1.24 (2012).

[130] D. E. Krause, R. S. Decca, D. López, and E. Fischbach, *Phys. Rev. Lett.* **98**, 050403 (2007).

[131] D. Kupiszewska and J. Mostowski, *Phys. Rev. A* **41**, 4636 (1990).

[132] A. Lambrecht, M.-T. Jaekel, and S. Reynaud, *Phys. Rev. Lett.* **77**, 615 (1996).

[133] A. Lambrecht, M.-T. Jaekel, and S. Reynaud, *Eur. Phys. J. D* **3**, 95 (1998).

[134] A. Lambrecht and S. Reynaud, *Euro. Phys. J. D* **8**, 309 (2000).

[135] A. Lambrecht and S. Reynaud, *Séminaire Poincaré* **1** (Birkhäuser, 2003), p. 107.

[136] A. Lambrecht, P. A. Maia Neto, and S. Reynaud, *New J. Phys.* **8**, 243 (2006).

[137] A. Lambrecht and V. N. Marachevsky, *Phys. Rev. Lett.* **101**, 160403 (2008).

[138] A. Lambrecht, *Nature* **454**, 836 (2008).

[139] A. Lambrecht, A. Canaguier-Durand, R. Guérout, and S. Reynaud, in *Casimir Physics*, D. A. R. Dalvit, P. W. Milonni, D. C. Roberts, and F. S. S. Rosa, eds. (Lecture Notes in Physics **834**, Springer-Verlag, 2011), p. 97.

[140] A. Lambrecht and S. Reynaud, *Int. J. Mod. Phys. A* **27**, 1260013 (2012).

[141] S. K. Lamoreaux, *Phys. Rev. Lett.* **78**, 5 (1997).

[142] S. K. Lamoreaux, *Am. J. Phys.* **67**, 850 (1999).

[143] L. Landau, E. M. Lifshitz, and L. P. Pitaevskii, *Landau and Lifshitz Course of Theoretical Physics: Electrodynamics in Continuous Media* (Butterworth-Heinemann, 1980).

[144] D. Langbein *Phys. Rev. B* **2**, 3371 (1970).

[145] E. M Lifshitz, *Sov. Phys. JETP* **2**, 73 (1956).

[146] E. M. Lifshitz and L. P. Pitaevskii, *Landau and Lifshitz Course of Theoretical Physics: Statistical Physics Part 2* (Butterworth-Heinemann, 1980).

[147] B. A. Lippmann and J. Schwinger, *Phys. Rev.* **79**, 469 (1950).

[148] M. Lisanti, D. Iannuzzi, and F. Capasso, *Proc. Natl Acad. Sci. USA* **102**, 11989 (2005).

[149] A. Liscio, V. Palermo, K. Müllen, and P. Samori, *J. Phys. Chem. C* **112**, 17368 (2008).

[150] A. Liscio, V. Palermo, and P. Samori, *Acc. Chem. Res.* **43**, 541 (2010).

[151] P. A. Maia Neto, A. Lambrecht, and S. Reynaud, *Europhys. Lett.* **69**, 924 (2005); *EPL* **100**, 29902(E) (2012).

[152] P. A. Maia Neto, A. Lambrecht, and S. Reynaud, *Phys. Rev. A* **72** 012115 (2005); **86**, 059901(E) (2012).

[153] P. A. Maia Neto, A. Lambrecht, and S. Reynaud, *Phys. Rev. A* **78**, 012115 (2008).

[154] S. de Man, K. Heeck, and D. Iannuzzi, *Phys. Rev. A* **79**, 024102 (2009).

[155] S. de Man, K. Heeck, R. J. Wijngaarden, and D. Iannuzzi, *J. Vac. Sci. Technol. B* **28**, C4A25 (2010).

[156] M. Masuda and M. Sasaki, *Phys. Rev. Lett.* **102**, 171101 (2009).

[157] T. Matsubara, *Prog. Theor. Phys.* **14**, 351 (1955).

[158] J. Mehra, *Physica* **37**, 145 (1967).

[159] R. Messina, D. A. R. Dalvit, P. A. Maia Neto, A. Lambrecht, and S. Reynaud, *Phys. Rev. A* **80**, 022119 (2009).

[160] P. W. Milonni, R. J. Cook and M. E. Goggin, *Phys. Rev. A* **38**, 1621 (1988).

[161] P. W. Milonni and M.-L. Shih, *Am. J. Phys.* **59**, 684 (1991).

[162] P. W. Milonni, *The Quantum Vacuum* (Academic Press, 1994).

[163] K. A. Milton, *The Casimir Effect, Physical Manifestation of Zero-Point Energy* (World Scientific, 2001).

[164] K. A. Milton, *J. Phys. A* **20**, 4628 (2005).

[165] K. A. Milton and J. Wagner, *J. Phys. A* **41**, 155402 (2008).

[166] K. A. Milton, *Am. J. Phys.* **79**, 697 (2011).

[167] K. Milton, *Nat. Phys.* **7**, 190 (2011).

[168] U. Mohideen and A. Roy, *Phys. Rev. Lett.* **81**, 4549 (1998).

[169] R. S. Mulliken, *Nature* **114**, 349 (1924).

[170] J. N. Munday, F. Capasso, and V. A. Parsegian, *Nature* **457**, 170 (2009).

[171] W. Nernst, *Verh. Deutsch. Phys. Ges.* **18**, 83 (1916).

[172] E. D. Palik, ed., *Handbook of Optical Constants of Solids* (Academic Press, 1995).

[173] V. A. Parsegian, *Van der Waals Forces: A Handbook for Biologists, Chemists, Engineers, and Physicists* (Cambridge University Press, 2006).

[174] W. Pauli, *Handbuch der Physik* **24**, 1 (1933).

[175] S. Perlmutter, G. Aldering, G. Goldhaber, et al., *Astrophys. J.* **517**, 565 (1999).

[176] L. P. Pitaevskii, *Phys. Rev. Lett.* **101**, 163202 (2008).

[177] L. P. Pitaevskii, *Phys. Rev. Lett.* **102**, 189302 (2009).

[178] M. Planck, *Verh. Deutsch. Phys. Ges.* **2**, 237 (1900).

[179] M. Planck, *Verh. Deutsch. Phys. Ges.* **13**, 138 (1911).

[180] G. Plunien, B. Müller, and W. Greiner, *Phys. Rep.* **134**, 87 (1986).

[181] S. E. Pollack, S. Schlamminger, and J. H. Gundlach, *Phys. Rev. Lett.* **101**, 071101 (2008).

[182] E. A. Power and T. Thirunamachandran, *Chem. Phys.* **171**, 1 (1993).

[183] S. J. Rahi, T. Emig, N. Graham, R. L. Jaffe, and M. Kardar, *Phys. Rev. D* **80**, 085021 (2009).

[184] S. J. Rahi, T. Emig, and R. L. Jaffe, in *Casimir Physics*, D. A. R. Dalvit, P. W. Milonni, D. C. Roberts and F. S. S. Rosa, eds. (Lecture Notes in Physics **834**, Springer-Verlag, 2011), p. 129.

[185] R. D. Reasenberg, E. C. Lorenzini, B. R. Patla, et al., *Class. Quantum Grav.* **28**, 094014 (2011).

[186] M. T. H. Reid, J. White, and S. G. Johnson, *Phys. Rev. A* **88**, 022514 (2013).

[187] M. J. Renne, *Physica* **56**, 125 (1971).

[188] S. Reynaud, *Ann. Physique* **15**, 63 (1990).

[189] S. Reynaud, E. Giacobino, and J. Zinn-Justin, eds., *Quantum Fluctuations (Proceedings of Les Houches Summer School, Session LXIII)* (Elsevier, 1997).

[190] S. Reynaud, A. Heidmann, E. Giacobino, and C. Fabre, *Prog. Opt.* **30**, 1 (1992).

[191] S. Reynaud, P. A. Maia Neto, and A. Lambrecht, *J. Phys. A* **41**, 164004 (2008).

[192] A. G. Riess, A. V. Filippenko, P. Challis, et al., *Astron. J.* **116**, 1009 (1998).

[193] N. A. Robertson, J. R. Blackwood, S. Buchman, et al., *Class. Quantum Grav.* **23** 2665 (2006).

[194] R. B. Rodrigues, P. A. Maia Neto, A. Lambrecht, and S. Reynaud, *Europhys. Lett.* **76**, 822 (2006).

[195] R. B. Rodrigues, P. A. Maia Neto, A. Lambrecht, and S. Reynaud, *Phys. Rev. Lett.* **96**, 100402 (2006); **98**, 068902(E) (2007).

[196] R. B. Rodrigues, P. A. Maia Neto, A. Lambrecht, and S. Reynaud, *Phys. Rev. A* **75**, 062108 (2007).

[197] A. W. Rodriguez, F. Capasso, and S. G. Johnson, *Nat. Photonics* **5**, 211 (2011).

[198] F. Rossi and G. I. Opat, *J. Appl. Phys. D* **25**, 1349 (1992).

[199] S. E. Rugh and H. Zinkernagel, *Stud. Hist. Philos. Mod. Phys.* **33**, 663 (2002).

[200] E. S. Sabisky and C. H. Anderson, *Phys. Rev. A* **7**, 790 (1973).

[201] V. Sandoghdar, C. I. Sukenik, S. Haroche, and E. A. Hinds, *Phys. Rev. A* **53**, 1919 (1996).

[202] S. Saunders, and H. R. Brown, in *The Philosophy of Vacuum*, S. Saunders, ed. (Clarendon Press, 1991), p. 27.

[203] M. Schaden and L. Spruch, *Phys. Rev. Lett.* **84**, 459 (2000).

[204] S. Scheel and S. Y. Buhmann, *Acta Phys. Slovaca* **58**, 675 (2008).

[205] O. Schröder, A. Sardicchio, and R. L. Jaffe, *Phys. Rev. A* **72**, 012105 (2005).

[206] J. Schwinger, *Lett. Math. Phys.* **1**, 43 (1975).

[207] J. Schwinger, L. L. de Raad, and K.A. Milton, *Ann. Phys. (NY)* **115**, 1 (1978).

[208] D. W. Sciama, in *The Philosophy of Vacuum*, S. Saunders, ed. (Clarendon Press, 1991), p. 137.

[209] M. J. Sparnaay, *Physica* **24**, 751 (1958).

[210] C. C. Speake and C. Trenkel, *Phys. Rev. Lett.* **90**, 160403 (2003).

[211] R. Sundrum, *Phys. Rev. D* **69**, 044014 (2004).

[212] A. O. Sushkov, W. J. Kim, D. A. R. Dalvit, and S. K. Lamoreaux, *Nat. Phys.* **7**, 230 (2011).

[213] V. B. Svetovoy, *Phys. Rev. Lett.* **101**, 163603 (2008); **102**, 219903(E) (2009).

[214] V. B. Svetovoy, P. J. van Zwol, G. Palasantzas, and J. Th. M. De Hosson, *Phys. Rev. B* **77**, 035439 (2008).

[215] G. Torricelli, P. J. van Zwol, O. Shpak, et al., *Phys. Rev. A* **82**, 010101 (2010).

[216] Q. A. Turchette, D. Kielpinski, B. E. King, et al., *Phys. Rev. A* **61**, 063418 (2000).

[217] S. Weinberg, *Rev. Mod. Phys.* **61**, 1 (1989).

[218] A. Wirzba, *J. Phys. A* **41**, 164003 (2008)

[219] F. C. Witteborn and W. M. Fairbank, *Phys. Rev. Lett.* **19**, 1049 (1967).

[220] P. J. van Zwol, V. B. Svetovoy, and G. Palasantzas, *Phys. Rev. B* **80**, 235401 (2009).